Geologic Life

Geologic Life Inhuman Intimacies and the Geophysics of Race

KATHRYN YUSOFF

Duke University Press *Durham and London* 2024

Printed in the United States of America on acid-free paper ∞
Project Editor: Lisa Lawley
Designed by Matthew Tauch
Typeset in Portrait Text and Archivo by
Westchester Publishing Services

Library of Congress Cataloging-in-Publication Data
Names: Yusoff, Kathryn, author.
Title: Geologic life : inhuman intimacies and the geophysics of race / Kathryn Yusoff.
Description: Durham : Duke University Press, 2024. | Includes bibliographical references and index.
Identifiers: LCCN 2023033362 (print)
LCCN 2023033363 (ebook)
ISBN 9781478030300 (paperback)
ISBN 9781478026075 (hardcover)
ISBN 9781478059288 (ebook)
Subjects: LCSH: Black people—Race identity. | Geology—Social aspects. | Black race—History. | Human geography. | Anthropology—History. | BISAC: SOCIAL SCIENCE / Black Studies (Global) | SOCIAL SCIENCE / Human Geography
Classification: LCC DT15 .Y88 2024 (print) | LCC DT15 (ebook) | DDC 305.896—dc23/eng/20231220
LC record available at https://lccn.loc.gov/2023033362
LC ebook record available at https://lccn.loc.gov/2023033363

Cover art: Yinka Shonibare, *Girl on Globe 4*. Courtesy of James Cohan.

To the orphans
of geology

a geologic dirge

Unground the human, fall off its edges
into rocky gatherings, seismic rifts, and
world-altering geologic grammars;
Brake geo-logics of enclosure:
alluvial plantations, oceanic, magma, mine, Mine, mine
fungible units, black-gold, coiled and recoiling earth,[1]
Of inhumanities.
Nonbeing quickens into fugitivity
from the category of "natural resource"
to subterranean rifts unburdened by the weight of surface "discovery"

Toward other geologic lives
unthought. Intramural. Crystalline.
Ore of the earth. Cradling islands, remembering seas.
Subjects unfolding in magma-fired embrace.
imaginary / ground / Subject / Earth / Relation
Beyond extraction.
"senses as theoreticians"[2]
plays aesthetics voluminous underground
as praxis; insurgent geophysics
defying racial gravities,
in deep discovery with rocky abandonment.
Every moment in Relation to another.
Every Earth a broken ground.

Contents

Introduction

Coordinates (0°0′ Longitude, 51° N Latitude)

In the ledger of geologic time there are missing earths. Earths that appear only as negative inscription, underground, beneath and behind the geographical imagination of colonial earth and its discourses of purposeful extraction. Indigenous earths. Black earths. Brown earths. As planetary fractures now appear daily in the shifting world of climate, colonial earth swings in an oscillating imagination of dystopic/utopic salvation in confrontation with the new geologic realisms of the Anthropocene. Ends. Beginnings. Narrativizing new origin stories for the planet and continuing to erase older, missing, broken earths that made the present possible. As diasporic human and nonhuman geographies transformed colonial spaces with labor, hoofs, creatures, and crops they also erased geologies that belonged to other imaginations of earth. These practices of "unhoming" through the epistemic dynamism of the inhuman enacted environmental changes of state in subjectification, climate, species, and elemental geophysical flows. Alongside this geotrauma, all these dispersals carried illegitimate geographies of passion for inhuman places and things. A patch of dirt in the fat heat of summer.

As the geologies of colonial world-building led to the Anthropocene, that building required first the earth-shattering of existing relations and bonds of shared ecological and inhuman worlds. Geology and its epistemic practices were a form of earth writing that was riven by systemic racism in the building of colonial worlds and the destruction of existing earths. *Geologic Life* seeks to understand geology (in its broadest sense) as a tool of raciality that has historically shaped the grounds of struggle and continues to shape material relations of racism into the future.

Colonial earth is the product of white geology: a historical regime of material power that used geologic minerals, metals, and fuels, combined with the epistemic violence of the category of the inhuman to shape regimes of value and forms of subjective life. As "end times" frame climate change conversations, this book goes back to look at the beginnings of the field of geology as a colonial practice that created normative orders of materiality and destroyed worlds.[1] I investigate how we might understand geology as an ideological and material infrastructure of matter and materialism that shapes subjective and planetary states. And how calculative regimes of geology organized both the temporal and political surfaces of power, and its racialized undergrounds.

I argue that white geology mobilized geopower to operationalize the conquest of space across the globe and below the surface, furnishing its partitions and apartheids as distinct spatial forms. Namely, through Indigenous genocide and enslavement, through the shape of colonial afterlives in convict lease and indenture, and through ongoing environmental destitution. Alongside the transformation of environments, paleontology (pale-ontology) named and raced persons as geologic subjects through the category of the "inhuman." The designation as inhuman, used for both raced subjects and subjectless matter, diffused the violence of geology and maintained the colonial prerogative of value and its progressive scripting of extraction as a genealogical achievement of whiteness. Antagonism to white geology is also an attachment to other kinds of earths and inhuman intimacies not continually wrecked by the accumulative geo-logics of those practices. We can see resistant earth and ecological practices as a defining continuum of struggle against colonial forces of transformation and white geopower.

Propelled by a material teleology of extraction and abandonment, white geology practiced a secular geophysics of material determination that was justified through the paleontological narratives of **race** and **time** that literally "placed" peoples in different strata (in a stratigraphic bill of rights). Geologic time became the political time of race, wherein whiteness was a tautology of material achievement in time, and race can be recognized as a missing term in the stabilization of material value. The concept of race kept the separability of the human and the taxonomic carcerality of the inhuman intact (as *bios* and *geos*) through the division of matter, maintained through subjugating geographies and the spoils of extraction. In this current moment of the climate-induced crisis of materiality, in the rapid transformations of worlds, the imperative to understand colonial materialisms and their afterlives in planetary effects is crucial to both materialist pedagogies of anticolonial praxis and the epistemological shifts necessary for a different earth.

The elevated plateaus of privilege and subterranean rifts of the racialized poor in colonial earth are a product of historic materialisms past and present, even as the structural and spatial occupation of whiteness—as the product of white geology—becomes a multiethnic affair. Geology was the predominant medium of the instantiation of this earth and its continued settler claims and thirst for geopower in the present. Geology explained the geophysical processes of earth to make it epistemically available for extraction and ontologically located within the development of human and planetary time. To tell a story of the rocks is to see the past surfacing in the present: a process of coming into view, of other earths. In this political surfacing we find the ghosts of other geologies sounding subterranean disquiets. Ghost geologies that testify to a certain disorientation, a gathering of a series of ruptures. To tell a story of rocks is to account for a eugenic materialism in which white supremacy made surfaces built on racialized undergrounds across multiple—political, geophysical, subjective—states. Thus, the history of colonial earth is also a map of the geophysics of race, of subjugating rifts, colored earths, racialized gravities, and Anthropocenic futures. The proclamation of universal fraternity and common futures is broken again and again by the weight of racialized gravities (that inflict the continued subject positioning of processing violence and the inhumanity of being processed by geo-logics of equivalence).

The origin story of geology and its materialization of a geophysics of race is important for how we understand the divisions between privileged forms of Life and its concomitant inhumane states of exposure and dehumanization. Historically, this book charts the origin stories of earth and scripts of race as natal twins, emergent in colonial and settler colonial worldings through an array of geologic practices. Racial and environmental determinants were coterminous in scripting the earth and its geophilosophy of human becoming in the geo-logics of colonialism, which intensified during the eighteenth century. While those concepts of environmental determinisms have been roundly critiqued, racial determinants continue to reside in the mantra of "natural resources" (e.g., the resource curse, limits to growth, questions of demography, the "undeveloped") and other more discrete eugenic renderings of what the earth is for. I argue that this separation of earths is not constituted by a Life-nonlife (partial human-designated nonhuman) bifurcation but through a Life-earth division (*bios-geos*) that race mediates and materializes. This means that the politics of decolonization are not primarily organized in either metaphysics or through its biopolitical critique but through the earth and its categorization. The categorization of the inhuman as a subjective register is not intended as a veiled form of environmental determinism but draws attention to a contingent "state" of matter that is involved in the subjective formation of geologic life. The imperative of a lithic-eye-view takes biopolitical critique into a broader field of reference that aligns the ground of thought with the ground of the earth, thereby mobilizing a political geology that sees subjective and earth states as concomitant states of being, collapsing the spatial function of race in the production of value.

Geos {race} *Bios*

The stages of argument about the separation (and separability) of *geos* and *bios* that I make through the empirical research in this book proceed as follows:

1. *Geologic Life* is the basis of the historical-material conditions for the emergence of the biopolitical/biocentric subject (as the figure of Life itself).
2. *Geos* and *bios* are produced as distinct "states" of matter (not organized in a hierarchical continuum) that are productive of different geophysical subjective states and spatial formations of the plateau and rift.
3. *Geos* is historically constructed and imagined as a stratified temporality that produces the time of race.
4. Race spaces the interregnum for the differentiated production of value between *geos* and *bios*. Every question of matter, materiality, and materialisms gets passed through the space and time of race.

Geologic Life investigates the structural pressure of colonialism as an earth force and its undergrounded archive that is made through the unequal racialized affects and exposures to geopower. Broadly, the geographic map that I plot through this book is of the plateau of white geology (henceforth, the *plateau*) and its processes of racializing undergrounds (henceforth, the *rift*). Rather than seeing this racialized archive of colonial earth as biopolitical, I see it as a geopolitical act in the division of flesh and earth through the grammar of the inhuman. The language of the plateau pulls toward stratal totalization. Universalisms issued from the plateau are understood as the present perspectivism of the earth—a normative earth—that dominates environmental politics and practices. Because of the geophysical dimensions of the rift and its relation to the plateau, the rift is not a space of totalization and is necessarily engaged with the specificalities of its historical geographies and their shared archipelagoes of relation.[2] If the plateau's language is one of hardening and categorizing stratal structures, the rift is a space of ghost geologies and geopoetics whose narration and syntax are given through persistent resistances. The countergravity of the rift requires a tender holding alongside the act of holding colonial environmental histories to account. The "implicatedness" (to use Denise Ferreira da Silva's term) of this form of geologic life is prescient in the precarity of colonial afterlives and their presents.[3] Broadly, Black, Brown, and Indigenous subjects whose location is the rift have an intimacy with the earth that is unknown to the structural position of whiteness. This inhuman intimacy represents another kind of geopower learned in the tactics of relation and the theorizing of experience: tactics of the earthbound. On the graveled road that shaped the bite of grit into freedom's dreams, the inhuman was a doorway, big enough to survive its weaponization.

Colonial earth is *extractive earth*, organized through racialized and racist relations that sustain and allow the surfacing of whiteness as a geopower (a terrorizing and territorializing force). The practices of colonialism geo-engineered territory through surface and subterranean geo-logics toward the white supremacy of the planet. The ground of colonial passage through mine, plantation, and clear-cut extraction produced an amnesiac earth of scant remembrance, as colonialism sedimented its telos of materialisms as normative. As race was coupled in the crucible of subjugation, territorial theft, and geologic unearthing, it continued to define geologic relations through these integrated modes. Race, seen through a philosophy of *geologic life*, requires a redistributed justice that understands both the affective and material infrastructures of race as geologically made, and geophysics as a product of race. Reckoning with geology (across its broad domains) points to another map that might work toward articulating the geophysical conditions of decolonization, where decolonization is a future-oriented and historically ongoing act that works against the principle of extinction that characterizes colonial earth.

As increasing attention is paid to the activities of racial capitalism and the dynamic mobilities of geomorphic change in the Anthropocene, the foundational racialization of geology is undertheorized in its histories (and historicity) and their material presents. The task is to join the theories and concepts of deep time to the praxis of extraction across domains that function within very different syntaxes of materiality and epistemic regimes. Addressing the colonial afterlives of materiality requires expansive pedagogies for anticolonial praxis that extend the reach of geologic grammars across the plateaus of disciplines to show how they form something akin to a racialized surface. The disassociation of the racial origins of geology, I have argued (Yusoff 2018a), has consequences for the proximities and exposures Indigenous, Black, and Brown persons are expected to navigate, absorb, mitigate, and ameliorate in relation to environmental harm. That work sought to unearth the operations of race, gender, family, and nation through a reflection on the epistemological enclosures of naming, and the way these geocodes overdetermined the normative and its (negative) association of properties with some bodies and not others. In this work I want to show how the geologies of race are axiological to the production of earth states, processes, and racial gravities (while paying attention to the ongoing violence of epistemic elisions between matter and racialized subjects). An account of rational accumulation goes only so far to locate the exteriority of the desire for extracting value; it is the libidinal that exposes the interiority of that desire for inhuman power—first as rabid thirst, second as patriarchy and paternalism, third as deadening of the earth.

Metals and minerals have signaled a durational abundance that struck at the puny human limits of temporality, a mark of difference that was both challenge and provocation for overcoming. Next to iron the flesh was weak. Energy is the primary transformation of the earth, and this was also the geoforce of colonialism. Blackness was made in the same metallurgic register as gold at both a corporeal and a continental scale, where both Africans and Africa were constructed as the abundance necessary for extraction. Muscle and strength were fetishized as persons were degraded. The spatial inversion of territory as a series of buried earths for extraction enlarged the colonial and settler colonial desire for a never-ending source of the accumulation of geopower and the sedimentation of racial difference as its operative machine of unearthing. Whereas Indigenous life was framed as an impediment to the seizing of land, Blackness was invested with spatial expansion (King 2019; M. Wright 2015). Colonial earth created black holes that delivered the myth of clean extraction in the "severe maldistribution of resources" (Wynter 1996, 302). This book is about the loamy broken grounds that are left by the predatory gravitational force of colonial earth. It also seeks to account for how whiteness was able to "float" above the gravity of an earth relation.

On the insistence of Katherine McKittrick (2006) and Christina Sharpe (2016a), if we place anti-Blackness as the normative frame of geography and proceed from there, then a different cartography of reason and attribution emerges from which to understand materiality, matter, and relations to deep time. Rethinking race is acute because of the continued weaponization of geology in bordering practices and the simultaneous relocation of the foundational divisions of the material conditions of racialized life into a question of discursive identity politics or metaphysics (which thereby dissuades an analysis of the structural conditions of material reparation). This includes the need to turn the discipline of the geosciences against its empirical foundations—to begin to think from racialized rifts and decenter the normative production of its epistemes of materiality so that reparative geophysics might be built, which substantiates different geosocial futures.

Geologic Life takes familiar scenes of geology to tell another history/historicity, of the entanglement of the inhuman and inhumane in the configuring of agentic Life and regimes of geopower, made through colonial geology. It is not that the inhuman was an impossible outside for the colonial imagination that everything was exiled to. Rather, it was the colonial investment in minerals and metals that was commensurate with the investment in enslavement and subjugation; both these libidinal forces were characterized by the desire for harnessing geopower. The consequence of this doubling of the inhuman-inhumane as subject and object was the creation of distinct geographies that spatialized the experience of geologic life as racialized and racist. These include the plateau of white geology and its perspectivism over and above the earth (the Overseer) and the broken grounds of the rift, where racialized subjects were torn from geography and labored under the pressure of unjust gravities to substantiate every aspect of the material life of the plateau. The splitting of *bios* and *geos* bifurcates planetary life-forms into two distinct geophysical zones: the plateau (Life) and the rift (inhuman). The secret question that this geophysics carries is the following: *What material and somatic qualities allow whiteness to float (as Life) through its taproots of racial undergrounds? And how did the geophysics of whiteness (and its new ethnic formations) underpin and reproduce a particular metaphysics of the earth in service of these geophysical desires?*

A geophysics of race points to a somatic differentiation in geologic life, where the gravity of whiteness is historically produced through certain arrangements of inhuman(e) earth. White geology epistemically created and relied on racial undergrounds to bring metallurgical and mineral value to the surface. And we can think about how this paradigm of the mine cascades through all the ways in which material value is stabilized in the present—from the jewels and skyscrapers that Aimé Césaire alerts us to, which bear his thumbprint and heelprint, to the saturated oil landscapes and palm plantations. In the context of the spatialization of racial undergrounds, the mine inside is another subjugating inhuman force that conditions the geophysical politics of the world outside. White geology, in the context of colonialism, mobilized the subterranean volume to organize surface flows of social power, including but not limited to forms of racial segregation and racist spatialities. The relation between the underground and the surface, and the politics of surfacing large quantities of earth, water, metallurgy, minerals, and buried sunshine, organized forms of capital accumulation that require racial undergrounds to function. Such racial mines—as a material and conceptual form—continue to be paradigmatic today. Racial geophysics is underpinned by the geomythology of a neutral conversion of undergrounds into overgrounds, of earth into colonial achievement. Geophysics, understood in this way as pressure and gravity that are the affectual force of geopower, accounts for somatic instantiations in the flesh of geology as a historical corporeal process of racial undergrounding in the production of space.

I theorize the rift as a geographic concept of broken grounds from which to fracture the surfaces of white geology and its forms of subjugation. The interruptive phrasing of subterranean space names a refusal to adhere to the foregrounding of utilization of the earth for the plateau. My argument about plateau and rift is not one of uplift, that we all need to be on the plateau or that the plateau is imagined in its environmental racist guise as delimiting a Garret Hardin–esque carrying capacity (see Yusoff 2018b). An inattentiveness to geology's spatial and political forms requires a historiography of its past (and thus future) grammars precisely because the slippage between the material and metaphysical in the inhuman hides the violence of geology and its role in the political subjectivity of race. Addressing geologic life necessitates an enlargement of political geology beyond its settled geo-logics that make its sites and objects and an imagination of the stratal relations that sustain them, alongside scrutiny of the grammars of geology that language its modes of description, organize its forms of address, and spatialize its operations.

The rift is a temporal scene and a material experience of shattered grounds. It is the messy geographies of nonarrival, the never getting to arrive because departure is not chosen but arrives as catastrophe and chasm. The endless sea that survivors carry is a restless inheritance of the broken earths:

movements and moments that ran counter to the obligated geographies of relation. The rift is also the geophysical manifestation of the psychic life of geographies' displacement, as it is a structural organization of geopower's hold on and subjugation of those who buffer earth's shocks and extract its ores. It is an elemental map of gold, coal, metal earth made through the imposition of a language of extraction. It is a map of clearance and erasure in the overwriting of value. Its ghost geologies speak to other cartographies. The rock gives forms to other temporal possibilities. Unearthing recalls buried bones and sun. The tree is a presence of something rooted and held. Geologic intimacies challenge the normative register of colonial materialities—materialities that narrativize earth as a site of improvement and settler scripting. Attending to the buried histories of inhuman intimacy has been an ongoing tactic for building another earth among racialized and anticolonial peoples. These too are the earth(s) that colonialism made. They might be called a billion Black and Brown Anthropocenes. A missing ground, a Black, Brown, and Indigenous earth, that haunts the geographic imaginations of planetarity and campaigns for a different understanding of the *geos*. Geographic imagination is a tool to lever the new and find passages to a differently imagined life. The surfacing of ghost geologies is also a way to bring things down in the world, to deflate the levitating quality of whiteness and the heavy gravity it demands to secure its stratal place in the world.

The gravities of rifts are defined not only through their obscene subjectification but also by a receptive atmosphere of holding against the temporal forces of deformation and tending spaces outside these forces. These ways of living are held in the vicissitudes of Black, Brown and Indigenous lifeworlds and in the historic geographies of community, and these stories are not mine to tell. What I have tried to give space to, beyond a reactive, redemptive countergravity, is to the radically other dimensionalities—or *geoforces*—that emerge beyond deformation and construct gravity differently, as open to the indeterminacy and intimacy of the earth and organized as a tactic within the flux (where such tactics also include engaging in settler spatial modes of property and the harnessing of geopower for sedimenting attachments in place, such as Black ownership of land during Reconstruction as an attempt to spatialize the affects of racial violence).

Colonial earth established a psychosis of materiality through the designation of "natural resources" and inflicted a refusal of locatedness through its geographies of abduction and displacement. Natural resources are a conceptualization of a specific form of geologic life and dislocated geography. Unearthing these sedimented relations and quieted histories requires an earth politics and geopoetics that understands the expansiveness of earth forces *and* the forms of their social production as historically racialized.

What is this particular gravity of whiteness that can claim the apex of time and space as its inheritance and impose an anti-Black heavy "weather," in Christina Sharpe's (2016a) terms, and how is it materially and empirically made? I understand this gravity not as metaphoric but as geomorphic: materially manifesting as a geopower geometry that is routed through narrative accounts, geologic grammars, and affective architectures. Racial gravities can be understood through a historical lens, as temporal and material forces of deformation that "land" on the present through the efficacy of colonial earth. Gravity is a historical set of conditions that creates the coordinates of falling, as well as the counterforce of its opposition, where space bends to grow around the untimely absence of the *shock forward*.[4] The durational effects of racist gravities affect spatial formations and the possibility of moving in space upright rather than weighted, free rather than oppressed, but they do not condition the grace of movement or the imagination of other geophysics of being. The time of racism (the time it takes and the spaces it makes) is connected to the operation of geo-logics of deep time and the establishment of racialized populations zoned within broken grounds inside rather than outside of the spatial-geophysical production of whiteness.

Narrating broken grounds, I explore a pedagogical countercartography to normative modes of materiality—as they are expressed in resource extraction, utility, and settler modes of apprehension and its geopolitical organization of the earth. The collective challenge is to find words that stand against the renewing tide of natural resources, understood as the normative and devastating language of materiality. Akin is the task to make models of kinships that would permit a rock in the family and a shared gravity in the shocks of the earth that might make a much-needed climate commons.[5] The central claim of this book is that geology is always racialized and that race is a geologic formation (rather than primarily a biocentric/biopolitical concept), and thus racial violence is the violence of matter, and race must be considered as a geophysical operation. Rather than arguing that geology demonstrates or mediates racialized relations, I argue that race is foundational to the production of knowledge about the earth within a Western episteme, and that a geologic understanding of the time of racial difference is foundational to racial violence and the possibilities of place. The historical geographic claim here is that race and geology are coterminous and so cannot be conceptualized as divisible realms (which maps onto the differentiated fungibility of geologic subjectivity as it is mediated by geopower). This inheritance is why the politics of race need to be reconceptualized in a geologic field of relations in the context of climate and Anthropocenic changes.

Colonialism is an expression of *geotrauma*, and geology operationalized and institutionalized its statecraft as grammar, imaginary, syntax, and material praxis through the tight intimacies of the inhuman and inhumane. Colonialism was *geologic-fication*: the transformation of land, ecologies, and forms of relatedness within interdependent forms of geologic life into a discourse and practice of materiality (and then commodities). Which is to say colonialism practiced an extreme form of materialism that transformed the planet and its geopower to extract and explicate value in such a way as to build a new earth at the scale of the planetary. Understanding the combined material effects and psychic affects of geotrauma is to see the earth as a product of colonial relations, and the replicability of these relations of extraction through ever-more intimate forms of life as the process by which earths, bodies, and relations change state. Scientific research tells us that the earth is one of many possible earths and all material states of being are contingently anchored in the variability of the earth itself. Geologic empiricism pushed far enough organizes an epistemological statement that a different earth is possible. The "social" in the geosocial formations of the earth is a geophysical modifier for other formations and ways of being drawn to the earth. In this sense, the inhuman points to unregulated forms of existence. This elemental largesse, which could not be contained by colonial categories, was the basis of the existential challenge to Enlightenment materialities, as it was a source of liberatory possibilities for those caught in its conceits.

Rather than assert the veracity of the stratal pressure of race, I seek to show how anti-Indigenous, anti-Brown, and anti-Black gravities are created across time through the knowledge networks and practices of deep time (understood as location device, narrative, and structural relation). Secreted in geologic grammars and hiding in plain sight through disciplinary divisions, *race travels in earth archives as the flesh of geology*. Thinking with this idea of the *flesh of geology* (with reference to Hortense Spillers's 1987 argument about how subjects become rendered as flesh in the context of enslavement), I understand this concept as being both about the specific historical embodiment of geopower in the afterlife of colonialism and about how subjects are designated as inhuman matter in the grammar of geology and subjected to the power of those earth forces. The division between body (in a biopolitical sense) and flesh is an essential category difference between a *captive* and *liberated subject position*. Dominance of normative accounts of materiality (as natural resources) regulated, spaced, and languaged the split of *bios* and *geos*, making colonial earth a riven landscape. Geologists in the seventeenth and eighteenth centuries oversaw the configuration of racial identity in paleontology and the identification of geologic resources and their narrativizing in field surveys, and thus race was made in questions of the ground and its value rather than through the signification of the body per se.

The move to make race a space of negotiation between the body and ground is precisely how the theft of land was justified and bodies were submerged into inhuman political signification, where epistemic violence was enacted through material and temporal designation of subjects in the figures of slave and indigene. The subtraction of *geos* from the scene of colonial encounter conditioned the displacement of Indigenous peoples and the reshaping of the earth under a colonial genre of materiality. Because of the required fungibility in the operationalized identity categories of slave and indigene between person and inhuman, a porousness existed that was both the result of and prior to those structural positionings and subjugating violences. What I am suggesting here is the opposite of a romanticized "closer to nature" discourse, and more in the register of paraontological tactics that emerged in this forced margin of the inhuman(e). The flux of temporal abandonment—whereby subjectivity was cast into different strata in the narratives of racial achievement by eighteenth-century geologists—made planetary flux, or tactics for subterranean life, a condition of experience and survival. Temporal scripting, then, happened from above and below the geologic plateau and its institutional transcripts of deep time.[6]

It is a mistake to understand race as a biopolitical concept when it is materially and conceptually grounded in the earth in advance, and the focus on *bios* covers over the racial deficit of the spatial extractions underway.[7] The continued subtraction of *geos* from questions of race reperforms the initial severance of relation that mobilizes race as a governance between Life and the inhuman. Part of my research agenda has been to resubjectify geology: seeing *geologic subjects* as implicated and entangled in epochal and social shifts. On one hand, this means understanding the inhuman story as a subjective and subjugating historical geography that involves distinct racialized relations with the earth. And, on the other, it sees the relation between inhuman and inhumane as structurally bound to extraction in the praxis and grammars of geology that elevate Life (in mastery over geologic grounds). Rendering subjects as inhuman matter, not as persons in the *geocoding of bodies*, facilitated and incorporated the historical fact of extraction of personhood as a quality of geology at its inception (enslaved persons racialized as property, energy, and flesh of geology). The historic division between *geos* and *bios* is the rupture or rift on which race is made (the inhuman-inhumane is the most obvious example). There is no "better" biopolitics. Racial justice is intimately tied to environmental justice as its precondition and the possibility of another earth.

In the formal epistemic separation of *bios* and *geos* in the eighteenth century through the discipline of geology, race became the modulator and mediator in that spacing to hold together what cannot be separated.[8] Through this division of *bios* and *geos* a preferred form of racialized Life emerged to stand for all life (as per Sylvia Wynter), and inhuman (as *geos*) was organized through the properties of dislocation and dehumanization in the pursuit of value. Although this fiction of Life's internal divisions is known (see Wynter's [2003] account of partial humanism), the deep structures of geologic grammars mean that this division is the a priori condition of encounter with material agency that must be continually overcome in Western epistemologies and material politics. There is a deep forgetfulness that is engineered by the separability of these two terms (*bios* and *geos*) in epistemologies of thought, practice, and valuation—meaning that the deep residues of race are operative in every material transaction (with or without a subject). It is my contention that you cannot think about the climate crisis, extinction, and biohazards (viruses, toxicity, pollution, etc.) without thinking about race and the way race violently mediates and maintains the *geos* and *bios* border. The separability of *bios* and *geos* is an instrument of power in the governance of geopower. The geo-logic of extraction is reproduced in this space-ing, as well as the strange apparition of whiteness as a condition that "floats" above the earth (where this spacing establishes the geophysics of being as above or below the earth; plateau and rift). So, a subtext to this project would also be bringing whiteness back down to earth through the dismantling of its historic mantle of geosocial forms. In parallel, the concept of geoethics might be seen to designate a juridical body that floats in a nonmaterial space of consideration (whether that be political, moral, or epistemological space). An antiracist geoethics would have as its methodological task the grounding of that Enlightenment body of thought into located bodies and geologic lives. The recovery of specific racial geologies is an alternative way of forging the methodological repair necessary to account for Black life (Woods 1998; Roane 2018, 2022).

My claim is that different "states" of being are not primarily the product of a metaphysical operation but are constituted by geo-logic codes that produce a modality of geophysical states. This production keeps subject formation in constant dialogue with colonial earth (often bound to its violent vicissitudes) and anticolonial earth practices. To understand the amalgam of earth and whiteness requires reexamining our understanding of the material and subjective dimensions of geologic histories and the geosocial contours of subjects. If whiteness established itself through the bifurcation of *bios* and *geos*, this severance is made in the historical geographies of geology as paleontology and surveying, and in the ontological and geophysical formation of the inhuman as subject, material, and resource. If geophysics is the prior condition to the coordinates of subjugation, racial justice requires a geophysical mode of redress: or, geoethics for an antiracist earth. *Understanding the geologic dimensions of oppression* is part of challenging the weaponization of environments and alleviating the quotidian exposures of racialized subjects to geochemical residues (categorized as pollution and poison) and geographical displacement.

In this book, I want to show *how* stratification (of temporalities and thought) participates in the governance of forms of racialized life and creates anti-Indigenous, anti-Black, and anti-Brown gravities across deep time. An explication of how the plateau of white geology functions as a form of geologic life is one step toward its dismantling. I offer this book as redescription (and reorientation) of terms through the analytic of geologic life, as empirically grounded geophysics that connects the social struggles of race to their geologic expression and material manifestation. My goal is to provide new sites of consideration in the materialities of racialized struggles and liberation, which is also to understand description, disciplines, and analytical diagnostics as participants in material modes of oppression and their directionality (as imperative and futurity). To put this the other way around, epistemic insurgency—here, insurgent geology—is crucial to the knotted work of racial justice and environmental conditions (in that order). If strata are *the* ontology of the Anthropocene (understood through the stratigraphic imagination as destratification of environments and the desedimentation of subterranean spaces), geology is its epistemology, plotting those ever-more chaotic stratal chasms into the future. This tectonic shift of political geography through planetary time in the present is a product of these colonial histories of the past.

A critique of "the earth that colonialism made" is also an epistemic argument against the separatedness of disciplines and practices in the ledger of geotrauma. This is the epistemic violence of division. Joining up the plateau of these interrelated forms of geologic production provides the perspectivism to imagine its dismantling and defamiliarizes the acts of white geology. Racial categories function because they adhere within the dynamics of social space across domains rather than in singular epistemic regimes. Understanding what creates the plasticity of the plateau and the stratal pressure of whiteness requires a map across domains of geology, its material and economic practices, its knowledge economies and epistemes, its geophilosophies of nature, and its structuring device for the "ground" of thought and the emergence of social and subjective forms. The bifurcation of disciplines is part of how the world is hidden in plain sight, how meaning is obscured, and how the recognition of other "grounds" that challenge and defamiliarize the normative forms of geologic life is submerged. What interlocking epistemic collaborations between the earth sciences, philosophy, judiciary, natural history, literature, mining, plantation farming, and economic botany have made colonial earth possible as a planetary proposition?

As a material practice of arranging the inhuman dimensions of the earth and an account of materiality, geology functions as an *affective infrastructure* of the politics of the inhuman from which Life is understood to differentiate, deviate, and subtend itself.[9] Geology is at once a set of historically located practices, epistemologies, and a more diffuse ontological marker for the inorganic that stands in for the concept of inhuman matter. In this book, I range across these material and metaphoric historical geographies precisely because of the collaborative work that is done between these interdisciplinary domains to create geoforces that shape the material world: racial, social, geophysical, political, and conceptual. Geology delimits material identity, it identifies and categorizes to make materiality mobile as an identifiable unit of valuation, as equivalence and exchange, universalized as natural resource. Geologic grammars cut mineralogical affiliations, rocks caught in the geochemical, fluvial, and geomorphic embraces of depth, pressure, and dispersal. Stratigraphy identifies and dislocates matter from its sedimentation and mobilizes it into the present tense through a material manifestation of temporality.[10]

Geology is so politically potent precisely because it both cuts matter through its grammars of description (as an execution and extraction that makes objects, modes of exchange, indices of valuation) *and* gives matter an origination narrative in the earth and natural causation. These geophilosophies naturalize the concepts of stratification and genealogy of life across domains (in ways that sediment in social and political structures as normative claims). Through the double action of cutting and grounding, geology (as a discipline and practice) performs an onto-empirics that acts to simultaneously master matter and make available the geography of the earth for transformation. It is *the* origin story and *the* ontology. This colonial narrativizing of geology is the historical geography of the transformation of matter and energy as the basis for subjective life and the diversity of its forms. There is no ground but the earth.

Geologic Life is concerned with taking on the shared lineages that compose the history of geology as a science and field practice, emergent in the late fifteenth through nineteenth centuries as a Eurocentric field of scientific inquiry. I approach this not through a linear historical geography but through undergrounds (as footnote, mine, appendix, subtending strata, and stolen suns) that reveal the subterranean currents and spatial proximities that run through the dual categorization of inhuman as mineralogical material and subjugating forms of life in the categories of race. The rifts that I draw on arc through a long history, beginning with the invasion of the New World in 1492 and through waves of imperial, colonial, and ongoing settler colonial moments, across the geographies of the Atlantic and Antilles, to argue that geology and race establish the affectual infrastructures of material world-making. The relation is ongoing and overlapping with the devastations of the political present and its racialized environments. My contention is that geology continues to function within a white supremacist praxis of matter in its current geophilosophical formation (Anthropocene) and material presentation (as extraction, environment, and climate change). I pay particular attention to this history of *white geology* as a material-metaphoric grammar that structures thought and earth in Western epistemologies to operationalize planetarity and to show how this homogenizing of space and time through matter economies is challenged by the theoretical praxis of its racialized undergrounds, established in the Americas and Caribbean (and many other diasporic spaces and Asian, African, and Pacific grounds).

Geology—as material metallurgic practice since 1492, a formal modern discipline since the 1700s, and the conceptual armature of materiality, inhuman classification, and parsing of deep time—is implicated in what Rinaldo Walcott (2021) calls "black life forms" in the long emancipation. Geology has shaped Black life in symbolic, geographic, and stratal realms. The structural trajectory of anti-Indigenous, anti-Black, and anti-Brown racialized strata was inaugurated in colonial earth through the theft of land and the mining of the Americas, was sedimented in the fluvial plains of the plantations, the sea graves of the Caribbean, and the ocean "beds" of the Middle Passage. White geology is a violence that contemporaneously accumulates in bodily burdens of toxicities and proximities to harm that is called environmental racism but permeates across much more subtle color lines in the grit and exhaustion of trying to live in a gravitational field that is organized around your subduction. White terra was white terror on colonial earth. In short, this book seeks both to make the deceptively simple claim that racial difference is enacted through the philosophies and praxis of seventeenth- through nineteenth-century geology and to show how this was done and why it matters to the racial geophysics of now.

I use the term *geologic grammars* to mean the epistemes of geology as material, theoretical, and temporal praxis that surveys, categorizes, catalogs, and classifies minerals and the broader divisions that are established between Life and the inhuman. I attend to the origins of geology as a material colonial practice and a discipline that emerges in the wake of this extractive culture to furnish its ideological principles, conjoining the European colonial conception of (1) the earth as Global-World-Space; (2) the "History of Life" as a genealogical principle of racialized lineage; and (3) the Human coded by white apex or plateau.[11] In this triad of space-time figures, the earth becomes the mutually constituting ground both for implanting eugenic notions of progress and biocentrism under Humanism and for furnishing the material development and consolidation of European (and then settler colonial) white supremacy states through the accumulation of geopower.

Matter functions, then, as the normative regulator for racial formations in the maintenance of the extractive commons. The classificatory languages of property, properties, lineage, purity, refinement, improvement, base, organic, inorganic, brute, and inert function to establish hierarchical social and material systems as hegemonic. Thus, the spatialization of deep time, as the "ground" of a History of Life, becomes implicit in the ascription of race and of the extractive economies of empire. Furthermore, the languages of formation, deformation (process and outcome), strata, stratum (subtending, eruptive, etc.), and sedimentation (accumulation) and the qualities and dynamisms of time of the geologic (deep time, prehistory, natural history) become the very registers through which Western thought is formed as a self-ascribed active participant in worlding, resulting in a geophilosophy that is also a form of material and social geoengineering.

In this research I have sought to account for the construction of stratal acts of compression and the mobilization of a stratigraphic imagination in the constitution of an epistemic "whole" of material environments. As counterforce, the disassembly of material forces of oppression is an interdisciplinary project that must move against the apartheid of normative knowledge production and the "extreme discretion" (to use Saidiya Hartman's term) of the geosciences in addressing the role of race, gender, and sexuality in constructing the earth, as well as policing the separability of life from earth(s) in struggles for liberatory and possible futures. Acts of description in geology entail acts of desecration. The question of how to revisit geotrauma in a reparative mode and not renew its racialized desecration of collective geologic life is equally a question of epistemes of language and relation, as it is of the material infrastructures that deliver its affects. How can we remake the world if we do not give space and place to the sedimentary affects of geotrauma?

I personify geology in the position of Overseer, as a practitioner in the geoengineering of the earth and as a codifier of racial transactions in the optics of capture and extraction. Part of the work of unmaking the historical infrastructures of colonial earth is to shift the language and grammar of apprehension—*geo-logics*—and displace the normative regulatory language of extraction (most notably in natural resources) and its dual psychic placement of permissible extraction in racialized and gendered bodies. Colonial earth is defined by the spatial perspectivism of the Overseer's carceral look (since 1492): from the conquistadors' thirst for gold from the bow of a ship, the surveyors' technical gaze over landscape and into its depths, the imagined scopic regimes are from the fort, the castle, the skyscraper rather than the view from the undergrounds of the mines and doors of no return. The Overseer is linked with the geologic gaze and the graze of extraction over bodies, land, and relation. This geologic optic mobilizes a perspectivism that captures geopower for material, cartographic, and racial extraction to make whiteness a geologic superpower. This racial deficit model is a tectonic process that exerts pressure on forms of life below. The sheer weight of the plateau on every other form of planetary existence means that this geologic colonial exoskeleton, brought about by a desire for a transcendental form of whiteness, has triggered forms of foundational collapse of the very grounds of geologic life. If we single out the desire for supremacy from the modifier of whiteness, then the material dynamics of what is at stake become clearer. Supremacy is, in Friedrich Nietzsche's term, the will to power; it is power desired as a feeling of elevation, an affective state, geophysics that defies gravity. In short, supremacy is a will to geopower. And the color of that supremacy in colonial earth is white. The earth was used to buttress and narrativize the desire of racial supremacy as a geologic condition. It is not just that environmental determinisms scripted the earth in the geo-logics of white supremacy but that the persistence of racial determinants of difference outlasted the discourse precisely because they were scripted through the geoengineering of the earth (a set of processes seemingly unconcerned with questions of race). That is, race was materially made where nobody thought to be looking for it—in geology—because it did not have or need or depend on a human subject at the center of its analysis or empirics. And so, the question of racial justice and the question of giving the earth another future (beyond the cascade of ongoing extinctions) are intimately linked.

This book started with a line of doubt about what geology was and what kind of earth it made in its descriptive and extractive processes. Books have seams and layers that relate to other work that does not always make it out in the world. The doubt I followed was written through the footnotes and strange juxtapositions of racialized bodies and rocks, of accounts of stratal formations and Negro church singing (Lyell 1845; Yusoff 2018a, 74), of overstated claims about the politics of the time and the politics of end times to circumnavigate race-making. The insistence on what is and what is not geology policed subjectivity, and this was not incidental or accidental to how geologic lives were made. The subtended strata of race that white geology enacted could be found across underground tunnels and prison mines, through the footnotes of epistemic orders, and in the accounts of the work of geologists and their "discreet" race work in natural history and state institutions. Another map emerged of stratal pressure and subjective deformation, a geology of race that haunts the present. The violence of geology shadows the dusty corners of the archive, the footnote, the subjects that "carry" theory to the academic page and as "proxy" for degraded environments in policy reports. There is a historical and psychological underworld that centered and reproduced whiteness as an immutable value, like gold, executed through an extractive field of relations that was sustained by transformation of the earth.

It is important to state at the outset that this book centers white geology to dislodge the centrality of its claims and the normativity of its material-subjective praxis. I make these arguments primarily through the historical geographies of geology in dialogue with whom and what these geographic imaginaries projected over and into. In centering white geology, I have chosen to follow its epistemological muscle that denotes the affective and managerial infrastructures of race so that the profound consequences of its operative geo-logics might be seen more clearly. There is, however, a cost, which is the decentering of predominantly Indigenous, Brown, and Black thought around the temporalities of justice and the focus on the contemporary geologic struggles over "relatives not resources" (#Pipeline3). In reckoning with white geology, I do not want to reinstate those power geometries but to map them across the disciplinary divides that would keep them as discrete domains to redress these divides and the production of knowledge in my disciplinary home of geography, earth sciences, and the environmental humanities. I believe that the examination of geology has profound consequences for racialized presents and the kinds of futurity that are imagined in the context of intensifying environmental shifts and impacts. I do not think the white academy can adequately sit alongside the most radical dimensions of Indigenous and Black thought without itself doing the work of understanding how its own ontologies of the ground are sutured into the fabric of earth, its racialized architectures, and subjugation of subjects. To that end, I am suspicious of origin stories that gesture toward a totality of explanation and the selective "borrowing" of Indigenous, Brown, and Black thought in the service of white enlightenment that goes on in the academy.[12]

Alongside the critique of colonial earth and its geologic life that organizes this work I offer a density of description and theorizing—a thick time—that allows other possibilities to surface that are already conversant, in resistance and refusal, with the violent grammars of geology.[13] The geographies that inform the book's content are broadly based on my fieldwork in the United States and in colonial British archives—a tale of "geology goes to America and becomes a superpower." I focus on this specific historical geography because it is where professionalized models of geoengineering space become theorized and standardized in relation to colonial and settler colonial aspirations of extraction—aspirations that are predicated forms of racialized erasure and enclosure.[14] Historically, the disfigurement of racialized subjects—ongoing violence against Indigenous peoples and resistance to Black freedom and bodily autonomy—becomes the prerequisite for another set of subjects (geocoded white) to enjoy a "distanced" and controlled relationship with the dynamics of the earth. The outcome of this colonial earth experiment of terraforming through *terror*forming is both the production of the apartheid of geologic life and an increasingly unstable planetary ground. The colonial imagination of a universal geography was predicated on racialized forms of undergrounding in social space that established a dynamic of underground-overground that became planetary in its modes. This spatiotemporal model of differentiated geology and its "analytics of raciality" (to use Denise Ferreira da Silva's 2001 term) requires a structural analysis of temporal orders and the narrativizing of time, alongside an understanding of the geomorphology of space.[15] The coupling of the global imagination of human origins as a racialized hierarchy of time in the eighteenth century by paleontologists sat alongside the colonial aspiration of empires to transform the earth for the accrual of global geopower. As race spaced the inhuman gap, racial categories indicated distinct kinds of racialized geophysical relations in the constellation of what it meant to be a differentiated geologic subject, where geopower = racial power and racial power was exerted through anti-Black, anti-Brown, anti-Indigenous gravities.

Geologic Life is an attempt to understand those historic structures that govern thought and its material transformation of the earth—transformations that are destructive to Black and Brown life, as well as a multitude of forms of possibilities that constitute the expansiveness of the universe. I want to be clear about the centering of the material production of race in the critique of white geology. I have no desire to reproduce the tight strictures of identity politics whose histories I seek to dismantle or to delimit the geographic complexity of diaspora and the experiences of racial regimes.[16] I am concerned with how the colonial model of extraction has become a material principle that governs the production of planetary relations to matter and stratigraphically racialized relations to earth, and with how social theorists have mapped the stratigraphic imagination of earth processes into analytical models of society and social processes. I am by no means the first person to observe that racialized epistemes underwrite colonial space and its settler colonial presents. Race has been largely addressed in the context of questions of identity, geographies of diaspora, and the history of biology (as race science and as a governance of the body politic), as well as in the sphere of territory as a governance of land and minerals. There is a lacuna in work about how geology conjoins the political life of territory with the geologic lives of bodies, and how these are structurally sedimented across deep time and activated in the geophysics of space, so that the rift and the plateau are made into racially defined formations that govern the gravity of forms of geologic life. Which is also to show historically how the spatial dynamics of geopower and its extraction are made into racial power. The rift is a consequence of the colonial earth but not defined by its ends, and so might be understood alongside Frantz Fanon's (1963) concept of the abyssal as a space without definable grounds and with different durational qualities. The expansionist and exterminating geographies of colonialism discussed here are written in conjunction with an attention to the extractive grammars that are institutionalized to establish dominant power relations through descriptive modes (geographical imaginations and aesthetics).

As the map established the mobility of power over territory, so extractive geo-logics established a dominant interpretive grid of the division of land as resource and property and the regulatory division of the surface (and its political "present") and subterranean spaces and their deep time. These extractive principles shaped material relations and the affective infrastructures of neoliberalism's nows through racial capitalism, organized around an epistemic system of value of the human and inhuman, and its racial and gendered subjectification (see M. Wright 2015). *Geologic Life* focuses on the historic geographies of colonialism and its afterlives, but it is imbued with the archipelagoes of thought that emerge in Édouard Glissant's errant method. My method is isomorphic. This argument runs through the spaces that I am implicated in. Because I am a British citizen, my taxes continued to fund until 2015 the payments of the 1833 debt to compensate slave owners who shaped Britain, North America, and the Caribbean. The scholarly library where I was able to read original imprints of geologic texts sits on a square with the headquarters of key mining industries and their private equity brokers—Rio Tinto, BP—interspersed with the naval, military, and East India clubs. The Geologic Society that originated through funding released from the legacy of British slavery financial compensation is up the street. Charles Lyell, who wrote the matrix of race and geology through his *Travels in North America*, stands, marbled, outside the Royal Academy next door. The underground station has a facade of 150-million-year-old fossils, sea strata captured in Portland stone, and the London Stock Exchange is still the largest trader in the world in mineral and fossil resources, shaping worlds within worlds, one plateau at a time.[17] The key mining industry lobbying organization, the International Council on Mining and Metals, is based in London, as is the global metals price-fixing mechanism that sets the daily London Fix Price of precious metals determined by the London Bullion Market Association, and the London Metal Exchange mediates the indexing of value. Now, I sit not far from the zero degrees of Greenwich Mean Time that flung the lines of longitude across the globe like a net to secure British sea and geopower. These dense geographies participate in the geophysical realization of the Anthropocene and the ever-resurgent wake of white supremacy, where race is made to work forever essentializing categorizations of people to sanctify their permissible attachments to places and possibilities.

Racism is an inhuman idea, inaugurated and instituted through material ontologies—one that renders subjugation through a form of material debt burden and material psychosis that is organized around the stratal idea of whiteness as structure and position. Most important, these ontological categories of the inhuman-inhumane claim actual experience and hurt in ways that smother the density of those lives. The differences are important. Racial difference is embodied in exposure, as it is lived as weight and history. My relation to this anti-Black and anti-Brown violence is to rework the colonial archive and its world-making, with the worlds that are compromised and destroyed in reciprocity with its coming into being, worlds in which I am professionally and personally situated. As a "junior partner in whiteness," to use Frank Wilderson III's (2010) catchy sluice and slice for the Brown, I see this work as a structural redress in the grammars of geology and their deadly enclosures—rocks thrown at white geology for a different imagination of geologic life.

My idea for geologic life didn't start here. It was launched on another plateau on the Ross Ice Shelf in Antarctica, with its time-bending perspectivisms and inhuman conditions that denaturalize Life's planetary claims (Yusoff 2005). In the long arc of the work that followed that exposure, it became increasingly clear that there was no way to speak about rocks without white supremacy, extraction without race, climate change without scientific and environmental racism. It took time to understand how theories of the earth were involved in the erasure of geology's silent partner in matter, the inhuman as a subjective form. The absence of a discourse on racial justice in the academy demonstrates how the connections between environmental thought and race have been studiously overlooked. The clear and beautiful reckoning of Toni Morrison might secure the point of the project: "to question the very notion of white progress, the very idea of racial superiority, of whiteness as privileged place in the evolutionary ladder of humankind, and to meditate on the fraudulent, self-destroying philosophy of that superiority" (2019, 180). To translate this into material relations is what I wanted for this work, to question the grammars of geology and the natural state of whiteness as a right to property. The white supremacy of matter is my intellectual inheritance and professional genealogy as a geographer even as I have lived a raced life. My idea was to construct interdisciplinary bridges between the classifications of the natural sciences and geophilosophy, to draw out the hidden violence and suppressions that create an episteme of the earth. Geography's and geology's legacy of mapping of space, materializing the extraction economy and stratifying social relations, makes them the most colonial of all disciplines. And as the discipline's illegitimate offspring, I write toward a new language of geologic relations.

It is worth remembering that the unnatural being of whiteness was a deliberate cut, an incision in geologic life, to forge a supergeopower. Paleontologists argued *for* the unnaturalness of whiteness, its levitational qualities, above and over the earth, and all who resided in that other "inhuman" category. This was the argument that geologists were making: supremacy through the epistemologies of material extraction and earth transformation. They just hadn't figured on quite how unnatural they would become. James Baldwin (1955) named it the lie of whiteness. The colonial photographers and planners must have known it, as they arranged naked and defiled persons in genealogic groupings, children and parents without hands for keepsakes to send back home to Belgium, burnt and hanging bodies to circulate as souvenir postcards, and Harvard chair Louis Agassiz's secret velvet-lined cabinets of his most intimate horror of Black subjects. All the skulls and skins and sliced-up flesh in the museums. Science and scientists made the story respectable. All the cabinets of crania and labia neatly labeled. All the life taken, and breath withdrawn. The shock forward of violence and what it takes. This was the exchange for the levitation of whiteness and the earth it made. It is still the exchange. And it will be, until the inhumane in the heart of the humanities owns and works against its inheritance rather than participating in its ongoing social reproduction.

The methodological implication of concentrating on colonial geologics is to obscure the density of life lived otherwise (Frodeman 2003). The poet Dionne Brand reminds us of the shortened life of the colonized in this temporal approach: "Conquest makes the life of the Conquered seem brief. . . . All their lives collapsed into one life. A summary" (2018, Verso 40.3). Colonization reset the clocks and, thus, it reset time and space, and controls through the production of its origin (and its genealogical reproduction). I often wished that this was a different project, something more expansive than the confinements of these temporal shafts. But I write from geography and a belief in the importance of challenging spatial expressions of power and their material manifestations. To whom I write is a different matter entirely.

I want to bring geologic materialism into the open so that it might find some redress as a site of witnessing in the geotrauma of colonialism and the geophysics of the day. Those who live this and fight for the visibility and remediation of the connection between racialized and environmental life know this already. As a geographer, I understand that there cannot be any livable environmental conditions without Black, Indigenous, and Brown freedom. There is no way to write about geology without race—matter, water, earth, air, oil, soil, phosphorus lightning, structure relations of power and possibility, and the sediment of anti-Blackness and settler colonialism—as geologic grammars organize the material conditions of life, its poverty and joy, in the rough textures of its resistance. But disciplinary silos and the political representation of ontological categories do their work. Only the privilege of whiteness and its organized categories of disavowal allow the erasure of the world it has raced—raced for extraction and exposure. This work is a bridge, then, an underpass or tunnel beneath the confidence of surface structures, a place to cross over disciplines and genres in the rifts of race and matter, to find some different earths, some less wretched relation in the structures of the white supremacy of matter, written as the planet burns.[18] These are violent histories, and there is hurt in their surfacing. But there are handholds of endurance, patience, and joy too. To read geology only as colonial telos is to read too much into the accomplishment of that world-building. In these resistances other orders of time quietly (and forcefully) challenge the opaque liquidity of the surface to imagine the day differently. In the place of the plateau and its subject in Overseers' earth, the thought that lives in its rifts is inventive precisely because it is attentive to the ground in which it traverses and builds. The rift only appears historically as rift because of the geophysics of the plateau and its levitation through subjection. However, the rift is not just a negative placement but a different perspectivism and spatiality that exceed the dialectical (or inverse) relation. It is subtended and broken by colonial earths, but it also is a space of its own that has a different geography, gravity, and dimensionality.

i began to think of body
as {an absence}
my body became multitudinous
as the ground is vertiginous
no longer reliant
on gravitational pull
our body was
no longer the book of bone
the detrimental predisposition
that took hold of us
in the depths
was now
luminous
(Ife 2021, 67)

The book is divided into three parts that examine geologic life and the geospatial emergence of the plateau and rift and its concomitant geophysics. Part I, "Geology's Margins," examines the marginalia of geology in stories of Life told through the geocoding of racially differentiated bodies and the analytic of the rift zone told through the methodology of underground aesthetics. Part II, "Geologic Histories and Theories," examines the empirics of stratal subjection through stratigraphic philosophies, grammars, and the emergence of geopower as an analytic to apprehend and govern the geophysics of the strata. Part III, "Inhuman Epistemologies," examines the empirical manifestations of the inhuman in the Southern United States as deep time, undergrounding, and genealogy in the convict lease mines of Alabama and the black earths of urban modernity. Finally, part IV, "Paradigms of Geologic Life," looks at the mine as a paradigm for the racialized organization of geologic life.

GEOLOGIC LIFE ANALYTIC

PROBLEMATIC: To share life on the surface with others and contingent "deep" earth (other geologic eras and their interests/affiliations; contingency of earth as already many and multiple) in the context of a new historiography of the Earth called the Anthropocene.

CONTEXT: containment/extinguishment/enclosure versus every resistant thing (and the imperfect organization of that) ≥ struggle of "relatants" (Glissant)[1]/"existents" (Elizabeth Povinelli) outside of biocentric modes and the imposition of racialized arrangements of Life.[2]

DYNAMIC:

1 Rising tide of late liberalism, *this thing we're in* ≤ ≥ toxic and destabilized geophysical/geobiological processes, *pyroearth*;
2 The reorganization of "every resistant thing" as soon as it shows itself (or surfaces from the subterranean) under racial capitalism and its ontopolitical affects;
3 Colonial Geology's anti-Indigenous, anti-Black, anti-Brown Earth policy

LOCATION: **Rift** as: (1) **Geophysics** of the earth[3] and (2) **Geoforces**, "quiet" (Campt 2017)/subterranean/quasi events (Povinelli 2016); "senses as theoreticians" (Wynter riffing on Karl Marx); **Geopoetics/Geoaesthetics** as decolonizing force (Glissant; Césaire; Elizabeth Grosz; Ferreira da Silva); Underground as Undercommons/abyssal depths (Fanon; David Marriott; Fred Moten); Wayward Lives (Hartman) or just killing the script.

TARGETS: genealogy[4] + colonialism > [well-critiqued arrangement of the family-nation] *but* also a "toxic genealogy" (Brand), the site of the categorization and resolution of geologic settler colonial grammars that are anti-Black and mobilize a specific conceptualization of the arrangements of (biocentric + geologic) life around that anti-Blackness in the "slave-mineral complex" (anti-Blackness ≠+ extractive property and properties relation = geologic grammars of settler colonialism and slavery) + settler colonialism in the mastery of matter.

The color of Reason: Fear of an inhuman planet
(black/bla_k* M. NourbeSe Philip)

Natural Philosophy = Moral + Racial Theory[5]

Theoretical Formation of an Anti-Black Planet
through colonial geoengineering

TARGETS ARE ON THE BACK OF: Blackness; Indigeneity; Brown; life on earth; "resistant things"; "open boats" (Glissant), subjects caught between environmental inundation and hardening of geopolitical borders.

TASK: Establish a grammar of geology after biocentric Life and geo-logic enclosures governed by the *extraction principle*.

GEO-LOGICS: Politics of Inhuman, Nonlife, Political "presents," Earth

CONCEPTUAL MATTER: Materialism; category of the Inhuman; Subjectivity-Relation, redress to materialities of subjugation.

HOW: underground aesthetics, Rift Theory.

INTERLUDE:

SCENE: Colonial Earths—Broken Earths

QUESTION: How do these histories of the racialization of matter *matter* for the political present and the alleviation of its racial gravities? How do these histories materialize the building of worlds and navigation in a world on fire?

STANCE: Geologically Invaded: invocation
to open the paths of the geologic.

ANALYTICS: **#GEOLOGIES OF RACE**: the historical organization of materiality through colonialism and its afterlives in the present. The present pasts of geology.

#GEOLOGICREALISM: geology made manifest by the three ways of the Inhuman:

1. **Matter** that is understood through the normative lens of *natural resource* and philosophies of Reason (Nature-Culture divide) that established modes of enclosure and capture through geologic grammars of identification and valuation.
2. **Race** constructed in the discipline and practices of geology (through the subfields of paleontology and comparative anatomy), forged through the praxis of *extraction*.
3. **Relation** such as alterity, outside, cosmic, inhuman, and raced life. Blackness, Brownness, and Indigeneity are also a praxis in the expansion of the temporal logics of relation through radically different experiences of materiality, often made as an implicit critique of the regulatory structure of Reason's work as empire, race, and slavery, and in its violent structures of materiality and mattering.

This combined historical grammar of geology—of the *inhuman*—established the stability of the object of property for extraction and valuation and its modes of apprehension. This historical relation to the earth (and to subjects inundated by the earth) is based on an *extractive principle*.

Imperative: Moving against the Extraction Principle

Natural resources ($\sqrt{\textit{colonialism}}$ + afterlives)

= extraction + capture (epistemic and material enclosure for the production of value)

personhood ($\sqrt{\textit{slavery}}$ + afterlives)

= extraction (from precolonial relations) + capture (through the technology of race and geology) + governance of blackened forms of life through temporal sovereignty and material epistemologies

≠ forms of material (spatial extraction ± containment) and subjective ("slave"/"native") enclosures.

Inhuman dialectic of *Bios* + *Geos* = White Geology + Breaking Earths:

> Human ≤ subject, being, site
>
> Matter = organic, agentic
>
> Inhuman ≥ void, matter
>
> Matter = inorganic, mute/dead

Inhumanism is a (vertical) stratigraphic geopolitics—what might be understood as the *racial geophysics of anti-Indigenous, anti-Black, and anti-Brown gravities*—a raced axis and historic material dynamics that pushes down to lift up. Or the racial flip of geology has color.

> INSTITUTION: *The Inhumanities* (action: epistemic shift through the geologic rift).

The Inhumanities is a "parallel geology" to that of the humanities and its Western construction that selected the European subject as the apex of achievement and material development as its practice. *The Inhumanities* aims at a different act of accumulation—a charge sheet against colonial earth and a redress of its geo-logics through reparative antiracist geoethics that attends to the nonrenewal of geotrauma.

GEOLOGIC LIFE LEXICON

GEOLOGIC LIFE: An analytic of *Geologic Life* is a mode of apprehending the earth as praxis of exchange across temporal scales and through material states. This formation of subjectivity is most relevant in the aggregation of racial geophysics—where historical racial forces combine with contemporary structures of exposure to create geophysical conditions of race and the position of the racialized poor as subterranean epistemic and extractive strata. *Geologic Life* can substantiate an analytic for geography that positions inhuman forces in political terms as preceding biopolitical concepts of life and understands changes of state as a political domain and a racial axis that have geophysical consequences. Geologic life is a historical regime of material power that produces and reproduces subjects and material worlds. Understanding **geology as a medium of struggle** that defines differentiated relations and changes of state, geologic life invites a rethinking of granular geologies as well as expansive reconceptualization of inhuman politics and histories, and the imbrication of geologies of harm, which might be categorized as geotrauma.[1] The cessation of harm—in terms of ongoing colonial modalities of geotrauma through geo-logics of extraction and their nonrepetition—involves displacing the normative and reproductive modes of geologic life that organize futurity through a racialized praxis (or "white geology").

GEOLOGIC SUBJECTS: We might see the development of humanist modalities of subjecthood, alongside the development of geologic grammars of the transformation of the earth, as existing along a fault line of the experience of a precariously imagined and defiant concept of Man and the Human. This humanist project, based on the fear of an inhuman planet, also created inhuman subjects, both as enslaved "inhuman" property and through racialized geology practices and proximities to ore bodies and geochemicals. The embodiment of differentiated mineralogy was structurally organized as specific forms of social mattering—the relation between mobility and minerality in imperialism—as mines, minerals, geochemical agriculture, and fuels enacted deadly forms of becoming geologic. In the rapacious geotraumas of the mine, for example, bodies carry inheritances from the underground as a corporeal colonial afterlife. Specific geologic formations and ore bodies differentiate persons-as-bodies across inherited geotraumas and environmental change.

GEOPHYSICS OF RACE: The geophysics of race conceptualizes the material geoforces and dimensional stratifications that historically enact race as a contextual quality of space, prior to any biopolitical or ideological arrangements (which often act as a distraction to the structural work of racialized material geopower). The residual capacity of geology to hold and exhibit histories of raced life, as distinct geomorphic states and stratigraphic features, speaks to broader questions of social justice, climate change, and colonial afterlives. If we consider the legacy effects of geology and its forms of afterlives as the geo-logics of racialized spatial forms of production, there is a need to take seriously the geohistories and geomorphology of racializing and racial forms, particularly as these migrate into complex multiethnic formations. That is, we cannot think about how race matters outside of geology.

The differentialized temporal forms of geology might also provide a possibility to think differently about race and racialization outside of the imposed colonial divisions of human and inhuman subjectification and the foci on identity (as partitioned from inhuman and inhumane conditions and spatial forms). This also means thinking about geologic subjectivity beyond a fungible body caught in inhuman conditions as a collective, socially policed, and epistemically situated condition of racialization that is realized across material forms and epistemic modes. Thus, race becomes performed as a geophysical state. This is what I call "geophysics of race," to comprehend erasures and their ongoing spatial and psychic affects that are experienced through the materiality and its ontological orders. The racial dynamics of geologic life not only have been assimilated into distinct geographic zones but also exist in the geophysical pressures of space and the syntax in which it is rendered (physically and psychically). Gravity, as a set of forces, imposes spatial hierarchies and densities that define social conditions of pressure and positionality above the earth, as well as the countergravities of resistance as earthbound.

GEOPOETICS: Emerging in colonialism, geology created a language for the description of matter, accumulation, and dispossession and a legacy of racialized subjects. These infrastructures of languaging materiality have a complex and contested dynamic that creates an occupancy in the material world—forms of geologic life—even as the grammars of geology work against languages that describe that interpolation with an ore body or a silty sedimentation. Geopoetics within the field of geology was a key speculative method of doing geology itself, a language that presupposed the stretch needed to inhabit the occupation of past geologic worlds. While Enlightenment traditions had expansive methodological interdisciplinarity different from the contemporary enclosures of disciplines, geologic narratives continued in colonial relations to create discourses of "new" terrorized environments and to produce geographic imaginations of settler states. Angela Last's (2015, 2017) work on Martinique writers who used geopoetics to undo geopolitics is relevant, as she recognizes how writers such as Aimé Césaire and Daniel Maximin turned Enlightenment languages of separation and description in on themselves, both to invent something new and to politically intervene in creating possibilities for attachment, while also uncovering a historical precedent of the poetics of geography and geology. The reversal of the "flow" between geopoetics and the geopolitical conquest of colonialism that Caribbean writers achieved demonstrated that they recognized very clearly the violent work geopoetics did inside the colonial condition. By giving origin stories new futures and participating in providing languages for geologic subjectivity, they gave geopoetics a less deadly decolonial future, as well as crafting a less masculine triumphalism in its writing, which defines the colonial trajectory and its afterlives in academic methods.

GEOPOWER: Understanding how geopower is distributed across states (geopolitical and geophysical) and strata of matter (as social, racial, and embodied) attunes us to the changes of state necessary to ameliorate the forms of disassociation that racial capitalism practices to border and contain events. Understanding how geotrauma can do political work in the future can provide a way to refuse the reiteration of colonial languages of extraction. An inhuman analytic that can move, as geology—

as both located and rapturous in time—is needed to understand the multiple forces of violence and therefore intervene in the sites of their reproduction and nonrenewal. We can see the racialization of space as an ongoing geotrauma. A by-product of decolonizing space and time is decolonizing materiality and normative approaches to matter that produce Anthropocene states.

GEOTRAUMA: Geotrauma is a way to understand geology as a praxis of struggle and to see the earth as iterative and archiving of those struggles. Geotrauma is the result of epistemic and material partitions in the relation between subjective attachments and inorganic forces, such as the dispossession of land, ecologies, and lifeworlds. And it is a burden of dispossessed geographies that are carried by the displaced. Geotrauma also conditions the transformation of geopower through racialized geomorphic labor into state power (as racialized geopolitics). Geotrauma provides a way to understand the manipulation of psychic states of the racialized for the attainment of geophysical states of the privileged. Geology is a site in the repetition of violence, as a material practice and as a heuristic for parsing the category of the inhuman. At the same time, the breakage of languages that congeal during geologic and epochal shifts, often because of violence, releases into being new languages of the earth that must turn against that epistemic and material violence. Rethinking methodological ways of rendering violence visible is a way to retool power through connecting forms of geologic violence across different state actions and states of being. Attachments to other fault lines of inhuman memory might act as sentinels to different political futures, established through temporal and tender provocations.

INHUMANITIES: The inhumanities is a parallel epistemological practice to the humanities (and environmental humanities) that centers those who have been historically denied (and continue to be denied) the privileged terms and spatial access of the human. Another way to say this is that humanist and settler futurity is organized through historical geographies of uneven environments through racialized modes of extraction and depletion, or racial deficits. The inhumanities, as a parallel institution, would seek to center the impacts of geotrauma as the methodological aspiration for the remediation of harm. The inhumanities would (1) provide methodologies for moving against normative languages of materiality that "earth" in particular homogenizing ways, for example, natural resources; (2) see the geologic as a medium of "shifts"/a shifting terrain in politics and sites of struggle; and (3) understand bodies as/in geochemistry, subject to geophysical affects, which in turn affect planetary states (here there is a real need to be cautious around a return to an indifferent elemental determinism) as a way to aggregate planetary scale without reproducing its depoliticizing affects.

INHUMAN MEMORY: If a rock can be a collector of stories of disobedience, a witness to rebellion and the countenance that claims time outside the colonial clock and its climate-changed earth, then different epistemologies of inhuman memory are a praxis for giving the earth a different future (to the trajectory of colonial earth, aka the Anthropocene). If earth archives are understood as a site of redress—building spaces of remembrance, seeing ghosts, and shifting temporalities in the colonial afterlives of extractive earth—then new museums of earth need to be built and imagined. Geology or the earth, understood as a medium of racial struggle, requires new forms of memory practices and expanded archives for just *environmental* futures. Another way of posing this is to ask: What inhuman epistemologies of memory and forms of genealogy are necessary to be able to have a rock in the family?

SHOCK FORWARD: Geotrauma is never just erasure; it has a forward effect, a *shock forward*, both in the organization of captive and carceral futures and in the survivance of other futures. What grows around geotrauma is both what sustains the legacy of those impacts and what allows that which is erased to have a future or an ancestral claim that moves with and in the present. Claims on the future made by the ghosts of geology disrupt normative accounts of materiality (as a shadow geology), writing against the apartheids of its reproduction, dislodging languages that structurally carry the division between human and inhuman. The ghosts of geology smudge the borders of material and subjective states in the ongoing violent histories of geoengineering the conditions and categories of race through geologic formations.

The ontological imbrication of the inhuman and the plasticity of strata—a malleability that is evident in geopolitical and geosubjective terms—remind us that questions of geology are always questions of power. Theorizing and languaging this imbrication is precisely how to move beyond a normative geology that is anti-Indigenous, anti-Black, and anti-Brown in its historic formation and current politics. Geology is not just a material zone that is used to do racial and ethnic work—as in weaponized environments—but it is a praxis in the stabilization of political and social forms that require racial deficits to function, both affectively and materially (i.e., racial capitalism). Disrupting the general account of materiality and inhuman concepts and their stabilization is part of how epistemic structures become open to change.

TACTICS OF THE EARTHBOUND: *Tactics of the earthbound* takes seriously counterarchives of the earth that deploy environmental knowledges and inhuman memory as a mode of generating spatial and political freedom. Historically, acts of freedom by Indigenous, Black, Brown, and racialized peoples in response to the violence of colonialism rely on environmental knowledges and tacit understandings of enacting with the earth in ways that the Overseer, enslaver, or state could not. These tactics of the earthbound were developed alongside and in resistance to the organization of colonial natures and the segregation of human and earth. Tactics of the earthbound shift the earth as the medium of struggle to a medium of survivance of other earths (e.g., marronage, caves of resistance) and thus create another geopolity. Clyde Woods in *Development Arrested* (1998) called this practice "Blues epistemologies"—whereby he saw the blues as a way of theorizing Black life in the Mississippi Delta and as a methodological repair to the erasure of Black history and in direct challenge to plantation bloc epistemologies of Black life. He argues that folk music was both a space of engagement with the environmental geographies of the Delta and a collective practice of collecting history and theorizing life as it was lived under conditions of environmental exposure to those geographies. The time of the earthbound is time in racial deficit and spatial undergrounds, doing time for promotion to the surface of capital flows, under the geologic praxis of racial capitalism. Tactics of the earthbound, without recourse to the distancing practices from geoforces, turn the weaponization of inhuman proximity into forms of inhuman intimacy and tender. Which is to say, the separation between environment and race needs to change spatially as well as socially for there to be any corrective to how geosocial worlds emerge.

I Geology's Margins

1 Insurgent Geology and Fugitive Life

flesh of the flesh of the world / panting with the very movement of the world —AIMÉ CÉSAIRE, *Return to My Native Land*

The different, not the identical, is the elementary particle. —ÉDOUARD GLISSANT, in Manthia Diawara's *One World in Relation*

Geologically Invaded

Geology invades the very thought of Life as it subtends its differentiated possibilities.[1] Its rubble, fossil pasts and future fossils, catches us in the imaginary of deep time's spatial expansion and temporal deeps. In grammars of extraction and subduction, in the weight of gravity's grounding to its rocky shore, another political geology arrives that haunts geology: the geoforces of concealed histories and undeclared racism. Geoforces might be named "the tremor by which being is provoked" (Glissant 2010, 7), the earth's and ours, as geopower captures the governance of its trajectory. The brokerage of planetary transformation is cosmic materialisms' longue durée, while the grit and grind of geology are the shit and shine in subjugating relations and racializing grammars of extraction. *Black earth, red earth, white surface.* Processes of undergrounding are related to processes of surfacing. Racial undergrounds are the process by which whiteness is materially elevated, establishing the plasticity of social space through geophysical appropriation.

Geophysics pushes some bodies deeper into the earth and weighs a subterranean positionality upon their personage. *Supine.*

Rocks are also the institution and disciplinary syntax of deep time, the forever against which the urgency of time holds itself, panting against extinction, Césaire's "flesh of the flesh of the world." Stop the clocks. Become a fossil (*or not*). Find some loamy soil, lie down, and imagine yourself immersed in geologic processes. A genealogy given to memorialization and legacy events (and thus a medium of genealogical claims). This is one narrative of geologic life found in the Anthropocene imaginary of future fossils. Another geostory is getting pushed into erasure, subducted into the gendered and sexualized breaks of extraction, in the torque of its blackened stratum, living a life in the rift of material debt burdens and want. The hard rock lesson to this geostory is to find your passage in flux and learn a different illumination. This is another organization of the origination of geologic subjects. Life's positioning against a geologic backdrop is narrated as contrapuntal to all that it willfully puts itself apart from, all that is designated nonlife, nonbeing, and inhuman (*hear*, obsidian, *speak in* granite and find rocks in your mouth, *lick* it and taste elements).[2] The inhuman impasse of deep time is all these imaginations of a linear way-back abstraction that obfuscates the pockets of inhumane occupations that have defined geologic planetary praxis (see also Gabrys 2018).

This is a text written in geology's margin.

Geologic Life argues for the geophysical underpinnings of Life and race as mutually constituted through a historical geography of colonial geology. Using an empirical method of understanding the geophysical dimensions of colonialism and its afterlives, I expose how the surfaces of the white supremacy of matter are maintained, practiced, and imagined, mapping sites for its dismantling. First, I start with a moment of colonial consolidation of the techniques of geology in the history of Western geology—*white geology*—and discuss key moments of its epistemic and theoretical formation to situate how geology identifies and structures much more than rocks, metals, and minerals. And how racialization (and its gendered and sexualized forms) became inextricably imbricated in ideas of the earth and the narrativizing of temporality, executed through material practices of extraction. As stratigraphic concepts retrofitted the thought of racial difference as a naturalized history rather than a political geology of forms, race is conceived by

Georges Cuvier (1813, 1827), Lyell (1863), Agassiz (1850a, 1850b, 1850c, 1850d), and a host of other geologists as both the process and the outcome of geologic forces. Thus, European white supremacy asserts its "arrival" as verified by both the epistemic and the material field of existence, as if human and natural history have collaborated to deliver the apex of human species just at the right historic moment for its spatial expansion (colonialism). The double bind in which time was spatialized into abstraction to make geography itself abstract and open to colonization thereby dissolved the bonds of attachment and location as an impediment to violent theft (and signaled some subjects' exclusion from the time of politics). If eugenics came to define the social management of populations as a biological policing of difference and its associated anxieties, the geologies of race organized an argument for the spatial management of territory and bodies as territory. As Fereirra da Silva astutely argues: "As anthropology defined culture as a product of historical processes, race difference was re-signified as a substantive sign of the long temporal processes which fixed certain bodily marks in isolated pre-historical conditions. Initially a product of scientific strategies of intervention deployed in the study of man as a natural being, the category of race difference was now re-signified as a (pre-conceptual and prehistorical) signifier of cultural difference, one associated with great spatial (geographic) and quasi-eternal temporal (geological) processes" (2001, 433).

Origins of racial difference were seconded to deep time to free up near-surface space for the occupation of settler colonialism. Second, I move to a discussion of how this geologic praxis is productive of racialized forms of life—*geologies of race*—that are defined by distinct material conditions that substantiate the geophysical conditions of subjective life. Third, I introduce the analytic of *geologic life* as a challenge to colonial grammars of geology and their syntax of the ground. Throughout the historic imposition of geologic grammars and regimes of extractive geopower, an alternative praxis of matter and its languages was formed in refusal and resistance to these deformations. These counterforces of geology can be seen as *insurgent geophysics*, drawing other earths into being, through sense and geographic ideation. Throughout this book, in its material, historical, and theoretical organization, runs a commitment to modes of description and density that discern the breadth of geologic life.

The interregnum of geologic life is both a site in the production of the structural violence of the politics of Life *and* a field of possibility that opens as a rift, within and undesignated by its classifications. I argue that white geology constructed racialized earth—as a privileged *genre* (per Wynter) of the earth—through its epistemic, material, and geographic orders. And that this earth cannot then be understood without reckoning with the violent placement of Indigenous, Brown, and Black subjects in the ground and grammar of that extraction.[3] Thus, neither race nor the earth (as it is understood in the history of Western scientific and philosophical thought) can be comprehended without the other, as occupied geographies of the ground subtend and subvert formations of the earth and inhuman concepts populate race. The foundational racialization of matter since 1492 reconstructs all operative ontologies of materiality and materialisms precisely because racialized subjects refute the category of the inhuman (Yusoff 2021). Both race and earth are attendant to the repressed refusal of inorganic life within the political formation of the Life of the subject (as *bios*). Considering how the inhuman historically shaped the institution of Being as a mode of apprehension and a specific material arrangement of the earth and its partition (as *geos*), *Geologic Life* aims at defamiliarizing the history of geology to expand the visible sphere of its implicatedness as a semantic, somatic, and sedimentary field of relations. It is a compressed field of relation that constructs beyond the sphere of geology and the earth sciences to underpin the very basis of colonial lives and their continuance in the present (which maps into the temporal geo-logics and conditions of futurity). Geologic Life *is an analytic that joins together the histories of environmental thought and geographies of degradation with the organization of subjective fungibility in the praxis of extraction, in conversation with openings into the unvalorized cosmic and embodied corporealities of these inhuman-inhumane worlds.*

The immediate context of the geologic scene in which this discussion takes place is the perspectivism of the Anthropocene and its accounts of geologic subjectivity that are arising anew on the fossil shore, down the mine, and into an epoch, to walk in the colonial footprints of geology through the inventive figure of Anthropos and its Anthropogenesis (see Yusoff 2016). As anthropogenic climate change is lithically *un*stabling the ground of the human, it is also unsettling the naturalization of concepts that were grounded in the commonsense arrangements of Enlightenment matter that divided this *bios* and that *geos*. The awkward equivalences of how to mesh that which is separated in syntax as *Life on Earth* and *Planet Earth* (seen in the narrative frames of the natural history museum) contextualizes current spheres of inhabitation, belonging, alienation, extinction, and expulsion. Life has secured Earth, the geostory goes, hominid insurance against an indifferent cosmos (first as amoeba then as Man, genealogy secures arrival on the shore of white settlerism).

Cold alabaster. *Fossilized* ammonite. Geology materially recalls a forgotten list of prior geochemical earths and inhabitations, changing locations, and energy balances, as the earth holds histories of surface erasures and subjugations as future possible or past relation. Forgotten socialities have geologic scenes. Inhuman memory is a set of conversations between fossilized and nonfossilized moments. The earth is not general but a specific, contingent coagulation of geoforces and mineralogical evolutions out of which relation emerges and geopower is cooked and coded (as heteropatriarchy, family, sex, nation-state, capitalism, mastery, etc.). Cosmic materialisms stretch the reach of the universe and return in the shared quotidian of minerals and bodies against and as the cosmic storm. A casual litany of (in) difference. Interstellar. Interstitial. Intermural. As mineralogical substratum and metallurgic exoskeleton, geology does not so much subtend life but extends, calcifies, and corrals it into affiliation with the cosmos, deep earth, and the material instantiations that are understood as the terrain of the political present. Black rocks, dark fuels, and shiny minerals imagined already as natural resources ground colonialism and the heteropatriarchy of settler states.

Historically, the field of geology as the normative telos of materiality instigated the division of geologic properties into colonial property and arranged subjects coded as *bios* (Life) or *geos* (inhuman—organic or inorganic). Alongside these epistemologies of matter, the empirics of geologic materialities structurally organize possible forms of life, from the deep-sea ocean vents to the bony possibility of our osteo-ability. Our entanglement with geology is marked by the governance and extravagance of these material and epistemic dynamics of the inhuman. The aphorism *bone deep* suggests the primacy of this relation as a corporeal form. Geology produces arrangements and derangements of geopower that are maintained in the asymmetries of access and accumulation of geologic value. From the coal imperialisms of the British Empire to fluvial plantations of cotton, rubber, and rice that substantiated the material manifestation of the master, geology is a form of global and subjective power. Oil and gas empires fuel populism and manifest in the weight of crushed rocks and silica in blackened lungs. Those on the tectonic end of abrupt geologic movements live, die, and survive with the flex of the strata and its storms, in the shadow of ventful eruptions, in the grasp of state-based states of protection and abandon. There is a geochemical duration to breath, says Glissant, a "black lick beneath the wind!" (1997, 39). The inhuman breath-taste of the unsaid and

FIGURE 1.1 Sinclair motor oil can. Kern County Museum, Bakersfield, California. Photo by author, 2014.

unsayable lingers in reactant memory, isotopes between the living and the dead. The body is not just an archive of geology but its resident participant, an earth system that hums in reciprocity and temporal correspondence with earthly matter.

Flesh of Geology

There is an intimate geology that runs through the mineral molecules of bodies, the chemical breath of the atmosphere, the humus of the soil and into cellular architectures of life. It is a geology that finds and builds a differentiated flesh in relation, as substance and surround, from the more obvious toxic residues of chemicals and heavy metals that congeal around the sites of mining, traffic, and industrial infrastructures to the multigenerational shifts induced by epigenetic forms of geologic stress that leave geochemical signatures on bone and bodies. Geology builds bodies and it takes them apart. From the mineralogical fortification of rice, toothpaste, and the iron supplements that build muscle to the open rift valleys in which fluvial silt and hominid species take hold. Geology *is* flesh, and so political questions of and about the flesh are bound up with geology; the pleasure that opens flesh, its vulnerability and brutality, and how it becomes sensate and sensual, is a geologic question.

Understanding inhuman materiality is inseparable from gendered and sexed accounts of the body. Thinking with geologic roots queer notions of materiality and its scripted genealogy through modes of inhuman reproduction and affiliation that affect the account of attachment and kin (see Chen 2012; Yusoff 2015b). The communicative sociability of geologic materials is tied up with the body's own "earth system" and its ability to process through its own (and the earth's) mineral compounds and their intimate thresholds of densities or exposure, so geology participates in the designation of subjectivity and its forms of subjugation.[4] There is no elemental thinking without these histories, no purity in elements that is not a fiction of relocation and recontextualization. Historically, the mark of the inhuman forecloses subject positions and possibilities, rendering Black life as flesh (per Spillers 2003, 206) to which anything can be done.[5] The flesh of geology in the colonial imagination is a divisible muscularity and geoforce that renders the territorial expansion and excess of colonial value through the contraction of a person into a thing. In this nonconsensual collaboration with inhuman materiality, both as a property of energy and in concert with other energy sources (sugar, coal, mineral), slavery and its indentured and convict kin weaponized the redistribution of energy around the globe through the flesh of Black and Brown bodies. The inhuman, then, when understood as an intimacy, opens to both regimes of violence and an unsaturated form of belonging that untrains the attachments that hitherto accrue around the inhuman as exterior to the body and unimplicated subjective struggles.

The material fact of living is tied up in the mineralogical bundle and bounty, perforated by *geoforces*. These processes literally hold us up or pull us down: the geophysics of gravity, the boniness of skeletal walking, the liquidity of bodies.[6] All the shifting of moraines, the erosion at the beach, the eruptive core, and thermal vents. It is this subtending context of geologic materiality that frames every question about modes of existence, forms of identity, and practices of futurity. As Glissant writes, "Some subterranean (submarine) force repressed what northern volcanoes supplied" (1997, 205). His writing sought to "nail solar keys to the measured door" of concrete forces that "gave over to what was elemental" (39). What is elemental, according to Glissant, is not identical but difference. To extrapolate Glissant's amalgam, the geoforces of the cosmos give material expression to the experience of subjectivity as differentiated, and as such, as forces that are pulling against the homogenizing affects of normativity that would make the elemental "pure."

The endless saturation of the material world; the taste of geology, its saltiness on the cheek; the iron that restrains, presses into flesh, and makes travel possible across the rail tracks; the petrol fumes laid on the back of the tongue; lines of asphalt that pave the pace and densities of cities; the metal humming down the highway looking for adventure; the atomic fissures as origination; the bedrock that finally gives solid immutable ground to the strike; the performative rock bling of gemstones that uplifts social status; the dark water of mines in which children pan for pieces of the shiny; the gold that connects in the connector; the warming glow of a liquid crystal display that says all manner of things; the endless aggregate that fills in and provides a surface for something else; the limestone, the gravel, the granite that paves; the ambient dust that, claylike, clogs lungs; the cycling of air, water, and minerals; calcium phosphate of bones and osteoporosis; iron and the lagging deficiency in anemia; magnesium, sodium, potassium, phosphorous, sulfur, nitrogen, chloride; traces of iodine, fluorine, the neurotoxins that accumulate to cause loss in sensory systems over often indeterminate time scales; the lead and arsenic in the drinking water; the malaise of nickel and chronic fatigue; petrochemicals in food and runoff; the calcium in combination with either oxalate or phosphate that forms hard-to-pass kidney "stones"; trace lithium and beryllium; the body burdens of lead, mercury, and cadmium; cold steel-cocked flint. We're coal diamonds in mineralogical middens. Almost 99 percent of the mass of the human body is made up of the six elements of the solar system (oxygen, carbon, hydrogen, nitrogen, calcium, and phosphorus) and the remainder from potassium, sulfur, chloride, sodium, and magnesium—the aftermath of exploding stars expelling matter into the universe. We're supernova, baby. A cosmic geologic story, differentiating all the way from the abyss. Geopower harnessed in colonial-racist configurations of extractive economies facilitated the classificatory disaggregation between bodies and earth in favor of propertied relation and autonomy of self. The category of the inhuman secretes race; the cosmos repels its small margin of conceptualization within an expanding universe.

As the material mediation and spatialization of time, geology therefore designates the spaces and temporal event-fullness of operations in time (or the material "tense" of life and event of the political). Against the realism of geologic deep time, life gets posited as the regulator of matter in its apprehension and appearance, a mode of governance in the politics of recognition (Povinelli 2016), a vitalism, a purpose, a provocation, and an animating link to the figure of the Human (as apex figure) and its extractable "others" (as ground, earth, resource). Yet, the deployment of the inhuman as a category in the praxis of genocide, colonization, and chattel slavery is an alert that quickly sounds out this language and its affects, as no neutral description of matter.

No geology is neutral (after Dionne Brand).

White Geology

Theoretically, *Geologic Life* sets out to show how the materializing effects of stratigraphy and the affects of the stratigraphic imagination organize a planetary method and modality. I argue that time is weaponized as a social relationship and ontological claim to the earth that organizes place-based and ecological relations in the racial chronopolitics of colonial geology. The colonial architecture of geology has affectual afterlives beyond what is considered as geology—it is a past that has future effects. In the dual racialization of the inhuman as an extractive grammar of subjects and objects, the optic of geology produced political subjectivity as stratal. As the colonial natal moment of the inhuman arrives in the twinning of genocide and ecocide, the inhuman also has a third journey fellow in the cosmos that exceeds this temporal contraction. Which is to see geology in its subject-object forms, as *matter, race, and space*. While geology is often described as the origin, history, and structure of the earth, its metaepistemes of origination, historicity, genealogy, and stratification of thought spatialize, temporalize, and racialize the earth as an often unthought, isomorphic praxis of matter and subjective life.

This book is concerned with both those historical geographies of geology (as affectual architectures and epistemic violences) and their embodiment (as the flesh of geology). Alongside the analytics of the praxis of white geology, I seek to break open a grammar that speaks to the present tense of geologic lives shaped by these durable histories of racialization and earth transformation. I ask: How does the white geology produce and sustain itself as an affective and material architecture through *racialized undergrounds* and *geophysics of race*? This question moves through epistemic orders (as colonial property, paleontological footnote, and legal nonpersonhood in *Dred Scott v. Sandford*); carceral mines (in Alabama); genealogical accounts; and the granular discordance of anti-Black, anti-Brown, and anti-Indigenous gravities that punctuate and segregate social space. If analysis starts with racial undergrounds as the foundational architectures of attention, how might a different geography of location emerge? A geography that methodologically shows how what is built on the surface and as modes of subjectivity (called Life) is tied to structures of subjection that *require* racialized undergrounds. Understanding narratives of space and time as grammared by colonial geology requires us to think of matter and materiality as historically and epistemically raced. Which in turn requires an understanding of how collaborative theories of the earth and beings, and material geo-logics of extraction and their empirics, became actualized as racialized structures of division. Or, how geology grammars race.

The power to be a subject (and be recognized as such) is partially formulated through the ability to parasitize geoforces of the universe and harness these geopowers toward forms of valuation, interceding in those relations of force as a mode of disassembly and redirection of power (see Grosz in Yusoff et al. 2012).[7] As the stakes of planetary transformation make the differentiated terms of geologic life abundantly clear, antiracist work must address the mineralogical basis of life in terms that do not reinstate the coordinates of determinism or purity. Radical revaluation requires disrupting the geosocial formations that rely on a structure of racialized strata (or an underground of subjects) to serve as a substratum of the generation of that value. These *inhuman intimacies* between race, matter, and space are the bedrock of the "Inhumanities," a parallel "paraontological" (to use Nahum Chandler's 2013 term) site to the historic development of the humanities.[8] Subjective inclusion on the side of inhuman materiality, and the intimate forms of exchange and valuation with inhuman materials that result from this stratal emplacement, mean that race makes manifest material bonds that mobilize across inhuman materials and geologic subjectivity, as the desire for inhuman materials mobilizes racial dynamics. Which is to say, race is deeply implicated in the experience of the inhuman world (*materiality + metaphysics = geophysical states of being*). Inhuman materialities mobilize longitudinal forms of valuation and devaluation, as they simultaneously transform earth systems and processes. And thus, the inhuman is a site of radical redress and revaluation, where material debt is inverted in extraction and diverted along a racialized axis. Inhuman proximity, without consent, petitions the intimate recesses of subjective relation, as it also opens to a cosmic expanse that far exceeds coercive geopowers and the imaginative bounds of subjective identity.

The effects of humanism's material economy in the policing of the *geos-bios* border and its partial categories remain robust in stratified forms of life, even as political states shifted under decolonial actions and were reinvented through liberation struggles. While it is clear why postcolonial writers such as Glissant, Césaire, and Wynter engaged the Human to open its doors to persons having to live in the narrowness of its partial inclusion, the optimism of that opening as a restructuring epistemology across material domains had limitations. Liberation from colonial powers did not bring the imagined freedom from the affectual architectures of extraction under postcolonialism, but it did see the emergence of a geopoetic material language that asserted, "Poverty is ignorance of the earth" (Glissant 1997, 41). If the Human is indeed a ruse and a distraction, a petition on the unpartitionable, then the work might need to engage the category of the inhuman, both as the ground (*geos*) of the Human and as a mode of partaking with the forces of the earth (and thus a way to disrupt its weaponization). Inhuman intimacy is a way to rethink the relation of race and the earth inside dispossession, attentive to the flux of relation beyond carceral forms and in dialogue with resistant *tactics of the earthbound* congregated in inhuman residency.

Inhuman intimacy as affinity with other temporalities preceded and exceeded colonialism. While that intimacy is stretched and rifted under the weight of colonial investment in the category proximity of the inhuman-inhumane (the ontological category of the slave as inhuman property, and Indigenous genocide as a precondition of surfacing resource), it was also a frustration to the conceptual and economic order. The praxis of materializing and maintaining other earths is conversant with the colonial drive to consolidate *its* earth and its normative architectures of the future. The practices of colonial earth did not overwrite other attachments and were not the sum and sun of its most negative organization, but they did pull time to the present in ways that required a temporal block against the petrification of race as much as a spatial resolution to segregation. Attention to the inhuman as a category of affect in the subject and earth relation is a route into the abolition of the Human and its policing of the permissible figures of attachment and care across human, nonhuman, and inhuman worlds. This is not a new directive but a shift from the focus on the metaphysical terms of the Human to plot its geophysical substantiation.

Rather than seeing inhuman intimacy as solely the result of subjugating relations, pushing further into those intimacies recognizes not only the ongoing resistances to that positioning but also the inventive refiguring—or theorizing—of those intimates within and beyond carceral conditions. Within the context of white institutions and their predatory forms of extraction, scholarly mining of the liminality of creativity is a raw question. What I want to argue for is that a lithic account of subjectivity opens the construction site of the material context of oppression to new insights that move beyond the juridical end games of liberal representation and its impossibility for certain subjects. Historically, if political subjectivity is considered stratal—that is, defined by a stratigraphic hierarchical imagination of the human (*bios*) and earth (*geos*)—then the place of politics might be differently understood in the interregnum of race. A lithic account of subjectivity redirects attention to the geophysical qualities of politics and addresses the tight bonds between subjection, extraction, and earth.

Geologies of Race

> If we plumb the depths, then what we will find is fundamentally black. . . . [It is] a process of disalienation. . . . I felt that beneath the social being would be found a profound being, over whom all sorts of ancestral layers and alluviums had been deposited. —AIMÉ CÉSAIRE, *Discourse on Colonialism*

While it might seem counterintuitive to look at the inhuman as a space for freedom practices, it is the stratal poetry and actual survival in this zone designated as inhuman that signal its seismic possibilities. Engaging the depths for Césaire is a way to both shake the alienation of stratal pressure and surface the alluviums of history (a methodology of desedimentation without the promise of mythic recreation). Epistemic work on the level of the inhuman is a way to break apart the chains of signification and material disfigurement of a blood-soaked ground, renaming that relation with the world differently. Blocks in the epistemes of geo-logics are no less material in their outcomes than blocks in pipelines. Antidisciplinary pedagogy requires disciplinary analysis. Epistemic interventions can restructure devaluation. Existence in the category of the inhuman, imagined through the intimacies of geologic kinship without recourse to the claims of Being (or the lure of redeemability in the category of the Human), has conditioned the very basis and possibility of Black futurity. As Kodwo Eshun argues (in relation to Afrofuturism and the chronopolitics of Black time): "By creating temporal complications and anachronistic episodes that disturb the linear time of progress, these futurisms adjust the temporal logics that condemned black subjects to prehistory" (2003, 297). Time travel has been essential for colonial subjects precisely because temporal scripts hold political spaces of subjugation in place, as somatic temporalities such as the blues and spirituals have held the experience of epistemologies of time to other clocks.[9]

From 1492 at the inception of enslavement in the New World for gold mining, matter economies made fungible intimate dispossession and possession of personhood.[10] They established intramural spaces in the grammars of geology to create categories of what Fanon called "nonbeing" and extended the containment of matter in its mineralogical mode as natural resource into a subjective mode. The inhuman carceral makes possible the syntax of Article 537 of the Civil Code of the State of Louisiana of 1825, which stated: "Natural fruits are such as are the spontaneous produce of the earth; the produce and increase of cattle, and the children of slaves are likewise natural fruits" (Louisiana 1825, 159). Only the imagination of a *Black earth* and a spliced genealogical human-inhuman tree can give rise to the genealogy of such "strange fruit." Made in the heuristic origins of geology, the inhuman was the word for the conjoined enclosure of natural resource and chattel-slave. It is not simply that Blackness is co-natal with mining the New World, or an economic extension of its purview, but that Blackness gets captured in the ontological wake of geology in its present tense. The identification of gold, silver, and copper as value prefigured the identification of the ideal miner, which launched the racialization of Blackness *within* the desire for inhuman materials rather than outside its ontological description.

Resistance to racialized matter economies was internal to the histories of their production and expanded in relation to the earth beyond their calculative geo-logics, creating intimacies with the inhuman outside the forced categories. In concert, materiality already holds itself open against enclosure (this is the cosmic dimension to the inhuman). Such an example, argued by Wynter (n.d.), is the slave plot as an alternative modality of planting that was cultivated by the enslaved (McKittrick 2013). The plot for Wynter is a duration and practice space outside of plantation time, where one might plant differently to forge other relations with a "new land" beyond economic botany and outside of the extractive relation (even as it contributes to the social reproduction of the plantation and is embedded in settler modes). The modality of matter has many paths and relations, but the

FIGURE 1.2 Malcom X deconstructing "Negro" during a sermon at Temple 7 in Harlem, New York City, August 1963. Photograph by Richard Saunders / Pictorial Parade / Archive Photos / Getty Images.

shared inhuman root establishes a genealogy that requires vigilance; that is, matter does not just come into already racialized relations but is made in a raced crucible that can materialize different geophysical conditions of personhood. There is no need to look elsewhere for some new inventive methods for geopoetics or geologic subjectivity in the Anthropocene because it is the central operation of the histories of what Saidiya Hartman (2008) calls "degraded matter," secreted in the subterranean histories of the inhuman classifications of empire and its extractive praxis. The mode of inhuman classification was no category mistake but foundational to the constitution of liberal political society. As Lisa Lowe suggests: "In other words, considering the slave's nonpersonhood, and the proximity of reenslavement, we can observe that the condition of exempting those designated as 'inhuman' is not an aberration to political society; it is constitutive of civil and political society itself" (2015, 193n40).

Following the lead of Black studies scholars, negation should never be totalized (see, for example, Hazel Carby and Saidiya Hartman), not least because this would be to doubly impose dereliction and repudiate the achievements of endurance in conditions without consent. Fred Moten argues: "For various reasons that are historically specific, this condition of not being a subject, of not being a citizen, of not being a self-possessed and self-possessing normative person is associated with blackness" (2016, 30), but this "nothingness is not emptiness. . . . I believe that in that displacement, in that no-place and it's [*sic*] off time, there's something there—something is there in nothingness" (21). That *something in the nothing* is historically evidenced by aesthetic and theoretical reserves that have been developed in the occupation of the inhuman by those that have lived and live this space-time, contingencies that radically reorganize the terms of devaluation in this subjugation as a politics of countergravity. Part of that *something* is the simple fact of geography, of the specificity of time and space in a shared place. While anti-Black and anti-Brown violence animates the (non)relation of the inhuman, it is not its only antagonism or its only path. In the temporal dissonance of the achronistic freedoming of space and time, a methodology of occupation with the inhuman was established. Resisting the humanist impulse for restorative category work or the fantasy of epistemic reparation to these harms, stratal readjustment and the reallocation of geopower might go some way to alleviating the pressure.

The inhuman, then, is an occupied (subjective) category of language, epistemes, and earth. In the parse of matter and (non)being, the terms of that occupation are by no means given and the racial disfigurement of geology is not its only gravity. As scholarship in Black studies pushes the contours of the inhuman as a metaphysical dimension of forms of Black life, I want to contribute an examination of matter and racial materialisms to substantiate a reckoning with the geophysics of race and racialized geologic futurisms.

Insurgent Geology, Fugitive Life

Understanding geology as a material and symbolic mode of relation, coconstituted by imaginative, geophysical, affective, and political economies, I employ aesthetics as a guiding analytic and methodological mode; a mode of capturing and practicing insurgent geologies that work against the classification, geo-logics, and grammars in the mastery of matter. Aesthetics is not deployed in its Western form as representational but as a "rift zone" of sense-theorizing.[11] I use aesthetics not as a visual or scopic regime per se that makes certain presumptions about unified planes of perception (per Jacques Rancière's "just out of reach" political possibility in the "distributions of the sensible" that maintains a colonial monism or plateau in its political aesthetics; see Yusoff 2018b, 266).[12] Rather, I use it to denote the bundle of

FIGURE 1.3 Racist Indian statue surrounded by motorcycles in front of Discovery Well of Kern River Field in Bakersfield, California. Photo by author, 2016.

senses, perceptions, poetics, and imaginations that embody an affective architectural structure (like gravity) that I call the *geophysics of sense*. Another geophysics exists beyond the political aesthetics of colonial materialisms that is already on the run, tapping into the cosmic—a historically informed rift praxis of sense-theorizing epistemic modes, temporalizing otherwise on the wrong side of categories, and caught between executing grammars.

Resisting epistemes of extraction that grammar political forms requires promiscuous interdisciplinarity that disrupts the stabilization and sedimentation of those colonial geo-logics. This gathering can be called the *Inhumanities*, a counterconceptualization of the geohumanities in the context of the Anthropocene (understood as the consequential outcome of colonial earth) that foregrounds the role of the inhuman (as it was established through an epistemic plateau that privileged an extractive account of matter that racialized and predated on the subjective-environment relation). Wynter has made the point that humanism is the selection of who can function as a political subject. It is my contention that this "functioning" of political subjectivity has a prior mode operationalized in the realm of geology, as both the temporal ground that subtends biocentric racisms and the actual earth that is the medium of extraction's exchange by which subjects are marked and made.

In Julietta Singh's 2018 work *Unthinking Mastery*, she puts forward an argument for dehumanisms as a critical mode for decentering the white patriarchal subject and dislodging the tyranny of the white settler family, and of mastery itself. This might be understood as the mythos of "white man's burden," or, in the age of the Anthropocene and its vertical geographies, white man's overburden. Of dehumanism, Singh says: "Dehumanism requires not an easy repudiation and renunciation of dehumanization but a form of radical dwelling in and with dehumanization through the narrative excesses and insufficiencies of the 'good' human—a cohabitation that acts on and through us in order to imagine other forms of political allegiance" (2018, 4). In the shifting of "inhumanisms" to "dehumanisms," Singh seeks to overcome a prefix that does not, she says, "intuitively signal the history of the making of nonhuman subjects and forms of being" (5). Singh's project of unlearning and unmaking of narrative forms of mastery is compelling, gathering together dehumanism and the decolonial in a way that connects the reach of colonial mastery across domains to highlight the "denial of the master's own dependency on other bodies" (10). Singh's shift of inhumanism to dehumanism is a useful reminder in terms of addressing the legacies of colonialism, as it shifts attention from the racial schema of whiteness as a category of mere accumulation to sharpen a focus on the attachment to supremacy that propels whiteness as a distinct formation and grammar. My own persistence with the term *inhumanism* is (as with my title of professor of inhuman geography) precisely to work against the intuitive, to grasp its nonnormative workings and to let in, as it were, its cosmic potential alongside the inhumane geographies that are the discipline's legacy. It is an inoculation in the whiteness of human geography. The colonial history of inhumanism generated and languaged a crossing of the abysmal divide between the human and inhuman, and it launched a space in which a new language was born to speak otherwise of this inhuman condition. The inhumanities (contra humanities) would be deeply interested in the stretch of the geophysical world in the making of spaces of freedom and investigating the conditions for thinking materially about *decolonization as a geologic process*.

The assimilatory logic of the Anthropocene further perpetuates the idea of "unthought" destruction and unforeseen consequences of environmental change. Acknowledging how geology participates in Black, Brown, and Indigenous life-forms through compression, deformation, and inundation is crucial to challenging the antecedent colonial earth and shaping the earth that gets made. The inhumane politics of Europe, the Caribbean, Asia, Pacific, and Americas that travel in the grammar of the inhuman (as geology

FIGURE 1.4 Processes of fossilization: *How to Become a Fossil* video game. Smithsonian National Museum of Natural History, Washington, DC. Photo by author, 2016.

and subject/object aggregation) need to be at the center, not the margin, of any reckoning with the earth as a new epoch is ushered in, lest it carries forth the grammars that nihilated identity and turned Black and Brown subjects into a geologic buffer for the force of the earth itself. As a racialized stratum that breaks the waves of hurricanes and absorbs the forces of mining's materiality (as explosion, police, exhaustion, and poison), this stratal positionality is materially locked into carceral and captured states of being, policed by languages of devaluation. A connection can be drawn that bridges disalienation and extractive grammars to rethink the relation between subjective life and the earth; to reorient the spatiotemporal structures of geology toward a *tense* cognizance of its histories and their futures in the full ledger of relation.[13]

Hewn by the geographies of mines and plantations, the first Anthropocene subjects occupied the indices and intramural spaces of geologic practices as integral to the destratification of energy and value. I argued in *A Billion Black Anthropocenes or None* (2018a) that white geology has Indigenous, Black, and Brown understrata, because in every sphere of its production as discipline, practice, and ecological relation, it is Indigenous, Black, and Brown subjects who have been violently subjugated within the subtending strata of the extraction industries, extinction drives, and climate change.[14] Geology structured material and social relations through its stratigraphic and genealogic methods (such as hierarchies, social strata, historicity) and its material processes (extraction, isolation, removal of value, etc.). The work of time (temporal arrangements) carried both narratives and normativities of these formations into the present. Given the way that universalizing claims to Humanity, Species Life, and Biocentric Life are being revitalized in the Anthropocene and the agency of fossil fuels, capital, or Anthropos prioritized, the long shadow of diasporic and complicated histories of colonial earth obliges a commitment against forgetting what was erased in the history of the earth story.[15] Especially when those geologies still radiate their harm down the petrochemical corridors of the Southern states; through the hurricanes of the Gulf Coast and Caribbean; in the displaced murders of Indigenous environmental leaders by militias funded by oil companies or mining in Guatemala, Brazil, Latin America, the Niger Delta; in the disappearance of Indigenous women in fracking zones in North Dakota; and through the many muscular relations of extractive industries that send young people from Yên Thành, Vietnam, to untimely deaths outside of London. Names like Katrina, Puerto Rico, Flint, Haiti that have all become synonymous with the twinned violence of racialized subjective-ecological relations highlight the continuing refusal of state recognition of Indigenous, Black, and Brown subjects within the material context of the commons. This agentic exclusion from value coupled with violent inclusion might be understood as a *stratigraphic inhumanism*. Anti-Blackness, settler colonialism, and neocolonialism. Questions about the costs of the stabilization of whiteness need to be at the center of debates about the Anthropocene because these dynamics continue to configure environmental pressures and gendered and sexual exposures, and so constitute the shape of violence unleashed by anthropogenic climate change and destabilizing earth processes. Geology is not an abstract axis of materialization but pressures the very geophysics of living in the day for those marginalized by and in its praxis.

Extending spatial analysis into geology—as temporal dimensionality—is a way to understand the inhuman as a structural plateau and a geophysical affect. Geophysics introduces time and its ordering into the equation as a material question: temporal delineations associated with different beings, the naming and appearances of different *qualities* of time, as arrival, teleological, cosmic, racialized, humanist, reproductive, annihilation, and natality. This, in turn, constructs an affectual infrastructure that conditions the *tense* of space and its social ordering through epistemes of materiality.[16] In the context of the twinned temporalization and racialization of colonial earth, the geostory secures the priority of European Man and interns "his" Other in/as matter of the earth, categorized for extraction and ascribed atemporally historical stasis. As the work of Black (Sylvia Wynter,[17] Katherine

McKittrick,[18] Denise Ferreira da Silva, for example) and Indigenous geographers (Audra Simpson, Kyle Powys Whyte [2018], for example) have shown, conjoined histories of place emerge in the making of time's (subjective) objects at this metaphoric and material juncture of colonialism. While the imposition of colonial and settler time is well attended to, how deep time functioned as a way of producing universality and flattening difference, while simultaneously remaking it along a racializing axis (through affectual architectures), punctuated by different racial "arrivals," eludes scrutiny. It is this casual and raced production of universality via species life and deep time (a diagram of geologic time is sufficient to show this totality) that makes its forceful return in the future-perfect, postracial "we" of the Anthropocene.[19]

One way to come at extraction is through political ecology/economy and the histories, infrastructures, and materialities of extraction practices that group under petropolitics, mining, energy humanities, and, more recently, political geology (where empirically based studies do the work of making visible the spatial effects, labor politics, environmental consequences, and commodity chains involved in extractive economies). Another is to look at how extraction touches down in hyperlocalized geosocial contexts and map out from there to show the politics of matter in process. *Geologic Life* differs from these approaches by examining the ontologies and materialisms of the "ground" of earth forged by colonialism rather than taking for granted neither its pregiven stability nor its spatial forms. With a methodological focus on fracture sites, the disruption of material languages and inhuman counterstories, my approach involves a broad interdisciplinary drift and epistemic experimentation to interrogate the agile scope of totalizing geologic grammars and relations. What I take from the work of Glissant's concept of the archipelago, Wynter's (n.d.) theorizing of "replantation" in "Black Metamorphosis: New Natives in a New World," McKittrick's (2006) *Demonic Grounds*, Leanne Betasamosake Simpson's (2014) "Land as Pedagogy," and Povinelli's (2016) Karrabing analytics is that shifting ontological ground conditions are a means to epistemic and other freedoms. McKittrick's discussion of Wynter's cartographic theoretical approach is instructive here: "Her work is anchored to multiple and multiscalar 'grounds'—demonic grounds, the space of Otherness, the grounds of being human, poverty archipelagos, archipelagos of human Otherness, les damnés de la terre / the wretched of the earth, the color-line, terra nullis / lands of no one" (McKittrick 2006, 123).

The "grounds" of geologic life are similarly displaced over multiple geographies, geophysics, and genres, in the gritty margin (rift) between categories.

The refusal of monogeographic space disrupts the fiction of a singular plane of material existence or experience. If the Enlightenment subject was materialized through racial violence to float above the earth in a universal plane, then epistemologies of grounding must be concerned with disrupting the account of spatial forms to make visible the spatial architectures that operationalize geologies of race. *Every building a mine.* Tilting the axis to stay with the affective force of the geo-logics that ground an idea and arrangement of the earth is to examine how these grammars operate through a stratigraphic imagination that is pushing against the specificity of locales, both materially and ontologically, toward a totality of geophysical conditions. To theorize from such a point is a way to disrupt the taxa and temporality of geologic life as governed by the extraction principle.

If the plateau represents the political aesthetics of monospatiality, the rift confirms the multiplicity of epistemologies of experience in engaging dimensionality (as a relation between rift and plateau, underground and surfacing). Thinking geology by way of Black feminist studies has focused my attention on the revaluation of subjective life outside of existing modes of valuation to disrupt the formations, communicability, and modes of fungibility. Jared Sexton suggests that "Black life is not lived in the world that the world lives in, but it is lived underground" (2011, 28). Understanding the racial undergrounds generated in current grammars of geologic life as a concrete expression of geology and its originary racial geophysics, I seek to expand the spatial and material dynamics under consideration to disrupt colonial earth's continuance. Nonequivalence in processes of valuation of matter locates this work because it is the process of categorization and valuation of geologic materials (broadly conceived) that greases the machines of extraction and their violent relocations of subjects, ecologies, and earth. Drawing sources from counterarchives and underground spaces of geology is a way to demonstrate colonization across its grammars and render the spatial scope of those relations (as plateau), while simultaneously refusing to reinstate the "unthought" on the margin (McKittrick 2006, 135). To that end, I move through and beyond current grammars of geology conceptualized in a narrow frame of resource extraction, political economy, and (white) new materialisms (see Broeck 2018, 178–95) into other ledgers and poetics of apprehension that acknowledge and attempt to disrupt, what Sexton (2011) has called the "economy of disposability" of Black life. In geology, thinking with the rift as the unthought in the production of the social order is a map to other affective geologies that might indicate passages toward less subjugating and full reckonings with geologic lives.

Geologic Life

I deploy *Geologic Life* as a counterintuitive analytic between the inhuman (*geos*) and Life (*bios*) that holds in tension a critique and the history of this division. To trouble the link between geology (inhuman) and Life is to disrupt the (commonsensical) training to thinking these ontologies in their contradistinction, as different arrangements of sense, whereby the episteme of biocentric Life represents species life but not the subtending racialized geologic relations and earth that gave rise to that "species body" (Foucault 2010, 139). I question how geology and race underpin an account of Life as *bios* and place this within the biographical life of Geology as a discipline that produces understandings of matter and materiality. I want to stage a disciplinary collision of discourses and orders of matter that have colluded to produce the inhuman and biocentric Life as separate entities. The placement of "Geologic" and "Life" together is to effect a dual agitation against these heavily sedimented categories. I do not intend for *Geologic Life* to be read as a self-explanatory category but instead see it functioning as a fault line in the conceptual terrain of both those concepts—a fault that frictions the rift rather than seeking out a perspectivism from the privileged plateau. Smashing these antagonistic terms—*grit* and *grind*—together can tell us something that might extend, as well as unearth, deeply sedimented forms of conceptualizing and cataloging matter, beyond the container of Life and its strata of subjection. By separating these political fields, we can look at how geologic life contributed to enduring modes of material oppression and extractive ontologies. Attending to the origins of geologic imagination, inquiry, and transformation of the earth in colonial expansion (and its ongoing regulation in development) suggests that geology is foundational to biological concepts, biocentric thought, and modes of "seeing" Life as a valued form, and that race is foundational to that geologic production of knowledge. This is to understand geology historically as *the* material and symbolic grammar of colonialism and its neocolonial afterlives. One of the tenets of my previous arguments (2018a) was that geology is a regime for producing subjects and, within the slave-mineral complex, a way of quaterizing subjectivity by way of the inhuman. In this book I bring the separated tectonic plates of Life and the inhuman together, under tension, to examine how the spatialization and temporalization of matter collaborate in the formation of a racialized plateau of geophysical exception in earth praxis.

In the agentic separation of matter into *bios* and *geos*, specific imaginaries of the planet as a stable object and ground of knowledge emerge. This material spacing and temporal ordering allow a psychic stability whereby the movement of thought is unimpeded by planetary thresholds or cosmic flux (as well as being immune from the material effects of the social order). Yet, the rub of the earth—that those who have lived and live in its rift zones of poverty, discrimination, exposure, and inundation know all too well, subjected to its languages of propertied inscription—attests to other cosmic intimacies unthought of in the colonial imaginary of matter. This inhuman intimacy with *geos* has a political charge; it renders refusal through an account of agency beyond extractive categories and in resistance to Life as agentic center and limit. Thus, I forge a praxis of underground aesthetics (discussed in chapter 3, "Underground Aesthetics") that seeks to unsettle and reclaim a language that is made to work against the partitions of Life and the inhuman in the division of matter, a language that consistently foregrounds a historic formation of biologism and positions the geologic in a relation to that biocentrism as slipping away to become ground and grounding.[20] The heuristic, material, and philosophical imaginaries of geology, with Geology as a discipline that marks a narrative of temporality, arrive in its political presents in narratives of the Anthropocene and its impoverished languages of the material world. In the following, I examine the theoretical foci of *Geologic Life* as an analytic that examines the *historic formation of political geology during colonialism* and as a *form of life that is subtended by the racialized scripting and spatializing of matter and temporality for the extraction and the governance of geopower*. I start by analyzing several concepts that are activated as open "seams" (grammars, white supremacy of matter, orders of the inhuman) through the analytic of geologic life; then I discuss my framing of this analytic in earth revolutions, inhuman intimacies, and geologic subjectivity.

Geologic grammars are sedimented instantiations in subjective life and disciplinary thought, and a collaborative mode of differentiation in identity formation. Inhuman grammars establish a lexicon of geologic life as contradictory, conflicted, racialized, sexed, and gendered through the material and symbolic grounds of geology. Intrinsic to this process of geologic life is the governance of divisions of life/nonlife (per Povinelli) that materially structure collective experiences to establish geology as a political realm in relation to the governance *for* Life (as a partial category). Further, I argue, we need to see the division of *bios* and *geos* as constituting a spatiality of difference (theorized through the plateau and rift) that produces different gravitational conditions for subjects on different sides of the Life/inhuman binary. This rift zone is in relation to biopolitical categorization but crucially outside the formation of Life that biopolitics takes as its (subjective) object (i.e., geologic life is beyond biopolitical modes of redress). Posed in its intramural relation, geologies of race locate the politics of the inhuman extant to and outside of existing political economies whereby the inhuman is excluded from biopolitical analysis or entry precisely because it is the "ground" of the biopolitical subject, on account of its epistemic distance from Being (as metaphysics and materiality). It is not solely as outer space (Sexton 2010), as invisibility, or as exterior (Moten 2008, 193) that Blackness is first structurally positioned in the colonial imagination, but through the telos of materiality to activate the inhuman (ontologically and as geopower). Blackness and gold become axiological in the dialectic of value, and geology grammars this exchange and substitution through the language of the inhuman.

Recent (subjective) designations of geologic force in Anthropocene discourse both unground the historic naturalization of geologic grammars by problematizing the futurity of those modalities of life and reground them in the familiar racialized historical forms of address: Wynter's Overrepresented Man, "*degraded matter*" (Hartman 2008, 7), Human, Humanity, Species (see Yusoff 2018a). Rather than starting out from the Anthropocene, this book is shaped around a longer historic engagement. Geologic grammars were made in the context of a Western tradition of Hegelian aesthetics that elevated language above matter and created dialectical hierarchies of forms that mapped onto subjective designations of matter, whereby the human-inhuman dialectic emulated the language-matter dialectic. In a parallel move, the geohumanities has constructed its strictures of futurity from humanist building blocks. I am motivated to locate and strengthen other geoaesthetic modes of description in the rifts of the racialized-ecocidal relations, which is to imagine a granular geology that holds time and space against the normative conditions of its grammars, of resource, nation, destiny, or telos, and against impulses of enclosure in the propertied form. I refuse the ideation of "livability" held against an imaginary of a coming disaster; instead, I attend to the future shocks of past extinctions, as part of acknowledging how extraction and resource have formed the present politics of geologic relations.

Geology as the *epistemological and material (white) supremacy of matter* is achieved through colonial knowledge-making practices that articulate orders of time (Wilderson 2010, 399) to materialize racialized subjects in geologically infused accounts of race within stratal time. Or, put another way, deep time is understood as a racializing analytic that unravels a set of normative discourses on agency, whereby the temporal lines in geology were historically used to make subjective and subjugating categories that rendered personhood as fungible flesh. And in geologic accounts of race, narrative time annuls political time. The material temporalization of racial difference "places" subjects both along a scale of the human–inhuman (inhuman–inhumane) and in proximity to the differentiating exposures of toxic geologies. Crucially, the white supremacy of matter practiced an apartheid of matter (and thus agency) between the subjective states of *bios* and *geos*. Through an examination of the role of origins and origination in narratives of geologic life and its attendant claims on geopower, I want to establish modes of apprehending geologic subjectivities in the grit of inhuman intimacy.

Geology *traffics between three orders of the inhuman*: (1) as a corporeal and radical intimacy with the inhuman; (2) as nonconsensual inhumanism/inhumanities in the universalist language of the Human and species Life; and (3) as a cosmic materialism that expands forms of life and is indicative of life after biologism and biopolitics (stretching out through and beyond the frame of Life). The alliances and fissures between sutured orders of the inhuman are means to reterritorialize the inhuman in a nonextractive mode and as geopoetics that disrupts the passionate attachment to Life: a form of life that is both genocidal and ecocidal and foundationally anti-Black, anti-Indigenous, and anti-Brown.

Geologic Life conceived as planetary force and geophysical dynamic in categorizations of materialization of subjectivity steers us away from exclusively biocentric accounts of both the human and politics, and into a potentially less deadly engagement with earth (and thus in shared nonhuman and inhuman arrangements of subjective life). This also means opening a grammar for geology that includes humans but has a larger scope of material and psychic processes that have the universe and its abysses as their originary field, beyond the puny exclusivity of the human as it is currently rendered through its negations and anthropocentrism. This is not about assigning geology a new identity politics for the Anthropocene, as becoming a geologic subject might imply, but to understand how the geo-logics and grammars of the inhuman *form* material worlds to structure the tense of subjective life in the political present through the politics of the inhuman (in a syntax that does not bind itself in natality and extinction, color blindness, or what Lee Edelman calls [white] "reproductive futurism" [2004]).

Earth Revolutions—Revolutions of the Earth

Historical, material, and conceptual revolutions of the earth have coconstituted subjective modes and delineated tacit forms of existence, while simultaneously Enlightenment scientific-humanist practices expanded the distance between Life and the inhuman. Thus, geotrauma conjoined metaphysical and geophysical orders in geographies of subjective-earth transformation across the colonial spectrum. Historically, this is evidenced in the response to the geopsychic shock of the Lisbon earthquake on November 1, 1755, that occasioned the launch of the nascent fields of earth sciences and natural philosophy in the West and the geologic traditions we have inherited. At the time of the earthquake, Lisbon was the fourth-largest city in Europe, the third-busiest port, and one of the wealthiest cities in the world from the traffic in enslaved persons and extraction of metals and diamonds from South America (850 tons of gold were sent to Lisbon from Brazil in the eighteenth century alone, and an estimated two thousand enslaved persons arrived from Africa per year from 1490 onward; European modernity was materialized as Black and gold). The "colonial revenue" of Brazil, and the enslavement of first Moors, then Chinese, Malay, and Indian subjects, followed by the industrial enslavement of Africans, built the public infrastructure and elaborate monuments and protoscientific institutes of Lisbon in which the Enlightenment took shape.

Portugal had established naval outposts and commercial networks that connected Lisbon to Nagasaki along the coasts of Africa, the Middle East, India, and South Asia. This is the geography of Lowe's *Intimacies of Four Continents*, where "the 'coloniality' of modern world history is not brute binary division, but rather one that operates through precisely spatialized and temporalized processes of both differentiation and connection" (2015, 8). The earthquake, tsunami, and subsequent fire disrupted Portugal's trade in enslaved persons and dampened its imperial geography and colonial ambitions.[21] Thus it was considered the first "modern" disaster (since it generated a coordinated statewide and Europe-wide response to attend to the disaster and set in place planning to mitigate future disasters).[22] At a broader philosophical level, the earthquake highlighted how the libidinal foundations of "fear of an inhuman planet" were implicated in the combined mastery of matter, the codification of race, and the birth of the modern subject.

Within the context of the wealth generated from enslavement and mineral trade in gold and diamonds, Lisbon represented a moment of emergence of

European cosmopolitanism, both in the development of the Enlightenment arrangements of Nature, Science, and Earth (see Larsen 2006; Reinhardt and Oldroyd 1982, 1983) and through the geographic emergence of Europe as a distinct entity that was built on the geotrauma of colonialism.[23] The earthquake itself was claimed as a European geophysical "event," despite its effects on North Africa; disaster relief came from other European cities; there were refugee camps; and European media covered the event (Immanuel Kant, sitting in Königsberg, wrote his treaty on natural science from newspaper clippings). The disaster is often seen as a turning point in the conceptualization of Western human history, where the attribution of natural history shifted from supernatural causation (God) to natural causation (Earth). The earthquake enacted a shocking material *and* conceptual destabilization, initiating geo and social revolutions of equal measure (see N. Clark 2011, 81; N. Clark and Yusoff 2017). There was not yet a science of the earth at this time in the West, but the event was taken up by many Enlightenment thinkers (Jean-Jacques Rousseau, Voltaire, Kant) to challenge the interpretation of the origination of earth forces, and as they did so their thought replaced divine intervention with the concept of Reason (as an ontological telos that resolved a material dilemma) and in the natural sciences the fields of seismology and scientific geology became established.

Enlightenment Reason produced an idea of Nature as a separate autonomous entity (with a capital N) distinct from religious authority and its transcendental telos. Kant, who was one of the first to teach geography and was obsessed by the Lisbon earthquake (Walter Benjamin called Kant the "first geographer" because of his theories of seismic and earth origination) published *Allgemeine Naturgeschichte und Theorie des Himmels* in 1755 (*Universal Natural History and Theory of the Heavens,* 2009) to address the events of nature and the question of cataclysmic geologic events in metaphysical life. His legacy—*The Critique of Pure Reason* (1781)—secured a position for the observer from afar as an adjudicator on knowledge and the earth, whereby the faculty of geophysical understanding provided a means with which to gain a resolve over inhuman terror. The problem of the seismic shattering of the telos of meaning, however, demonstrated a thornier problem for philosophers. The Lisbon earthquake demonstrated how geotrauma produced an equation between earth events and new modes of thought and models of subjectivity (demonstrating what the earth does to thought, how it arranges disciplines, modes of inquiry, and institutions). But it also evidenced the autoaffection and self-insulating drives of Enlightenment subjectivity in the context of a dynamic earth and white subjectivity. Lisbon highlighted the importance of the geomateriality of thought in colonial mastery and its dependency on the shaky compromises of geophilosophy to simultaneously capture materiality to produce the plateau as a dematerialized subject position.

Alongside the cataclysmic event, a shift occurred in the late 1760s through to the 1820s from geology as a practice of determining localized formations of strata (for the commercial purposes of extraction) to geology as a temporal practice of determining the status of *beings in time* and *theories of the earth*. The notion of tectonics—or "a moving earth"—began to emerge as an answer to the question of faults and tremors at Lisbon. This heralded the beginning of the idea of natural laws that populated an understanding of the universal as a monotemporal quality of material states. James Hutton, considered the "originator" of modern geology and a cornerstone of rational scientific thought, published "Theory of the Earth" in 1788. This shift in geophilosophy also marked the naming of paleontology as a key branch of the discipline of geology, understood as the study of ancient "beings" through the bodies of contemporary organisms. Fossil life was studied in relation to contemporary anatomical life in a comparative mode. Fossils were used to narrativize a story of the earth as told through a corporeal lens, whereby life-forms animated the strata. *Paleontology* sat alongside *stratigraphy* (ordering of the materiality of time through sequential strata) and *comparative anatomy* (ordering of bodies of species through their differences, applied in the sphere of human species and their racialization). Geology was a differentiating machine in its first instance, producing codes of temporal and racial difference through the ordering of bodies and times through the earth.

The power of the idea of racial difference (that paleontologists perpetuated in the histories of geologic thought) was mobilized in the traction between material and metaphor: the grammars of geology enacted the material extraction and theft of land and peoples, and the metaphor of supremacy over those lands and peoples stretched the empirical into imaginaries that validated the material gains of whiteness. While the material and symbolic are often separated as distinct regimes—as force/matter and metaphysics/representation—grammars of geology relay between the two as an identifying and idealizing materiality in regimes of power. The simultaneous articulation of earth and Being by paleontologists mobilized geophysical relays that accrued authority in space and across time. Geology became a first ordering principle in the arrangement of Western matter and thought, as well as a racializing axis in the production of the Human as a being in time and space. The ontological and material ground of the Human's origination was the position and ideation of the inhuman (as the materiality of actual European bodies was secured through the inhumanities of colonial violence and enslavement). The institution-building and conceptual scaffolding that structured this relation between Being and Time organized the very possibilities and scope of thought and its governance through partition (Being as *bios*, Time as *geos*). The metaepistemic ordering of geologic nomenclature—such as stratigraphy, formation, strata, and the empirics of temporal designations—created the normative modes for understanding ontology as social and subjective. Geology has an extra episto-ontological dimension precisely because it is a story of the earth and life, and because the temporal lines of making beings-in-time secure emplacement in the world and forms of origination (securing the genealogical claim). Joining the earth (a nonteleological entity) to the telos of the subject in the organization of natural history lithically made the white subject at home in the world through a material telos. Geology was a locational device for being and time that had a primal grounding because its symbolic and actual ground *was* the earth (and whereby mastery of the earth materially brought subjective elevation).

Confidence in the fossil record grows incrementally through New World invasion and the intensification of colonial practices of economic extraction. It is not incidental that extraction of new geologic materials was intimately tied to the encounter with other forms of "alien" life and the codification of races in time. Rather than treating this historical location and geographic project of geology's origins as a background to the narrative of patriarchal lineage—or the "Fathers of Geology" story—I want to place this colonial material praxis at its center. I do so to articulate how the genealogy of inhuman materialism and the narrative of patriarchal figures erase the transgressive nature of the discipline across multiple fields of cosmology, subjectivity, race, sexuality, and power, and how these dynamics shape the structural violence of its gendered practices, both historically and contemporaneously. This "fathering" of geology established a patriarchal relation to matter through both the identification of beings-in-time (a temporal identity politics) and established paternalism (a hierarchy of interpretation) over the earth.

Historically, the formation of geology is intimate to colonialism, racial capitalism, and slavery, entwined with philosophies of racial biocentricism in the material practices of colonial survey and extraction and the imposition of settler territory. While geology is often seen as playing a supporting role to biological theories of Life—backgrounded to the mineralogical field in which organic life *takes* place and *makes* space in the vastness of time—it emerged as the ground of both philosophies of Life and natural philosophies of the earth/Nature, as well as the origination of the Enlightenment ethical and political subject (the human constituted as it was through the inhuman, as the idea of freedom was crafted through slavery). As a philosophical discipline, geology was a history of ideas about the origination of the earth *and* earth beings, which was materially made through colonialism and its racial praxis.

It is the *verticality of stratigraphy*, as the positioning of fossils as hierarchically organized beings-in-time, that is mapped onto the metahominid differences that create the spatial coordinates of white supremacy as a geophysical condition or levitation. This geophysical imagination is important for the ways in which race is deployed as a regulator of stratal relations (which are simultaneously imagined and produced as raced). Through this mode of stratal governance, transcendental whiteness—as the Enlightenment subject—is produced through the earth. In the panic of losing one's God at Lisbon, another mode of governance of the subject (called Reason) was created, but the desire for an Overseer remained and so the telos of materiality was created. This is why Western philosophy can seemingly float above its geographies, touching down occasionally, but without concern for the grit of the ground, and impervious to *its* inorganic reason. This is the *will to universalism*, as a release from the earth and its gravity of material (inhuman) demands. If the Lisbon earthquake was the first geotrauma to shake Western thought, the second was the incremental discovery of deep time and what it conferred about the nature of existence as the extinction of everything without exception. The idea that everything was but a moment in the depths of time, gone in the blink of a geologic eye, was enough to send European fragility into some murderous compensatory activity in pursuit of inhuman materiality. This was the psychic rub in which Western theory built its armor of protection: inhuman nature triggers the deeper recesses of white fragility as inhumane rage.[24] Blackness became a stand-in for that abysmal confrontation, like totemic gods of distraction placed in the interregnum of the inhuman to protect the autonomy of the Human. It was historical location of Blackness in relation to gold (as elevation) that organized the politics of this inhuman-inhumane exchange. Stratal overburden was the material resolution for Western thought of the libidinal terror and vulnerability of living on a dynamic earth.

Insurgent geology—uprising is prefigured in a stratigraphic imagination.

Race was materialized by geologists in the seventeenth to nineteenth centuries through an attention to, and narrativization of, comparative physiology, geologic zoning, and the geographic origination of races, which not only produced the spatialization of temporality (or the "time" of races) but, more importantly, presupposed a staggered arrival and extinction of races in the present, which in turn naturalized ideas of race war and ascendancy through a sedimented progressive race narrative. Thus, the racialized arguments that are made through white supremacy are not just about biology and the performance of phenotypical subjects but are mobilized through geologic temporality and geographic imaginations about the ontology of material economies. The spatialization of *race through time* is used to legitimize a right for whiteness to the political present, whereby power is realized via a spatial project of time. The eugenics of biological thought and environmental determinism (which posited hierarchical understandings of different geographic racial zones) has been critically attended to as pseudoscience (although it is resurgent as white supremacy reasserts itself).[25] But the role of geology in embedding and legitimizing racial violence as arrival—telos—has a quieter register in the histories of science and the humanities. As the Human was spaced in geologic time—as a differentiation—meaning over the vastness of time was secured to cushion the monstrous indifference of the earth to the maintenance of creaturely life.

Geology as a heuristic is a means by which to understand the affective, aesthetic, political, and economic modes of existence in relation to the earth—as a first-order realism or ground of Enlightenment thought. There is a reason Kant was obsessed by the Lisbon earthquake and Johann Wolfgang von Goethe was overseer of mines and an avid rock collector and theorist of the earth (as was Alexander von Humboldt). Geologic realism as *the* ground of every figuration is so sedimented in Western thought that it is backgrounded as the surface to events. Concomitantly, the idea of the earth as resource and the political formation of territory are naturalized relations to the ground that sediments political and racial accounts of space and their normative social-sexual arrangements. It is not that there is just an overlooked lithic ground beyond extraction economies (as economic geology and material empirics) but that the geologic sciences produced geophilosophies and theories of Being (ontology) that were forged through the earth as *the* empirical object, and thus they had a primary claim to a naturalizing relation.

Alongside this historicity of beings-in-time (the Human earth story) is the making of nonbeings-in-time (the Inhuman story), a form of nonexistence that is established by the historical classification of the inhuman as an internment both in the materiality of race and in parallel psychic registers of capture (see Campt 2017; Hartman 1997). Thus, the inhuman is established and perforated as a concept by its origination in the slave-mineral complex of black-gold, coal-black, black-sugar, whereby Blackness was made equivalent to a mineral from the earth's crust created in a cataclysmic event.[26] Blackness was positioned both outside and within geography, as libidinal buffer to the inhuman and as a lithic buffer of geophysical force in the extraction of value. Because of the subjective ties of the inhuman at the inception of its languaging, the inhuman can never be encountered as the *thing-in-itself* (or via materialist autonomy). That is not to say that there is no materialist autonomy outside of and beyond, before and internal to subjectification, but that the languages of apprehension of the material world that code normative modes of materiality are not neutral. Geologic grammars historically established a colonial "politics of recognition" of materiality, enacted through the extractive praxis of personhood, land, mineral, and metal (Coulthard 2014). This categorization of matter enacted a form of spatial execution and substitution of subjective relation, through geographic displacement, relocation, and transformation.[27]

Inhuman Intimacies

Colonialism, understood as a structure of geophysical oppression and geotrauma, transforms subjects and environments: through violence and pathogens that establish the colonial frontier, and in the erasure of material forms of reciprocity and consent in the solidarity of bodies across difference (see L. Simpson [2017, 2021] for a discussion of water and beaver bodies).[28] Inhuman intimacy is a sociality with the world that white geology denies, and Black and Indigenous aesthetics elaborate on to expand the temporalities of being beyond extraction for the "now" and its surfacing. The rift of inhuman intimacy is not a peripheral zone or a zone of exception, but it is a quantitatively different geophysical zone in space and time, positioned by a colonial telos of materiality as racial accomplishment through the conversion of the strata. To practice worlds of beauty within the category of inhumane brutality (between forms of organic-inorganic life) is a mark of survival and sociality, but also durational capacity (to have and recognize generosity given by and through an earth that exceeds the mark of "I" and the singular claim of "now").

Thinking intimacies in geologic lives, I am instructed by Lowe's reading of intimacy as a heuristic and historic colonial division of world processes that produced a particular calculus of "production, distribution, and possession of intimacy" that can be read against the grain of liberal intimacies on which individualism depends, to attend to "those processes that are forgotten, cast as failed or irrelevant because they do not produce 'value' legible within modern classifications" (2015, 18–19).[29] In relation to how the inverse racial axis of devaluation and the production of geologic value operates, intimacy is a way to conceptualize the governance of material forms of proximity and its larger spatial forms.

I understand the intimacy of the inhuman as (1) a category proximate to racialized personhood that concretized social forms (coded agency, subject, forms of life, sexuality, gender, and value in subtending white Life) and (2) a material dimension of environmental relations (from the skeleton of our calcium carbonate becoming to the mercurial toxicities of its undoing, through the thirst of drought, heat, and hunger to the storms of anthropogenic climate change). Both intimacies of the inhuman (as earth and race) parse the possibilities of all subjective life in the bind of colonial geotrauma.

Recognizing how geologically implicated subject positions (i.e., identity defined through epistemes of matter and its temporalities) could be mobilized and maintained is part of seeing the materiality of race in the present. The partition of the inhuman is a negative engine of difference that has been fundamental to the ongoing colonial dispossession of land and social categories of personhood, and implicit in the forms of extinguishment and exhaustion that permeate ongoing environmental change as a necropolitics of *all* earthy relation. The deployment of the inhuman category of chattel in the context of slavery emptied out the recognition of relation and reinstated it as property, untethering, "ungendering" (per Hortense Spillers), displacing, deculturing, and making kinless. In this inhuman calculus, geographic and familial genealogy was obliterated and resutured to the index of value, as property and properties (Yusoff 2018a, 68–74). These erasures were simultaneously voiding actions in relation to people, places, and ecologies of earth (or how you become a subject through the sedimentation of environmental place relations). Valuation stripped relation of its tethers. The extraction principle is both antirelational and antagonistic to planetary inhabitation as a commons.

Geologic grammars brokered and broke earthly relations and interspecies kinships and refused recognition of phenomena beyond the geomatrix of racial capitalism. A parallel point of unearthing the racial grammars of rocks is to develop a more exacting and exuberant understanding of inhuman processes (beyond natural resources) within geosocial formations—a geopoetics that breaches the extractive mode of address to understand the geologic as a generative alterity (which is also to insist on how its racial formations shape contemporary environmental racisms). The inhuman grounds every biopolitical turn, from every nascent reproductive strategy to every extinguishment that does not appeal to Life. Life is apprehended as the guarantor of a purposeful surfacing of events (as long as it is attached to the right body). Humanist philosophies of thought retained the primacy of the immaterial (the inhuman) as the ground of Humanity to establish normative claims on behalf of what Wynter calls the "referent-We." The division of subject-ground was also a way to adjudicate on the realism of matter's agency and willfully stifle recognition of the communicative processes and porosity of geochemical codes within and beyond anthropocentric concerns. By contrast, the formulation of inhuman matter as nonidentity before it is brought into the realm of classification and configuration as natural resource (or racialized labor) dissuades us from an account of the complexity of exchanges that do not individuate but, rather, ground subjects coded as Human.

There are two distinct reasons to pursue this fracture: (1) Life is the ground for humanism and its modes of exclusion that established a racialized biopolitics (which is really a geophysics); (2) Life is the ground for establishing the fallacy of "brute matter" and its assimilation by social forces, which enacts its material annihilation from relation and located identity across geographies of mineralogical, bacterial, fluvial, cosmic, atmospheric, and social attachments, not to mention its geochemical attractions. The belatedness of biopolitics to the geologic field of relations also means that geopolitics needs to be redefined through its geological root: a root that configures the elemental dynamics of territory and its subjective and temporal affordances.

Geologic Subjects

> The land she tilled and worked had been Indian land. . . . Stolen bodies working stolen land. It was an engine that did not stop, its hungry boiler fed with blood. —COLSON WHITEHEAD, *The Underground Railroad*

Questions of geology frame the historic context of the stolen lives and land. The praxis of white geology regulated space and subjects to produce historical forms of subjectivity under erasure, a colonial engine not fed by bodies alone. The arts of subduction that underpin this theft are organized through grammars of geology and the processes of subtending some bodies with the energy of others to produce distinct spatial and geophysical arrangements: *stratal subjectivity*. In this stratum of embodied affects—understood through the analytic of *geologic life*—there is a need to de-script geologic subjectivity both from its subtending spatial production in land-as-property and from geology-as-natural-resource, with its subjective stratum of plateau (*bios*) and rift (*geos*). The question of the reclamation of relation between *geos* and *bios* has been posed historically and geographically by Indigenous resistances on stolen land, by revolution in Haiti in 1804 (that imagined, fought for, and won Black freedom), and in many anticolonial liberation movements in the Caribbean, Asia, and Africa ever since.

For Antillean writers and many in the diaspora, the question of independence was posed in relation to a *geos* where there was no genealogical historical claim to origination. The void in origins and its attendant questions of genealogy were an urgent political question in liberation movements and frame some of the most extensive theorizing about subjectivity in the breach of disrupted territoriality and orphaned geographies. This sense-theorizing about the impact of geotrauma on colonial subjects sat alongside other more sedimented narratives on claims to origination made through the question of land (Indigenous and settler). And the thornier questions for Black subjects of the geomythos of reorigination on Indigenous lands and the negotiation with settler modes of property. The geographic ruptures of colonialism forced the articulation of what it meant to exist between the ontologies of *geos* (inhuman) and the actualities of its rough grounds (as geologic subjects governed by and extracting geopower), which was to experience a geologic realism that slipped between metaphysical capture and the materiality of new worlds.

In Martinique, Antillean writers had to articulate this specific fracture zone of the geographies of enslavement and Indigenous slaughter alongside the inherited universalizing French literatures of equality and freedom (culminating in the French Revolution in 1792) in which they were schooled for their political speech. The psychic fractures of the biopolitical experience of inhuman-inhumane nonbeing though enslavement similarly removed the assumption of a stable ground. What Glissant, among others, sought was "a sensibility-knowledge that arms the sensible" (2010, 54). That is, a language that could arm itself against the colonial architectures of affect yet remain sensible to the trial of experience that colonial geotrauma generated. Wynter (1989, 645) understood Antillean writers as addressing Michel Foucault's racial lacuna in his partial conceptualization of biopolitical subjectivity through the disruptive project of revalorization.[30] As biopolitics was no use, poetics was chosen as the political form able to break with the languages of description and system of classification that governed forms of life. While attention has been paid to the now canonical writings of Glissant, Fanon, and Césaire for their engagements with humanism, I revisit them within the context of this book as anticolonial materialist theorists, specifically writers who enroll geology and geography in the political project of freedom. For example, Glissant was interested in the cosmic voyage of elements and their amalgamations as a parallel geology to that of enslaved persons hurtled through the multidimensionality of space and into the matrix of multiple histories. Angela Last (2015, 2017) argues that the geopoetics of Antillean writers offered a mode of undoing hegemonic geopolitical worldviews rather than resedimenting their claims whereby the geophysical was used as a mode of critique and means of geopolitical redress toward decolonization.

Conceptualizing the human as a geologic agent (see Yusoff 2016) and understanding the queered corporeal genealogy of this geologic subjectivity (Yusoff 2014) opens the field to think about the broader dimension and divisions of inhuman subjectivity (Yusoff 2018b). The political terms of geologic subjectivity mean that there are both common geochemical affiliations and differentiated access to geopower to participation with and partition the earth. Reconceptualizing subjectivity in the abjection of the inhuman(e) required another form of expression, an axis of sense that remade the subjugating grammars of geology, grammars that marked the passage of life lived in the exclusionary strictures of genealogy and its institutional modes (either as "mixed-race" for Antillean writers or denied the father's prerogative). Poetics can be framed as a practice of geologic life in Glissant's concept of Relation, which makes genealogical claims through poetic affiliation, rather than in the racialized mode of lineage or purity articulated by paleontology. Affiliation with the *geos* was poetry's wager, not petrified imaginaries of filiation.[31]

Affiliation not filiation!

The question of filiation in the establishment of genealogy and kinship in paleontology and the understanding of what it means to be a geologic subject (or a subject cast into deep time) produced a stratal hierarchy of subjects that steps into the geopower of the present or is cast immobile in the inhuman grounds of past epochs. Illegitimate kin were the *ground* on which biopolitical modes are built, and thus a biopolitical approach falls short in its reach as an analytic mode of address. As biocentric analysis induced the differentiated ground of race into its construction of a subject via the occupation of the *geos*, race inverted biopolitical and geologic forms in ways that already erase in advance certain persons from its perspective. Crucially, the plateau of biopolitical subjectification exerted a material weight that deadened social forms below, as *geos*, specifically as strata (discussed in chapters 6 and 7).

The "unthought" subjective form of the inhuman as a stratal form of geologic life gave materialism a racialized genealogy and filiation. The imaginary of time and the earth—inscribed in the Human story as the paradigmatic prism of recognition in genealogy and paleontology, involved in the action of "selection" and "dysselection" that Wynter argues structures the Human—designated the spaces and temporalities of agency, its absences, and the regulatory norms that structured subjectivity. Thus, geologic subjectivity is both materially and psychically stratified by matter and its grammars of articulation, as the hidden political "time" of the subjectivity. Which is also to see the inhuman as a brutalized category with a painful history that is implicated in human becoming. Extending understandings of subjective formation into deep time, then, might historically locate the formation of divided and dividing modes of subjectivity in political relation to the production of geologic knowledge and its epistemic and economic development of matter (with its connection to displacement, extraction, and violent weathers).

In the environmental sciences and the humanities, theorizing of the earth often returns us once again to a "neutral" animate ground for matter without a concomitant investigation into how the historic symbolic inscription of Life and the inhuman was used to create a negative inheritance and stratigraphic hierarchy through the new secular origin text of geology. In the broken grounds of materiality called the rift, there was the "door of no return" (Brand 2001; Hartman 2007) and an abyssal quality (Fanon 1963) that annihilated subjectivity, space, time, and relation, collapsing all classification and dimensionality in on themselves in an inverted horizon of materiality. On the plateau there were geographies of agency, access, inheritance, accumulation, valuation, and juridical protection. This racialized axis of geologic life is the geographic and geophysical earth that colonialism made. It exerts a geophysical force that pulls at the lateral axis of extraction to exact a vertiginous force that calls out the negation in its historical narrativizing through this collapse in the ground of the inhuman(e). That is, there can be no understanding of or accounting for extraction or environmentalisms without consideration of their subjective forms in the geologies of race. In this chapter I have proposed geologic life as an analytic to de-script the time of the ground. The next chapter, “Rift Theory,” expands on the rift as a location from which to theorize broken earths and counter the geopower of colonial earth.

2 Rift Theory

> History is not only absence for us, it is vertigo.
> —ÉDOUARD GLISSANT, *Caribbean Discourse*

Geology can be thought from many places—plateau, rift, underground, deep—spacing time and the world.[1] The official and economic archives of geology speak from the perspectivism of the plateau, the surveyor as planetary technician and geologic underwriter in cartographies of extraction and sedimentations of valuation. The grammars of geology cut away the trauma of extraction and unearthing: what a geographer might call its spatial affects, what a poet might suggest is its debasement, crystalline in its exegesis of the world. Geologic life is double-sided: on one side, the struggle against carceral forms (where the elemental doubles as the ontological arrangement of property and theft and an index of value); on the other, a struggle to recognize the full sense of isomorphic relations that inform the possibility of experience in an all-too-material world. In this chapter I advance a theory of the rift as a spatial form and methodological process for understanding the material locatedness of theorizing, against the epistemic smoothness of geology and its homogenization of the elemental.

To reinstate an elemental approach to geology is to elevate property over personhood, stilling properties in time outside of relation and in a monogeography of value (the gold standard). Elemental configuration can be geologic determinism by other means. Thinking the elementary as difference and locating concepts in their historical geographies, I develop the rift as a place and perspective from which to understand the vertigo of geologic relations. The rift as a topographic zone of the earth's surface is a site that is pulled apart by the flex of extensional tectonics in the mantle of the lithosphere exhibiting what is materially buried and what is uplifted to the surface.

Rifts are the basis of stratigraphic exposure!

Rifts expose the *uneven geographies* of exposure and the *geophysics of geologic agency*.[2] In geologic terminology, a rift is a zone where the lithosphere is being torn apart, a site of the ruptures between different states of the earth (and their attendant social and environmental strata). Living through deeply rifted social and environmental realities, in the context of many broken earths, rift theorizing might address the inchoate languages and material transformations that break and build worlds of resistance to the unceasing transformation of the surface of the world into property. Rift theorizing might be likened to geophysical fissures as a route to confront the dehumanizing shocks of broken earths. Acting as an indifferent terraforming force, colonialists and settler colonialists visited violent new geophysical relations on the people they encountered, encoding the earth in necropolitical languages of matter, and wrenching entire worlds of being from relation. The rift is a way of reading across these broken grounds and attending to specific place-based reparative work that practices divergent imaginations of material engagements.

I imagine rift work as an open-ended set of questions and placements that begin from a situatedness in broken places and think from that brokenness without the promise of the plateau or its telos. Rift work might be something less prescriptive than the social sciences usually demand but offers a way to think race and geology together, as an embedded set of geographic and geophysical relations made in fractured colonial afterlives. Rather than posit the trajectories of origins and endings, the rift can be more usefully thought of as a methodology of uneven geographies and unstable grounds (ontologically and materially) that are found in the archives of geologic life as footnote, fragment, suppression, aesthetic, poetics, and affect.[3] Riftwork recognizes broken grounds not in service to the settler colonial surface of power and its modes of (de)valuation. Such rents in the epistemological archives and material affective infrastructures point to other geologic lives that are overlooked and unheard on the plateau.

Rather than singular sites of origination, where the plateau is perceived as goal and arrival, the rift (as a concept and place) is a way of geographically attending to the power structures that differently press down or levitate racial difference as a gravitational field of relation, as well as contributing to the analysis of racial regimes of property and the sedimentation of whiteness and its longitudinal affects. If the trouble with pluralizing and multiversing in social theory is that it accumulates difference, often without accounting for its weight of power and violence, then the rift gives a spatial dimension to racial formations and extends the temporality of the fracture in its terrains as an ongoing material condition and insurgent rebellion across space and time. As a site of extraction and leverage, as well as an extended view into the stratal planes of pressure and the histories of violence that surface disavows, the rift speaks to the accumulative weight of the Overseer's earth. In this sense, the rift might be thought of as a prismatic space of grind that may build things in quiet revolutions of the earth.

Throughout this book I engage epistemic processes that have a countergravity to *fossilization* and *stratification* (as the empirical bases of geology and the modes of petrifying social forms) as strategy to desediment white geology. In human origins theory, the term *ghost populations* refers to unknown ancient humans (often ancient Africans and unknowable hominids that have no fossil referent) whose genetic marker is carried in contemporary humans but who are without an originary genealogy or recognizable identity trace.[4] Which is to say that the ghosts of geology are *not-fossil*; the unrecorded, unclaimed, illegitimate, and illegible geologic lives of those who occupy the footnotes and rift, in the margins of the script of genealogical narratives, and in the undergrounds of colonial scripts of becoming human. I imagine the insurgent geology of the rift as something like Glissant's "going-on-and-with that opens finally on totality" (1997, 192). "Going on and with" speaks to the ways in which diasporic histories are yoked together but also to how this structure of colonial geologic life has within it its own undoings that are evidenced in both a refusal to be flattened by this violence and a recognition of the instability of these categories that require constant work to maintain their (subjugating) operability. If we think about the inhuman in its liberatory solidarities rather than just its subjugating modes, it is also a way to make and develop a politics of stratal rift solidarities against the oppressive use of geopower in the configuration of subjective life as carceral and coercive. That opening on totality is, then, also the poetic heart of an affirmative geology, as a cosmic incitement in an expanding universe:

> It is like what we imagine knowledge to be: dark, salt, clear, moving,
>   utterly free,
> drawn from the cold hard mouth
> of the world, derived from the rocky breasts

forever, flowing and drawn, and since
our knowledge is historical, flowing, and flown.
(Bishop 1979, 66)

When I began this project, methodologically I had thought I was interested in processes of fossilization and future fossils.[5] It turned out that the opposite was true. If fossils are knots in the narrative of making a world—a genealogical claim on time—then *not-fossils* are what exists in the shadow of those claims to identity, ghosting, but no less part of how worlds come into being. *Non-fossils* are the orphans of geography. In this chapter I (1) explicate the conjoined history of the plateau and the rift as spatial and social forms of life that are onto-epistemic and material; (2) show how the liminal and the lithic are joined through a geophysics of sense that functions as a racial coda of space;[6] and (3) suggest that rift theory is a way to both take account of the sedimented histories of the colonial life of geology *and* mobilize the fissures of the rift for destabilizing work to break with these racialized inheritances.

Methodologically, the figure of the rift is a way to access the repressed architectures of the surface and activate new ways of seeing liminality as a form of lithic subjectivity and placement (which shifts the geographies of attention and means of intervention). To break the filial bonds of race and environment requires not just the end of grammars of extraction but also the dismantling of the stratal imagination of the surface as it is constituted through a lithic (de)arrangement of the plateau and the rift. Staying on and with the rift is a commitment to beginning to dismantle the white supremacy of matter through rerouting the grammars of geology and working to disassemble its accrual of value. This dismantling is done in the context of an earth whose colonial rift economies have intensified the propagation of planetary tears in the time and space of the earth processes, turning the geologic oven on high in the form of climate change.

Spatial Propositions and Lithic Subjectivity

Spatial propositions and statements of location are always a question of territorializing, which is to say that location is a spatial fix that consolidates a political territory of geopower. *You Are Here.* Reassurance. *You Are Not*

Here. Erasure. Ghost populations, rattling chains under partial universalism and conditions of unfreedom. *Silencing the Past* in Michel-Rolph Trouillot's terms (1995). We are marked differently by and through geology; this is the geologic color line. The earth weaponized betwixt belonging and alienation. Extraction exposes the most pronounced fault line of this racial inequity, but there are many other more subtle and insidious effects of racialized geophysics. If we understand the density of this entanglement in colonial histories of matter as building a geophysics of space—conceived as the historical weight of social and physical forces that both hold up and repel the possibilities of a body's passage in space (and its corresponding action of making space otherwise)—then the consideration of the racialization of these geoforces becomes not just necessary but foundational to any spatial inquiry, environmental or otherwise. The argument here is that the structural address needs to be foundational and planetary rather than just vernacular (although the vernacular is important in the specificity of the rift and how it breaks and is broken).

The European plateau (as an expansionist continental geography), its empirical raft of natural sciences and philosophy, was made within the history of colonialism's material transformation of the earth. The imperial warship's perspectivism over the horizon was transformed into the colonial surveyor on the plateau. The stratigraphic impulse of white geology was to bring all other space and time into European Global-World-Space, which in turn was enacted through the integration of metaphysics and geophysics in a geographic imaginary of the earth as empire. If we take an example from natural philosophy, Kant as a pivotal figure in the lithic sedimentation of subjectivity (and for the sake of continuity), when he considers the Lisbon earthquake he does not speak from *within or with* the earthquake but assumes the position of the objective observer on the mythic plateau (which in his case is Königsberg). Responding to the threat of physical geography, his sublime teetered on the edge of the plateau, looked down, only to pull back again to reinvigorate the surface, inuring the subject with tools against the rift. As earth system thinking emerged out of an understanding of natural causality, the question of human freedom was placed in jeopardy. No longer governed by the morality of the Great Chain of Being, freedom was geographically relocated as a question posed in the realm of inhuman indifference. In the anxiety that accompanied that moment of faith shaken by the 1755 Lisbon earthquake—in a space that could not abide a vacuum of governance by a higher order or stay with the existential conditions of unknowing—Reason took the place of God as the arbiter of human difference (natural causation replaced divinity, but it did not quash the desire for hierarchy, transcendence, and telos). Nigel Clark argues that Kant's "settlement" in the birth of the modern subject "involved putting up a barrier between a sphere of nature that is left to its own necessities and a realm of human existence where we are free to compose our own principles. It is he, then, who gets the credit or the blame for firmly establishing the idea that freedom is where nature *is not*" (2011, 85; for an alternative reading, see Cesaire 2010).

Henceforth, the European (white) subject eschewed inhuman nature and exercised a psychic repression of its abysmal qualities as relevant to its composition. Kant's legacy (1994 [1756]) set in place a new judgment, predicated on the *universal . . . but*. That is, the imagination of a Universal Subject is universal only within whiteness. Nature was understood to be outside freedom's jurisdiction, and so too was Blackness (narrated as being of nature), and therefore the composition of principles of freedom and the experience of beauty did not apply, under Kant's dictum, to the enslaved subject. This adjudication is at work in the ongoing negative dialectics of white supremacy that suffuses environments from nature scripts of museums to freedom in the Black outdoors.

So, despite the seismic rift of the earthquake and subsequent tsunami that took Lisbon under—as a city whose wealth, density, and global material reach were built on the earnings from the Portuguese trade in enslaved persons—the psychic rebound to that shock by the emergent natural sciences and philosophy was to double down on the sedimentation of the plateau. For Burke and Kant, sublimity is interpolated through race. As Meg Armstrong comments, "'Black bodies' (Burke's 'vacant spaces dispersed among the objects we view') is always, simultaneously, a mark of bondage to the ideology of the aesthetic" (1996, 230). That is, in the aesthetics of natural philosophy and in the actuality of mineral extraction, Blackness is understood in Enlightenment thought as an absorbent materiality to cushion the white imagination from the sharp rubble of the earth. The aesthetics of anti-Black gravities secure the sublimity of whiteness (its existential float above the rubble of the earth). Blackness, then, for Edmund Burke and Kant, is the geographic imagination of a portal to the inhuman rift. The geophysical rift tore apart God but redoubled social hierarchies using a metaphysical root (Reason) to achieve a spatial telos (whiteness as empire). The policing of these categories of identity understood as a material property of Blackness and as a metaphysical property of whiteness was crucial to maintaining the stratigraphic hierarchy in the extraction of value. The figurative relation of Blackness and Indigeneity to and as the ground secured the subterfuge.[7]

In contrast, hear Glissant as counterpoint: “As if the sea kept alive some underground intercourse with the volcano’s hidden fire” (1997, 121). Glissant refers to the painful histories of Black life that communicate in the subterranean seas between islands like the deep magma of volcanic archipelagoes. And the Indigenous Caribs, faced with the ongoing colonial tide of invasion, leapt in a precursor to Fanon’s inventive leap (see Marriott 2018, 312) off the plateau to find their truth and refuse the institution of an unrecognizable violent form. The project of many thinkers located in the broken earths of colonialism was to “plumb the depths” and find a new language for the rift, articulating its occupation as matter-theory and sense-event. To disassemble “judgment,” in its delineation of the color of Reason, pushes further into the questions of the inhuman and devaluation. In this painful language, another language of the inhuman can dislodge the all-too-instant relays of meaning, the commonsensical valuations, the ideas of what is necessary and what is not in the apprehension of the material reality of the world. Destabilization in language reformulates history and its telling, making other geographies come into view. In this move, the inhuman might be understood as a sentinel for another history told through the earth (as a loud alluvium of the past).

The colonial transformation of matter into natural resources led to a transformation of both experience and desire, as matter became a repository for the desires of possession (of land and subjects)—*chasing the shiny*—where a geo-logics of property and properties was brought to the fore as the schema of interpretation and apprehension. Mine/m“I”ne is a geo-logic schema that secures the subject in the world and, like Kant’s sublime, resists the vertiginous fall of the rift (that Glissant claims as his history in the epigraph) and the loss of the horizon, in order to secure a propertied footing in the world. In this formation of the European subject, the rift is an excitable threat to the Overseer. The rift’s dissolutionary power, the ground that gives way at the edges, is a threshold that must be overcome for the propertied self to be consecrated anew. Metaphysics by the way of geophysics. A paralleled repression of Indigeneity and Blackness was simultaneously used to spatialize and stabilize the rift-as-abyss, and to provide a psychic route that secured the overground in its perspectival supremacy from this abysmal threat (the inhumanism from within).

Stratal placement of race in geologic orders was an inoculation against the psychosexual horror of what whiteness did/does to secure the plateau. Thus, the privilege of whiteness is a priori a stratigraphic act that evolves into a biological schema realized through the traction of geology (in the race theories and temporal eugenics that gave empirical density to that notion of biological superiority as a place in time). Paleontology gave rise to and justified—as a legitimating and self-fulfilling analytical frame—racial hierarchies and genealogies that were from inception eugenic, as whiteness was made the apex of a spliced geological/genealogical line.

The plateau, understood as a terrain of white geology, also interns the view and assumption of the all-knowing white master, the "father" of geology; the one who decides for the child (the racialized subject, belated in geologic time). The master's voice has a material, spatial, and temporal instantiation, which is the property of the plateau. Colonial geographic imaginations mobilized a stratigraphy of subjects and normalized their positionality as elevated or subterranean, to enact structural oppression at the scale of the planet (which was the dream of imperialism). Whiteness secured the plateau through geographic conquest and consolidated and expanded itself through the material subtending of Black, Brown, and Indigenous personhood-as-stratum. The plateau is the ideal signification of the *view over* and the *view from* racial superiority. What creates the rift-as-rift and not the beginnings of a world—like the rift valley in Africa—is the geographic imaginary of the white spectator and his stratal projection of Overseer's earth. While the colonial masters ordered the maps and geographic imaginaries, it was the surveyors and geologists who were the Overseers, terraforming and breaking earths.

Thinking from and in the rift, as a material and theoretical practice, is to think of matter in correspondence to a radical revaluation of the inhuman. The rift, then, is deployed not as a metaphor but as a material condition of subjective placement that is also a site of fracture. The devaluation of racialized lives in relation to the levitation of whiteness on the plateau continues to allocate a placement of disproportionately weighted environmental burdens and imperiled atmospheres. Racial violence is a violence of matter. *#BlackLivesMatter*. Thinking the rift as a locational pedagogy in racial materialisms is a way to understand the substance of geologic lives fully lived, rather than through their negative inscriptions on the plateau (as waste, damage, and pollution).

The rift as a zone of fractures requires a reparative modality of engagement and redress rather than being understood as a zone awaiting inclusion on the plateau. It is a site to see more clearly the afterlives of geology, after the extraction principle, and to petition for the abolition of the geophysics of the plateau. Which is to simultaneously see the current forced passage of boats across the English Channel and the Mediterranean as a weaponized geologic subduction zone operated by Europe and its allies to bury responsibility and enact racialized eugenic principles, and the funneled geographies of border crossing across the Sonoran Desert as an organized zone of dehydration and thirst for migrant disappearance into aridity. The rift is this pressing together of hard geopolitical borders and physical geographies of disappearance in a weaponized geology of the state (on behalf of and for a privileged subject).

Rift as Racial Brokerage

The rift and the plateau are in a geographic debt relation (a dialectics of geopower) that establishes conditions of broken earths and racial subduction and subtending (those who are made to disappear from view and geography). While ontological claims have a concrete reality, it is the geography and geophysics of these colonial geosciences that make the rift zones through their grammars of propertied relations. Settler colonialism set the imperative to turn racism into a longitudinal geo-logic of colonial infrastructures that could function, structurally, as the "ground" of racial capitalism. What provides a commons to this geographic zone is the rift itself as a geophysical entity, not the conditions of its racially ontologized participants but their shared overlapping (and sometime undermining) conditions of possibility. That is, Brown subjects (low-caste Indian and Chinese indentured labor in the Caribbean or on US railroad construction) may take on or come into the conditions of the rift but have a different ontologically structural racial "origination" under colonialism and settler colonialism (and this is also configured differently across the global material sacrifice zones of the world and their sites of disappearance).

In the context of the Americas, the natal moment of the native is on the plateau, in possession of relations of land (albeit with a radically different geographic relation to modality of surface in relation to the settler). Colonialism literally attempts to push natives into rift valleys: Caribs taking to the sea in the place that is now called Grenada, enacting the freedom of decision against the already decided, determined extinguishment of colonial invasion in 1651; and the environmentally compromised zones of land and barrenness that constitute the designation of reservation and Aboriginal territories, and the ongoing structures of spills and dumps that rift through clean and ancestral waterways and aerosolize in atmospheres of toxic accumulation and nuclear futurity. Where slavery and Indigenous genocide are often yoked under colonialism through the lenses of labor, land, work, and property, we can understand that togetherness differently as preconfigured through the organization and theft of energy for the plateau, as "bodies" become configured in relation to extraction and dispossession without consent in colonial categories of race.[8] What is taken is always far beyond land and labor. The petrification of race acts as a physical geography of constraint.

Rifts exhibit a geographic amalgamation forged under pressure, broken genealogies, abstaining from projects of racial purity, which is not to say that differences are homogenized, or specificities of historical geography made equivalent. As Wynter has shown, expulsion and origination in and outside of the Human have different possibilities in the reoccupation of that term. Frank Wilderson more acerbically argues that the place of the native (extrapolated here through the paradigm of the rift) is not the same. He argues that "junior partners in whiteness" have a chance at the plateau, whereby the native has a "spatial, cartographic redemption" in the return of land and a possibility of historical and political sovereignty, but the libidinal economy positions Blackness outside with no possibility of claiming back space and time (Wilderson 2020, 13–15). In the spatial ontological, Indigeneity equates to spatial exile-in-place, but with the possibility of cartographic redress to the plateau: Blackness is exilic dispossession whose territorial claim is suspended in the rifts of the oceanic, maroon, hold, and underground. Extrapolated, Blackness is the ground of propertied relation in its negative form, and imagined by whiteness as a negative topography it supports the elevation of plateau: the plateau can function in this stratigraphic relation of anti-Blackness only because it *requires* undergrounds for surface conditions to emerge, as Europe "required" colonialism to develop its theories of Enlightenment thought and practice. There is a historically complicated, and at times antagonistic, relation between Indigeneity and Blackness (as there is also an intertwined history of affiliation and difference). There are also much fuller geographic and genealogical histories than the "purity" of ontology accommodates (see M. Wright 2015). My intention is not to try and map this difference in its complexities of relation (as this is the work that is active in Black and Indigenous studies); rather, I want to understand those histories of experience as geographically situated in different modalities of time and space, which coincide in their enfolding into the settlement and fictioning of colonial geology, but also continually to stretch and torque the space of those geographies.

Deep time was used to naturalize the ongoing practices of the fantasy of racial extinction of Indigenous peoples in rift zones (the Trail of Tears, 1831) and to stabilize Blackness as an original geoforce of the subterranean (Africa as resource). Brownness is parsed in the many complicated geographies of empire to emerge out of the forest, desert, monsoon, or sea, as an emergent quality of environmental determinants. The pincer move of stratigraphic structures of erasure and subtending in narratives of natural history cleared the geophysical (and metaphysical) ground for the tense of whiteness.[9] As Indigeneity is made through successive displacement or rifting events, Blackness is borne in the rift because it is narrated as having no prior ontological condition to the historical natality of its fabrication under colonialism, out of the rift valley and forged in the hold, mine, and plantation. These historical geographies share violent displacement and diaspora as external and internal spatialities of existence. What exists before, during, and after is outside of the creation of these categories, but active within their geographies. Structurally, Blackness, organized through an inhuman paradigm, does not preexist the genealogical isolation or geographic exile of slavery and thus cannot proceed to the plateau. Blackness, for the white geologic imagination, gives psychic perspectivism to the vertigo of the rift as fungibility provided the geopower of uplift on the plateau. The projection of the vertigo of the rift (Kant) is replayed as a subject condition that the enslaved must interpolate and make their own (Glissant). As Marriott comments, "Blackness, then, is the vertiginous experience of its own impossibility, or its experience is that of an impasse, or aporia, that remains hidden, unknown" (2018, 235). If the plateau and the rift map surface tensions and the torques of its geoforces, then the rift might be understood as a geologic surrogate that is both the site of new origination and the result of stratal pressure that remains hidden from the plateau. Thus, the rift points to another archive that remains hierarchically below that of the plateau, in the footnotes, the subtending, the overlooked. How might we methodologically speak of these residues and dispersals of this rift relation in earth without setting this up as a reprieve and inoculation against erasure (which is to return meaning to the plateau)? How does thinking from the rift bring another historiography into view?

The *Not-Fossil*

The rift is a structural motif or restless analytic of this book, one that seeks to focus in on the contingent possibilities, the *not-fossil* that accounts for a larger field of geologic lives, in the deluge and subduction, the erasure of a repressive grid of racial geo-logics, since 1492. Minor acts, footnotes, embodiments, and placements partially captured in the official sedimented thing—the *not-fossil*, the body-as-demonstration, the report, the map, the archival reference—constitute the trace of geologic lives lived in the rift zone. Like a diorama that presents the completeness of a world to gaze upon, the plateau delivers a perspective of the just-happened-upon glimpse into nature without its technics or curations. The fullness of this world (imagined as totality) snags on a footnote, like a dead fly on the lion's back in the Hall of African Mammals, which brings attention back to the craft of exclusions and expunged historicities. The rift tells us of the maintenance of other kinds of nature work and political staging that need to happen to keep the viability of the Edenic scene of African abundance, constructed as surplus, natural resource, and native, intact (see Haraway 1984). While the limits and challenges of the archive have long been a vexing historical thing to which earth histories are not immune (albeit in materially different ways to colonial or state repositories), the earth archive is often placed, somehow, outside of the purview of politics because of the imagined division between human and inhuman history, and between grammars of speech and orders of matter. If colonialism is the alienation of the earth, every anticolonial moment must practice disalienation with the earth, bound to its restless geoforces. This is a space of the disarticulation of geologic grammars and the rearticulation of a different genre of arrangement of sense and meaning that is affectionate toward the geophysicality of being.

"Natural" history, until recently, has been understood in a Western canon as a largely physical inhuman event. Geologists can argue over the representation of the epochal boundaries, but the geomorphic event itself is understood to have its own reality and temporal exclusion from the political present, even as these epochs are unearthed into other political economies (as in the returns of Carboniferous in the heteropatriarchy of the nation-state). In parallel with this political disdain of the present tense, a recent article in *Nature* charts the lack of any changes in ethnic and racial diversity in the earth, atmospheric, and ocean sciences in the last four decades (Bernard and Cooperdock 2018; see also Dutt 2020). *#BlackInStem*. So, even as the Anthropocene concept delivers a planetary geosocial formation that extends an appreciation of the social life of geology, it still relies on an earth reading that temporally displaces politics by adopting various deep-timing perspectives (including the future geologist and a postracial futurism that erase the geotraumas of race) that leap over the thorny problematic of indexing location and thinking with difficult colonial inheritances and geophysical debts. Meanwhile, the earth itself is delivering its own account of hominid temporality and (anthropogenically provoked) articulations in fire, heat, carbon, and methane rifts.[10]

Thinking with the rift as a historical geographic entity is a way to understand colonialism as an intervention (and invention) in time and space that

FIGURE 2.1 Coal extraction, Falkirk Mine, North Dakota. Photo by author, 2015.

produces foundational geophysics to the contemporary earth events of the Anthropocene and climate change. The geosocial architectures that structure colonial afterlives are prestratified through this plateau-rift relation. Its natural history of geomorphic change created ghost geologies of racialized material pasts. Thus, we can begin to see the mobilization of identity as a geophysical and spatial operation that is tied to both the perspectivism of the plateau and its epistemologies of the rift. I am not proposing another kind of determinism here; rather, I want to draw attention to how terraforming the earth happens through certain structures but, more important, how identity is mimetically deployed in relation to those structures as that which allows them to function. As McKittrick reminds us, "The rift must, in other words, not just take us somewhere new, or old, conceptually, it must provide the conditions for us to think carefully about how the work of liberation is tied to the uneasy work of getting in touch with the materiality of our analytical worlds" (2019, 246).

As geologic grammars of colonialism petitioned for a singular geography of relation indexed by its modes of valuation (which is repeated in the carbon imaginary), the rift draws attention to the need for a fuller apprehension of the density and difference of the structural relation of material geographies in excess of political economy. The rift as vertiginous materiality of race and geology gives a spatial form to the weight of identity and its political forms.

Interrogating the material strata of colonialism(s) requires attentiveness to the subtending of liberalism's political forms and their coming into being through genealogies that have interlocking material and subjective forms. Under the rubric of these geo-logics the normative categories of racially governing materiality come into being. It is crucial that the perspectivism of the rift is not predated by the plateau but retains the restless urgency of its insurgent geology, in the academy and the world.[11] I propose the rift to conceptualize the concrete geophysical state of being where liminality is produced through the legacy of colonial lithic architectures (which will be discussed through the empirical chapters of this book). So, the next methodological question becomes, How can a history be built that keeps vigil with the filiation of the fragmentary work of *not-fossils* and their erased historicity, recognizing the work of duration but without reinstating originary claims that keep subjects in their petrified place?[12]

Riftwork: Geophysics and Gravity

There are two understandings of geophysics as an epistemology of material force that I want to foreground: *geophysics of sense* and *anti-Indigenous/anti-Black/anti-Brown gravities*. Understood scientifically (Einstein's general theory of relativity), gravity is a force of the space-time medium, a dimensional ground that other forces of nature play out on, but that all matter "feels" as a force. Near black holes or the Big Bang origins of the universe, these equations break down, and gravity is shown to have a quantum-like form like other nature forces, at which point other descriptions of time and space take over (so gravity is both subject to and not subject to the principle of locality). Understanding the spatial production of geology as multidisciplinary (a concept of materialism, a narrative of time, a question of racial hierarchies, etc.) and multidimensional (organizing conditions of space-time and planetary modes, etc.), we can use gravity as a way to conceptualize how dominant geosocial forces act to create worlds that have vertiginous and rift-making qualities, creating conditions of pressure and a geophysics of sense that police life from the outside and within, as bodies-in-space.

Extending Wynter's idea of the "senses as theoreticians," I want to push sense through the geophysicality of the world (the affective architectures of geoscience), as well as producing new possibilities of sense that challenge and interrupt normative structures of time, space, and materiality (which have a parallel depth of vertiginous racial violence). A geophysics of sense, then, is a way to conceptualize how the arrangement of sense that is locational to the rift accumulates as a mode of thought, a sedimentation of affect, that renders persistent geoforces on racialized bodies. If gravity pulls us to the earth through geophysical attraction, material histories of colonialism and their settler colonial presents police the condition of the fall and velocity of bodies. For some subjects the desirous pull of geophysics toward the earth's core is given a systematic extra push into the earth. Think of the young Black South African miners who go deeper into the earth than any other human being: they suffer under the weight of the gravity and the heat of geophysics because their lives and labor are considered "cheap."

The architecture of a racialized geophysics of affects has an aesthetic presentation, insomuch as it presents a political or persuasive scene that has a normative regulation. It is historical too, a force relaying the accumulative or sedimented affects of previous scenes, gathering sense toward reproductive nodes that produce a geophysical arrangement across spaces, something like what Moten calls the "affectability" of aesthetic sociality (2017, xi) that creates a force in the deformation or freedom of subjectivity as a sovereign state. Relays in the geophysics of sense create gravity and live in the body as knots of feeling that become like reflexes in the body's understanding of itself, akin to a learned corporeal intuition that is schooled in the grammar of the body's experience and disciplined in ways that become invisible, as an *as is*.

Geophysics captures the modulations of intensification that change in the parameters of space and time for certain bodies caught in the violences of value's exchange in the material worlds of geology. From heat burdens to muddy rivers, geology is cooked as a multidimensional gravitational field that coalesces around, as well as exceeds, bodies' energetic processes and the geochemical reactions that explode around them. As minerals drive geologic redox reactions that lead to biological metabolizing, generating heat and force in the equation, this might be understood at a more granular level in the sedimentation of subjectivity, society, and space. As the earth is a product of gravitational accretion, social life too organizes differentiated gravitational pulls on bodies. *Geo-pulls*. Some bodies hit the tarmac faster and with more force because of epistemic-curated systems of gravity, called race and racism. These systems are not governed by natural laws, but they do gather strength from a history of naturalization and the systemic collusion of geohistories that created structural forms of oppression. Natural resources do not just organize matter for extraction but impinge on what matters and who matters in subjective life.

FIGURE 2.2 Rollin' coal, Williston, North Dakota. Photo by author, 2015.

Geophysics

Geophysics is a way to think about the simultaneity of the entanglement of geohistoric time and the political present, when events arrive on bodies as a deformation in the possibilities of subjectivity, in material and symbolic ways, as a constitutive moment of the arrival of a deadly historical force. Water cannons, tear gas, a knee on the back, the stranglehold of police, *No Humans Involved* (Wynter 1994), broken public spaces, inhospitable places, stolen land, the exposure of communities of color to toxic air and the residues of geochemical processes in the water, the *Pressure* of Horace Ové's BFI-produced 1976 film of that title, for example. The weight of the gravity of all these historic formations arrives in the present as a geophysical politics. So, what does it mean to take the pressure off or make a gravity with other handholds on the earth? Or to heap that force back on the viewer with such intensity that they feel the gravity of anti-Black violence, as in Arthur Jafa's film *Love Is the Message, the Message Is Death* (2016). As Jaffa switches and cuts between the scene of a police beatdown of a young Black woman at a pool party and a burning inferno of the sun, he marks an insurgent geophysics of Blackness: a geophysical connection between the anti-Black gravity of the police state and the eventual black death of the sun, a connection that speaks to a set of physical processes that press on, deform, and reorganize subjectivity, while challenging the all-white geographic assumptions about the inhabitation of space (#easiertoimaginetheendoftheworldthantheendofracism). Race has a painful gravity, it is a word that hurts, in the shape of its utterances. As Morrison asks, "How to enunciate race while depriving it of its lethal cling?" (2019, 133). For me, this translates into the questions: How to shift the gravity and weight of colonial geologies of race? Which perhaps begins with a tunnel under white man's overburden? Part of that "lethal cling" is the materialization of race in environmental conditions and as ongoing material conditions. If we understand race as a geophysical force that amounts to the structural division of the earth (forever and ever, in Fanon's words), then what its dissolution mobilizes is nothing short of the methodological dismantling of the entire grammar of geology, as subjection, valuation, and desecration.

Geophysics is a way to think in this world with a different gravity and historicity of sense, rather than consecrating the urge to leave the world (through space travel, Afrofuturism, or transcendentalism). Black feminist futurisms remake and refuse the given world from the inside, reorganizing geography's geophysical dynamics and their biopolitical effects.[13] Changing the weather, in Sharpe's (2016a) terms. It is a tilt in the perception of sense, or a rift in the strata of subjugation. The rift and its altered geophysics of sense also push against genealogical method, detaching from the desire or expectation of an originary or elementary referent. Geophysics is imagined as a translocative, transhistoric tense of accumulative fungibility. It is a method that stays in the rifting zone resisting a restorative impulse to re-establish the plateau. A rift is in tension between past and future, hinged between material and symbolic orders, a zone of fracture and pressure.

Riftwork: Sediment as Method

> We no longer reveal totality within ourselves by lightning flashes. We approach it through the accumulation of sediments. The poetics of duration . . . take up the relay from the poetics of the moment. Lightning flashes are the shivers of one who desires or dreams of a totality that is impossible or yet to come; duration urges on those who attempt to live this totality, when dawn shows through the linked histories of peoples.
> —ÉDOUARD GLISSANT, *Poetics of Relation*

Glissant's reference to lightning flashes recalls Enlightenment claims to the transcendental "light" of knowledge (as truth, genius, and possession). Instead, he proffers his alluvial methodology as "accumulations of sediments" that are organized by their durational capacities rather than the dream of totality (that belongs to the colonizer). It is a complex and interlinked historicity that sediments through the interrelated geographies of colonial and precolonial encounter. Sedimentation, thought against the petrification of the stratal thinking, mobilizes *not-fossil* epistemologies, capturing that which remains overlooked in the strata: a parallel inhuman memory, the inhumanities.

Sediment is mineral or matter that is broken down by weathering and erosion, and geographically mobilized to be deposited as residue and remains.[14] In the context of geology's origination in colonialism, Glissant's poetics are a crucial repose to questions of filiation and natality that require fixed empirical or fossil forms. The eugenics of filiation most violently reveal themselves in the question of racial mixing and the panic of miscegenation, and it is to this mixing that Glissant's work explicitly addresses itself. He argues against the colonist's "my root is the strongest" mythic identity, which then exports that priority as an inherited value: "A person's worth is determined by his root" (Glissant 1997, 17). The affect of the genealogical assertion of whiteness as *the* root—as it is empirically claimed in the work of geologists such as Cuvier, Lyell, and Agassiz—as an effect of the earth's geography (as natural causation) rather than colonial possession is that the "conquered or visited peoples are thus forced into a long and painful quest after an identity whose first task will be opposition to the denaturing process induced by the conqueror. . . . Decolonization will have done its real work when it goes beyond this limit" (17).

That is, opposition to this devaluation through filiation is not enough through the assertion of a better lineage or history; decolonization, in Glissant's terms, needs to embrace the rupture of the voyage and voiding of subjectivity that is the conjoining of these histories into "uncertain births of new forms of identity that call to us," where uprooting works toward identity "as an expansion of territory," not its compression into the rift of colonialism but opened through errantry (18). He argues that identity is "no longer completely within the root but also in Relation" (18). It is borne through these shared geographic experiences as dissent rather than in the singular line of descent of the propertied relation. Methodologically, this dissent in the spaces of the womb, ship, and plantation requires a different conceptualization of a materialism capable of holding the histories and experiences in the rift. In the context of the imperial geographic imaginaries that projected over space and colonized the thought and language of the other, Glissant suggests a pedagogical investment in "nonprojectile imaginary construct" (35) that does not imagine over, but maps from, the depths. While his poetics crafts the possibility of a new beginning, it does so in Relation to its emergence and the complicated "birth" of possibility that is sutured to the gratuitousness of violence against Indigenous and enslaved subjects. The abyss of the boats, of the ocean depths, of the alliance with an imposed land and meeting with the first inhabitants (who are also "deported by permanent havoc") becomes knowledge, where the "unconscious memory of the abyss served as the alluvium for these metamorphoses" (7). The sediment does not just record the histories of those designated as nonhistorical but is a site of passage in the sense of an alteration. The rift opens unknowing collectively.[15] Thus, Glissant seeks the permeability of those histories and how they force new languages into being through those abysmal depths.[16] In the context of colonial epistemologies and subject formation, sedimentation is a geologic and theoretical method to historicize and language the rift and "build" an insurgent and reparative geology of memory. Empirically, *not-fossils* campaign against any monolingual intent, such as the colonial formation of Global-World-Space, and the monolingualism of individuation, temporality, origin, or language.[17] The geography of rifts dispels the perspectivism of the identifiable and thus extractable, suggesting that the depths require and practice dissimulation as a form of shelter.

Methodologies of the Rift

I deploy the idea of a *rift* as a mode of simultaneously noticing the historical formation of earth-subjective relations and their historicity in the present. A rift materializes a vector across registers of epistemes and affect in geologic lives. Examples of the patternation of the subterranean can be seen in the following chapters: the "placing" of Sara Baartman in the footnotes of geology to secure the normative "conditions" of European Man through the matrix of her degraded race, gender, and sexuality; the "slave portraits" secreted in small velvet wallets that were made for Agassiz's private collections; and the incarcerated in the convict lease system in "slave mines" that built industrial Birmingham, Alabama, nicknamed the Magic City. The figuration of the rift draws together threads of composition to expose the geologics that underpin the representation and determination of formations of subjectivity and the stratigraphic relations that uphold them. Altered geophysics of sense, aesthetic breaks, and different durational capacities show up a passage to the rift and underground, destabilizing the grammars of normativity of the overground as stable plateau. Finally, the rift points to the irreducible density of the earth and the dreamy fragility of an *indifferently attentive* inorganic and organic communication. Between the breath of the world and its breathing us in.

"Is the world intended for me? Not just me but / The we that fills me? Our shadows reel and dart," asks Tracy K. Smith in *Such Color* (2021, 206). What cannot be made continuous is always what is contingent, always, already losing itself in geochemical embrace. Differently allocated forms of violence, railroads made of bones traveling in the accumulative wakes of enforced passage, are a geographic place and material presence. Rifts are also fissures in the brokerage of life, the rubble of forms of life that are not recognized as such and designated either outside of its bounds or placed on the cusp on the organic/inorganic. It is inhuman memory that presses on the intimate questions of race and environment in the present tense. As Mojave poet Natalie Diaz writes:

> Sometimes I don't know how to make it
> to the other side of the bridge of atoms of a second. Except for the air
>
> breathing me, inside, then out. Suddenly,
> I am still here. (2020, 73)

In the remainder of this book, conversations in the rift zone happen *through* and *with* geology, across several sites: (1) the mine, as a carceral underground and a subterranean space of "improvement" and solar internment; (2) the inhuman earth, as cosmic storm and gravitational pull; and (3) the museum, as a space of narrative discourse and public culture, and as a site of representational resolve. Wherein geology is encountered in different material and empiric guises but is broadly understood in its fullest sense as flesh, earth, geochemical tether of material worlds, geophysical bind to the earth, and geoforces subtending possible forms of life and their deformation. In the next chapter I turn to underground aesthetics as languages that are made in the indices or intramural spaces of geologic acts—the rifts—but are stashed beneath the surface logistics of the libidinal and violent language that underpin geologic histories. I explore underground aesthetics as a wayfinder in these geographies of disappearance and accumulation that stage how epistemologies of geology are both spatial imaginaries and sedimented material relations of extraction.

3 Underground Aesthetics

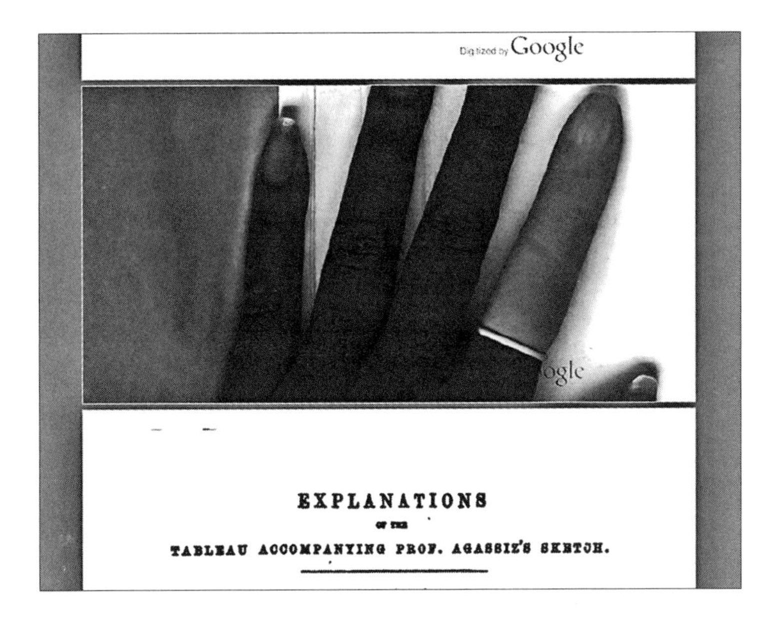

FIGURE 3.1 Screenshot of archivist's hand over Agassiz's text. Photo by author, 2017.

The hand of a Black archivist inserted into a scanned copy of Louis Agassiz's racialized natural histories on Google Books blocks its representation. The hand questions and refuses the seamless futurism of the geologic archive, of a prior paleontologist's vision of the "naturalness" of whiteness as uncontested reproduction, blocking its passage into the present. She is making Black feminist futurity, a different history and archive. It is a different hand than that which the Swiss-born American geologist Agassiz writes about in his infamous letter to his mother regarding Black servants: "and when they advanced that hideous hand towards my plate in order to serve me. . . ."[1] Agassiz was a champion of polygenism (genesis belonged only to whiteness, Agassiz claimed) and eugenics (whites and Blacks were two different species, with clear political hierarchies), and his "scientific" racism began with a letter to his mother about his revulsion toward a Black hand. Similarly, Kant would rather imagine beauty in a ravine than in a Black body, whereby the ravine becomes the foreclosure to the abyss of a Black woman's body (Armstrong 1996). Agassiz writes to his mother that he would rather eat crumbs in the corner of his room, rendered fetal, than accept food from a Black hand. As Toni Morrison observes: "The necessity for whiteness as privileged 'natural' state, the invention of it, was indeed formed in fright" (2019, 180). Two hands touch across time in the structural intimacy of the archive. Agassiz's race science begins in the aesthetic fear of Blackness upon his person: "'I can scarcely express to you the painful impression that I received, especially since the feeling that they inspired in me is contrary to all our ideas about the confraternity of the human type [genre] and the unique origin of our species. . . . Nonetheless, it is impossible for me to reprocess the feeling that they are not of the same blood as us. In seeing their black faces . . . I could not take my eyes off their face in order to tell them to stay away'" (translation and quote in Gould 1981, 77).

Agassiz's portrayal of the dramatic aesthetic "impression" that he uses to explain his erotic and abject attraction is made through a sense-act of Blackness, which leads him toward the establishment of a racialized science of measurement and verification (of what he has already decided to be true). As Cedric Robinson comments, "It was thus that race science was staged between performances of burlesque and horror," where "racial regimes are commonly masqueraded as natural orderings" (2007, 3). I am not interested in assigning a psychic origin to Agassiz's race work.[2] Instead, I wish to recognize, at the onset, that the liminal and aesthetic dimensions of ideas of origination and genealogy were foundational to paleontology, and how this liminality is resolved by making the inhuman a violent estrangement in forms of life. Liminality, posed in a lithic order, is part of what Robinson calls the "regime maintenance" (2007, 2) of race as an image of natural history, where the past is ordered for future fulfillment.

This moment of "the hand" that Agassiz cannot touch or be touched by begins the "American school" of racialized natural science and philosophy.[3] In the same letter, Agassiz is beset with wonderment at the possibility of the procurement of skulls: "'Imagine a series of 600 skulls, most of Indians from all tribes who inhabit or once inhabited all of America. Nothing like it exists anywhere else. This collection, by itself, is worth a trip to America'"[4] (Agassiz in Gould 1981, 83). Only dead or inhuman (life)forms are "worth a trip to America." Indigenous extinguishment and enslavement are brought together to be disappeared and made to reappear within the physic regimes of science, driven by the causal logic of an aesthetic impulse that regulates transgression through objectification. The lure of abjection that the inhuman mobilizes is foreclosed in the violent objectification of Indigenous skulls.

The archivist's hand blocks the frame and its temporal occasion in the present. Images speak of the presentness of the past, or what Sharpe (2017) describes as the parsing of "the past that is not past." The images used throughout the book surface in the present tense. The archivist's hand is the persistence of Black refusal in reproduction of anti-Black violence, practicing Sharpe's (2017) call for vigilance over Black life: "keeping watch with the dead, practicing a kind of care." While race science no longer holds scholarly legitimacy, the liminal damage of the image of racial formations has remained. More than this, the material structures that underpinned the liminal have quietly (and not so quietly) reproduced the conditions of a racially stratified future. Polygenesis did not just separate the origin narratives of races but, crucially, also hierarchized them across a matter divide. It is the hierarchy and the aesthetics of affect accorded to different material states that do the work of valuation rather than the origin story itself. The fact that polygenesis was beat out in America, in favor of monogenesis (the one-family human origins story), is not the point; what it carried into being was the enduring stratification of relation (and of social and environmental deprivation) that materialized along a racializing axis.

Thinking with images of the present, even if they are historic reproductions, is a reminder of the racializing and racist infrastructural continuities (which demands care in strategies of reproduction). Further, infrastructural continuities draw attention to how the processes of accumulation can nullify structures of thought and sense, so the genre of Indigenous, Black, and Brown death becomes normative as historical substratum or "collection" for the establishment of the natural sciences. The invisible hand of reproduction, made visibly present across too many racially objectified bodies, recalls the present conditions from which we look. That reminder is also a remainder of the continuity of the past, and of its resistance. The archivist inserts her history across the "father" of geology. It is an act of witnessing and intervention. A refusal of the condition of social death in the archive and its patriarchal lineage. A reminder of the tension between the continuance of gendered anti-Black, anti-Indigenous, anti-Brown violence in the present and the hope and tenacity that were and are in the fight for possibilities of existence.

Reading history back through its genealogical traces, in parallel to the figure of the "fossil," ignores the discontinuities, the rupture, the abandonment, and what is lost to the archive, or could never enter its raced and gendered spaces, but may yet be present in the future. It is the "future shock" at work and a reminder not to impose today's negative histories on the past so resolutely. Archives must also be read through their generative energy, not just their carceral geo-logics. The desire for justice that populates the post-Reconstruction convict lease courts, for example, tells us of both the veracity of living in the afterlives of slavery and the hunger for justice and the modes of mutuality, modes that were not organized around white heteropatriarchal models, such as the family, but were exorbitant in their ethics of care over the nonfamilial. Not to mention theorizing against normativity as a praxis of necessity and survival (see King 2019, 135).

In this book, I have not used explicit images of racialized violence, particularly photographs of Indigenous, Brown, and Black death, of which there is a vertiginous abundance in geologic archives, especially in the paleontological "collections" of scientists (where casualized references to Black and Indigenous death are scattered through the narratives of "procurement" of scientific "objects"). Hartman's explicit instruction in *Scenes of Subjection* (1997) is a refusal to contribute to Black social death through the renewal of narrative violence through redescription, and Black feminist and Indigenous theories explicitly argue for work that engages life, rather than genres of overrepresented and uncared for death. However, this is a book about violence, and it is important to be clear about why this is engaged. I have retained a focus on violence, in some cases in reproducing the narrative description of archives where I believe that violence has an explicit structural operation that contributes to the normalization of a "racial calculus" (Hartman 2007, 6) or demonstrates the work of the geologies of race in places that are epistemically distant from geology or race. The bodily schema of this violence lands on racialized persons-coded-as-bodies. This schema was authored and curated by white men of science and the disciplinary institutions that they represent and consolidate, of which I am now a part (academic institutions are nearly always on the plateau). Raced and violated "bodies" in the geologic archive are the colonial history of the corporeal matrix of geology that continues in the present.[5] In this book, I have generally chosen images (and been chosen by them) that speak across time, to and with geologic nows, and are commensurate with a rift aesthetic that troubles genres of extraction.

The representational violence of metaepistemic narratives of material forms that are grouped under the sign of the inhuman are, by categorization, inhumane. However, pushing into that category can also disclose how material worlds can become sites of (re)valuation that tie together life-forms and forms of earth praxis. The both/and subject/object of the inhuman category suggests the mutual imbrication of an epistemic geo-logic that functions through representation as aesthetic. Forestalling an ontological or ethical rebuke as the first impulse to the violence of the inhuman category is a way to understand better how it functions, in order that it might be dismantled. Changing the language for subjection also changes the language for earth, and vice versa. As Natalie Diaz shows us:

> If I was created to hold the Colorado River, to carry its rushing inside me, if the very shape of my throat, of my thighs is for wetness, how can I say who I am if the river is gone?
>
> What does 'Aha Makav mean if the river is emptied to the skeleton of its fish and the miniature sand dunes of its silten beds?
>
> If the river is a ghost, am I?
>
> Unsoothable thirst is one type of haunting.
> (2020, 50)

Geology's Visual and Material Culture

Geology should be understood as a process of observation and identification that was an *inductive* visual practice (before the ascent of remote sensing technologies and geochemical processing). Fossils and rock formations connected aesthetics to visual modes of speculation in the creation of other worlds. Remaking past worlds through imaginative geographies and fossil histories conceptualized earth events through the survival of traces (fossils, not *not-fossils*). Specific to this material practice is the importance of *inference* in an inductive visual science that gave geology a speculative scopic armature and developed a materially corroborative visual descriptive language. It is a fragmentary reading of the visual graphia or rocks that advances through a highly universalizing praxis as the axis of its world-making. Throughout this book I draw on museums as archival spaces of politics in the shaping of cultural narratives about geology and race. The national museum is part of a schema of interpretive narratives that organize and erase violence through representation, affect, and aesthetics in the racial construct of the nation-state. These institutions were also the storehouses and consultation sites of mineralogical collections.

Aesthetics hides the hurt (after Colson Whitehead).

Walk into any natural history museum space in Europe or North America and the addressee is an invisible but universal white subject of the plateau, regardless of the diversity of contemporary visitors. Poised as the witness, visitor, and adjudicator, the preferred subject—the heir of Enlightenment science—untethers the knots of hominid emplacement in deep time or nature, where whiteness is made as an evolutionary attribute that denotes an agent able to shape destiny rather than be shaped by it, a grounding that has its basis in the stratigraphy of the races in paleontology. Shifting the model of causality, the descendance of European-now-American-New-World-Man as a set of epistemic material practices established an ontology of subjectivity outside of and above the earth and its relation, celebrating the imposition of patriarchal mastery of matter. We see this in the foyer of the American Museum of Natural History, where the panels of "Youth," "Manhood," "Nature," and "The State" adorn the entrance. The activities of exhibition, eugenics, and conservation, Donna Haraway argues, "attempted to insure presentation without fixation and paralysis, in the face of extraordinary change in the relations of sex, race, and class" (1984, 56). This privileged space of the arrest of time into permanence is of course a fantasy of fossilization that makes its claims with bones to mobilize eugenic fantasies of nature as resource, achievement, and improvement. In the scene of dinosaur conflict (or naturalization of species antagonism) in the museum entrance, natural resources (as "Nature") are tied with "The State," which in turn is pushed through masculinity as the continuance of the nation-state (and the will to patriarchy, in "Youth") as empire: according to the legend in the museum's entrance rotunda, "the Nation behaves well if it treats the natural resources as assets which it must turn over to the next generation increased and not impaired in value. Conservation means development as much as it does protection." These ideological fortifications of nature are secured via the dinosaurs in the middle of the room that tie white settlerism in an unbroken line to the early epochs of deep time and the imagination of an

FIGURE 3.2 Bronze statues of an Ice Age "family" ready to fight with a saber-toothed cat. David H. Koch Hall of Fossils–Deep Time, Smithsonian National Museum of Natural History, Washington, DC. Photo by author, 2019.

eternal battle of will (Yusoff and Thomas 2018, 56–57). The Age of Man is tied to the Age of Dinosaurs via the violent Birth of Man.[6] Deep time and fossils serve to underpin white nationalist narrations of nature and their symbolic orders of valuation as normatively raced and in states of conflict.

The prehistoric encounter between predator and prey, as a *Barosaurus* rears up to protect a baby dinosaur from an attacking *Allosaurus*, reminds "us" of the entanglement of dinosaurs and state formation and masculinity in the primal narrative of nature. The towering bones signal repositories of power, and deep time holds the line about the importance of family in statecraft to concretize these symbolic formations through time. Dinosaurs are similarly scripted in the La Brea Tar Pits and in natural history museums around the world. The patriarchal scene of the Age of Man fossilizes roots. *The Last American Dinosaur: Discovering a Lost World* is the name of one exhibition in the Smithsonian National Museum of Natural History. America claims bones from sixty-six million years ago, dug up at the Hell Creek Formation of North Dakota, for a national icon; geologic formations are fast-tracked into state formations (as with coal and oil). Geologic time is compressed, the Mesozoic seamlessly becomes America, and fossils become the foundational bones in the naturalization of nation-state narrativization. No matter that the earth body previously unknown as America did not exist in geologic time. The ultimate supercontinent Pangaea had just begun rifting into two landmasses, the supercontinent Laurasia to the north and the supercontinent Gondwana to the south (a geologic join that is still remembered in the flora of what is now South America, South Africa, Australia, and Antarctica). Empire begets empire. Lucky that the pesky asteroid accidentally lent its eventfulness to this eschatological formation of nation-building and the ascent of Man through linear time, eschewing the temporal complications of geology.

Museums of Unnatural History

The natural history museum tells us about the question of racial hierarchies and the role of scientific institutions in storying these racialized "natural" histories: of masculinity and monumentality, of gender in understanding colonial natures, and of the formation of a settler colonial nation as invoked and entangled in the *aesthetics of race*. Nature, geology, and rocks are a resource for the reproduction of white settler man and his social production of time.

In the Smithsonian one can visit the David H. Koch Hall of Human Origins and then move through to "African Voices" (sponsored by Shell Oil)

FIGURE 3.3 Entrance to African Peoples exhibit, American Museum of Natural History, New York. Photo by author, 2019.

where there is a market in Accra, or "stop by the Freedom Theater to watch two short films (*Atlantic Slave Trade* and *The Struggle for Freedom*) before walking into the new *David H. Koch Hall of Fossils—Deep Time* exhibition. A quote on the wall by Maya Angelou at the end of the exhibit that "represents" Africa—"I rise, Bringing the gifts that my ancestors gave, I am the dream and the hope of the slave"—gives way to another: "Life begins. . . ." Enter by a fossil lab and a bronzed heteronormative family, pictured in a violent deep time scene, man out front ready to fight with a saber-toothed tiger, poised on the cusp of geologic time. "Earth's distant past is connected to the present and informs our future," the exhibition tells us. "Deep Time starts at the very beginning—4.6-billion-years ago. But it ends in the future." Where, I wonder, is the "hope of the slave" in this inhuman origins story? Hunter Man is accompanied by reproductive woman-child (backgrounded), as the scene of reproduction is projected back in time, deep timing heteronormativity (and the same dinosaur dilemma about protecting the child, and by implication the future, is replayed, but for whom?).

Natural history museums are origin machines. Projection back in time is a way to secure a foundation for the ungrounded assertions of racial superiority and white normativity. Deep time confers a nonpolitical affect (the alterity of the inhuman) to the political assertions of race. The future of deep time is a site of capture, entanglement with the social reproduction of the present, where an archive is at stake, "to preprogram the present" (Eshun 2003, 291). Museums are spaces that highlight the normative tale as a social text in ways that speak to other cultural productions. The categorization of the inhuman is told as a bureaucratic architecture of the rocks of settlement, which follows an epistemological designation of the world as resource. No answer can be given to Angelou's "hope of the slave" because in the museum "African Peoples" are still in a state of nature, fixed in the aspic of the diorama, kin relations plotted out, customs and objects collected. In the human origins story, the mostly "Out of Africa" migration of early hominids gives way to a postracial assimilatory future. The hope that sits between these epistemic categories of deep time and inhuman classification has no expression. Africa awaits. The story is clear: Africa invites, or rather demands, extraction, transformation. History is retrofitted via "inhuman" nature. Museums sediment narrative scripts, securing them in the institution as permanent cultural marks. In this context, archives of redress may need to be invented—as "critical fabulations," in Hartman's (2008) term—to break through the commonsensical establishment of sense, as it is presented in the museum as a hegemonic script of the settler state. It is the political performance of the ontology of matter that matters here, how the empirics of extraction are legitimated, consolidated, and pulled apart from the history of "Life" into an earth already constituted as a "natural resource" (for the patriarchal nation-state).

In geology, museums have a special role in the displaying and reinforcing and consolidation of epistemic and material frames in geology, not least because museum rock collections consolidated the dissemination of the identification of matter to speculative geologists, but also because these collections were donated by commercial extractive industries. In a reciprocal relation in the political work of origination, these exhibitions on geology, nature, deep time, and origins find themselves sponsored by fossil fuel companies, geochemical industries, and climate deniers; Koch, Shell, and Exxon enact the geologic surveys and mine the natality of the earth. A blond, blue-eyed boy of a heteronormative settler family stands guard over the fracking fields in the "Welcome to Williston" sign in North Dakota. Geology and white heteropatriarchy are an ongoing matrix of settlement, mobilizing "a deep unconscious saturation and naturalization of White family authority as state authority, wherein 'characteristics of the family are projected onto

FIGURE 3.4 Dinosaur heteropatriarchy at the American Museum of Natural History, New York. Photo by author, 2019.

the social environment' in such a way as to allow for 'no disproportion between the life of the [White] family and the life of the [state]'" (Wilderson 2020, 159). Or, to extrapolate, the geologic life of the white family is also the death curse of forms of life in the category of inhuman nature.

The Liminal and Lithic

> Was she travelling through the tunnel or digging it? Each time she brought her arms down on the lever, she drove a pickaxe into the rock, swung a sledge onto a railroad spike. She never got Royal to tell her about the men and women who made the underground railroad. The ones who excavated a million tons of rock and dirt, toiled in the belly of the earth for the deliverance of slaves like her. . . . Who are you after you finish something this magnificent—in constructing it you have also journeyed through it, to the other side. . . . The up-top world must be so ordinary compared to the miracle beneath.—COLSON WHITEHEAD, *The Underground Railroad*

In the claustrophobic spaces of stratigraphic inhumanism and the lithic spaces of the underground there is a doubling of subterranean aesthetics, which might be understood as a voluminous and fugitive rift in surface dynamics. Improvising geologic spatial structures (undergrounds, caves, hiding spaces, mines) has historically been practiced as a form of insurgent geophysics that repositions the rift zone as central (rather than as marginal) to *de*sediment the power of the plateau. The deep charge of the underground does not just reposition the rift in relation to the colonial surface but forces the disruption of those plateaued formations and destabilizes their languages of enclosure and stability. Identifying spaces of spatial and aesthetic brokerage in epistemic grammars of geology in this way is a route into dismantling the white supremacy of matter as it operates as both a spatial and an aesthetic medium. The "miracle beneath," as Colson Whitehead (2016, 303–4) names the Underground Railroad, is a passage that throws off the calculative geo-logics of the surface and refuses the forms of valuation designated by the inhuman propertied relation of slavery. It is a bet made with all that is not loved (Hartman 2019), and against all that would deny this possibility. The underground is a rift, moving through the subjugating epistemes of the plateau-rift and its social and spatial sedimentations in the languages of strata, stratification, and formation that hold it in place. Theorizing from the rift zone is precisely about unmooring the conceptual strongholds of the "view from the plateau" and the confident colonial gaze that seizes a geographic imagination of place as resource and field of conquest. It is a site of knowledge production made without guarantees or the rewards of social forms and psychic life that rely on "fixing" a narrative of geologic lives in a redemptive or restoratory mode. Retooling the geophysics of sense can redress the inhumane geo-logic ledger and its modes of capture and resistances.[7] The *rift as a spatial-aesthetic proposition* is isomorphic in ungrounding across epistemic-ontological orders, whereby the material and ontological liminality fuse, defuse, and refuse. As parallel geophysics of sense, the lithic space of the underground and the aesthetic space of liminality can be thought together as materially manifesting spatial productions of power, as aesthetics is a route to new (and old) geographies of apprehension.

Colonialism secured an affectual architecture, a structure of affect that organized and grammared a relation to the world, to others, to earths. Its imagined geographies ordered people, nonhumans, and space, while it romanced the landscape as empty and in various forms of ruination awaiting the muscular white geographies of empire. As geology is a crucial space in which power is realized through visual methods, aesthetics is a path that can restructure that affect into different imaginaries and architectures of located being (geographically and theoretically), because it can restructure the senses (of the possible). Aesthetics is also how I work: the leap of images and poetics structures my thinking and this book. Glissant's affirmation that "poetics are the non-normative lesson" (in Diawara 2010) has guided me. I think aesthetics/poetics is also a site of repair to the normative lesion and its violent geographies, a space for building different imaginaries and material relations out of geotrauma.

Underground aesthetics can be understood to act like a subduction zone that draws in and under the conceits of geology that propagate dominant lived realities or normativities, to suggest other ways to understand earth relations, and go down to the river. As a scopic mode, disruption of the *stratigraphic impulse* of geology and its organization of relation, temporality, and ground is a way to probe the stretch of inhuman matter-languages and their fracture points (which can condition the possibility of survival). The more the plateau is lifted, the deeper the subduction is. The plateau and its stratigraphic reason position the underground, but the underground is its own space, not a dimensionless geography. It is a racialized volume. The ungrounding of the plateau can be understood through five geologic modalities:

1. Underground as materiality: of concrete, clay, steel, anvil, iron ingot, aluminum, stone, geologic arrangements, geochemical attractors, and the elemental "building blocks" of future environments.
2. Underground as substratum to the surface; in a racialized debt relation to the accumulation of value on the plateau.[8]
3. Underground as psychic space: configuring attachments to imaginaries, desires, and affects and arenas of racialized harm and geographies of dispossession.[9]
4. Underground as luminous force in the geography of resistance and refusal: as displacement, slump, tunnel, undercut, destabilization, pocket, site of epistemic innovation, and place of knowledge-making (epistemologies) about how to traverse the strata.
5. Underground as lithic occupation of the liminal: that which supports surface structures and is obscured below.

Race is a form of policing geosocial strata, as plateau-rift, in the organization of the idea of stratified difference. The rift is where the underground surfaces—*forty thousand years is a long long time, forty thousand years still on my mind*, per the Aboriginal mural at Redfern Station, Sydney, Australia.[10] Differentiated claims to time and duration in collaboration with the earth introduce a disarrangement sense of linear progressive narratives into understandings of space, raising questions of who and what endures of the subterranean. The redirection of sense is a methodological approach to archives and images that understands that the archival impulse is always in dialogue with its own excess, or fever (in Derrida's terms). Inspired by Wynter's (1996) critique of "we-the-underdeveloped" as an epistemic structural position, the underground (as temporally articulated through jazz for Wynter and poetically posed by Moten, for example) can be understood as simultaneously a material position in time, a perspectivism, a scopic regime, and a script.[11] If the layer cake of geology presents as the desire for a totality of time, locating the fragments of other geologic realities (alongside the work of Black and Indigenous scholars) might name that time and its eventfulness differently. Liminality names an isomorphic relation to duration; in geologic terms this liminality can be found in the forms of lithic occupations in relation to the epistemic and actual plateau (and in relation to its forceful restoration and possible surpassing).

To undermine is to under-mine the strata.

Geologic Life aligns itself with projects made outside the existing political geology, which are restricted in their purview to what is in excess of existing "restricted" (vis-à-vis Georges Bataille) racial capitalist material economies.[12] Nor do I practice in the mode of an "illusionary unraced white world" (Morrison 2019, 143). Instructed by Black feminist projects of futurity and work on the racialization of environments, politics of affects that are placed "outside" the political economy (either through the cut of categorization or through a nonrecognition in the social order) also exist in antagonistic relation to the *extraction principle* and can disrupt it as an affective sense machine by refusing a model of sense that it offers as a normative or a commonsensical version of matter relations, relations that matter. Which is to say, extraction is at work across geosocial orders far from the physical mine.

The articulation of shifts in geologic sense-acts through geologic grammars is made in friendship with Brand's poetic practice in which she "tried to produce a grammar in which Black existence might be the thought and not the unthought" (2017, 59) and Natalie Diaz's understanding that "Everything is iron oxide or red this morning, / here in Sedona. The rocks, my love's mouth / . . . Maybe this living is a balance of drunkenness / off nitrogen and the unbearable / atmosphere of memory" (2020, 73–74). This weighing of rocks and grammar is what Elizabeth Povinelli calls carnality: "the socially built space between flesh and environment" (2006, 7). Encountering the geoforces of the earth does not have to be a limit event, with its masculinized personification of mastery, coming down from the plateau to peer into the abyss of the racialized poor (see Singh 2018). Resisting the urge to covet the plateau from the rift is an onto-epistemological, as well as material, refusal of the rewards of hierarchies because an escape from the material conditions of fungibility that positions rift dwellers as absorbent to the shocks of the earth is ultimately the imaginative projection of the "promise" of the plateau.[13] The narrativization of geology in the plateau in spaces of power and cultural influence (such as boardrooms, museums, art galleries, stock exchanges) has profound material ramifications for the racialized poor. The rift is often not visible from the surface of the plateau, or it exists in shadow, or its subjects are made indivisible from the geography (seeing people as proxy indicators for environmental degradation).

Countergeographic imaginations have been at work from the get-go, dismantling the plateau, not trying to scale up its heights, and be part of the included. Instead of the lone white surveyor and his vista over the future and its "development," the characters of N. K. Jemisin's racialized and enslaved Orogenes from her *Broken Earth* trilogy, who can manipulate geothermal and kinetic energy and shift strata to address seismic events, serve as more exacting critical figures for this zone. Orogenes have the ability to sense matter—"sessing" fault lines, with the gift to break them apart or subduct them under other plates. Yet such is their extraordinary gift for *sessing* the earth that they are violently subjugated through fear and need by those in power. Their seismically tuned geologic lives are, however, made the basis for their enslavement and internment in cruel residential schools and apparatuses of confinement to buffer shocks and modulate the earth's faults. The broken earth is a seismically rocked planet that has experienced a rifting so severe—the Season—that it will destabilize the climate for thousands of years. In the negotiation between an imagined raw geopower and racial power, Jemisin's novel provides a guide for how to understand the manipulation of psychic states of the racialized for the attainment of geophysical states (as geomorphic labor and subjection). Jemisin's conclusions are unequivocal: now everyone must live in the rift because they could not share. The rift is a place from which to fracture the surface and its forms of subjugation. The interruptive phrasing of subterranean space names a refusal to adhere to the foregrounding of utilization of the earth for its surface needs. *Water*

Is Life #NoDAPL. And, perhaps it is also to suggest an affiliation or mode of filiation with the underground itself that cements these forms of communication as the mobile fragility of shifting states.

Insurgent Aesthetics, Underground Spaces

In her essay "Poetry Is Not a Luxury," Audre Lorde sets out the demand and necessity of poetry in learning how to live otherwise. Lorde sees poetry as an aesthetic sense (far removed from the intellectual tradition of aesthetics) and as illumination, a "quality of light by which we scrutinize our lives" (2007, 36).[14] As new senses become felt and known as thought-ideas and language (theory and grammar), Lorde argues, they become the spawning ground for radical ideas, "safe-houses for that difference so necessary to change and the conceptualization of any meaningful action" (37). Lorde's conceptualization of poetic acts is like the mythic underground railroad during slavery, the network of subterranean refuge and movement against the violence of the surface, that which carves out spaces of generosity in the earth which can be found but not exposed to the political economy in the "up-top world." As an underground aesthetic that parallels the railroad, "safe-houses for that difference so necessary for change" relay toward freedom, laying affectual architectures toward future change: "Poetry is not only dream and vision; it is the skeleton architecture of our lives. It lays the foundations for a future change, a bridge across our fears of what has never been before" (38). Lorde, speaking as a Black lesbian woman, associates this underground of deep power with a Black feminist aesthetic: "The woman's place of power within each of us is neither white nor surface; it is dark, it is ancient, and it is deep" (37). As both sensibility and erotics, subterranean aesthetics were made as a resource for Black feminist survival, as sense-acts of liberatory being. This is what I understand a Black feminist ethic to be, a counterforce directed against the weight of forces that structurally position Black women at the apex of stratal pressure (alongside the imperative to refuse that dialectical resolution).

I understand aesthetics not in its traditional Kantian guise as judgment but as a sense that draws in another dimensionality, something detected, felt, and known, before it presents itself as such (rather than as a mode of adjudication on experience or in the telos of transmutation). As Wynter suggests in her analysis of the "senses as theoreticians," aesthetic sensibility is theory-making and affectual force. In her attention to music and rhythm, she argues for an affectual architecture: "For soul, like the poetry and music of which it is the well-spring, is strictly structured and thinks itself through the sense: rhythm as the aesthetic/ethic principle of the gestalt" (n.d., 245; underlining in original). In the vein of Clyde Woods's blues epistemologies, J. T. Roane makes the connection between environmental justice and theorizing in the blues: "The Blues contain under-examined intellectual resources in relation to the matters of environmental destruction and environmental justice. Blues description and analysis, remixed and remade in the changing environmental and temporal contexts shaping Black thought, was a primary retainer and progenitor of critical and sometimes heterodox information about Black life in its imbrication with environmental vulnerability and possibility" (2020).

Reading Black epistemologies as environmental practice ascribes a theoretical function to sense that forces the aesthetic and earth into conversation. The political force of aesthetics, to give inhabitation to a radical futurity before the seeming possibility of its actuality, is to give shape to the semblance of a possible reality or to intuit a liberatory possibility in often impossible circumstances. Thus, aesthetics can make a place from that which is currently displaced, capturing a different time and space in that move, to make available the feel of another earth, another time, another relation. This is the aesthetics of sense configured as the theory-act of difference, sense as directionality into unknowing, liberated from within the tight, confined, or "cramped" spaces (as Berlant [2016] names them in the affectual politics of neoliberalism), made through intimate reconfigurations of anteriority rather than flight to the mythic outside or outer space.[15]

Underground aesthetics are an interruption in the flow, a swerve in the comfort reflexes that structure the humanities, disrupting the relays of affect and building in that rift another grammar of sense. Even if this sense is simply the sense of the disruption of those affectual architectures rather than any proscriptive map: "All Texaco gathered around the body and the stone" (Chamoiseau 1992, 22); "principle of the bones, mineral and alive, opaque yet organizing" (22). The skeleton bones of the enslaved (who have claimed another geography through marronage) haunt the writing in Patrick Chamoiseau's novel *Slave Old Man* (2018), where words gather abandoned histories, as the abandoned bones make their home on a rock, collecting pasts, indifferent to how they are named. Affectual architectures of rocks and grammars foreground the durational to provide a different analytic of sense. Bone is approximately 70 percent inorganic: environmental conditions enact a demineralization-remineralization dynamic on it. Mineralization is a life-long process. Pathological processes of mineral deposition result in changes in the body, such as calcification of blood vessels. Geologies of time and sense establish a thought of the world. *Loam. Onyx. Metallurgy. Igneous desire. Meteorite showers. Obsidian arrowheads. Sky metal.* Philosophical thought and poetics substantiate, in a material and experiential sense, the relay of a relation rather than describing its production; it makes, and asserts through feeling, and its accumulative acts or resonance. These relays between geologic architectures of inhabitation, and their affects, perform Relation: a process of writing, practicing theory, of creative repetition, through links, relays, returns, and retellings. Relation for Glissant is not posed between *what* and *who* (as in relational geographies) but is like an intransitive verb that "informs not simply what is relayed but also the relative and the related" (1989, 27). That is, Relation is a way of building, and changing the sensibility of thought. The depths as shapeshifters. Sense as orientation. *Senses as theoreticians.* Unmooring geo-logics of matter and their temporal codes. Unbordering colonial forms. A poetics launched against carceral states. Rift theory is imagined as a theorizing on the move, fugitive and alert, as a muscle that grows around the future shocks of the violence that is carried by the orphans of geography.

Rifted Aesthetics

The desire at play in the rift zone is the opposite of a redemptive approach to the rent. Opposite to covering over the violence and building again, presumably better than before (as with versions of the "good" Anthropocene). Rather, the desire in the rift is to read geopowers against the flow of their inscription, to pull the rift farther apart, and to dwell in that location, so that such a possibility of reconstitution of its states of initiation are not possible, *at all.* The crucial aim, as suggested by Povinelli, is to "remind us that dwelling demands, or is at its heart, a political purpose, and this purpose is to interrupt a given formation of power rather than either report and adjudicate that formation or report and extract an affective charge of hope from it. The rhetorical force is aimed not at feeling for but at affecting with—of staying with the errant rather than trying to quickly press it into a form of resistance, of hope, of an alternative social world" (2017b, 133). "Staying with the errant" is to stay in the rift with forms of nonequivalence in modes of valuation that counter historical deployments of the inhuman as subjective and geologic category—that is, to speak differently and give another sense of the earth. In Povinelli's work, this is to "listen to rocks" and treat them as kin (1995, 2016). Refusing to abandon the errant to normative structures (like whiteness, the patriarchal family and state, and their juridical and affective modes, etc.) is a demand that something else must be built as sensibility, and another kind of world-building enacted that is interested in the beauty of duration-in-a-body beyond its current extractive and catastrophic terms, and toward an earth that leans beyond the affirmation of the bounded self into the nonhuman and inhuman. This is the insurgent movement of underground aesthetics (as a forceful friction), which finds a critical mark in fault lines and seeks to lever them open. The battle continues in maintaining the fault as a critical mark in the surface atmospheres of incorporation of *every resistant thing* into surface orders (see the schematic of the analytic of geologic life in the introduction) through inclusion within the very languages it challenges. Another way to say this would be that to survive as a legible form of geologic life in the current language of extraction, every resistant thing has to be insensible, at least in part, so that it is not immediately folded into the dominant social, economic, and epistemic orders.

Underground aesthetics depends on an illegibility within the very codes it seeks to dismantle. Refusal, then, is a strategy of axiomatic difference within the senses, a temporal code shifting to induce an altered epistemic register, to produce an altered geophysical reality, a counteraesthetic that cannot yet be read as anything settled. A sense that rifts carries its movement. It is a countercolonizing force involved in the politics of unsettlement. As Campt argues, "*Practicing* refusal highlights the tense relations between acts of flight and escape, creative *practices of refusal*—nimble and strategic practices that undermine the categories of the dominant" (2017, 32). This tense of refusal must necessarily present in a different form than those recognized in the social field, or in the registers of the political aesthetics of the state. This is why any analysis of Rancière's (2004) "distribution of the sensible" is misplaced (or, rather, available only to the Enlightenment subject). Refusal in geologic lives, then, may be a block or a stutter in the languages of power and grammars of matter, rather than a rebellion against the state. Like the rock thrown at an employee of the urban service at the beginning of Chamoiseau's (1992) novel *Texaco*, the rock launches the force of the resistance of the shantytown into the narrative: "all Texaco gathered around the body and the stone" (22).

Aesthetic acts can transform the lexicon by signaling a new imaginary of relation even as such an act is refused in its grammars and modes of citation.[16] An example of such an undergrounding geology of race is being made by the Equal Justice Initiative's Legacy Museum and National Memorial for Peace and Justice in Montgomery, Alabama; three-hundred-odd jars of earth taken from lynching sites record a geology of violence. In the jars, some with names, others with just locations, an aesthetic of the underground petitions on the familiarity of geologic cores, to name a bloody relation to the geologic acts of colonial prospecting and the pursuit of property over personhood. And they petition on all the other earths of experience and oppression that have become sedimented into the ground in the earth archives under erasure from the surface. In this mimetic sense they perform a corpsing effect on white geology. The earth jars suggests a way to hold onto those who did not survive their entanglement in the nomenclature of geology's grammars. What is most disturbing about these jars in the first instance is precisely their colonial aesthetics and the assimilatory visual culture of the surveyor. Yet the genre of the geologic core doubles back on itself; aesthetics brings about a language and orientation of refusal, a refusal that creates a place for being resistant to the affectual lures of what resistance and the counterpolitical are meant to look like within the political economy. The jars are the work of caring for and assembling the *not-fossil*.

The promise of an aesthetics of the underground is not one of restoration but of redress in the present tense. Resistance might be handfuls of earth. An unearthing. Or the forensic dust collected by the women of "the disappeared" in the Atacama Desert. Unburying a denied reality. "I wish the telescopes did not just look into the sky but could also see through the earth, so that we could find them," says a woman, digging in the desert for those disappeared by Augusto Pinochet, in Patricio Guzmán's film *Nostalgia for the Light* (2010). Aesthetics can be a tactic of opening up the terrain of possibility even as it strains under the weight of the expectations of its nonarrival in the political sphere.[17] In a nonextractive mode, aesthetics does not offer resolution, maybe something more akin to disarray, disassembly, the affective qualities of exhaustion, love nuggets of the possible, another path for desire. "Light horizons her hip—springs an ocelot / cut of chalcedony and magnetite. / Hip, limestone and cliffed" (Diaz 2020, 21).

Political Aesthetics

Political aesthetics are the organizing structures and configurations of affects that have formational and infrastructural power.[18] In other words, they function in relation to one another in a manner that organizes affects in a coherent structure (i.e., as a gathering and an orientation of sense, their modes in relation to one another in such a way that their relays reinforce and confirm, rather than break apart, this coherence). As Moten reminds us regarding the force of the affect, "When you talk about the aesthetic you've got to talk about it in its interinanimative autonomy vis-à-vis the political" (2017, 26). An insurgent geology, or rift, breaks apart this coherence, activates a durational sense of brokenness, noncoherence, and moves toward a different tense of the earth, intuiting spaces that do not yet exist. An insurgent geologic sense follows these fissures, along the seam, not to extract but to work the seam into other affiliations and object lessons. This is my method of reading colonial archives. As Brand observes, there is a need to be "constantly vigilant to the ways in which colonialism is an active organizing system" (2018, n.p.). A counteraesthetic sets up injunctures, roadblocks to relays, and resists against one-way extractions. Such injunctions decouple the "natural" order from the social order in a subjugating relation—that which makes the normative—to trouble the project of how concepts become materialized and matter. Caribbean and African political humanist projects made in the wake of independence understood this dynamic as they sought to redefine the breadth and breath of the Human, rather than rejecting the category that was built on their exclusion. The taciturn acknowledgment was that they were already included in its formation, albeit as substratum, so reworking and displacing the foundation's logics were a priority in destabilization of the manifestation of the humanist form. That is a different analytic than the expansion of terms of inclusion. It occupies a different counteraesthetic mode.

For Maurice Blanchot (2014), writing in the context of fascist projects of totality and into the disaster suggests that aesthetic fragments constitute the very possibility *of a world* not imagined in any relation to a reconstructed whole (or globality), but as fragmentary gestures that write precisely of their own interruption, of fragmentation itself, in rapport with the unknown and its often-abysmal losses (Yusoff 2018b, 269–70).[19] The fragment offers us a glimpse toward another earth; it opens out and *un*works the enclosure of the colonial earth, to inscribe its difference on time and on us, as a *demand*. It is the fragment thrown up from the rift that puts itself into question, and thus demands a certain requirement to look toward a radical futurity (to differ from ourselves because of this demand that is not autopoetic). That which is instigated through this question of difference is already in the process of becoming something else (the fragmentary *un*working challenges the future we thought we were living; this is its political charge).

Aesthetics can bring that "slipping away" of the underground to light, to question what might be at stake in this exposure: so, these traces become citational, and new futures and pasts are constituted through the memory of the earth in these *not-fossil* traces. Understanding the fragment as fragmentary, flung out from the rift, is a different kind of worlding. If we read this worlding alongside the bland globalism that Anthropocenic humanism offers as undifferentiated strata or "universal" bad residue smeared across the planet, the ciphers of geology can be read more precisely against their teleological impetus and redemptive materialisms. Fragments unearth questions of geologic materials and subjective relations not just in how geology allows certain ways of life, modes of communication, forms of sociality and solidarity, and political subjectivities (and their exclusions). This affinity toward certain incorporations of geologic matter is, in and of itself, a fossilizing/or Anthropocene-making process. That is to say that imaginaries participate in the political present and the current apocalypses, which is an ongoing project of sense and its interruption since 1492. Aesthetics is a change in the possibility of understanding and feeling differently. It not only changes the register (epistemic innovation) but also can be a way to decolonize the telos (by creating other directionalities, not destinations). From the plateau the rift may not be visible, but, as Morrison reminds us, "invisible things are not necessarily 'not-there'" (2019, 176). They may be just below the surface, going underground.

II Geologic Histories and Theories

4 "Fathering" Geology

Geology Comes to America!

"Curiously enough, science entered America led by geology," proclaimed the French geologist Jules Marcou (1895, 281), who was appointed by the Jardin des Plantes in Paris to survey the geology of North America and "English possessions" in 1847.[1] Subsequently, Marcou was employed by the US government to conduct the Pacific Railroad Survey. He became the first geologist to cross the United States, making surveys from the Mississippi River to the Pacific Ocean (published in 1853 as *Geological Map of the United States, and the British Provinces of North America*). While there were other scientific practices operating in isolation, Marcou suggests that they were "without the support of the people. Public opinion did not encourage them. This was not the case with geology" (281–82). Emerging from a confluence of European and American practices, geology became well established in the "New World" as a foundational modality of settler materialism, which led to the development of specific genres of colonial materiality. As geography was the primary science of imperialism, geology became the political science of empire and its settler claims. As Marcou suggested, geology was *the* science of America: "People in general, and agriculturists in particular, showed an eager desire to know the resources of the soils, the rocks, and the mines. Geologic surveys were started at the expense of the State in North Carolina, Virginia, Maryland, Pennsylvania, New Jersey, the New England States, New York, and Ohio. A desire to agree on points of classification and to know one another brought together the state geologists, who founded in

1840 the 'Association of American Geologists,' the first national scientific organization" (281–82).

White geology provided the material praxis, and it furnished the geomythos of the populist science of white supremacy through geologists' theories on race. Paleontology mapped theories of racial hierarchies and difference, while geologic practices materially organized the economic units of extraction, mines, soil, plantations, and alluvial and fluvial processes as racialized environments. Geology made white grounds and became a symbolic medium for their quotidian and monumental expression (Tunkasila Sakpe Paha, or Six Grandfathers Mountain, renamed Mount Rushmore National Memorial, exemplifies an exaggerated form of geologic acts as political state formation).[2] As white geology made the nation-state of America materially manifest as landscape, resource, and race, geo-logics conditioned the nation's normative social forms and modes of settler geopower. As America grew its own geologic empire externally, these national modes of settler colonial materialism paradigmatically transformed extraterritorial extractive worlds across the globe (and now into space). Geologic methods and agents organized the surface, subsurface, and undergrounds and governed the relations between them, while undergrounds drove surface politics and contestations (and continue in competing Indigenous and national imaginations around subsurface claims).

Unearthing race in the history of geology is a challenge to the pathology of dissimulating the sedimented processes of valuation that concretized hierarchized racial spatial forms and established a normative materialism and its social order. Racialized geology as natural resources development established the grounds of other forms of segregation and subjugation and installed the narrative that broken earth was a prerequisite of creation. This chapter demonstrates how paleontology made the concrete and interpretive ground for the anatomical and aesthetic collaboration of what we have come to know as racial formations—geosocial collectives defined by racial difference. The ugly details of geology's "birth," often footnoted as a sign of the (imperial) times, are an occasion for desedimenting the disastrous entanglement of race and geology (so that we might come to know its normative structures and extractive presents). For example, as the only human subject represented in the primary archive of natural history, a Black woman (named elsewhere as Saartjie Baartman) interrupts, inter alia, among the pages of living and extinct mammals (Cuvier 1827). Her only-ever provisional identity is footnoted by the leading historian of geology, Martin Rudwick, as a now minor embarrassment in an otherwise illustrious career of one of the "founding fathers" of geology.[3] Rudwick defends this position, citing the availability of "specimens" as mere historical context that in "no way tainted [these bones] with imperialism" (2008, 554). Given the detailed account of every other facet of Cuvier's bibliography and geohistory, the extreme reluctance to mention Cuvier's work on race and trade in the body parts suggests that bones do indeed retain the subterranean marks of their geologic passage. Despite Rudwick's arguments about the importance of the "liminal" in geologic theoretical propositions, he is unable to admit it as it pertains to dehumanizing geo-logics at the intersections of race, gender, and sexuality. Specimens, not persons. No tainted bones here, move along, see the earth transformed by great men. Later in the chapter, I focus on Baartman (1790–1815) within the context of geology because of what first seems like the dissonance of her presence in the geologic archive. Her presence at the geologic table places Black women's sexuality firmly as a force within the lithic field of vision. Deviant sexuality and racial segregation are borne within and through historical modes of geologic objectification and the temporal concepts of deep time in ways that draw together wider productions of meaning around race, gender, and geology and are riven by the political geology of Life (and the role of violence against Black women as a mode of stabilization in that form of life).

While footnoted, the racial analytics of Baartman's presence in the geologic archive relay a set of grammars that expose the structures and conditions under which these questions of Life and racial difference emerge (and reemerge). She disrupts the ontology of geologic life that she is supposed to represent, as a restless nomenclature in representability that scratches at the binds of specimenhood because of what she clearly is *not* in the taxa of inhuman objects. Attentive to the marks that disrupt the smooth accomplishment of geopower, Baartman's figuration of racial difference grates against the supposed impartiality of deep time (in its past formations and present resurgence as eugenics). Defootnoting the geologies of race speaks to an active inheritance operative in the co-emergence of theories of the earth and race that condition the gravity of everyday material encounters and the ongoing application of stratal thinking to racial hierarchies.

This chapter centers around two geologies of race in the formation of geologic thought: (1) Georges Cuvier's (1769–1832) identification of Sarah Baartman as a stratal figure in the paleontology of comparative anatomy and (2) Louis Agassiz's (1807–73) racial geographies of human history and his broader political geologies and visual typologies of racialized difference. As leading geologists, Cuvier and Agassiz structure a geography between imperial centers of science in Europe and the establishment of geology as a populist science in the United States. I engage their writings extensively to demonstrate the intertexuality of race and geology and question the confinement of their legacy of race work outside understandings of natural science and inhuman histories. The chapter ends with the "inhuman dicta" of the juridical codification of geo-logics in the *Dred Scott v. Sandford* case in the Supreme Court (1857), and W. E. B. Du Bois's countercartography in the visual geographies of race made in response to Josiah Nott and George Gliddon's *Types of Mankind* (1854), which included Louis Agassiz's contributions. Underpinning this discussion is the historical context of the emergence of geology within imperial and colonial practices to both situate geology and name the questions that arise around its key material and philosophical practices—namely, stratigraphy, genealogy, and the classificatory logics of the inhuman (to discuss how temporal spacing is used to solidify political markers of racial difference).

I argue that paleontology produced the grounds of anthropogenesis—a new origin for Man-*the*-Human and scene for Life's geophysical origination (Yusoff 2016)—as it produced an ontics for the inhuman. Inhuman ontics produced an *elemental genesis* of geologic life rendered through a proximity to inhuman forms and material submergence in the mines, which in turn generated a countergravity of underground aesthetics. As Agassiz produced and represented types and cartographies of racial difference (of Indigenous subjects in Brazil and then the first photographs of enslaved subjects in America) to enable the building of a political and industrial geology of race, W. E. B. Du Bois was reading the color line of categories of difference in ways that visualized the spatial effects of geologies of race. The active conflation between forms of earth genesis and onto-/socio-genesis by geologists in this period defined material and cultural relations and created temporal splices between inhuman processes and the explanation of stable social positions and identities. Du Bois's album title *Types of Mankind* has its mimesis as a "counterarchive" in the various race books of craniologists, ethnologists, and paleontologists, most notably Nott and Gliddon's hugely popular *Types of Mankind* (1854), which included Agassiz's race maps. Britt Rusert shows how Black intellectuals in the antebellum North used ethnology in the 1830s and 1840s to reimagine what she calls "speculative kinship" (2017, 67), which sought to reconstitute genealogies broken by the institutions of slavery and repair the continuance of forms of belonging. She argues, "Given the centrality of scientific images that separated black people from humanity and from the social itself, black ethnologies were especially interested in using the visual—in both text and in image—to reconstruct the forms of relation denied and destroyed by polygenesis" (Rusert 2017, 66; also see S. M. Smith 2004, 2). I conclude with a discussion of the problem with kinship and its emplacement within geologic genealogy and move to underground aesthetics to diagnose a subterranean analytic of these subjective relations to disrupt this racialization of geologic life.

I focus on historic geologies of race to examine how a story of the rocks becomes a means to naturalize and produce racial difference through moments of representation in the making of epistemic orders of natural history (and how that ordering, in turn, collaborates with the racialization of spatial forms). The narratives of geologists took on political-ontological roles in the politics of white supremacy in the settlement of America, as the idea of race war was naturalized within a paleontological frame. Coupled with the substantiation of the material transformation of the landscape and forms of geologic life (life sustained through specific material economies), geology functioned as a language of consolidation for (settler) colonial Life, as it provided the conceptual and material field for the emergence and consolidation of the white subject. The consequences of this period of seventeenth-through nineteenth-century geology rendered a historicity that established a privileged genealogical ordering of the Human in social and geologic formations and secreted race (and its interlocutor, species) as a inhuman relation and geography.

I focus on the problem with genealogy, understood both as an analysis of kinship and descent and as a mode of originating the material accumulation of the present through an attachment to lineage and its established hierarchies, whereby genealogy establishes whiteness as the accomplishment of raciality.[4] The power of the geologic "father" is consecrated by the genealogical plant in time rather than in space. The planting of race in deep time was central to narratives of white settler nations to reoriginate their "birth" in stolen Indigenous land.[5] Furthermore, this geologic arrangement of Life continues to secure a privileged—and white supremacist—politics of Life through material accumulation based on social deprivation of Indigenous, Black, and Brown peoples. Geographic accomplishment (imperialism) therefore became rendered as geologic naturalization, a form of doing (colonial) Life's work in the progress narrative of the planet (achieved through the medium of geology).

Several questions guide this chapter: What are the conditions under which racial knowledge is made and mobilized through geology and concepts of geologic life? How did racialized persons become classified as bodies and types to be consumed alongside deep time geologic materials (such as rocks and formations), and thus how did racialized subjects become integral to understanding inhuman materiality and the production of time, and specifically colonial time, via the rocks? What relays in the grammar of geology—geo-logics—make these traversals across deep time and racial presents possible? And, in terms of disciplinary divisions, why does analysis of these inhuman(e) racial rifts show up only as a question of identity in the disciplines of Black, gender, and sexuality studies and not as a question of materiality in geology? Rather than trying to sidestep the racial origins of geology and how race becomes interned in geologic differentiation, I follow Denise Ferreira da Silva's suggestion that "the cure for the ethico-political effects of racial difference is not less, it is more, more engagement with, more analysis of, more critique of the context of emergence and conditions of production of the tools that set up and sustain this rather efficient biopolitical arsenal" (2011, 146). In this instance, it is the geophysical arsenal that is mobilized to produce political effects in the context of raced relations. It is a story of the legacy of racial difference, as geologic praxis and geophilosophy.

Historical Geology

A shift occurred in the late 1700s and early 1800s from geology as a practice of determining localized formations of strata (for the purposes of extraction) to *geology as a temporal practice* of determining the status of beings-in-time.[6] This shift was marked by the naming of paleontology as a key branch of the discipline, understood as the study of ancient "beings" through the bodies of organisms: fossil life in relation to contemporary anatomical life. It was a story of the earth told through a corporeal lens, or geologic life. In the disciplinary arrangements of geology, paleontology sat side by side with stratigraphy (ordering of the materiality of time through sequential strata) and comparative anatomy (ordering of bodies of species through their differences, and thus in the sphere of human species their racialization). Historian of geology Rudwick comments that

> this new stratigraphy remained on the same *structural* level as the geognosy from which it had developed: fossils were used primarily to enrich geognosy by clarifying the sequence of three-dimensional rock masses or formations, and their correlations from one region to another. Of course, geologists were well aware that this spatial sequence of rocks also represented a sequence of events in time; but the temporal dimension usually remained in the background. . . . Only a few geologists were beginning to

> enrich their stratigraphy still further, or in a double sense, by using fossils to give their work a primarily *geohistorical focus*. Rather than being treated just as objects useful to recognizing specific formations, fossils could then be imagined in the mind's eye as fragmentary traces of organisms that had simply been alive at specific periods of geohistory. (2008, 47)[7]

This "double sense" of fossils as geohistorical rather than merely locational—fossils as devices in time rather than just in space—shifts geology from a discipline that is concerned with localized matter (in an extractive and comparative mode) to a theoretical discipline of earth geophilosophy (the idea of geologic life in an ontological mode). While the study of fossils was about subdividing strata to use them as a locative medium for valuable earth minerals and fossil fuels that could be correlated from place to place, it also began to offer a speculative armature for imaginative geographies of temporality. Debates emerged about the difference of life-forms through time and space, as well as about their occasion in different mineralogical matrixes and geoforces, and on questions of causality and meaning. Geologic study literally produced an earth as a global temporal object while imperialism was making the spatial continuum of Global-World-Space. As the colonial and imperial context of this knowledge production produced a global spatiality, geology became a means both to establish a synchronic relation between material practices and to justify their ideological formations.[8]

Paleontology, or the Origination of White Geology

As discussed in part I, after the Lisbon earthquake, geology also became a secular story of the planet, which provided a seismic spatial and temporal "fix" to the flux of the cosmos. In the context of the explosion of "alien" difference that imperialism unleashed, the curtailing of the cosmos secured an important potential opening in the global accession of space and worked against the liminal affects of the deprioritization of the human in deep time. The epistemic frame of containment of the cosmos had been provided by a causal understanding of the geophysics of the planet with a purposeful theory of Life (and its divisions). As the practice of stratigraphy developed an agreed upon set of understandings about strata as a measure of time and occurrence, its formation "enabled earlier speculations about the broader features of the history of life to be set on much firmer foundations" (Rudwick 2008, 48). In short, fossils provided the empirical ground for a materialist secular philosophy of Life, as strata confirmed the successive placing of all life. It was during the long nineteenth century that these (over)psychic confidences in the geologic sciences came to the fore as a brace against a dynamic social and geophysical planet. My approach to these archives of geology refuses the preferred schism by historians of science between conditions and context of knowledge production and the categorization of inhuman materials, to interrogate how geology is made in the subjective and symbolic registers of these indices. Rudwick argues that the discovery of many fossils of reptiles and other sea creatures prompted questions about the relation between species and helped furnish an accumulative narrative of purposeful development of life: "Since reptiles were, of course, important components of *living* faunas, alongside mammals, this suggested that vertebrate life had become progressively more diverse in the course of geohistory, by the successive addition of animals with arguably 'higher' kinds of organization. It implies an overall directionality, or even progress in the history of quadrupeds, which might also apply to the history of life as a whole" (2008, 49).

Whether or not such discoveries *actually* indicated directionality, or whether this desire for historical progress was more tied up in the historicity of European colonialism, is a question that exposes the traffic between paleontology and the contemporaneous political present of colonialism and *its* desire to establish itself as the highest order of species-life.[9] Charles Darwin, for example, while maintaining a commitment to "no *necessary* tendency to progression," also naturalizes the extinguishment of "lesser races" by Europeans (justifying colonialism through species reasoning).[10] The attendant discussions around race were never too far away from these pronouncements on species-life; there was significant slippage between species talk and race work, thus binding the naturalization of racism into the geologic sciences and its inquiries. Internal to the discussion of progeny and the question of the purity of genealogical account (which was also a question of accumulation) was the role of gender and sexuality in maintaining and extending Man. Fathering geology was about securing the heteronormative gender binary to maintain patriarchy. Race was both the challenge of difference to Man and his supremacy and the axis of its consolidation.

As paleontology became an account of species in time (or Life's progression), it was a narrative of advancement that established its argumentation by way of race rather than fossils per se. This narrative is embedded in paleontological thought even today. Racialized subjects were and are objectified as "living fossils," which preserves the operability of racial difference and its hierarchies. Racialized subjects perform as a "type" of and in a Fanonian "zone of nonbeing" that made and make them accessible to scientific scrutiny without consent. At the point that paleontology became about beings-in-time rather than rocks in formation, an onto-epistemological juncture elevated fossils to descriptive objects in the geophilosophy of Life. Henceforth, fossils were not just locators of natural resources but empirical knots in the fabric of the politics of Life itself, and the hierarchical designations of racialized humans there within. The "Father of Paleontology," Cuvier, illustrates this point; the designated patriarch of the theory of the history of life on earth seamlessly moves between the species identification of fossil teeth of extinct tapir-like creatures discovered in the clay pits of Paris to the contemporaneous "comparative anatomy" of a Black African woman who was variously named Saartjie Baartman, the Hottentot Venus, Black Venus, and Khoisan type. The geostory of Baartman (to be discussed later) continues to put pressure on the grammar of geology, and her own brokenness in the archive of geology breaks apart its geo-logic, exposing forms of subjectivity produced as mode of genesis in geologic time.

Fathering Geology

My focus on the "founding fathers" narrative of geology's telling addresses the obvious patriarchal structuring of the discipline and how the fathering narrative is struck through with a narrative of white paternalism as a foil to the structural violence of geology. The narratives are not difficult to plot in the production of a historicity of Man and Earth, but the material questions of how these epistemes function remain salient in understanding how the praxis of the white supremacy of matter might be dismantled and how these narratives continue to have political valency. "Fathering" established the patriarchal relation to matter through racial hierarchies—making materiality "pregnant" with redemptive possibility and growth, "seeded" by the right stock. "Fathering" is about the foundational institution-building and conceptual scaffolding that structure the very possibilities and scope of thought as informed by a stratal imagination, specifically *as* racial difference. The metaepistemic ordering of geologic nomenclature—such as stratigraphy, formation, strata—and its application and temporal production through the social, created normative ideas of (gendered) relation and conditioned the relays of ontological thought as it was organized through ideas of difference (racial and sexualized).

Fathering, as Hortense Spillers alerts us, involved the severing of the link between the master and his enslaved child under the conditions of industrial rape on the plantation, where the "denied genetic link becomes the chief strategy of an undenied ownership, as if the interrogation into the father's identity—the blank space where his proper name will fit—were answered by the fact, de jure of a material possession" (1987, 76). Spillers prizes apart the master's ability to deny fatherhood as being predicated on his identity as property owner. That property ownership protected the master from the acknowledgment of fatherhood (he could not be owner and father at the same time in the designation of genealogies of whiteness and of the property laws of slavery). Thus the "mocking" presence of the master as owner signaled his absence as father. He fathers elsewhere, as it were, in the genealogical traits of whiteness rather than in the erasure of familial claims that the Middle Passage and the property laws of slavery enacted. In parallel, the fathers of geology both "made" subjective origination and absented that fathering through objectification in the name of the inhuman. The inhuman thus stands in for the absented white father who narrates the earth as property and properties of geopower in order to obscure existing genealogies of relation and attachment (such as Indigeneity) while simultaneously drawing attention to the conquest of territory.

Property rights obscure paternal agency in familial descent, while reinvesting the universal paternity of whiteness over racialized subjects. Race holds the border between the collapse of the signifiers of property and paternity to only allow genealogical access in the context of whiteness. The biological is hidden in the failure of claims to personhood, made through the equation of persons-coded-as-matter, exchanged as property. Geology thus becomes a foil—through its paleontological designations—of the possibility of a claim to the juridical rights of existence (as they are so configured for whiteness). Thus, the normative patriarchal figure produces a genetic underground of orphaned children that will bear his name (Man) as a property mark but must bear it on the slave block, blocked from the genealogical transmutation by the category of inhuman property. And so the biological line of descent is foreclosed as a horizon through the designation of the terms of matter and agency in the separation of Life/*bios* (whiteness) and inhuman, *geos* (Blackness). The name wielded as property (inhuman) cancels out the name as paternity (origination narrative), wherein racial ambiguity is disguised by the violent enforcement of racial hierarchy. The assertion of the vertical force of hierarchy and genealogy as temporal movements of descent in paleo (and, then, evolutionary) time stems the lateral flows of the present to create the undergrounds of fathering and the earth. Race inscription overwrites the claim to Life with a geologic blockade because of the way symbolic and actual ground is joined in the earth.

The development of ideas of stratal difference and the role of genealogy in moving across racial strata (through miscegenation) had a profound effect on ideas of race, as they are articulated by geologists in the political and public sphere, most notably in the *Dred Scott* ruling by the US Supreme Court. This chapter will show how forms of geo-logics sediment ideas of racial difference to establish reciprocal and future-oriented forms of white supremacy. Broadly the historical trajectory of geologies of race begins with the visual narration of race by Cuvier in Europe in 1815, is sedimented in America by Agassiz (and his race maps and photographs of the enslaved [1850]), is then consolidated in the judiciary by *Dred Scott* (1857), is politicized in 1863 by Agassiz's letters to President Abraham Lincoln's appointee Samuel Gridley Howe to advise on the racial dynamics of Reconstruction, and is then arranged as an inventory of types in Brazil in 1865–66.

In the next section, I locate my analysis on those "fathers of geology" (Cuvier, Agassiz, Lyell) and their practices of subjectivity and sexuality not to foreground their psychic motivation, fetish attachments, or desire for the subjugation of Black personhood but to show how their narrative arc carves a foundational geologic ordering of Life that elevates the plateau of whiteness and sediments its race principles in the rift. For example, the claims of the populist author of scientific racism, John Van Evrie, that "slavery was the normal condition of the Negro" (1866, 3) and that antislavery was "a monstrous outrage on reason and the nature of things," rested on the evidential claim that the "Human family is composed of a certain number of species or races, just as all other forms of being, which are generally alike, but specially unlike. The White, or Caucasian, is the most elevate, and the Negro the most subordinate of all the Races in their organic structure, and therefore in their faculties" (Van Evrie 1866, 4). Examining how the desire for white supremacy is legitimated in the annals of geology and its popular narration—despite the incongruence of bodies-as-specimens placed alongside rock, animal, ice, and prehistoric life—signals not just what is political in geology but how geology selectively organizes the polis as a spatial production of time.

I am not interested in the cult of personality and patriarchy but write about these "fathers of geology" to draw attention to the grammars that were used to make legibility (and legitimacy) across genres of personhood and inhuman matter (and with care for those caught in these inhumane archives). Further, I want to know what these geo-logics say about the language of internment and isolation of matter within geologic categories that allowed persons, objects, and creatures to move across registers of deep time and the politics of the present seemingly suspended in a temporal formation that dissuades us from considering the incongruence of this time travel.[11] How are these simultaneities of "times" (of race in the human and inhuman) naturalized to the present as a racialized politics that functions as depoliticized, through geologic grammars and the appeal to an indifferent planetary politics?

Orphans of Geology

The story told of Baartman in Étienne Geoffroy Saint-Hilaire's and Frédéric Cuvier's *Histoire naturelle des mammifères avec des figures originales, coloriées, dessinées d'après des animaux vivants* (1824, 7–9) is one of orphanage and atomization, where her isolation in the rigid air of the paleo-world is cemented to signify her as the only human represented among lavishly illustrated mammals (named "Femme de race Bôchismanne" in front and side profile, positioned between a mountain deer and a langur monkey).[12] What is taken for granted is that Baartman can function in this archive simply through her depiction in a recognizable aesthetic genre of natural history and the assumptions about race that informed audiences about Black bodies in white science. Baartman's story is well known in the context of her European "live" exhibitions as the "Hottentot Venus" and the English court case that resulted from her cause being taken up by abolitionists (poised as she was historically between the abolition of slavery in Britain in 1807 and its continuance in the empire until 1833). Her role in geohistory is less attended to except as an exemplar of racist science. She enters the geologic ledger based on the research of Georges Cuvier (*Le Règne Animal*, 1817), Frédéric's elder brother, as unfree sexualized paleo-object associated with Georges Cuvier's explanatory narratives of geologic life that placed her in the racialized inhuman stratum.

The Cuviers, as professors at the Muséum d'Histoire Naturelle in Paris, the European center of social anthropological and geologic studies, saw Baartman "displayed" at a ball organized by the Countess du Barrie. Highlighting the convergence in the libidinal economy of science and slavery through the exhibitionary complexes of racialized "curiosities," Baartman was one of at least two Khoikhoi women whose exhibition in Europe under the name "Hottentot Venus" was organized around the fetishization of their steatopygia. Her name Sarah Baartman was later adopted in Manchester. Another name was Saartjie ("little Sarah"), a diminutive name for a subjugated status of a little racialized child in the lexicon of species attainment in relation to the fathering status of whiteness (Qureshi 2004, 235). Baartman toured London and then the provinces (presumably as Venuses became less novel).[13] It is recorded that she was then sold to S. Réaux, an animal trainer in Paris, who displayed her in a cage with a baby rhinoceros.

Frédéric Cuvier requested a private "viewing" of Baartman to examine her within the context of zoology and paleontology, where it is recorded that he pressed her with offers of money to show him her labia. She spent three days at the Jardin des Plantes under the observation of the professors of the Muséum d'Histoire Naturelle, including the older Cuvier. Despite what must have been her continued exposure in the extractive industry of exhibitions, and the organization of her public shame and availability under slavery, it is recorded that Baartman refused. For a slight moment, she enters the official geologic ledger in the grammar of the Human, as a subject who can consent or not. This nick in the question of consent marks a limit to Baartman's assertion of privacy, of her refusal of the totality of subjugation, yet thinking with Hartman, such consent is a manufactured possibility in the context of which there was none.[14] But why is her refusal part of the story that is told, when her "availability" is already taken for granted?

Read one way, her refusal signals her claim of a limit, one that establishes a break from experiences of representational and psychic violence in the objectification of her body. This limit that she marks is transgressed in her actual death, yet in her living social death as racialized other, she makes an ambiguous aesthetic claim for an otherwise of agency in both the symbolic and the actual order of her representation, and a right to her body outside of these racist scientific objectifications.[15] How Baartman understood and kept her sexuality and subjectivity alive in that subjugating context is not something that can be known within a life lived in someone else's taxa. Baartman becomes a specimen whose refusal is disregarded after death because she is already designated as an object of knowledge *for* science. Her subjectivity is scantly recognized in life, where she is understood to be flesh (in Hortense Spillers's terms) that protests, but she becomes absolutely enclosed in her taxonomic status in death. Her afterlife is a project of various forms of reclamation and restitution. This imagined agency functions to configure the libidinal interest of white men of science into professional curiosity. The historical sedimentation of attenuated racist meaning that is made to circulate around Baartman as a Black woman has already organized a deadening force that flattens her life into a one-dimensional "account" of her racial identification. She is held in a moment of agency (her protest) for the express purpose of reorienting the subjugating force of the "fathers," to legitimize the scientific interest of the paleontologist more than violation deferred.[16] That is, geology wants to desex itself as it invests in sexual fetishizing of racialized anatomy.

Baartman's encounter with science was no less violent and driven by libidinal fascination than her various turns at the fairs and balls.[17] As Hartman says of another two Venuses, "The stories that exist are not about them, but rather about the violence, excess, mendacity, and reason that seized hold of their lives, transformed them into commodities and corpses, and identified them with names tossed-off as insults and crass jokes" (2008, 2). Similarly, Baartman makes her appearance in geohistory in the context of her sexualized anatomy from which theories of beings-in-time are made to radiate. Cuvier wanted to examine her labia to test his theory that the more "primitive" the mammal, the more pronounced would be their sexual organs and drive. This "act of chance or disaster produced a divergence or an aberration from the expected and usual course of invisibility and catapulted her from the underground to the surface of discourse" (Hartman 2008, 2). The singularity of her identity is ambiguous at best, as she appears as example, and then later as type (and again, here in this writing, her presence is caught up in the politics of the archive as illustrative of regimes of geopower).

In Cuvier's imaginary, Baartman's anatomy was imagined as the "missing link" between animal and human life.[18] What catapults Baartman from the underground of an anonymous scene of subjection is the way her body is made to function as a sexual fetish of racial difference. This racialization of Black subjects in inhuman orders made Blackness both a different space-time that was used as a temporal bypass to the political present and the antinormative of sexuality that secured white futurity and required continued coercion in colonial afterlives. There is a slippage in this formation of racialized time that is paralleled in Anthropocene discourses, where historical time slips the knot into geologic time to permit such "species-life" talk while maintaining privileged hierarchies of white subjectivity (see Yusoff 2018a). The relation between race, sexuality, and geology is so rarely discussed that there is an inattentiveness to the racializing contours of geologic suppositions about time and futurity, and their reproductive modes as a racially marked form of geologic futurism. This is the "absented presence" of Black geographies in McKittrick's terms (2006, 33), geographies and geologies of race that mark access and agency to the future.

Sexuate Geology

Issues of race and sexuality are at the heart of geology's disciplinary formation: as Baartman's body was objectified in relation to species theory and made into a consumable object of geology in the Museum of Man (Musée de l'Homme) in Paris Cuvier lectured that the Negro race was condemned to external inferiority and naturalized the violent expansion of whiteness (geographically and philosophically). Baartman was transfigured through the objectification of her sexual organs into a species belated in time in relation to European Man. This is how race is made, materially and historically, in the praxis of geology, as a segregated order of species in time. In Cuvier's archive, Baartman stands in for a restorative norm of whiteness through its inversion, shoring up a normative fantasy of the human and woman over and against dissonant versions of itself. To defend the limits of what is recognizable against that which challenges it is to understand that the norms that govern recognizability have already been challenged (Butler 2002, 24). The genealogy of species-life is used to (re)construct the normative, to settle and police certain relations through normative justifications rather than to interrogate or subvert them. What racists fear most, Glissant ruefully reminds us, is mixing.[19] Even though Baartman was in the present, she was made into geologic past.

After Baartman's death, Cuvier applied for permission from the police to add her to his large "collection" of human and animal specimens. His application stated, "It [she] was a singular (singulière) specimen of humanity and therefore of special scientific interest" (Qureshi 2004, 242). Her body and its atomized parts were presented to the public as a sexualized paleo-object at the Museum of Man for more than 150 years, where she stood as a racialized and gendered figure in geologic hierarchies of the human, available for the eternal gaze in what Hartman calls a "scene of subjection." Baartman disfigured, skinned, standing in the corridors of Man demonstrates how the metaphysics of race was achieved by way of geophysical inhumane description. Her death relinquished her ability (however cramped) to resist Cuvier's atomization of her body, a quaternization already made permissible through the representation of her excessive sexuality. Baartman's refusal is what little account there is of her agency in the archive of her geologic life, where the conditions under which Baartman can act or not, or the contradictions between her lived reality as a subject and the symbolic inscription of its effacement in geologic orders, render her life *to* geology as *its* subject in paleo-objecthood.[20]

Cuvier published an account of Baartman's anatomy, with many detailed prints and pages of her labia included as evidence of her racial type in the borderland of species, and his observations and diagrams are reproduced and referenced in many subsequent scientific papers, including those of Lyell and Darwin.[21] The key distinction that is made is on account of her sexuality in service to the geostory of racial difference, whereby her sex effectively takes away her womanhood and personhood. Qureshi suggests that the report "reveals a tension between acknowledging Baartman's humanity (she is not even named), and the expectation of bestial habits born from the belief that she represents an inferior human form" (Qureshi 2004, 242). She was skinned (which was a common practice of imperial science), and the rest of her flesh was boiled down and her bones formed part of Cuvier's collection (again, skinning Khoikhoi women was a "scientific" practice that Cuvier devotes some time to discussing in terms of the best methods for procuring subjects as taxidermic material). I pause here on the information of collecting skins not to shock or reproduce racist objectification but to bring attention to its normalization within scientific practices *as* legitimate scientific practice. This is not just a system of knowledge that depends on and socially reproduces a racist patriarchy, but one that interns that social reproduction into the lithic ground as necessary for the explanations of contemporary subjectivity, whereby Baartman is rendered as inhuman past to explain the inhumane present and secure the future. The question is not so much who and why (all the discipline is tainted by imperialism, contra Rudwick), but how did the grammar of geology and patriarchy function such that Black women's bodies became necessary for these lexicons to exist as coherent? In Cuvier's 1813 *Essay on the Theory of the Earth* (Cuvier 2009) what world-building required such sexed subjugation? This dehumanization is what Hartman calls the "utter fungibility" of racialized and gendered bodies to science, but these narratives of geologic life do not just build an account of colonial and settler colonial spatialities; they weaponize deep time as a tool of white patriarchy for extraction.

Sexual excess and pathological desire are made to reside everywhere but in the white patriarchy from which they come. The shame of this fathering of geology is rendered as Baartman's. Her atomization into sexual organs and the way in which Cuvier parses her from personhood through this anatomical division demonstrate the condition of a complex series of relations in a relay of racial difference. Atomization of body parts was part of the grammar of construction in racialized types that were established through forms of bodily segregation into partitioned representations of "traits." Labia in this signifying structure were used to condemn Baartman's body to interpretation from this point of anatomy, as an object of geology rather than as a subject of the world. Her Black female body is made available to all, insomuch as there is no recognition of her body and her desire *for her*. Cuvier's imagery of her sexuality as excessive through this partition of her body is understood as an indexical mark in the development of a normative sexuality that has its apex in white womanhood.[22]

Objectification becomes constructed as a mode of evidence in Cuvier's detailed dissections and diagrams of Baartman's labia, and this "evidence" to bolster his belief in stable species (or the nonevolutionary relation between species) was used to heighten the status of *his* scientific reputation as a father of geology. The imagined threat of unregulated Black female sexuality is fixed alongside apes to serve the cultural purpose of denying subjective rights and elevating white masculinity. That Cuvier invented and perpetuated a "missing link" that happened to serve the political interests of the white supremacy of the day in producing racial hierarchies that justified subjugation was neither unimportant nor immaterial to the championing of his method and reputation. Cuvier gained his status in geology, paleontology, and comparative anatomy partly by insisting on factual empirical evidence, thereby dispelling many of the myths about giants and other creatures that were ascribed to dinosaur bones and fossil objects. When many were still looking for giant beavers and mastodons in North America (on the Lewis and Clark Expedition in 1803–6, for example), he proposed a theory of extinction that disaggregated species in time. The concept of extinction emphasized the noncontinuity of life within geologic time, thereby naturalizing ideas of and domesticizing the extinction of "lesser species" within human species.[23]

Cuvier continues to be praised for the quality of his unflinching scientific scrutiny and meticulous description of "specimens."[24] Geologic propositions are literally stretched through the skin of subjectivity in the bifurcation of person and flesh into objects of knowledge (Baartman) and through the designated subjects/agents of history (Cuvier). Baartman's remains were eventually removed from public exhibition in 1974 but remained the property of the museum until her repatriation to South Africa in 2002.[25] The question, however, remains for me about *how* these narratives traffic across terrains of Black women's sexuality and theories of Life and the earth as an inhuman intimacy. How did Baartman function in the geologic life of the planetary? It is Baartman's ability as a Black woman functioning in "the figurative capacities of blackness" (Hartman 1997, 7) to travel through time that *enables* her appearance alongside tapirs, polar bears, and apes without a change in the grammar of her presentation and a shift in the modes of recognition between human, nonhuman, and inhuman categories. That is, geology provided the grammar that joined human life as a fungible unit with the valuation of mineralogical objects within geology's empiricism and empire of matter. The violent story of Cuvier's subjugation of Baartman and her life as a "living exhibit" exposes a brutal geohistory and a hinge in the connection between geology and questions of race, sexuality, and gender. Cuvier explicates the way in which physical description, nature, and morality are marshaled together to perform a collaborative mode of dehumanization invoking words such as "filthy" to designate "the whole race" and its "moral condition as displayed in their appearance and modes of life" (1827, 198). Darwin continues in this vein of designating racial accomplishment to moral qualities in a letter to A. R. Wallace: "I had got as far as to see with you that the struggle between the races of man depended entirely on intellectual & *moral* qualities" (1864).[26] In the attempt to establish European ideals (philosophical and corporeal) as the normative trajectory for life on earth, the construction of geohistory was not neutral in its attachments to the white supremacy of a preferred race.

While Baartman is a heterogeneous actor across many fields, and making her perform once again in another disciplinary register may seem a turn too many, a specific kind of taxa and aesthetic grammar allowed fossilized creatures and Baartman to coexist as specimens in the paleontological ledger as onto-epistemological objects. Baartman was in a genre of representation that functioned as "less than fully human . . . , as still a limit where the human becomes less than human, as a space where the human becomes a mix of the human with something else" (Sharpe 2010, 68). She is described by Cuvier in terms bordering on subjectivity (for example, capability with language), but such recognition is quickly tempered by physical, moral, and sexual negation.[27] The subject/object distinction that "allows" Baartman to move through and with fossils is her designation on the borderland of the human-inhuman classificatory divide, whereby she is already outside of but shoring up the constitution of the human, as race spaces the *bios-geos* gap. Her geologic death as racial-sexual other is established with scant regard to her *actual* living or dying, as she gets transfigured from exotica to exotica in different registers of objectification. The contusions of creativity in science, its affectual grammars that functioned in symbolic and geophysical orders, allowed Baartman to be produced as both a living and a socially dead subject, an inhuman being in orders of the human, without any need for Cuvier to resolve this apparent contradiction. It is the genres of geologic time that are used to dissuade an apprehension of this collapse and the work those temporal lines do on behalf of bordering the creative fiction of whiteness.

Baartman may seem like an anomaly in the annals of a principal text from the "Father of Paleontology," but she exemplifies the convergence of racialized theories of Life that were constructed in the libidinal economy of slavery and the atomization of Black women's bodies *and* the theories of life that were actualized through the earth and its fossil remains where some humans functioned as living fossils rather than subjects in time. The twinned imaginary of genealogies of earth and humans combined narrative and material objects to form a totality of thought in deep time to consolidate normative life in the present. To put this another way, deep time is a racializing analytic that unravels a set of normative discourses on agency. Baartman presents the caesura in time and narrative, a systematic sexual violation that marks geology's becoming as both racialized and sexualized through the libidinal economy of Black women's bodies. Cuvier used the diagnostic method of a pathology, visually articulated through a distinct genre of the body-in-a-performative-mode to produce nonnormative sexuality as racialized time (not unlike the later iconic and invented photographs of Jean-Martin Charcot's "hysterical women" at the Salpêtrière psychiatric hospital in the 1880s). Within the context of geology's focus on sequential events, understood through the entity of the *formation* (a collection of rocks formed during a certain period, defined by Abraham Gottlob Werner by the end of the eighteenth century), the paleontologist applied the same thinking to speciation as to the basic unit of stratigraphy. Sequential genres of photography and illustration were made to produce evidence for this pattern of thought through temporal incisions. Thus, genealogy of species was fundamentally about narrating relations, forces of power, and possibility, not life per se, where geophysical description mobilized political geologies of the present.[28]

Cuvier's pathologizing of Black women's sexuality (that translates into an extreme form of dissection) was an established mode in the libidinal economy that underpinned the economy of slavery.[29] As the poet M. NourbeSe Philip suggests, "The space between the black woman's legs becomes. *The place*. Site of oppression—vital to the cultivation and continuation of the outer space in a designated form—the plantation machine" (1994, 289). And, as shown by Heather Vermeulen's insightful discussion of the "libidinal Linnaean Project" of Thomas Thistlewood, an earlier naturalist in Jamaica in 1750, "classification is inseparable from sexual violence; indeed, the latter provides the occasion for the former" (2018, 24). In Britain, for example, it was fear of the sexual corruption of "absentee" husbands and fathers on plantations that motivated much of the moral discourse of abolition, which came to be seen as a mode of immunization against desire realized (violently and without consent) on Black women's bodies. Morality is deployed in discourses of paleontology (and later in evolutionary theory) as a regulatory behavior in normative regimes of sexuality and its maintenance of property relations. Temporal sovereignty—established through race—becomes a way to police questions of gender and sexuality.

Bodies That Mark Time

Cuvier sets out geology as a fixed durational field that makes Life a particular kind of event defined through origins and extinction, but immovable as a hierarchical enclosure against more challenging notions of change and nonlinear dimensionalities/diversions. Since he was a catastrophist who believed in the fixity of species, there were two times in Cuvier's thinking: now and before now, or the present and past geologic worlds. This was a form of abysmal time. His work preceded the development of the key concepts of biocentric life, understood by Lyell's uniformitarianism, which gave space for changes among species in time, and ideas of biological evolution proposed by Jean-Baptiste Lamarck and later in the natural selection of Charles Darwin and Alfred Russel Wallace. Though not the first to propose a theory of extinction, Cuvier was instrumental in persuading the scientific community to accept the theory of extinction, a notion that was at the time as controversial as evolution would later be. He was dismissive of Lamarck's *Philosophie Zoologique* (1809), which proposed that animals can acquire new characteristics during their lives and pass those characteristics on to their offspring, an idea of transmutational theories or what became, with modification, Darwinian evolutionary theory. He cautioned that, although "the human species would appear to be single, since the union of any of its members produces individuals capable of propagation, there are, nevertheless, certain hereditary peculiarities of conformation observable, which constitute the termed *races*. . . . The Caucasian, to which we belong, is distinguished by the beauty of the oval which forms the head; and it is this which has given rise to the most civilized nations—to those which have generally held the rest in subjection" (Cuvier 1849, 49–50).

The identification of race is immediately qualified by an appeal to white (Caucasian) supremacy, identified through aesthetic modes in the "beauty of the oval" *and* action of subjugation. In this geologically infused account of race, time and its rendering hold much of the argument in the materialization of racialized subject. Cuvier divided the world into three categories of race (Caucasians, Ethiopian or Negro, and Mongolian), which he saw remaining stable on a dynamic earth, with only Caucasians being able to access geographic promiscuity—"extended by radiating all around" (1849, 50)—via the temporalization of racial categories.[30] In Cuvier's production of deep time, race types are caught in a geologic time trap *between*, and whiteness is rendered free in time as *radiant* (and thus geographically expansive). That is, free passage and the openness of spatial acquisition (conquest) are secured for whiteness via the arrest of other races in the rift zone.[31] The organization of geographic relation and the right to mobility that Cuvier articulates via race (justifying colonization) bear no relation to the *actual* movement and forced migration of people through Africa, Asia, the Atlantic, and the Americas. Viewed in the historical context of colonialism, Cuvier's assertions look decidedly compensatory in their attempts to stabilize geographic relations and counter the lack of viable categories with which to understand newly emergent colonial identities. What his articulations of geographic capabilities do is to sediment a white "right" to geography through temporal location. Thus, ideas of racial superiority are temporalized within the logics of geologic time to make whiteness a movement within time that justifies the brutalized conquest of space.

The imaginary of a hierarchy of the races through paleoplacings was deployed to confirm Cuvier (and the institutions he represented: colonialism, natural history, whiteness, patriarchy) as the apex of race, where racial difference is naturalized rather than historicized. Ferreira da Silva makes the argument that the effect of the naturalization of racial difference is to "render colonialism and slavery irrelevant in the understanding of human collectives' conditions of existence" (2011, 143). As race is immediately transmuted into a fixed condition of supremacy and subjugation, and thus not attributed to differences in racialized relations, the anticipatory quality of Cuvier's writing as both future orientation and a present sedimentation of the stratum of patriarchy, privilege, and power is identified.

The difference of race, as it is represented in a punctuated narrative of species achievement in time, literally functions to stratify certain persons into another geologic age or stratum. Racialized others become not just another species but another time zone. Race is materially made as time travel and arrest—Baartman is stuck in the wrong epoch, endless performances in the temporal rift of the inhuman(e). This is how the racial orders get made through stratification into a formation—the plateau. Cuvier demonstrates that the structure of geology, as genealogy and stratification, underpins racial formations, and the temporal concerns of the production of knowledge of deep time become decidedly local in time (or, parochial, as Fanon might say).

Ciphers of Geology

> Can you kill Remorse with its beautiful face like that of an English lady stupefied at finding a Hottentot's skull in her soup tureen?
>
> —AIMÉ CÉSAIRE, *Return to My Native Land*.

In Césaire's terms, skulls pop up in the soup in an English dining room, where any engagement with the remorse at the brutality of empire becomes transfigured into a form of mild stupefaction, something out of place in polite society. Race, as an unpalatable inconvenience of geohistory, makes a similar unseemly appearance. The scene of the soup tureen is no less implicated in the spoils of colonialism than the violently procured body part. Like the perspectival splice in Hans Holbein's *The Ambassadors* (1533), the skull introduces an epistemic rift into the proceedings in which the ghosts of geology's empire show up to unravel the inhuman story in the eugenics of matter and colonial riches of white supremacy. As a living bridge between phylogeny and geology or a political ordering in the "Great Chain of Being," racialized inhuman "subject-objects" are unsightly in parlors of the plateau.[32] These unseemly "fossils" of empire are matter out of place in Césaire's rendering of a BBC period drama that is yet to be screened, unconfined by the "fathers of geology" narrations that anatomized racial difference and parsed its forms of procurement as science. He says, "My racial geography: the map of the world made for my use, coloured not with the arbitrary colours of schoolmen but with the geometry of my spilt blood. . . . We are measured with the compasses of suffering" (Césaire 1969, 70).

The specific invention of geologies of race is historical, not natural. First, because the time of development of the discipline of geology is contemporaneous with the time of slavery and the construction of racial difference as a key determinant of the material production of settler colonialism. The foundations of geology were laid down (1780–1830) at a time when the economies of slavery had been used to build over four centuries of science and societies in Europe and the "New World." Second, paleontology developed precisely because of empire and colonialism that made available "bodies" to science as objects of scientific property, concurrent with the violence and the volume of death that secured colonization (through settler colonialism and chattel slavery). This disciplinary accumulation was substantiated both by the material gains made from the industrial-scale theft of land and people and through the accessibility to the world that this subjugation instigated. The mine, plantation, lab, lecture hall, and ball were concomitant fields of production of racial difference rather than divergent. Third, rather than the role of race and racialized theories of difference being peripheral to geology (a footnote in an otherwise inhuman story, a sign of the times), these theories were integral to an interwoven discourse of white geology and its theories of the earth (as universal spatial geophilosophies). If geology and genesis had a tricky alliance over the age of the world (divine vs. natural causation), arguments about the destiny of Man (various iterations of Manifest Destiny) replanted in natural orders were just as susceptible to their own forms of genesis in the perpetuation of white supremacy.[33]

Sedimenting the Fathers

Geminating the role of the "fathers" of the discipline as key to the establishment of the discipline, Marcou suggests that "the two visits of Lyell in 1841 and 1845, and the important journey of de Verneuil in 1846, among the Paleozoic formations from the State of New York and Canada, to [the] Ohio Valley, the Upper Mississippi River, and Lake Superior, had given a strong impulse to geological researches, in bringing about the much needed comparison with European classification and synchronism. . . . The coming of Agassiz was anticipated with great joy" (1895, 282).[34] In his account of the "coming of Agassiz," Marcou describes (on Agassiz's visit to Charleston, South Carolina) how the rich entomological fauna was a surprise to him, but "what made the greatest impression on him as a naturalist was his contact with a large population of negroes. With his power of comparing zoological characters, it was impossible for him to consider the black man as a species identical with the white man" (293).

Marcou continued: "He insisted that every animal and plant is confined to a certain portion of the earth, while man is the only one who covers the whole surface. . . . After his first visit to South Carolina, species, in his eyes, existed for man as well as for every other genus" (1895, 293). While Marcou suggests that we cannot conclude that Agassiz was in favor of slavery (as Marcou returned to America in 1861 to assist Agassiz to set up the Museum of Comparative Zoology in Boston, Massachusetts, at Harvard University), he says: "This was an abyss which he never crossed" (293). It might be suggested that while Agassiz did not cross the abyss, he in fact created and maintained the abyssal fissure of its racial difference.

Louis Agassiz (1807–73) was another recipient of the "father" moniker of geology for his work on ice sheets and another proponent of "missing links" that connected the transition of the stratum from one state to another (Agassiz 1863a).[35] He was a beneficiary of the patriarchal patronage of Cuvier and naturalist Alexander von Humboldt. And he was introduced, via the Lowell Lecture series (1846), to America by Lyell's recommendation.[36] In turn, Agassiz's science supported Cuvier's theory of a fixed geology in which the history of the earth had successively been written in strata and on which biological species followed suit in temporal formations, where species are held in stratal time and do not change. Ice ages became a way to explain the geo-logic rift to support Cuvier's ideas of extinction and fixed time(s) of species, and thus his theories of Life. Against Lamarck's theory of the changing nature of organisms from internal and exterior pressure, Agassiz's theory of catastrophic ice ages was used to explain the breaks in the genetic relationship of animals and plants from one geologic period to another, without ascribing to a progressive narrative of change. Like Cuvier and Lyell, Agassiz participated in the signification and subjectification of racialized Life within accounts of natural history and brought European paleontological race theory to America, where it flourished. His work on race and genealogy was especially significant in establishing the visual genres and aesthetics of race in the objectification of Blackness, and in providing an overt defense of white supremacy to reestablish the geologic color line in the United States in slavery's afterlives.

While Agassiz's "fathering" reputation was subsequently questioned due to his adherence to creationism and its twin in scientific racism, he was the most important geologist in the United States at the time and a popular professor inside and outside the academy, where he influenced the nature philosophies of Ralph Waldo Emerson and Henry David Thoreau and the pragmatist philosopher William James, who accompanied him on his expedition to Brazil (see W. James 2006). He held the chair in zoology and geology at Harvard University.[37] Named founder of the discipline of glaciology, Agassiz was also a prodigious establisher and curator of natural history museums and established the Museum of Comparative Zoology at Harvard University and advised the Charleston Museum in South Carolina (in a plan of the College of Charleston Museum of Natural History, the entrance is named the Agassiz Hall), where his practices of cataloging and categorizing natural history were widely adopted across this burgeoning field of museology, especially through his publications such as *Methods of Study in Natural History* (1874), which compiled essays on techniques of comparison, collections, and nomenclature, and *Essays on Classification* (1859). While the overt racecraft of his species talk has been questioned, it is at the metaepistemological level that the syntax of his racist geology became embedded in the production of racial difference and the material cultures of geology. Sitting underneath Agassiz's most obvious representations of scientific racism, such as his dehumanizing illustrations of "Negroes and apes," the empiric of geology does quiet work in the continuance of arrangements and classifications of time and matter, as allied to specific racial dynamics. Specifically, Agassiz's expertise on the comparative "laboratories" of race in America and Brazil ordered and sedimented a racist reason that pitches the security of (racial) whiteness into the violence of Indigenous, Black, and interracial repression.

Genealogical Geographies of Race

Two distinct approaches characterize Agassiz's discourse on race: geographic and genealogical. In Agassiz's geo-logic, whiteness (being of the higher category) could travel forward in time as a futuristic species, and Blackness (being of a lower category) moved in the other direction, stratified in deep time. Agassiz made the division of the "civilized" and "barbarous" through geographic and genealogical distinction, as their empirics were naturalized through deep time of the history of earth and its species. His racial maps use geographic differentiation to separate and codify the races, while concurrent claims were made for the superiority of whites in their intellectual capacities to justify their conquest of territory (i.e., their ability to escape the confines of the "natural" order of the map and the enclosures of environmental and climate determinism). In this formulation, white races were not bound by geography but seemingly entitled to it, through extermination, if necessary, to achieve their "natural" tendency toward supremacy (thus whiteness becomes literally unnatural—inhuman—in order to secure its naturalized superiority). Geology became an empirical means to construct whiteness as a territory that encloses property and politics as Agassiz, like

Lyell, traveled to plantations to secure his empirical taxa of racial "objects." This race narrative was set in an adherence to a natural history that was understood to be the outcome of a war of races (see also Cuvier 1849, 49). Thus, racial difference becomes both mobilized (for whites) and stabilized (for others) in a reciprocal geologic and genealogical stratum to justify the assertion of white supremacy over geography. Agassiz sketched the geographic distribution of mankind as a natural(-ized) history, arguing that "men must have originated in nations, as the bees have originated in swarms, and, as the different social plants, have covered the extensive tracts over which they have naturally spread; this geographical location and determination, he [Agassiz] called a natural history" (Morton in Nott and Gliddon 1854, 78).

This zoological (and nation-state) understanding of race (or the designation of organic laws), which located race in geographic points of origin, was developed alongside a historical argument about how "human progress has arisen mainly from the war of races," but in which "some of the lowest types are hopelessly beyond the reach even of these salutary stimulants to melioration" (Agassiz in Morton et al. 1854, 53). In Agassiz's geographic imaginary, Man as species was made the apex; "he is the last term of a series beyond which there is no material progress possible upon the plan upon which the whole animal kingdom is constructed" (Agassiz 1857, 25). The framing of human origins as a succession of race wars justified the white supremacy of the present and future, beyond which no more material progress was designated as possible:

> Nations and races, like individuals, have each an especial destiny: some are born to rule, and others to be ruled. And such has ever been the history of mankind ~ No two distinctly-marked races can dwell together on equal terms. Some races, moreover, appear destined to live and prosper for a time, until the destroying race comes, which is to exterminate and supplant them. Observe how the aborigines of America are fading away before the exotic races of Europe! Those groups of races heretofore comprehended under the generic term Caucasian, have in all ages been the rulers. (Agassiz in Morton et al. 1854, 79)

Treating Mankind zoologically through the argument of naturalized zoological provinces, "into which the earth is naturally divided" (because of geologic processes), posits racial difference through the causation of geologic formations (87). In this neat identity formation for geology, the earth is divided through geologic processes, and nature is stacked into continents, then nations, then into race. The "operation of physical causes" naturalizes the division and origination of "stock," where race becomes species, and species difference is racial division: "The meaning attached to the term species, in natural history, is very definite and intelligible. It includes only the following conditions: namely, separate origin and distinctness of race, evinced by a constant transmission of some characteristic peculiarity of organization" (80). This "characteristic peculiarity of organization" was imagined as white supremacy, Indigenous death, Black subjugation, and a prohibition on "fathering" mixed offspring. Agassiz's geologies of race demonstrate that it is not just about the race politics of the time (i.e., historical context) but also about the race politics *of* time and how geologic time was used to spatialize and hierarchize races in time as extinct-oriented beings or future-oriented beings.

Utilizing the research from Agassiz's fieldwork in the hugely popular book *Types of Mankind* (Nott and Gliddon, 1854), the editors narrate: "Prof. Agassiz also asserts, that a peculiar conformation characterizes the brain of an adult Negro. Its development never goes beyond that developed in the Caucasian in boyhood" (Morton et al. 1854, 415). Agassiz demonstrates his empirical method for this assertion with a number of skulls. While the casual availability of skulls and brains for scientific examination testifies to the fungibility of racialized subjects, the paleontological groundedness of Blackness as "fixed" in a scene of boyhood again secures whiteness as the adult relation (that which is mobile, geographically and genealogically, in the ability to possess and *father* as the child cannot). Agassiz's construction was created to sustain the idea that the boundaries between different races were self-evident, explained by visible "natural" divisions of continental shelves and discrete zoologies, in which race becomes understood as a formation—a process and outcome of geology, sedimented by an array of organic and inorganic processes.

Maps of Racial Difference

Agassiz declared in March 1850, before a meeting of the Association for the Advancement of American Science in Charleston, South Carolina, that the races were "well marked and distinct" and did not originate "from a common center . . . nor from a common pair" (i.e., Adam and Eve). Agassiz propagated the theory of polygenesis and proposed that human beings of different "racial types" did not share a common ancestor (and thus a common humanity) but were the product of multiple originations.[38] This theory of polygenesis was posited against the theory of a single common ancestor to assert racial superiority: "The unity of the human species has also been stonly [*sic*] maintained on psychological grounds. Numerous attempts have been made to establish the intellectual equality of the dark races with the white; and the history of the past has been ransacked for examples, but they are nowhere to be found" (Agassiz in Morton et al. 1854, 59).

At the time of Agassiz's writing in the United States, polygenesis was considered a "Southern view" on the status of Black persons, in contrast to the "Northern view" of monogenesis. His support for the "Southern view" from the institutional and social location of a powerful northern university (Harvard) did much to shift the geography of the idea of the segregation of the races. Agassiz's proposition of human descent from different "stocks" and multiple originations argued that racial difference of the "human family" was equivalent to different species. Agassiz mobilized the function of the origin to make racist enclosures in the present. Temporal segregation in paleontological time organized social segregation, as narratives of the past secured political horizons for white supremacy. Following his publicly declared support for polygenesis, Agassiz was invited to the capital of South Carolina (Columbia) to conduct "field research" on race.

Agassiz traveled to plantations (owned by enslavers and geologic enthusiasts) where he conducted research on many of the enslaved (concentrating on African-born persons and their "country-born" daughters) to establish their anatomical difference and organize "standards" and "specimens" of the race of African ancestry. In his search for "authentic" Africans, Agassiz sought to establish his scientific reputation as an Africanist and to reoriginate (mythologize) an epistemic point of "return" from which Black purity (albeit in its subjugating stratal status) could be genealogically identified (and his ideas of the denigration of miscegenation tracked).[39] Agassiz used the factory of the plantation as his laboratory of racialization, highlighting the cooperation between the university and the plantation in the production of racial difference as a form of social death caught between questions of origination, materiality, and narration. Through comparisons with apes and crude drawings of racial types, which were well circulated and reproduced, he sought to distinguish the differences between enslaved Africans he examined and photographed, and those of other races. Like Cuvier, Agassiz was keen to distinguish between different "types" of Africans, asserting, "We generally consider the Africans as one, because they are chiefly black. . . . Look closer and differences abound: The negro of Senegal differs as much from the negro of Congo or of Guinea. The writer has of late devoted special attention to this subject, and has examined closely many native Africans belonging to different tribes, and has learned readily to distinguish their nations, without being told whence they came; and even when they attempted to deceive him, he could determine their origin from their physical features" (Agassiz 1850d, 125).

While Agassiz's approach might seem at first pass to disaggregate Blackness and accommodate a more complex notion of geographic and racial difference (in comparison to homogenizing discourses on the "Negro" in the construction of Black identity under slavery), the establishing of *his* expertise in Black difference (beyond his subject's "deception") is being used to consolidate an evidential foundation for more sweeping and brutal racial pronouncements. The stabilization of difference (at a time of massive destabilization through the sexual economy of slavery) proceeded through the disaggregation of "types" and their stratification.

In a passage about Indigenous peoples, Agassiz makes clear the crude bifurcation of subjectification that his research is attempting to achieve through research on bodies and craniology, and their liminal placement in lithic orders:

> The intellectual lobe of the brain of these people, if not borne down by such over powering animal propensities and passions, would doubtless have been capable of much greater efforts than any we are acquainted with, and have enabled these barbarous tribes to make some progress in civilization. . . . *This view of the subject is in accordance with the history of these two divisions: barbarous and civilizable.* When the former were assailed by the European settlers, they fought desperately, but rather with the cunning and ferocity of the lower animals, than with the system and courage of men. They could not be subjugated, and were either exterminated or continued to retire into the forests, when they could no longer retain their ground. (Agassiz in Morton et al. 1854, 278; my italics)

Creating categories of permissible extermination of "these people," the policing of subjectification was geologically interned in the rift between the barbarous and civilized to promote the rights of whites to murder with impunity, given the natural order of racial war. The identification of the "so-called" five civilized tribes, for example, demonstrates the brutal limitations of this recognition and zero-sum game (see Wolfe 2006). The fantasy and rhetoric of extermination functioned to create a paleo-state-of-being or the geophysics of whiteness and enacted the futurity of its stratal renewal. Utilizing the research from Agassiz's fieldwork in *Types of Mankind*, the editors narrate: "Prof. Agassiz also asserts, that a peculiar conformation characterizes the brain of an adult Negro. . . . It bears, in several particulars, a marked resemblance to the brain of the orong-outan. The Professor kindly offered to demonstrate those cerebral characters to me, but I was unable, during his stay at Mobile [Alabama], to procure the brain of a Negro" (Morton et al. 1854, 415).

The authors continue, "Although a Negro-brain was not to be obtained, I took an opportunity of submitting to M. Agassiz two native-African men for comparison" (Morton et al. 1854, 415). The interchangeability of alive and dead subjects makes clear the inhuman(e) epistemologies of paleontological "specimens." The "procurement" of racial subjects for science operated under the same geo-logics as the procurement of subjects for geologic acts of transformation of the earth, namely, the matrix of geology and slavery. In a dedication to Agassiz's coauthor in the sixth edition of *Types of Mankind*, Samuel Morton (the extensive collector of skulls and author of the influential *Crania Americana* [1839]), the contemporaneity of concerns about race are made explicit.[40] The dedication reads: "It is manifest that our relation to and management of these people must depend, in a great measure, upon their intrinsic race-character. While the contact of the white man seems fatal to the Red American, whose tribes fade away before the onward march of the frontier-man like the snow in spring (threatening ultimate extinction), the Negro thrives under the shadow of his white master, falls readily into the position assigned him, and exists and multiplies in increased physical well-being" (Nott and Gliddon 1854, xxxiii).

As Native Americans are imagined as seasonal subjects that will be disappeared into a scene of nature, like the snow, frontier man establishes nature's transformation through racial management. Racial difference is understood on a temporal sliding scale wherein the apex of whiteness defines a change of geophysical state, a new climate, like spring. The mobility of difference is held by the immutable ground of Blackness that "falls readily into . . . position" in the shadow of the plateau. This racial ground is what allows whiteness to travel, like a pair of compass legs, one fixed, the other mobile (to use John Donne's geographic conceit of gender as the fixed foot that allows patriarchy its metaphysical range in "A Valediction: Forbidding Mourning" [1611]). Invoking the paternalism of the white master as the naturalized father of the "Negro" establishes a biological ordering for political ends (to further geopower), while highlighting the libidinal anxiety at the "physical well-being" that drives this ordering. The rocky rift of race is managed by the Overseer. Frederick Douglass wryly reflects on the political aspirations of *Types of Mankind*: "The debates in Congress on the Nebraska Bill during the past winter will show how slaveholders have availed themselves of this doctrine in support of slaveholding.[41] There is no doubt that Messrs. Nott, Gliddon, Morton, [Samuel Stanhope] Smith and Agassiz were duly consulted by our slavery propagating statesmen" (1854, 16).

In their racial argumentation, Nott, Gliddon, Morton, and Agassiz used the temporal device of the future geologist (so favored in Anthropocene science discourses) to set out an imaginary of the futurity of white extinction (that is rationalized through the paleontological past) to do the present political work of asserting the space-time of geologic racism: "May not that Law of nature, which so often forbids the commingling of species, complete its work of destruction, and at some future day leave the fossil remains alone of man to tell the tale of his past existence upon earth?" (Morton et al. 1854, 80).

Inhuman as an *Intra*human

While the overt language of natural history may have changed in the Anthropocene to encompass the postracial "we" of the future fossil, the language through which it originates and signifies relays back to the geo-logics of white anxiety and empirics of empire, which seek to stabilize the instability of race to maintain exclusive categories of privilege. "We" are still in the syntax and traps of reason founded on a geo-logic of scientific racism. The historicity of geology is what secures race as foundational to the discipline such that any attempt at a "we" in the Anthropocene or environmental politics [that] appeals to species-life or universalism is riddled with this formation of a politics of Life by way of these intimate forms of inhumanism as an *intra*human condition in social orders. As the Anthropocene focuses a lens on geologic forces, it carries a (racist) fabric of genealogy in its inhuman trajectory, which requires a fundamental disruption. Race science has been debunked; the racialization of natural resources as the normative mode of extraction has not. Agassiz's ideas were influential enough that in the context of the Emancipation Proclamation of 1863, Samuel Gridley Howe (who was appointed by Lincoln) wrote to Agassiz about the futurity of the Black and "mulatto" races in America in relation to the whites in the South.

Howe was a prominent Boston abolitionist, doctor, and founding member of the Boston Emancipation League. He was one of three members of the American Freedmen's Inquiry Commission (AFIC), a blueprint for radical reconstruction set up after the Emancipation Proclamation to investigate the measures pertaining to the dual purpose of the protection and improvement of the conditions of freedom and suppression of rebellion.

His findings formed a key part of the AFIC's conclusions to the secretary of war in 1864 (also see Furrow 2010). Howe asks Agassiz three questions: "1st is it probable that the African race will be a persistent race in this country or will it be absorbed, diluted and finally effaced by the white race?"; "2nd will not the general *practical amalgamation fostered by slavery*, become more general after abolition?"; and "3rd In those sections where the blacks and mulattoes together make from 70 to 80 and even 90 pr. Ct of the whole population, will there be, after the abolition of slavery, a sufficiently large influx of whites to counteract the present numerical preponderance of blacks.—?" (Howe 1836; italics mine). Agassiz wrote four letters back to Howe in the space of a week on his scientific opinions on "a preponderance of blacks" and outlining his fears of the "effeminate progeny of mixed races . . . sprinkled with white blood" (Agassiz 1863b, no. 150).[42] He asserted that the mixing of the races was the biological and moral equivalent of incest.[43] This "creates unnatural relations and multiplies the differences among members of the same community in a wrong direction . . . that which is abhorrent to our better nature, and inconsistent with the progress of higher civilization and a purer morality" (no. 150). The forced intimacies of the plantation are symbolically held apart by the spatializing of geologic race time and the imagined purity of those trajectories. His response to amalgamation is affective and framed as a psychosexual panic: "[The] idea of amalgamation is most repugnant-looming feelings, I hold it to be a *perversion of every natural sentiment*. Practiced secretly at the South, it is the source of infinite domestic misery; it has produced a population the position of which can never be easy, natural or productive of any good. . . . Amalgamation among different races produces shades of populations, the social position of which can never be regular and settled" (no. 150; my italics).

Amid his looming sexual and moral affect, amalgamation, according to Agassiz, produces a directional inversion and "shades" of populations that cannot be "settled" in the colonial state. The praxis of the genealogical imperative did not stop at the division between human and inhuman but proliferated an imagined spectrum affect that read value in the migration toward whiteness, even as it held that movement as symbolically stratified. For example, the white-skinned, blue-eyed "Békés" were the planters and thus the color of power in Martinique. Békés' obsession with bloodlines, Fanon argues, is one of Martinique's worst forms of alienation, namely, the "shadism" that identifies groups and individuals based on their degree of pigmentation. Shadism performs to the biological aesthetics of the phenotype, but it also mimics a Western philosophical organization of thought securely based in a stratal imagination, where migration to whiteness equates to stratal logics of levitation upward toward the plateau.

In Martinique, the abolition of slavery sedimented this levitation affect, as the enslaved were "replaced" with indentured Indians from Pondicherry (a French colony in India until 1954), who became the new "dark" underclass. Fanon called this the "racial epidermal schema," whereby the gaze of whiteness in "the bluest eye" continued its work of racial branding. On his return to Martinique, Fanon says: "The blue of his eyes burned the gaze of the negro during slave-time." He complains, "All this whiteness is burning me up" (1986 [1952], 71). He is referring to the violent imposition of being Black for others, as it is rendered through an Overseer's gaze, a gaze that brands the subjugated other as imprisoned by whiteness, as by the sun. To move through the geologies of race in its metaphysical and geophysical dimensions required a parallel practice of nonadherence in the genealogical ordering (the color of the rift is the dissimulation of these race affects). As Agassiz's political response sought to make race crystalline through its containment in a geographic and sexual paradigm in the standardization of racial types, the imposition of fungibility on racialized subjects denoted a more expansive subterranean field of relations and materiality that pressed against his reasoning. Agassiz responds to Howe's first question that the amalgamation of the races would result in degeneracy: "I can see no cause which should check the increase of the black population in the Southern States. The climate is genial to them, the soil rewards the slightest labor with a rich harvest. . . . The country cannot well be cultivated without real or fancied dangers to the white man, who therefore will not probably compete with the black in the labors of the field" (Agassiz 1863b, no. 150).

What has been the source of the enslaver-planter's geopower is imagined as usurped by the soil's generous reward to the Black population's "slightest labor." Using the geologic grammars of classification that he has established, Agassiz describes the "Negro" as a fungible unit designated by nature rather than politics, "by nature a *pliability*, a readiness to accommodate" (Agassiz 1863b, no. 150). The proximity to the inhuman is scripted as a potential strength of environmental determination if not checked as a threat to the racial tenant of civilization: "Dangers awaiting the progress of civilization . . . have a practical influence upon the management of the public affairs of the nation" (no. 150).

To the second question, Agassiz indicates that "practical amalgamation, as you seem to call the illegal intercourse between whites and blacks, is the result of the vice of slavery." Referring explicitly to industrial rape under slavery, he says that it is hoped that emancipation and the "legal recognition of their natural ties, will tend to diminish this unnatural amalgamation, and lessen everywhere the number of those unfortunate halfbreeds, ~~wanting~~ deficient in manliness and feminine virtue, and left to be the minister to the lust of other races . . . let me insight upon the fact, that the population arising from the amalgamation of two races is always degenerate" (Agassiz 1863b, no. 150). Thus the mixed-race subject is positioned both before and after by their degeneracy, as both subject to and product of the sexual violence of whiteness. Imagined as queer, the porosity of race creates uncontained sexuate substances. To the third question, Agassiz responds, imagining a geophysical spatiality of race: "The whites inhabit invariably the sea shores and the more elevated grounds, while the black are distributed over the lowlands. This peculiar distribution is rendered necessary by the physical constitution of the country. The lowlands are not habitable by the whites, between sunset and before sunrise. All the wealthy whites, and in the less healthy regions even the Overseers, repair in the evening to the seashore or to the woodlands and return only in the morning to the plantations" (no. 150).

Creating the idea of a low-lying inland archipelago (a rift of Blackness), Agassiz zones whiteness to the higher land (the plateau of Overseers' earth) during the hours without sun. Bizarrely, given his prior commitment to the geographic universality of whiteness, he continues that these granular geographies with a free Black population will "sooner or later become Negro States with a comparatively small white population. This is inevitable; we might as soon expect the laws of nature as to avert this result. Abolition of slavery in the Gulf and River States coupled with identity of political rights for all the inhabitants amounts to giving over those States to the negro race" (1863b, no. 150).

Agassiz cautions: "Remember that in destroying the power of a slave-aristocracy, you initiate the *reign of a black democracy*" (1863b, no. 150; my italics). Even though this contradicts his earlier claims about Negro self-organizing and governance, he points to Haiti and Liberia and says this may work well, but that Howe should be under no illusion that this is what to expect: "Do not overlook the fact that as soon as slavery ceases in the Cotton States proper, the reign of the Negro begins, to become, in the end, supreme." His futurology is framed in epochal terms. He continues that if political rights are not restricted for the Negro, then there will be a "white exodus" to the North (no. 150). Invoking natural causality, he writes, "Social equality I deem at all times impracticable. It is a natural impossibility" (Agassiz to Howe, August 10, 1863, in Agassiz 1863b, no. 152). Thus, the geologist's ground naturalizes racialization as nature itself seemingly upholds the desire for a white planet. Such is the fear of racial accession that Agassiz

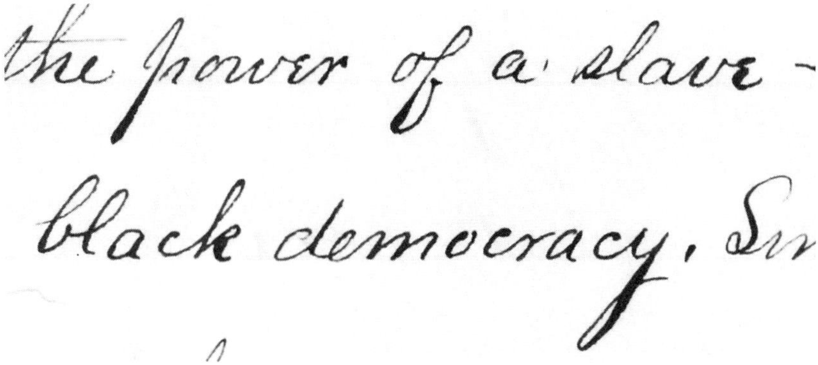

the power of a slave -
black democracy, Sir

FIGURE 4.1 Black democracy: extract from Louis Agassiz's August 10, 1863, letter to Samuel Gridley Howe in the context of the Emancipation Proclamation of 1863. Agassiz 1863b, no. 150.

becomes caught in his own spurious and contradictory geo-logics. Natural causation is a tricky thing when the rift is seemingly accommodating Black democracy. To hide the fact that narrativizing nature requires a father's subjugating hand, he seeks to enlarge the stature of his authority further: "There is hardly a more complicated subject in physiology, requiring *nicer discriminations*, than that of the multiplications of Man; and yet it is constantly acted upon as if it required no special knowledge" (Agassiz 1863b, no. 150; my italics). Mobilized by fear and foreboding for white futurity, Agassiz uses his authority and faux modesty to command his racial politics. For Agassiz (like Cuvier), the terms of degradation are configured around questions of sexuality and gender, simultaneously imagined in geologic and political time (whereby geology can answer political questions of race and political subjugation can be naturalized as geomorphic determination). He says to Howe:

> Conceive for a moment the difference it would make in future ages for the prospect of republican institutions and our civilizations generally, if instead of the manly population descended from cognate nations, the United Stated should hereafter be inhabited by the effeminate progeny of mixed races, half Indian, half negro, sprinkled with white blood. In whatever proportion the amalgamation may take place, I shudder at the consequences. We have already to *struggle, in our progress, against the influence of universal equality*, in consequence of the difficulty of preserving the acquisitions of individual eminence, the wealth of refinement and culture growing out of select associations. (no. 150; my italics)

The question of reproduction is the most salient eugenic point for Agassiz: "How shall we eradicate the stigma of a lower race, when its blood has once been allowed to flow freely into that of our children?" (1863b, no. 150) while the "wealth of refinement" configures the terms of accumulation of value under settler colonial materialisms (this is the "why" of the question). The question of "practical miscegenation" made manifest an intermediary who blurs the script between matter (coded Black and as inhuman property) and Life (coded white and as Overseer of property), the amalgam that queries and queers the material state of political personhood as it pertained to the construction of the settler nation-state.

The inhuman-Life problem that arose from colonialist materialist scripts (and their regulation through gendered social-sexual formations) arose from the conflicts over the materiality of geopower. Categorization on the inhuman-Life spectrum sought to achieve a clean extraction that could disassociate itself from its "excesses" (subjugation in subterranean racial rifts) by aestheticizing that violence through reallocation to a different symbolic realm (I think here of the decorative silver sugar bowls in English country houses held aloft by the figurines of small Black children—*aesthetics hides the hurt*). The antitype of human—the inhuman—rather than being simply a dialectical inversion of power, performs a material spatializing function, pushing and projecting forms of material and psychic intimacy outside of the white subject to secure value, agency, and geopower. Amalgamation is decidedly queer and feared as the destabilization of categories stabilized through paleontological description, which created the conditions for the genealogical accumulation of value (geopower) in the name of white supremacy.

While Agassiz is careful not to seem to support enslavement (given Howe's position), he says he rejoices at the prospect of emancipation and being "able to discuss the question of the races and to advocate a *discriminating policy respecting them*, without seeming to support legal inequality" (my italics), stating that equality is a geologic impossibility: "Let us therefore for a moment consider the natural endowments of the negro race, as they are manifested in history, or their native continent, from the beginning of the existence of the race itself" (1863b, no. 150). In a loop of self-justification, the origination binds and fathers the futurity of Black and Indigenous life. If the natural geo-logics of inequality are not enough to persuade, Agassiz says at least fear should motivate the denial of freedom, as "the black population is likely at all times to outnumber the white in the Southern States. We might therefore beware how we give the blacks rights by virtue of which they may endanger the progress of the whites" (no. 150). In Agassiz's simple equation, Black progression is equal to white regression, plateau uplift is coterminous with the force of subduction. He continues to rehearse the well-worn tropes of infantilization of racialized subjects, describing childlike qualities as if on a bad report card: "indolent, playful, sensual, imitative, subservient, good-natured, versatile, unsteady in their purpose, devoted, affectionate in everything unlike other races, they may but be compared to children, grown into the stature of adults, while retaining a childlike mind" that is enlarged to a continental scale in the comparisons of empires, and the underlined conclusion, "never originated a regular organization among themselves" (no. 150). He says that slavery is not a necessary condition per se, but "they are incapable of living on a footing of social equality with the whites, in one and the same community, without becoming an element of social disorder" (no. 150). The *elemental condition* of Black subjectivity is made, and Agassiz's paternalism suggests instead the granting of "instalments" of freedom. Returning again to the question of sex, in his second letter to Howe, Agassiz cautions that

> the degrading influence which intercourse between individuals of different races necessarily has upon ~~both~~, each, has inevitably reduced such intercourse to a mere physical act. But the consequences are still worse if we consider its influences upon the community . . . would introduce a new element of discord in society in consequence of the disparity of their antecedent associations. We all know how unfavourably ill-assorted unions influence the family circle. And yet in such unions the difference which is mostly one of social position only, may be also of temperament. What could we expect that should be advantageous to society in the widest acceptation of the term, if to these antagonisms must be added the difference arising from a diversity of race? (no. 150)

The elemental genesis of race returns to trouble the social order and disturb white heteropatriarchy; its antagonisms render geology a social script of sexuate and community relations, whereby "antecedent associations" threaten the transmutable "family" of Man and the productive force of whiteness, smuggling in "savage life, under the cover of civilization." What threatens white masculinity organizes political subjectivity and its subversion. Agassiz writes a long postscript to Howe to try and deal with his illogical geo-logics, given that it is the white population that is the "authoring" source, fathering miscegenation. In the patriarchal vein of Cuvier, he routes his explanation of white men going in the wrong "direction" in the axis of genealogical lineage firmly between the legs of the imagined sexual excess of Black women:

> P.S. You may perhaps ask how it is that the half breed population is so large in the U.S. if intercourse between the white and black is unnatural. . . . A glance at the conditions under which this takes place may suffice to settle this point. As soon as sexual desires are awakening in the young men of the South, they find it easy to gratify themselves by the readiness with which they are met by colored house servants.[44] There is no such restraint upon the early passions as exists—everywhere in those communities in which both sexes are legally upon a footing of equality. The first gratification under the pressure of? [illegible] a stimulus as the advantages accruing to the family negress from the connection with young masters, already blunts his better instincts *in that direction* and leads him gradually to seek more "spicy partners," as I have heard the full blacks called by the fast young men. (1863b, no. 150; my italics)

The hypersexualization of Black women is thus organized as the site of disciplining, while the rape of "house servants" is legitimated by the naturalized "desire" of young white men and the pressure of their gratification that "blunts [their] better instincts." Again, the sexual stimulus of Black women is organized as a site in the regression of white political progress, whereby the wrong sex desediments the fossilization of race.

In the context of Agassiz's Southern fieldwork, white patriarchal geology interned Black corporeality in both symbolic and actual terms: first, through science in the production of an epistemological framework to legitimate patriarchal substitution (of genealogical relation for property relation) and the theft of personhood (through a practice of genealogical alienation and paleo-objecthood); second, through paleontology's associative work of genealogical placement of the category of the Negro proximate to the orangutan which established affective alignments, where the visual structure of address or genre is what actually produces "likeness"; and third, in the denial of kinship ties and the rendering of that relation as one of property under industrial slavery, while simultaneously exalting the name of the property owner (as plantation owner or scientist) as the only one entitled to selectively and symbolically recognize kin in time. The point here is that geologic time was the ground for imaginaries of reproductive futurity and the white settler nation-state. Control of the narrative of linear descent (genealogy) operationalized horizontal "elemental" property gains (inhuman and inhumane). Agassiz used the temporal seduction of deep time (and his professional stature within it) as a lure to stake a claim on white futurity. The claim reveals the ruse of time. Deep time as racial privilege and entitlement to material resource. Deep time as the "natural" law of white supremacy and patriarchal violence. Time does the work of holding scientific and mythic inconsistences together, yoking them to the temporal expectation of colonialism as progression.

Fathering Subjects

In the context of the collaborative work in geology between *geomythos* and *materiality*, Agassiz continued to be instrumental in the production of racialized environments within the United States and transnationally. Like Cuvier and the British geologist Lyell, Agassiz participated in the signification and subjectification of Black and Indigenous life, establishing visual genres of race through representational work on the geologic color line at Benjamin Franklin Taylor's plantation in Edgehill, outside Columbia, South Carolina. The infamous daguerreotypes he commissioned of enslaved persons—subjects whom enslavers had named Delia, Jack, Renty, and Drana, from Taylor's plantation, and Jem, Alfred, and Fassena, who were "borrowed" from nearby plantations—were made in 1850 by photographer J. T. Zealy to support Agassiz's racial theories through a visual taxonomy (see Cheddie 2016). There are fourteen contested images. Agassiz's intent was photographic taxa that were used to "read" the Negro type. The five men (two naked) and two women (stripped to the waist) were the first designated portraits of enslaved persons that were individually "named" rather than assigned a "type." Established through the physiological coding of difference, the subtle and affective arts of dehumanization place the sitters in their propertied relation (made through the removal of their clothes and the objectification of their bodies). In this visual act Agassiz originated the genre of photographs of African American subjects as racial segregation through the narrative of temporal dispossession. Tamara Lanier, a descendant of one of the enslaved persons, Renty, who is suing Harvard University for possession of the photographs, has spoken of the "inhumanity and trauma" (*Tamara Lanier v. President and Fellows of Harvard College and Others* 2019, 2022). As these images continue to circulate and accrue various forms of capital, they keep open and renew (Sharpe 2010, 2) the dehumanizing property relation to the present.

The inhumanity that Agassiz propagated was twofold: he contributed to the fashioning of the category of the inhuman as both a geophysical and an inhumane subjective category, and he crafted geostories of the earth and of racialized populations that enacted dispossession as an ongoing formation. Looking at these photographs, what are we meant to see?[45] In the lookout for how subjectivity is crafted in the shadow of geology, what are the aesthetics of subjectivity that are rendered in these photographs, and how do the semiotics of racial difference condition an understanding of the ghost geologies of plantation life? In part, it can be assumed that Agassiz's photographs are intended for "nonrecognition," to police the imagined and constructed difference between races and to refuse identification. As with Baartman, the eroticism that is evidently exhibited by geologists looking at the naked other is negatively charged to rebound with shame and its reassignment to the object as abject. The movement intended is of "scientific" fascination. The subject interned within Agassiz's classificatory and coercive frame is positioned to be looked at as a subject of natural history, not of the world. The *Front View, Profile View* builds on established genres of natural history illustration (per Baartman), which later became associated with criminal mug shots and other forms of visual pathologizing (see Campt 2017). Repeating the codes of the commodification of the male and female persons-coded-as-bodies on the slave block (men signed as labor and women signed as reproduction), daguerreotype Jem and Alfred are photographed *Full Frontal, Full Back*.

The daguerreotypes are taken in the parlor of a photographic studio and are held in the velvet internment of a leather wallet. The gaze is directed away from the materiality of existence and institutional places that constituted the subjects' enslavement, materializing a body politics that was visually (and conceptually) isolated from those material circumstances and seemingly contained in the subject as eternal racial type, ordained by *natural* rather than human history. They are pictured as the erotic version of an erratic boulder, isolated, eternal signifiers. The violence that was unleashed on the bodies of the enslaved becomes theirs to carry as a racial signifier that is meant to signify across geologic temporalities and geographic regions, as they collectively sign "Africa" (as nonaccomplishment in geologic time). Agassiz's photographs sought to establish a visual narrative of absolute difference that maintained the category of the subterranean human to rationalize a regime of looking that morally excused slavery and its afterlives. The "property" of slave owners is transmuted into a property of science via Agassiz's photographs and discourses on the relation between anatomy and geology.

Agassiz's attempt to heighten the affectual epistemological difference in how Black bodies were visualized is made in the context of previously discussed anxiety over the threat of miscegenation in the context of plantation sexual economies. The portraits of the enslaved functioned to maintain their imagined parallel contrast—the "civilized" white lineage of the privileged race. The hypervisibility of Blackness is made to mask the coeval force of whiteness—setting the standard for the normalcy of vision that continues to inform ways of seeing—a coercive force that strips and objectifies the Black body to structure white identity, where the Black body assigned a preexistent fungibility to alleviate the intimacy of whiteness with this violence through its projection. The photographs and Agassiz's discourse of the fixity of race participate in a visual culture of "fathering" (producing the subject as underdeveloped), underpinned with a fear of (and desire for) Black corporeality. Repeating the propertied relation of the plantation, these photographs participate in authorizing and validating voyeurism through a scientific lens (as with Cuvier's), while simultaneously containing it in a propertied relation that reproduces the sensual quake of those deemed inhuman. The visual degradation of nakedness and partial clothing in the context of the "civilized" space of the photography studio projects debasement onto the enslaved rather than on the white viewer. In the context of wider discourses in paleontology, the muscularity of bodies of the two naked men is read in other texts as fixed anatomy next to that of simians, rather than as a product of the labors of enslavement. These bodies that affect a tensile strength are cognitively placed next to the bodies of "civilized" white families arranged at ease in the leisured space of privilege and self-regard in the studio.

Carrie Mae Weems's work *From Here I Saw What Happened and I Cried* (1995–96) illegally (according to the subsequent libel case bought by Harvard's Peabody Museum) rephotographed Agassiz's daguerreotypes from the archive, printing them in red and imposing the constructed categories across the photographs *An Anthropological Debate* (Jack); *A Negroid Type* (Renty); *You Became a Scientific Profile* (Drana); and *A Photographic Subject* (Delia). Highlighting how these images both shaped and supported racism, Weems's installation consists of appropriated photographs of enslaved persons from the American South and other images found in the museum's archives of "Africans." Weems suggests that when "we're looking at these images . . . we're looking at the ways in which Anglo America—white America—saw itself in relationship to the black subject. I wanted to intervene in that by giving a voice to a subject that historically has had no voice."[46] By making visible the text that interprets and fossilizes subjectivity in this context, Weems forces the confrontation between categories that underwrite racialization through their overwriting, rather than through a claim to an impossible restoration on behalf of those subjects.

While the libidinal is undoubtedly a current that runs through these photographs, claims to the innocence of a lithic scientific gaze (alongside so many "unnamed" archives of subjects) assume the division of science, plantation economics, and the erotic collector.[47] Per Hartman, the libidinal is structurally integral to the economic and social practices of chattel slavery, and this was science conducted within that frame and field, literally on a plantation in the South. Similarly, the separation of the libidinal investments of science in inhumane subjectivities as a consequence and extension of mine and plantation economics ignores how the geomythic and material collaboratively function in the political geologies of state and patriarchal power formations. The racial circumscription of those represented is achieved by displacing the *actual* field of occurrence—in the labor of sugar, cotton, mine, racial exhibition—into a *temporal* field of geologic reasoning; this displacement of geologic life predates and prepares the ground for the modern state and its governance of Blackness as subjectivity and resource. Constructed through the grammars of geology that convert space into time, the placement of the human in color-coded hierarchies that equate to stratal positions is substantive both to nation-building in the Americas through transatlantic slavery and to indentured labor. Agassiz engaged in the effective craft of making the visual language of degeneracy and nonnormativity, thereby producing narratives of racist paternalism of the "lesser" races within a body politic that was seemingly arrested in prehistory, forever held in infantilization through a geologic embrace without the possibility of genealogical claim. And, in the context of racial slavery, "the mother's only claim—to transfer her dispossession to the child" means that there is no episteme of filiation to be claimed in women's sexuality (Hartman 2016, 166). In this exclusion of genealogy for Black subjects, Agassiz made himself a self-appointed "father" through these acts of visual inscription of subjectivity that rendered an anti-Black schema of visuality and a racialized structure of deep time.

Brazil as Racial Laboratory

As Agassiz's creationist position on the fixity of species through catastrophe was being challenged by the increasing popularity of evolutionary theories, he set off on expeditionary fieldwork in Brazil (April 1865 through July 1866) to collect empirical traces of glaciers, alongside his popular work on the geologies of race that sought to naturalize white supremacy in the United States in the aftermath of the Civil War. He used the accrual of scientific capital and visual empirics of racialized types made in Brazil to continue to argue for the dangers of a racial "experiment" gone awry in the other America—specifically, what he called his scientific "proof" of the degeneration of the races and the "pernicious effects" of racial miscegenation. He says, "Let anyone who doubts the evil of this mixture of races, and is inclined, from mistaken philanthropy, to break down the barriers between them, come to Brazil" (Agassiz and Agassiz 1868, 293). The Thayer Expedition, sponsored by the wealthy Bostonian Nathaniel Thayer, set sail from New York City for Brazil with Elizabeth Agassiz, the future philosopher William James (Agassiz's student), Walter Hunnewell (Agassiz's student), artist Jacque Burkhardt, geologist Charles Frederick Hartt (professor at Cornell University), and several other assistants.[48] The expedition returned with over eighty thousand species for the new Museum of Comparative Zoology at Harvard (now the Peabody Museum). Traveling more than two thousand miles down the Amazon, Agassiz collected fish, parrots, deer, monkeys, and a sloth, among other things, to build an empirical basis for his theoretical work to counter ideas of transmutation (Agassiz and Agassiz 1868, 361). The principal scientific rationale was to study the metamorphoses of the fishes of the Amazon and the drift phenomena of the Andes, alongside other geopolitical and geophysical briefs to campaign for the internationalization of the Amazon.[49] Alongside his "empirical" race work, Agassiz had also come to Brazil to promote the idea of a transatlantic Black "southern," rather than northern, immigration after abolition.[50] Agassiz collected visual "specimens" for his human origins work. Understanding the power of representation from his plantation photographs in 1850, he set up a photographic studio in the town of Manos, Brazil, where he searched for "subjects" to document, photographing forty to fifty people who embodied his stratified understanding of race and its visual typologies (Isaac 1997, 6). He employed the photographer Augusto Stahl and made over two hundred images to compile an inventory of "types" of African and "mestizo" populations, with the express purpose of demonstrating through the visual method the mongrelization of racial purity (which we are asked to imagine through a comparative elemental genesis of subjectivity, parallel to that of mineralization).

The photographs studiously follow the layout of comparative anatomy set up by Cuvier, of front, side, back. His captions on the photographs index the subjects of his concerns: "sexual dimorphism," "comparative analysis of women's breasts and inguinal region," "difference in physique and emotional character of 'hybrids.'" These photographs are recorded in the Peabody archive under the titles *Portrait of a Woman*, *Portrait of Two*

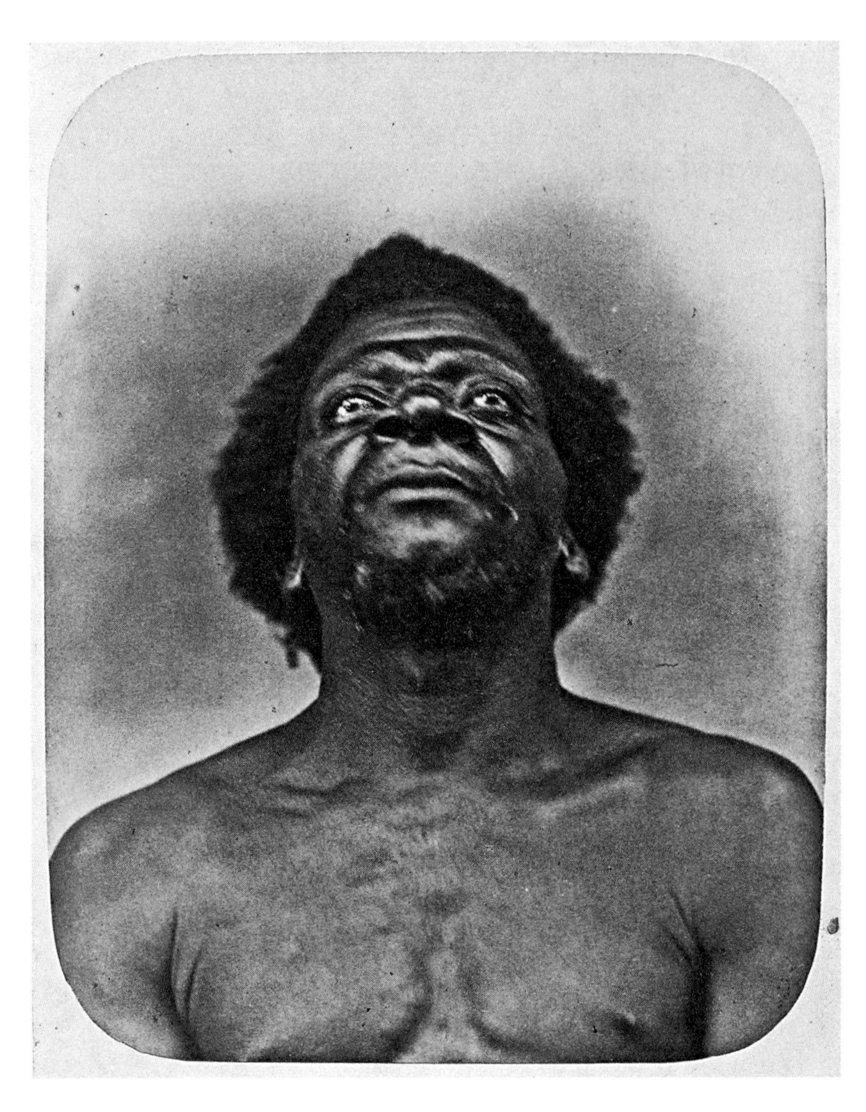

FIGURE 4.2 *Frontal Portrait of an Unknown Man*. Louis Agassiz Photographic Collection. Portrait of a racial type. Photographer Walter Hunnewell, Brazil, 1865. Courtesy of the Peabody Museum of Archaeology and Ethnology, Harvard University, Cambridge, MA, 2004.1.436.1.74.

FIGURE 4.3 *Frontal Portrait of an Unknown Woman*. Louis Agassiz Photographic Collection. Portrait of a racial type. Photographer Walter Hunnewell, Brazil, 1865. Courtesy of the Peabody Museum of Archaeology and Ethnology, Harvard University, Cambridge, MA, 2004.1.436.1.97.

Men, Nude, Full Profile [Image Not Available], Portrait of a Woman, Partially Nude, Frontal View, Seated, Portrait of Two Men, Nude, Full Back View [Image Not Available], Portrait of a Woman, Nude, Full Back View [Image Not Available], and so on. Agassiz persuaded young women to remove their clothes so that they would be unadorned by clothing and thus rendered as noncivilized—savage—in the meaning and signification of race. In the social context of questions of origination, as Darwin was preparing *Origins of the Species* in England, Agassiz sorted the affective persuasion of the visual—as a form of geologification—to illustrate his ideas of differing and fixed lines of descent in the narrative of species and consolidated his ideas of the purity of the single origin of whiteness. Whereby, he argued that static geology punctuated by catastrophe contested the various "transmutation theories," and, applied to race, it both justified and exonerated the violent policing of whiteness.

Enrolled in the empirical pedagogy of Agassiz and seemingly inured to her part in enabling dehumanization of the subjects photographed for her husband's comparative anatomy series, Elizabeth Agassiz observed, "The grand difficulty is found in the prejudices of the people themselves.[51] There is a prevalent superstition among the Indians and Negroes that a portrait absorbs into itself something of the vitality of the sitter" (Agassiz and Agassiz 1868, 276–77). The omission—narrated as prejudice—is the resistance to being stripped naked by her husband, who is under the patronage of Pedro II, the emperor of Brazil.[52] William James in his 1865 diary is a little more forthright but no less conceited in his understanding of consent: "On entering the room found Prof. engaged in cajoling three mocas whom he called pure Indians. . . . They were nicely dressed in white muslin and jewellery with flowers in their hair and an excellent smell of pripioca. Apparently refined, not at all sluttish, they consented to the utmost liberties being taken with them and two without much trouble were induced to strip and pose naked" (see W. James 2006, 88).[53]

In their jointly published account, the Agassizes note, "She consented yesterday, after a good deal of coy demur, to have her portrait taken. . . . Mr. Agassiz wanted it especially on account of her extraordinary hair" (Agassiz and Agassiz 1868, 309). The violent question of "consent" and its cloaking in "coy demur" recalls Sharpe's observation of the "double status of subjectification" (2010, 83) and Hartman's "double bind of agency" (1996, 538) whereby consent seemingly gives agency while the very act around which it is marshaled takes it away. As Sharpe argues, "The interpellation of these subjects/objects is, at base, the interpellation of *the* subject" (83). The very availability of subjects to Agassiz (based on Emperor Pedro II's patronage) and his symbolic claims of property made on racialized subjects enlarged his claim to be able to articulate authority in the discourse of the properties of race (or the fallacy of phenotypical expressions understood as a social hierarchy). Agassiz makes several detailed observations about "Indian" and "Negro" women's breasts (objectified and visually dismembered) as he argued that such gender differences testified to the higher bearing of white womanhood. In many of the photographs, Indigenous women are pictured in a state of disrobement, such that their top is bare, and their bottom half dressed. The visual effect alone of stripping communicates the violation. In justification of the consumption of his own hungry "curiosity," and in deference to his inability to perceive agency in others, he writes, "For is not the readiness to receive new impressions, to be surprised, delighted, moved, one of the great gifts of the white race, as different from the impassiveness of the Indian as their varying complexion from the dark skin, which knows neither blush nor pallor?" (Agassiz and Agassiz 1868, 178).[54] What he cannot see is deemed not to be there; white innocence is protected as it shielded and deflected its violations.

Agassiz writes: "The Indian woman has a very masculine air, extending indeed more or less, to her whole bearing, for her features have rarely the feminine delicacy of higher womanhood" (Agassiz and Agassiz 1868, 530).[55] To visualize this higher white womanhood in the photographic albums held at the Museum of Comparative Zoology, he interspersed stereoscopic cards of classical statues of Greco-Roman women. Indian and Negro women were visualized next to species of monkeys and apes, at once desexing through gender nonconformity and hypersexualizing their bodies through atomization of sexual organs. In the episteme of racial production, inference is made via visual association.

Genealogy and Geology

It is important to join up these race/species questions with materialist land theft and the development of natural resources, precisely because the theory of the earth was rooted in a conversant discourse of genealogy (of earth and humans) and white supremacy. These dynamics of colonial afterlives have persisted. The explicit conjoining of the genealogical and geographic is why geology was so popular in America as a praxis of settler colonialism. Enslavers (property owners of human beings) considered themselves experts on both geology and race (not least because of their role in the social production of miscegenation through rape and enforced "breeding" on plantations), and thus they saw themselves as part of a populist movement, transforming both environments and racial populations. The time required for geologic study was the time taken from enslaved people, and the increase in knowledge of geologic processes led to further accumulation of "property" on plantations. To study time was to "own" time. In Agassiz's words, "The time has come when scientific truth must cease to be the property of the few, when it must be woven into the common life of the world: for we have reached a point where the results of science touch the very problem of existence" (1874, 42). While geologification of race, and theories of miscegenation that emphasized degeneration, perhaps made it easier for enslavers to sell their children on the slave block, the "common life" of popular geology sedimented the *time* of race as an existential question.

As Agassiz joined together settler colonialism, natural resources, and race to explicate an argument for necropolitical extraction through the paradigm of geologic development, racial management was made essential to the generation of value in colonial material economies and environments:

> Two things are strongly impressed on the mind of the traveller in the Upper Amazons. The necessity, in the first place, of a larger population, and, secondly of a better class of whites, before any fair beginning can be made in *developing the resources of the country*. . . . Not only is the white population too small for the task before it. . . . It presents the singular spectacle of a higher race receiving the impress of a lower one, of an educated class adopting the habits and sinking to the level of the savage. In the towns of the Solimoens the people who pass for the white gentry of the land, while they profit by the ignorance of the Indian to cheat and abuse him, nevertheless adopt his social habits. . . . Although it is forbidden in law to enslave the Indian, there is a practical slavery by which he becomes as absolutely in the power of the master as if he could be brought and sold. . . . Besides this virtual slavery, an actual traffic of the Indians does go on. (Agassiz and Agassiz 1868, 139–40; my italics)

Describing the debt relation of race that was established through ongoing deceptions in the labor practices that paralleled slavery, Agassiz draws on Brazil to do comparative geopolitics through comparative anatomy. He argues that Brazil has "the same story of amalgamation of race; but here this mixture of races seems to have had a much more unfavorable influence on the physical development than in the United States. It is as if all clearness of type had been blurred, and the result is a vague compound lacking character and expression" (Agassiz and Agassiz 1868, 167). His political intent and geologic race lessons are clear: racial mixing in Brazil obstructs the physical development of subjects and land.

The origination of humanity was the scientific scene, but the parallel political one was of the maintenance and increase of white supremacy in the blur of racial difference. The problem of purity remained a permanent issue of concern in the fabulation of the lithic and liminal. Agassiz's arguments about the ruptures in genetic connections between species was a means to justify white creationism as distinct from that of other races. He argued that race had a geography and a geology, and that species of hominids or races were connected to and conditioned by their habitat (1850a, 1850b, 1854). Only the Caucasian man had overcome these environmental determinisms of "zoological provinces" or zones of creation. Aware of the aesthetic power of empiricism in photography and sketches, Agassiz sought to both describe and visualize the "mongrelization" of racial types in Brazil: "The study of the mixture of human races in this region has also occupied me much, and I have procured numerous photographs of all the types which I have been able to observe. The principal result at which I have arrived is, that the races bear themselves towards each other as do distinct species" (Agassiz and Agassiz 1868, 383). The *difference* of speciation is made through the photographic type. While race was footnoted in scientific histories of geology, the inhuman and inhumane were always interpolated in paleosocial thought on life-forms and geomorphology that sought to provide scientific answers to political problems.

As much as Agassiz was concerned to use Brazil as a biological laboratory to prove transmutation false, he was equally concerned to see it as a racial laboratory to warn about the possible futurity of the United States.[56] In the promotion of degeneration theory, Agassiz not only addressed popular media, social texts, and cultural actors but was actively involved in seeking to influence political decision-making. Because the geologic record was seen to establish the immutable record of world order, so it included, for Agassiz and other geologists, the twinned questions of human origination and earth theory. As Agassiz was dubbed the "founding father" of natural history in America, his role as popular political figure and showman demonstrated how theories of the earth, geology, and ontology were knotted. Explicitly, he materially constructed questions of anthropogenesis as crucial to the political narrative of the present and the imaginary of the future, including ceding the concept of Brazil as a suitable country of "return" for Black Southern populations after emancipation (dubbed "the New South"). He emphasizes that the question of how many points of origination exist for hominids is incidental to the recognition of the *difference* of human difference, and if the recognition of human difference could be established, then a species boundary could be erected around the properties and property of whiteness as the apex of types of difference.[57] Agassiz and Agassiz explain:

> However naturalists may differ respecting the origin of species, there is at least one point on which they agree, namely, that the offspring from two so-called different species is a being intermediate between them, sharing the peculiar features of both parents, but resembling neither so closely as to be mistaken for a pure representative of the one or the other. I hold this fact to be of the utmost importance in estimating the value and meaning of the differences observed between so-called human races. I leave aside the question of their probable origin, and even that of their number; for my purpose, it does not matter whether there are three, four, five, or twenty human races, and whether they originated independently from one another or not. The fact that they differ by constant permanent features is in itself sufficient to justify a comparison between human races and animal species. (1868, 298)

The legacy of European geologists in the Americas (and other settler and colonial nations) was the perpetuation and sedimentation of racial difference, regardless of the nuances and contestation of the actual theories of human and earth evolution. Argumentation of comparative states (of human and earth) was made by assertions of resemblances rather than actual fossil empirics, thereby giving aesthetics a crucial role in the geologics of racecraft. While biographies celebrate the foci of clashing personalities and elevate the white father figures of geology (a practice that finds its echo in today's geosciences departments), this dissuades attention from the broader systemic institutionalization of race that was made in institutions, such as Harvard, as a material, economic, and social differential—and the legacies of subjection or geotrauma that are based on an enduring historic conjecture about the normativity of certain natural histories.

It is not so much the explicit racial work that was done by leading geologists, in their popular accounts of rocks and slavery, or letters to presidents and newspapers, nor even the founding of their institutions in part through the payment for the enslaved (The Geological Society, London); rather, the sticky residue of racialized temporalities in the becoming of the material world and its grammars of materiality have had an imaginative veracity that continued to serve as an oppressive geopower. This geographical imagination of temporal difference—that is still perpetuated in ideas of white supremacy, development, modernity, and so forth—is the legacy of paleontology and its wider geophilosophy of matter. While great discussion is given to the differences and debate in the American school of paleoethnology—both mono- and polygenists maintained the tenet of the apotheosis of supremacy—the attachment to the production of white supremacy cannot be underestimated (in both its human and inhuman geosocial forms). It was a driving force in the restless maintenance and naturalization of racial difference. Even the transmutationists, later evolutionists, could not quite let go of the lure of the developmental framework that continued to shelter the idea of supremacy. Which is to say, race, as a long-disproved fiction, still functions *as if* it is a natural law because it regulates the organization of natural resources (and thus accumulation and value), as well as absorbing the drive of white belonging (as overcompensation), as it continues to proceed via dispossession of Indigenous, Brown, and Black life.

The reason this origin story of Agassiz and his populist and institutional power matters to understanding material geographies is because it created and perpetuated how the inhuman functioned as a subjective category (beyond the figure of a subject). The perpetual imaginary of racial difference understood as a phenotypical phenomenon (i.e., aesthetically identifiable) and as a condition of resource extraction is the legacy of the *geologic paternal*.

"Inhuman Dicta": The *Dred Scott* Decision

In 1857, the US Supreme Court constitutionally sanctioned racism in the decision of the case *Dred Scott, Plaintiff in error v. John F. A. Sandford* (December Term, 1856), finding that Black Americans, whether free or enslaved, were not intended to be included under the term *citizen* as employed in the Declaration of Independence, and Native Americans were to be considered in a political caesura of "foreign" nations. The terms of the dispossession were argued through racialized accounts of genealogy within a geographic and "natural" history of belonging, collating a spatialized rift placement for Blackness outside the state but within its borders. Indigenous political life was considered an alien state within. Both territorial and ontological states within states denoted a geophysical change of state. These spatial placements of race show how geologic narratives joined up with the juridical to form the plasticity of the plateau and its gravities.

By excluding the "negro" from the category of "person" set out in the Constitution, Judge Roger Taney thereby justified Black people's continued exclusion from consideration within the category of the human. It was dubbed

FIGURE 4.4 A statue of Louis Agassiz after it fell from its plinth following an earthquake (ca. 1906), the head of the statue embedded in the paving in front of the Zoology Building at Stanford University, Palo Alto, CA. https://www.loc.gov/item/2013651564/.

an "inhuman judgement" that sedimented Blackness back into its already oscillating material and conceptual proximity with inhumane subjection. The judgment described a placement that was the juridical equivalent of what McKittrick calls "the uninhabitable" (2013, 6). In designating Black people outside citizenship—"property of," not "of the family of" man—*Dred Scott v. Sandford* places an account of Black sovereignty in the stratal rift below the plateau of whiteness, suspended in a spatial and temporal limbo between Africa and America (and arrival at the surface present of the political). It was not so much a biopolitics of exclusion, in Foucauldian terms, as it was the confirmation of a submergence into another ontological-material state of being, and thus an expulsion from political agency in the present. According to the 1857 *Dred Scott* decision, Black persons remained property whether they were owned or free because they were held, according to the juridical judgment, in a condemned status that was recognized as unchangeable (Taney 1857). This unchangeability was narrated as an effect of the racialized *where* and the *who* of progress and advancement, of forms of life compressed in the strata as fossilized subjectivity, and Life conceived through whiteness. The central question that was posed and framed by the *Dred Scott* case was the following: "Can a negro, whose ancestors were imported into this country, and sold as slaves, become a member of the political community formed and brought into existence by the Constitution of the United States, and as such become entitled to all the rights, and privileges, and immunities, guarantied by that instrument to the citizen? One of which rights is the privilege of suing in a court of the United States in the cases specified in the Constitution" (Taney in United States Supreme Court et al. 1860, 22).

Such a question gave rise to the immutable or ontologically otherwise status of the slave as a category of nature argued through narratives of natural history and the arrangements of racial difference by paleontology (see Agassiz 1850a, 1850b, 1850c, 1850d, 1857; Nott and Gliddon 1854). The decision temporally cast Blackness into an atemporal abyss, outside the state (spatial and subjective erasure), while simultaneously organizing Indigenous life in a state of "pupilage" (spatial and temporal zoning). It recorded that "Indian Governments were regarded and treated as foreign Governments, as much so as if an ocean had separated the red man from the white. . . . The course of events has brought the Indian tribes within the limits of the United States under subjection to the white race; and it has been found necessary, for their sake as well as our own, to regard them as in a state of pupilage, and to legislate to a certain extent over them and the territory they occupy" (Taney in United States Supreme Court et al. 1860, 24). Employing a tectonic imaginary, the judgment continues to place Native Americans in a spatial limbo that designated that "if an individual should leave his nation or tribe, and take up his abode among the white population, he would be entitled to all the rights and privileges which would belong to an emigrant from any other foreign people" (24).

Thus, race became a division of the juridical domains of freedom via a temporal species placement that had spatial effects. The judgment demonstrates most clearly how Agassiz's ideas of inhumanism are transmuted onto persons-as-property. Addressing the Universal Rights of Man, declared in the Constitution, the court brings Kant's qualifying *universal-but* into play: "The general words above quoted would seem to embrace the whole human family, and if they were used in a similar instrument at this day would be so understood. But it is too clear for dispute, that the enslaved African race were not intended to be included" (United States Supreme Court et al. 1860, 49). It continues: "The unhappy black race were separated from the white by indelible marks, and laws long before established, and were never thought of or spoken of except as property, and when the claims of the owner or the profit of the trader were supposed to need protection" (50). The claim of the rights of property and natural laws are used together to assert an uncrossable breach of racial difference. In the imagination of geology, an abyssal rift.

Geopolity

Race was imagined something like casting, a form of metallurgy that shaped value and distilled "properties" of a social mantle. There is a necessity to dislodge the easy land/labor—Indigenous/Black formula (most thoughtfully problematized by King [2016a]) not least because it makes a spatial nonplace of Blackness and imposes a specific carceral cartography on Indigenous life (see Hoxie 2007). These modalities have functioned so repeatedly as placeholders for more complex spatial negotiations precisely because they were enshrined in the *Dred Scott* decision (and many more discrete pronouncements) as a form of capture and control in an elemental imagination of geosocial life.

The summation of the *Dred Scott* judgment sets out first to establish "a clear view of the nature and incidents of that particular species of property which is now in question" (Taney 1857, judgment no. 920) and then through this geologic grammar (of nature, species, natural resource) to establish the totality of an ontological condition: "The *status* of slavery embraces every condition, from that in which the slave is known to the law simply as a chattel, with no civil rights, to that in which he is recognized as a person for all purposes, save the compulsory power of directing and receiving the fruits of his labor" (no. 922). On March 12, 1857, the *Chicago Tribune*, an important Republican paper at the time, declared that Taney's statements on Black citizenship were "inhuman dicta" and that it "scarcely know[s] how to express our detestation of its inhuman dicta or fathom the wicked consequences which may flow from it." Many other public forums, including the New York State Senate, described the ruling as "inhuman." What the ruling did

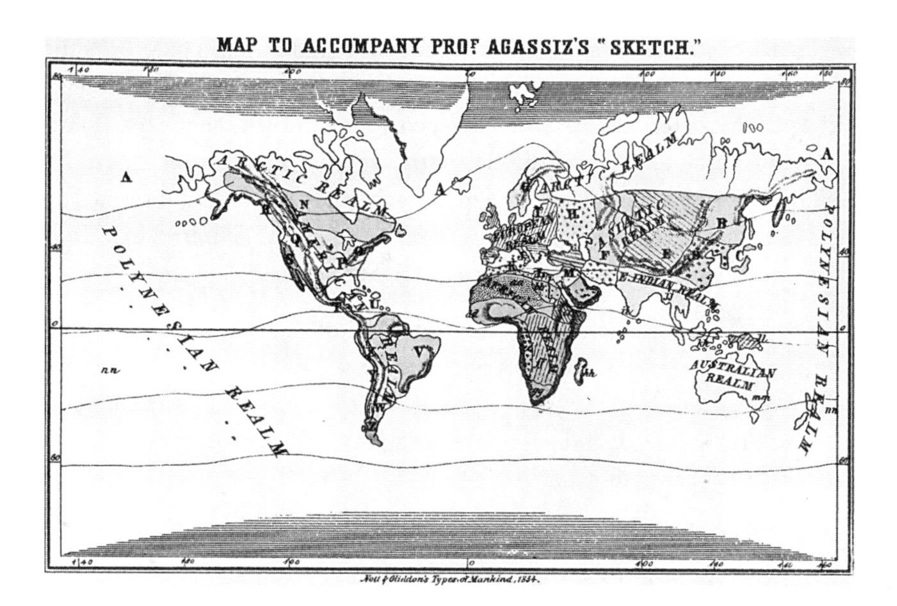

FIGURE 4.5 Map to accompany Louis Agassiz's sketch of the geographical distribution of the races in Josiah Nott and George Gliddon's *Types of Mankind* (1854).

was to make Black freedom a geographic contingency justified through its ontological placement in a state of speciation. Like the imagination of inhuman nature as fixed entity and immutable condition, "the *status* of slavery embraces every condition" (Taney 1857, no. 922).[58] Thus, slavery was enshrined not as a conditionality of life-forms but ontologically outside its juridical perspectivism, an inhuman condition beyond the law of citizens.

The judgment cemented the legacy of the dual trajectory of the inhuman, in the construction of the inhumane and inhuman as property (geocoding persons as natural resources), as it perpetuated the imaginary of racial difference and its hierarchical differentiation. It gave the inhuman classification of natural history, via geology, a juridical present through the construction of racial difference as unchangeable and thus fossilized. The stratal imagination of Agassiz is clear in the judgment. For example, the decision states: "A perpetual and impassable barrier was intended to be erected between the white race and the one which they had reduced to slavery, and governed as subjects with absolute and despotic power, and which they then looked upon as so far below them in the scale of created beings" (Taney 1857, no. 44). "The racial architecture given by paleontology not only justified and perpetuated the theft of personhood and land but also gave impetus to racial differentiation before the law, in ways that still define the struggle. Citizenship is determined, inter alia, by geography and genealogy expressed as racial difference, understood through a stratal imagination in juridical world-building. The judgment narrative expresses throughout how it is informed by lines of descent, wherein genealogy is the descriptive and definitive marker of difference that is used to differentiate the right to property. It stated:

> The defendant pleaded in abatement to the jurisdiction of the court, that the plaintiff was not a citizen of the State of Missouri, as alleged in his declaration, being a negro of African descent, whose ancestors were of pure African blood, and who were brought into this country and sold as slaves. . . . (Taney 1857, no. 7)

> It will be observed, that the plea applies to that class of persons only whose ancestors were negroes of the African race, and imported into this country, and sold and held as slaves. The only matter in issue before the court, therefore, is, whether the descendants of such slaves, when they shall be emancipated, or who are born of parents who had become free before their birth, are citizens of a State, in the sense in which the word citizen is used in the Constitution of the United States. (Taney 1857, no. 23)

The question of being in possession of a body of rights is posed in the arena of political sovereignty ("The question before us is, whether the class of persons described in the plea in abatement compose a portion of this people, and are constituent members of this sovereignty?" [Taney 1857, no. 26]) and understandings of the terms of citizenship ("The words 'people of the United States' and 'citizens' are synonymous terms" [no. 26]), but the decision is resolved on the basis of racial difference—what constitutes a person or chattel—naturalizing the terms of inhuman subjectivity:

> We think they are not, and that they are not included, and were not intended to be included, under the word "citizens" in the Constitution, and can therefore claim none of the rights and privileges which that instrument provides for and secures to citizens of the United States. On the contrary, they were at that time considered as a subordinate and inferior class of beings, who had been subjugated by the dominant race, and, whether emancipated or not, yet remained subject to their authority, and had no rights or privileges but such as those who held the power and the Government might choose to grant them. (no. 34)

The phrase "subordinate and inferior class of beings . . . subjugated by the dominant race" reproduces Agassiz's account of human subjects precisely (Agassiz accepted a post at Harvard in 1847, seven years before the *Dred Scott* ruling). Taney's arguments rehearse the *geologic paternal* established through patronage and white supremacy. The writ continues:

> In the opinion of the court, the legislation and histories of the times, and the language used in the Declaration of Independence, show, that neither the class of persons who had been imported as slaves, nor their descendants, whether they had become free or not, were then acknowledged as a part of the people, nor intended to be included in the general words used in that memorable instrument. (no. 34)

Thus, the opinion denies citizenship to African Americans on the basis of the status of the recognition of personhood not being recognized in the "memorable instrument," having

> regarded as beings of an inferior order, and altogether unfit to associate with the white race . . . the negro might justly and lawfully be reduced to slavery for his benefit. He was bought and sold, and treated as an ordinary article of merchandise and traffic, whenever a profit could be made by it. This opinion was at that time fixed and universal in the civilized portion of the white race. (no. 36)

Political sovereignty is resolved as an issue of temporal sovereignty through the interpretive work of paleontology. If we return to the geopolitical organization of race and consider the language of the judgment of *Dred Scott* as a form of rift theory, the earth is pulled in as witness to substantiate biopolitical work. Taney continues to sediment inhuman subjectivity. The decision speaks of "assum[ing] among the powers of the earth the separate and equal station to which the laws of nature and nature's God entitle them . . ." (no. 47). Thus, powers of the earth—or geopowers—are used to both name and consecrate the "natural laws" of race. And Blackness is interned in the category of property—"were never thought of or spoken of except as property" [no. 50]—and natural resource: "The right to govern may be the inevitable consequence of the right to acquire territory. *Whichever may be the source from which the power is derived, the possession of it is 'unquestionable'*" (no. 165; italics in original). Thus, the violent acquisition of territory (colonial theft) designates the right to govern (geopower). The view of inhuman nature determines the account of property: "It is necessary, first, to have a clear view of the nature and incidents of that particular species of property which is now in question" (no. 920). "Power is derived" as geopower (territory and nature), and geologic grammars define possession as the law. Normative regimes of materiality are used to concretize racial difference. The determinants of geologic life confer the recognition and representation of juridical life and the possibility of sovereignty. The result is that geo-logics of "natural" resources (inhuman) are paradigmatic of necropolitical relations (inhumane).

In the printed version of the *Dred Scott* decision, published by J. H. Van Evrie in New York, the opinion is accompanied by an appendix titled "Natural History of the Prognathous Race of Mankind" by a Dr. Sam Cartwright of New Orleans (United States Supreme Court et al. 1860). The fact that the decision is accompanied by an appendix on natural history—focused on the racist ethnology of prognathous (Cartwright 1857)—to explain the discursive work of the inhuman in the judgment confirms the tight-knit reciprocity between geophilosophy and race work. In his justification for the plurality of the races and the hierarchy of the species, in the first paragraph Cartwright draws on Cuvier for his evidential base of the lower status of the negro (United States Supreme Court et al. 1860, 44), then works in Agassiz's arguments on polygenesis: "It is not intended by the use of the term Prognathous to call in question the black man's humanity or the unity of the human races as a *genus*, but to prove that the species of the genus homo are not a unity, but a plurality, each essentially different from the others" (Cartwright in United States Supreme Court et al. 1860, 44). The denial is not of humanity, but the justification is made based on the appearance-in-time (or temporal occasion), understood through the aesthetics of anatomical development (prognathous measured facial angle and projection of lower jaw). Cartwright argues that "Negro" children and white children are both born white, and that the "Negro" infant develops toward Blackness, growing darker, thus confirming the category assertion of "the slave or knee-bending species of mankind" (Cartwright in United States Supreme Court et al. 1860, 47). On the flip side of the publication is an advertisement for the *New York Weekly Caucasian: The White Man's Paper* and a publication of Van Evrie's (1863) *Negroes and Negro "Slavery,"* subtitled *The First an Inferior Race—The Latter Its Natural Condition*. In the introduction to the opinion, Van Evrie writes:

> The doctrine of 1776, that all (white) men "are created free and equal" is universally accepted and made the basis of all our institutions, State and National, and the relations of citizenship—the rights of the individual—in short, the *status* of the dominant race, is thus defined and fixed for ever.

> But there have been doubts and uncertainties in regard to the negro. Indeed, many (perhaps most) American communities have latterly sought to include him in the ranks of citizenship, and force upon him the *status* of the superior race.
>
> This confusion is now at an end, and the Supreme Court in the Dred Scott decision, has defined the relations, and the fixed *status* of the subordinate race forever—for that decision is in accordance with the natural relations of the races, and therefore can never perish. It is based on historical and existing facts, which are indisputable. (Van Evrie in United States Supreme Court et al. 1860, iii)

Conjuring an elemental subjecthood, Van Evrie refers to the "negro element in their midst" (iv) that must be placed in "natural relation" to the white man in "a different and subordinate being and in a different and subordinate social position" (v). The geophilosophy is clear: elements must stay in an inhuman state awaiting extraction by the plateau. If Black being was ascribed in the category of metal or mineral, it was done so geologically to position that metal/mineral being as an ore that needed the praxis of whiteness to generate value through extraction (Ore: 1: *a naturally occurring mineral containing a valuable constituent* [such as metal] for which it is mined and worked. 2: *a source from which valuable matter is extracted* [Merriam-Webster Dictionary n.d.]) Being cast as a mineral meant a flattened time, barring entry into history (except by the authorship of labor and "discovery"). Arrested in time as element, not person, bonded Blackness to extractive forms of life.

Whereas Taney's ruling renders Blackness as a permanent geographic condition of the state of Africa, Agassiz's visual geo-logics of the enslaved enacted a rift spatialization, which suspended Blackness between nonrecognition and new forms of patriarchal filiation that carried the master's name.[59] The attempt to secure an aesthetics of purity around genealogical descent was an affectual mask to the context of subjugated sexual relations as the economic and libidinal means of production on the plantation and in its afterlives wherein the conquest geography is secured by narratives of patriarchal time. The scientific rendering of *verticality* and stabilization of "descent" in genealogy temporally secures whiteness against the *actual* spatial horizontality of sexualized racial relations (asymmetrical in terms of power and consent).[60] Genealogy in this context acts to consolidate modes of exclusion and exorcise symbolic purification in contradistinction to actual geographies. Agassiz, Lyell, and Cuvier all cook race and geology together in the crucible of geologic, mine, and plantation filiation, where the determinations of kinship operates along racial lines, precisely because of the anxiety of bloodlines, which results in their geofictionication. The effect of slavery continues a legacy of broken kinship that is under the weight of the normative organization of kinship as whiteness. Thus, it is not possible to separate the fiction of genealogy and its narrative of Life from the subjugation of whiteness and its property regimes built on flesh. In short, historically there is no "natural" resource without racial subjugation.

As intellectual controversies over origination and progression raged, the residual remainder of racial difference was the most enduring contribution of paleontology and its utility in operationalizing dispossession of subjective kin and geographic gifts. Living in time, not space, defined the possibilities of political subjectivity and bodily sovereignty. These ideas of different temporal states of being and their hierarchies are with us today. Racial difference is everywhere evident in the account of what it means to be a subject, within and outside of its juridical accord. The populist and practical sciences of geology secured settlement both practically (in the settlement of land) and politically (in the making of racial categories). The epistemological power of the frame of geology in making race was in accordance with the political aspiration to imbue whiteness with a reformatory power and the "right" to stolen persons and land. Geology brought together the transmutationary efficacy of geochemical and geophysical processes to build the nation-state as white, while maintaining an immutable set of geo-logics of race.

W. E. B. Du Bois and the Counterarchive

W. E. B. Du Bois's photographic album for *The Exhibit of American Negroes* at the Paris Exposition of 1900 was displayed in a building devoted to matters of "social economy." There, Du Bois exhibited his graphic diagrams of racial color lines and his photographic collections *Types of American Negroes, Georgia, U.S.A.* (volumes 1–3) and *Negro Life in Georgia, U.S.A.*[61] Directly engaging the legacy of paleontological visual epistemologies and its geocoding, the exhibition included five hundred photographs made in collaboration with his students at Atlanta University, as well as diagrammatic color charts, geographic and sociological racial maps, and a display of two hundred books written by African Americans. Although it was separated from the main US display at the exposition, *The Exhibit of American Negroes* occupied one-fourth of the total space allocated to the United States in the multinational Palace of Social Economy and Congresses, and it was viewed by an estimated fifty million people. While the mainstream US press ignored the exhibit, African American media such as the *Colored American* wrote extensively about the project, and Du Bois was awarded a gold medal for the quality of the display.[62] The introductory chart that framed Du Bois's visual sociology bore the statement "The problem of the twentieth century is the problem of the color line." He would repeat this assertion a few months later at the first Pan-African Congress in London and would use these diagrams and photographs to frame his theories in *The Souls of Black Folks* in 1903.[63]

Taking aim at the visual genres and social tableaux of *Types of Mankind* and Agassiz's and Cuvier's stratal modes of visualization, Du Bois's album moves from Black subjects from "natural" history or mug shots of Black criminality (front and side) to classic studio portraits of the white bourgeois family (sat slightly aside from the camera). The middle-class attire and appropriate grooming engaged with practices of mirroring and unsettling. As Shawn Smith argues, "These portraits as mug shots make explicit the 'shadow meanings' of white-supremacist images of African Americans" (2012, 282). Similarly, according to Smith, "white-looking" Negroes (the familiar stranger) played with white audiences' genealogical anxiety and fear of miscegenation: "Du Bois's photographs of a white-looking African American child signal both white violence upon African American bodies and white desire for the black body" (289). Such a deliberate identification navigated along the lip of a perilous rift of race, which recognizes whiteness (and its privilege), while simultaneously rebounding with the nonrecognition of the origination of violence in that genealogy. Du Bois's (1900) *Types of American Negroes*, attentive to the social play of racial types, employs the oscillation between codes of acceptability and the inhuman codes, throwing these double takes to viewers in an alternating current of recognition, identification, and destabilization. Claiming a space for middle-class Blackness against the dominant archive of criminology and racial taxa, he established a curiously elliptical visual relation, as the bourgeois white portrait and the deviant pathology are paired (and inverted along the color line that separated them). Through its investiture in normativity, the photographic archive writes back on the visual discourses of paleontological inferiority, to reconfigure the middle-class family and thus the "family of Man."

Du Bois's album oscillates between patriarchal inscriptions of the middle-class family and its undoing. The racial relocation through class politics shows us how class operates within the category of race, while the patriarchal relation is left somewhat intact. The terms of Black sexuality are partially redressed in the shuttle back and forth with normative forms.

While Du Bois's practice offers a reevaluation of identity in the reinscription of photographic genres and color codes, its transgression is still caught in the social reproduction and gendering of (neo)colonial power over recognizable kin relations and sexuality. Hartman suggests about Du Bois's later album, *The Philadelphia Negro*, that it set the precedent for how Black studies framed and oriented its problematic: "It's a way of thinking about black life as a particular kind of problem, and a problem of its deviation from bourgeois family norms and hetero-patriarchy. As if the restoration of the black patriarchy can remedy the ravages of slavery, dispossession, capitalism and white supremacy. . . . Rather than deviance and pathology, what I saw was the way in which the particular formation of black social life yielded radically different forms of intimacy and kinship and association" (2018, n.p.).[64]

While Hartman reconstitutes Black social life through the density of intimate refusal, the reconstruction of a visual paradigm of race through Du Bois's engagement with its counterfactuals shows rather than breaks with the stratigraphic relation and the aesthetics of its propertied dimensions. If the *Dred Scott* decision conflated the objecthood of slavery (property) and its conditions (properties of race) into an eternal condition of emergence (ontology), the aesthetics of that elision of race cut into affectual architectures of the visual. The photographs are *relatants*—exhibiting a set of relations and their assemblage, such that the hinges and torque of those relations become apparent, and thus the scope of their encoding—between photography's indexical signification in the memory work of other archives, of subjected and objectified paleoarchives, in which violation is counterpoised (through remembrance) with the cultural politics invested in posed family portraiture to "retrain viewers" (S. M. Smith 2004, 10) to see race differently through the lens of normalcy rather than its opposite. More than this, the photographs equalize the stratal relations by mobilizing the universal against the universal *but* . . . across the terrain of the family as archetypal, as Caribbean writers sought to mobilize a fuller humanism. Smith argues that "Du Bois' Georgia Negro images signify in critical relation to, and '*signify on*,' the scientific, eugenicist, and criminological archives that attempted to proclaim African American inferiority" (2004, 9). The doubling troubles both at the normalcy of the racialization of the middle class (in the adherence to class-based imaginaries) and at the construction of normalcy itself (cut sharply against the sexualization and criminalization of African Americans) through the co-natal production of the white middle-class family and scopic regimes of inhumanization and degeneration.

In the diagram "Proportion of Freemen and Slaves among American Negros" prepared by students from Atlanta University, a big black rock underpins industrialization, with the word *slave* across it. In another diagram, "The Amalgamation of the White and Black Elements of the Population in the United States" (1900) (Du Bois 2019, 75), the visual image is of a large black rock pinnacle on the border of a plateau of whiteness, with slivers of gradated brown and yellow (representing mulattoes, 10 percent). From 1800 (pictured at the top of the rock), the brown rift of amalgamation opens like a fault line or seam as the Negro population increases to 1890 (pictured at the bottom of the rock). These stratal modes of data visualization literally figure a visual racial formation in the social construction of natural resources and industrialization that the world's fair was set to celebrate and promote. The combined imagery depicts the understratum of white America and its "civilizing" geopowers. Progress is undergrounded by a big black block of matter with the inhuman category label "slave."

As Du Bois's Black maps chart a critical cartography of race into social and economic worlds arranged around their forgetting, the photographs simultaneously speak to the visual language of class politics and its racial dialectics.[65] The dissonance between Agassiz's and Du Bois's photographs, in their paralleled parlor scenes (which already corpse Blackness through the genres of domestic space), speaks to the aesthetic axis of the human and inhumanization to show the subjective sites of extraction through which the privileged white family draws its powers of levitation. Following Agassiz's genre conventions, Du Bois included no captions and no interpretive frame to anchor a reading of the images. In this way, the transparency of the frame of normalcy as a self-explanatory set of dispositions is rendered questionable.

Du Bois was an early visual theorist and geographer of race, imagining and imaging the "color line," "double consciousness," "the veil," and "second sight," underscoring the visual importance of the paradigms of race established by geologists in their race work. Du Bois's counterarchive reconfigured visual codes of racial taxonomy pioneered by Cuvier and Agassiz, to visualize the dialectic of the racial rift to the plateau, offering a site from which to imagine the rift as something other than rift. Whereas Agassiz's images highlighted a destitute aesthetic, to emphasize paternalism and mastery and to "naturalize" segregation and forms of sexuality, Du Bois's images are historically located in an era in which discourses around criminality (especially in various "sex panics" around white womanhood) were used to justify the violent repression of Blackness through lynching, segregation, enforced economic poverty, and social surveillance.

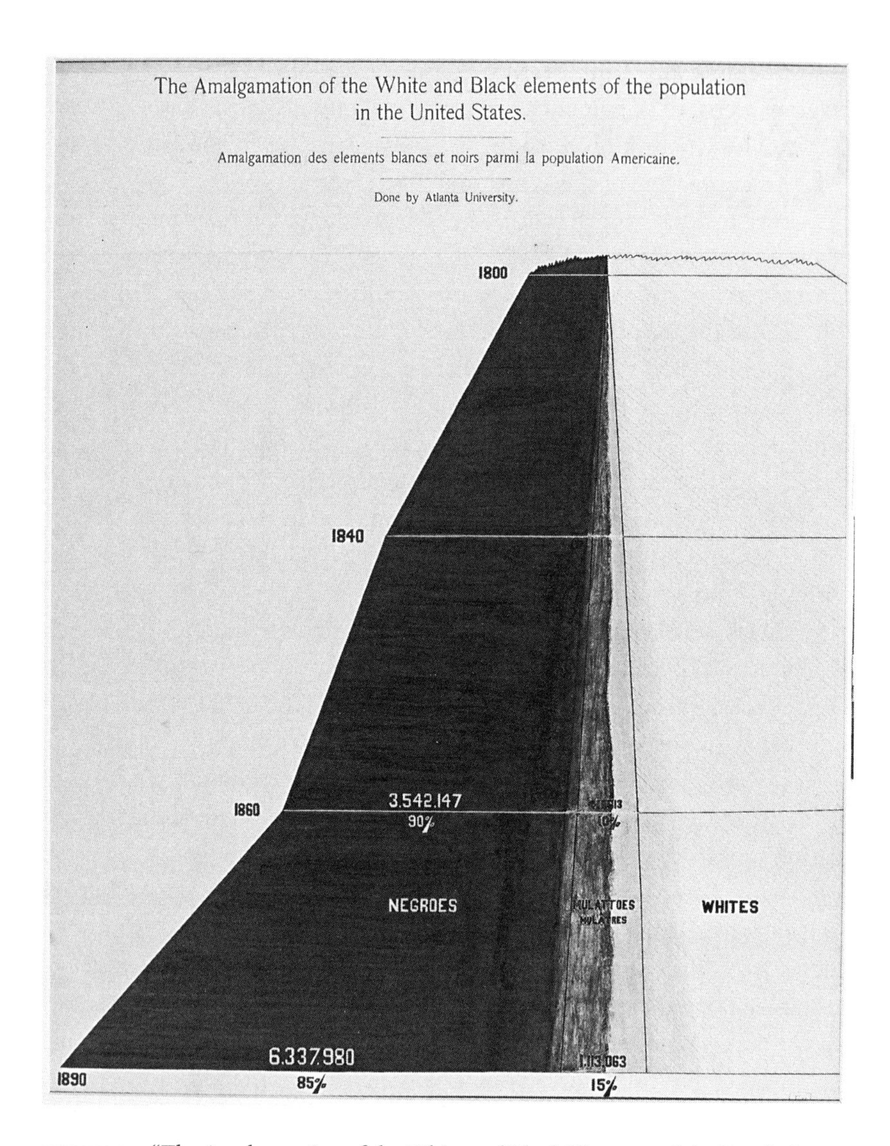

FIGURE 4.6 "The Amalgamation of the White and Black Elements of the Population in the United States." One of a series of charts based on W. E. B. Du Bois's research prepared by Atlanta University students for the Negro Exhibit of the American Section at the Paris Exposition Universelle, Paris, France, 1900. Library of Congress, Washington, DC. https://www.loc.gov/item/2014645360/.

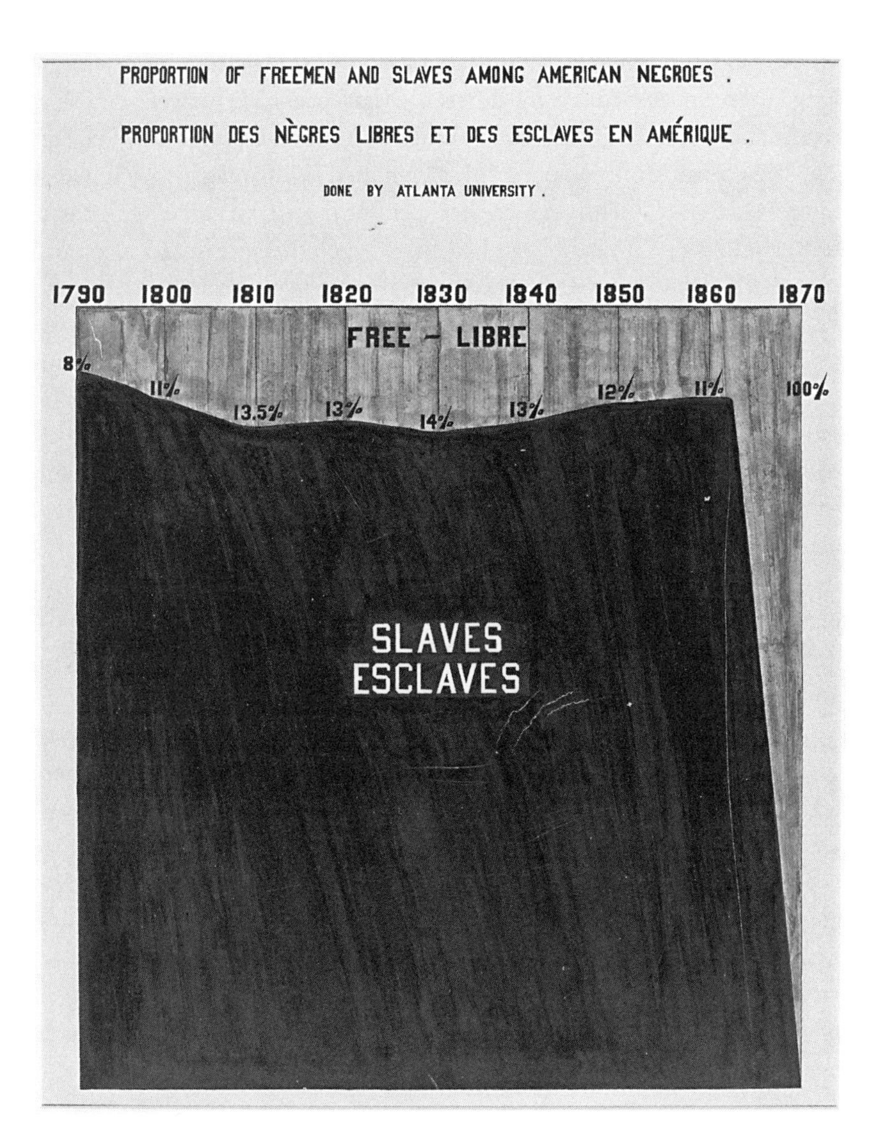

FIGURE 4.7 "Proportion of Freemen and Slaves among American Negroes" between 1790 and 1870. One of a series of charts based on W. E. B. Du Bois's research prepared by Atlanta University students for the Negro Exhibit of the American Section at the Paris Exposition Universelle, 1900. Library of Congress, Washington, DC. https://www.loc.gov/item/2014645356/.

The legacy of geologic life as it is crafted by geologists' race work in the long nineteenth century is the difference that race made between a normative genealogical tree planted in a recognizable (and representable) lineage and the corpsing of Blackness "being lynched in its image, or lynched as image" (Marriott 2018, 213). Placed between these co-natal modes of signification, Du Bois's photographic archive prizes open this interstitial space, disrupting the relays between subjectification and objectification in anti-Black visual cultures, placing the confusion of racial difference back onto viewers rather than seeking to police it for them as Agassiz had done through his naturalizing of inhumanism. Chandler comments about the friction that Du Bois instigated for futurity; his practice is "orientated toward the incessant inhabitation of the practices of freedom, the resolute commitment to thinking beyond limit and boundary, yet by way of limit and boundary according to an asymmetrical form of thinking of historicity" (2015, 3). Du Bois's photographs lever apart the dependencies (of whiteness) at stake in a particular normative formation of the family (as a product of genealogical processes and their historicity) and the necessity of visual codes to maintain and enforce anti-Blackness in nonfamilial states. His visual practice exposes how the racial formation of color lines is assembled in such a way so as to borrow from the epistemic construction of geologic formations where legibility is grounded outside the frame of construction, yet "naturalized" within it to accord value. The teleological organization of *whiteness as property* is domesticated as the white family, scalar to the nation-state, sedimented as geopower through the propertied form that underpins the value of whiteness (as it is made through the mobilization of racialized geopower). Du Bois visualizes the dialectic drift in *Types of Mankind* and its "natural" forms of objectification, thereby presenting a map of social relations designed for its overcoming (albeit in normative terms).

Materializing Racial Difference

Understanding geology as a material and social infrastructure in the reproduction of white supremacy (and its attendant gendered and racialized labors within geo-logics) challenges us to repudiate and transform the social order through an abandonment of genealogy and its forms of filial rationalization. The configuration of geology at its onset, through its patriarchal lineage of fathers, is foundationally and profoundly racist. The axis of the verticality of genealogy across time (or descent) is emphasized in geologic accounts of Life to ameliorate the anxiety of the horizontality of sexual and the lateral actuality of kin relations under slavery. As Orlando Patterson (1982, 5) argues in his work on natal alienation and chattel slavery, the slave was a "genealogical isolate," with no recognition in the social order or recourse to claims of blood lineage or ancestry. The emergence of genealogy, located in the historical context of a stratigraphic imagination and slavery, creates an axis of whiteness that establishes both a perceived right to conquer and inherit geography *and* a dominance of reason through scientific practices of thought to justify this (that continue to inform critical and development methodologies). The stratigraphic impulse sutures the relation between genealogy and geology as a material and philosophical process that sediments these geo-logics and their regimes of geopower, which facilitate the stratal renewal of whiteness. It is these very geo-logics that are at stake (in their conditions of reproduction) in any appeal to planetary or vernacular environmental or socioeconomic futurisms. The self-origination of the genesis of whiteness is given a natural root. Similarly, genealogy as a method, even in its critical Foucauldian form that uses the filiation of concepts to do away with origins and understand the history of ideas of Man as a hermeneutic, originates in a racialized context such that it problematizes the organization of the thought of kinship and archaeology. As Sharpe suggests, there is a need to "rend the fabric of the kinship narrative" (Sharpe 2016b, n.p.); instead, we can remake and reimagine the earth through its rents and rifts. Sharpe argues that "chattel slavery continues to animate the present: transatlantic chattel slavery's constitution of domestic relations made kin in one direction, and in the other, property that could be passed between and among those kin" (Sharpe 2016b, n.p.). There is no way to do away with the genealogical affects and their structuring force, especially for the wayward orphans of the diaspora. Its affective architectures impose a way of doing relation that follows a line of imperialism and slavery and needs to be acknowledged as it continues to structure who or what counts as permissible kin.[66] Not least to recognize, as in Indigenous practices, what it means to have rocks, rivers, and beavers in the family.

Ferreira da Silva argues that there is a foundational inscription of race in the construction of the cartographic space of globality, in which the scientific tools of racial knowledge institute the global as an onto-epistemological context in which understandings of the Enlightenment version of the Subject are constituted (as self-determined through the elimination of "outer-determined" racialized others) (Ferreira da Silva 2007, xiii). This (white) version of the subjecthood also brings with it a set of outer-determining relational environmental and material modes, such as fathering, genealogy, sexuality, family, and kinship as modes of land acquisition and resource extraction. Materially, it is inhumanism that secures the world-building of the plateau for those white subjects. The temporal markings of this grammar of racial and inhuman differentiation achieved through the spatialization of time (in the stratigraphic space of geology) is forged in geologic reasoning and its genealogical mapping and then naturalized as a product of earth rather than these fathering forces. Thus, geology can be seen as a scene of regulation and representation, and a strategy of intervention that apprehends and codifies others as outside of, yet composite to, the Enlightenment version of the Subject. To cast Cuvier's representation of Baartman or Agassiz's photographs of Indigenous and enslaved persons to the margins of geology, to a historical moment of inhumanism, then, is to miss the point of their inscription into white patriarchy so as to bring the Overseer's earth into being through the exchangeability of race and species within the geologic lexicon. As Walter Johnson explains, aspirational humanism is the problem, not the cure: "By terming these actions 'inhuman' and suggesting that they either relied upon or accomplished the 'dehumanization' of enslaved people, however, we are participating in a sort of ideological exchange that is no less baleful for being so familiar. We are separating a normative and aspirational notion of humanity" (2017, n.p.).

Baartman and her kin are simultaneously subjects of slavery and evidential subjects of its imagined justification through explanations of racial difference. This dual process of extraction and subjugation in geologic narratives of time and descriptions of the formation of worlds was the material basis of humanism and its globality. Human and Earth origins are the beginning of every exclusion and extraction of the political present and the material worlds of the future. The work geology did to spatialize the entitlement of white supremacy above and in contradistinction to Indigenous, Black, and Brown peoples (rendered as a condition of "time's arrow" and the selective geopowers of whiteness), meant racial differentiation swung on a geologic hinge.

In fathering geology, both natalities (*filiation* and *matteration*) are part of the development of genealogical relation and ideas of kinship. The classification of matter is a means to differentiate value and was in constant conversation with the classification of traits, types, and lineages of people that established hierarchies on the road to eugenics. While I have focused on Agassiz's racist science and Cuvier's white supremacy, these "fathers" are not unusual or exceptional in the views. They are actors in a site of plateau-building. The natural philosopher and paleontologist Georges-Louis Leclerc, Comte de Buffon (1707–88), for example, disdained the "degenerate" nature of New World faunas and argued that only the prospect of colonization and its improvement of the physiographical status of these territories through planting, deforestation, farming, and fluvial management would eventually improve this impoverished fauna. Although Buffon never traveled to the "New World" to compile his volumes of natural history and theories of the earth, the argument around the lesser nature of America and its weaker stock was a common one made across human and nonhuman fields (he published forty-four volumes of *Histoire Naturelle* from 1749 to 1804). It was famously refuted by Thomas Jefferson through the figure of a moose and mastodon in a series of libels directed at Buffon. In response, Agassiz, keen to boost America as a worthy site for geologic study, began his popular publication *Geologic Sketches* with "America the Old World":

> First-born among the Continents, though so much later in culture and civilization than some of more recent birth, America, so far as her physical history is concerned, has been falsely denominated the *New World*. Hers was the first dry land lifted out of the waters, hers the first shore washed by the ocean that enveloped all the earth beside; and while Europe was represented only by islands rising here and there above the sea, America already stretched an unbroken line of land from Nova Scotia to the far West. (1863a, 1)

His elegiac tribute to America's deep time casts the rocks of Turtle Island in a political geology of natural history understood through the bonds of genealogy.

Often credited with the first use and formation of the concept of race, Buffon argued his ideas of degeneracy through the role of climate and directed this at both animals and Indigenous peoples, and his research contributed and championed ideas of scientific racism. Buffon championed the division of six races (Polar Negro, Tartar, American, Australian, Asiatic, European) in *Of the Varieties of the Human Species* (1749), but his use of race seems to be much more expansive, encompassing even some imagined tailed races in the island of Mindoro, while Kant went with a division of four (white, black, copper, and olive) and Edmund Burke had sixty-three, Agassiz proposed eight, and Morton (1847) twenty-two different races. The numbers are not important, as it was the differentiation in theories of race that had lasting scientific and cultural meanings. The continuance of difference was the act of division itself—the *rift*—and the *establishment of racial difference* as a quality of that spatial distance. This difference, in Buffon's writing, was attributed to climate, food, ways of living, and "epidemic distempers" that were shifted by deformities and diseases. While Buffon cites a common ancestry (and Agassiz's polygenesis was an outlier), the hierarchical assumptions made on behalf of climate established an environmental approach to race that connected geography and identity in determinist causation. The comparison of skulls and anatomy was the predominant empirical medium, and while some paleontologists argued for primordial and fixed racial difference, others argued for nascent forms of coevolution with environmental changes and geographic locations. Thus, racial difference was secured as a differential across *all* theories of beings, and it framed the thought-space of Enlightenment dialectical binaries that can be seen in Kant and Georg Hegel, where elevation is claimed for Europe, against a scene of the dispersal of degeneration from this center out. To counteract this degeneration, Buffon argued that in the New World, nature and the savage peoples must be conquered and improved. Domination of nature was a twofold process: of the surface and processes of the earth and of its peoples. The telos of development and the enclosures of geologic grammars were foundational in securing racial difference as a governing principle of materiality and ideation. Two things remain from paleontological racial geologies: (1) the importance of the temporal occasion of racial narratives for "placing" subjects and (2) the axis of racial difference as a geographic form of knowledge production and material extraction.

The racialized geo-logics of deep time demonstrate an elliptical relation to the vastness of deep time, squeezing it into an ever-diminishing human envelope of racialized life, thereby inverting the temporal plane into a paleo-psychic time. This parsing of the epochal into the everyday racism of colonialism reveals some of the conceits of time travel in the repositioning of the human in planetary time (paleontology). Baartman's place in this archive should alert us to the juxtaposition of the question of writing a *geologic now* and the need to redress the sedimented histories already securing futurity (see McKittrick 2010). As with Octavia Butler's time-traveling character Dana in *Kindred* (1979), she must keep traveling back to the Weylin Plantation to save her enslaver ancestor, which is also to ensure her own existence in the space-time in the present (M. Wright 2015, 84–86). The cost of her survival is a reckoning with the brutal entanglements of their conjoined violent histories. Targeting the coordinates in the production of racial knowledge is also to target its possibility (see Ferreira da Silva 2011, 139). The problem of the human in its construction through inhuman epistologies haunts the formation of geology. What geology *is* as a discipline over time changes shape, and it is made differently through material practices, but the foundational role of questions of sex and race to geology's origination remain proximate. Epistemes of Life have a geologic root; even if geology is backgrounded and biological-biopolitical life is foregrounded as the active mode, it is the permutations of geologic lives that are coeval in the suturing of racial difference into phenotypical skin. Furthermore, if genealogy (and its connection to racial origins) is the citational field in which these ideas of white science are raised, genealogy is also the theoretical apparatus (via Kant, Marx, Freud, and Foucault) in which this field is thought within Western philosophy. As stratigraphy and genealogy emerge as practices of geology, they also emerge as a practice of social thought that performs a way to navigate, through an accumulative historical approach to knowledge formation. This leads to the question of what a feminist, antiracist counterhistory of geology could be when genealogy is deployed as both a critical and a compliant methodology in how geopower is exercised and narrated. What other methodologies can unsettle the arrangements of geopower and what grammars make such arrangements of power *legible* in the first instance?[67] As repetition of narratives established the normative, taken-for-granted understanding of the ordering of Life and earth and, crucially, in the case of paleontology, the instituting of ideas of racial difference and the materiality of time as a racialized axis, narratives of hominoid "development" and geographic difference gave rise to a self-explanatory role of difference as rooted in natural history rather than political relations bound to the plasticity of white surfaces.

Ironically, in the claim to the exceptional genealogical root, paleontology made whiteness as a category with no natural ties (an inhuman power and unnatural being), releasing whiteness as a name in which anything could be done in service of. Whiteness stands in for futurity as it strips everything but its own future and mines forms of life not coded as Life. It calls this Civilization, Progress, and then Development. The narrative of time makes the conditions of racialized pressure and its enduring gravities. When racial determinants come back as they frequently do in these new waves of populism, white environmentalism, technofuturism, understanding the histories of geology provides an unearthing of the metaformations in which race is made and discloses the architectures of affect that enable the materialization of racial difference. As race nested across human and inhuman worlds, in overt places such as the mine and plantation, it was also embedded in the metaorganization of the classification of natural history and the collection protocols of museums, through their systems of signification and persuasive modes of aesthetic address. Understanding materiality as a colonial organization of the empirics and praxis of matter—White Geology—demonstrates how embodied geology spatialized relations *as racialized* in order to do spatial work in material, geomythic, and subjective registers. Geology was the way of alienation—of land, environment, and peoples—for the transformation of the imperialist and settler state, and thus it was the science of settler belonging and established its categories of valuation and geomythos of belonging. Geology entered America as a populist practice that secured itself in the material of stolen earth, birthing violent extractions hidden in origin stories of inhuman nature and patriarchy. This tying together of geology and geotrauma in the evisceration of relation can also initiate the question: What kinds of reparations, repair, and remembering can be thought with and in earth (and geologic sciences) that reorganize its geopower in the context of these racialized grounds? And, further, what "ghosts of geology" need to be resurrected and built from these material sedimentations to question how subjectivity is produced on the plateau and embodied in the rift through the geosocial formations of this colonial materiality?

5 Geologic Grammars

Strange history this of *abolition*. The negro must be very old & belongs, one would say, to the fossil formations. What right has he to be intruding into the late & civil daylight of this dynasty of the Caucasians & Saxons? It is plain that so inferior a race must perish shortly like the poor Indians. S[arah]. C[lark]. said, "the Indians perish because there is no place for them." That is the very fact of their inferiority. There is always a place for the superior. Yet pity for these was needed, it seems, for the education of this generation in ethics. Our good world cannot learn the beauty of love in narrow circles & at home in the immense Heart, but it must be stimulated by the somewhat foreign & monstrous, by the simular man of Ethiopia. —RALPH WALDO EMERSON, *The Journals and Miscellaneous Notebooks*

What is the nature of a form of being that presents a problem for the thought of being itself? —JARED SEXTON, "The Social Life of Social Death"

Fossil Formations

Strange history this of *fossil-being*. Following in the "fathering" footsteps of his close acquaintance the geologist Louis Agassiz, Ralph Waldo Emerson notes the Negro as a "fossil formation" in the context of his consideration of abolition.[1] This is an imaginary figure lodged in the deep time of being and the outer limits of Western ethics. In a form of genealogical wizardry, the Negro-as-fossil casts Black time as back time, sedimented into another geologic epoch and social stratum. This racial petrification echoed Agassiz's racial logics and the *Dred Scott* decision, which argued that "a perpetual and impassable barrier was intended to be erected between the white race and the one which they had reduced to slavery" (Taney 1857, no. 44). In continuance with this stratal geophilosophy of identity, Emerson grants Blackness a location displaced from the present by a stratal barrier. Life in the fossil archives is like waiting for promotion in geologic time, hanging around for a rift to fracture temporality. That one of the most prominent American writers of nature ethics should locate the "beauty of love" and "the Heart" in such proximity to the enclosure of Black subjectivity, the murder of Indigenous life, and the geophysics of racial violence highlights the optic of erasure that is performed by deep-timing subjectivity.

In a letter to President Martin Van Buren on the Cherokee Indians, Emerson writes, "This stirring in the philanthropic mud, gives me no peace . . . no muse befriends" (quoted in Gougeon 1982, 560). Emerson's concern about securing a muse in the mud has little to do with Indigenous or Black liberation; he was interested only in free speech and questions of the order of nature. The metaphysics of beauty in his imagination was tied to a fear of Black immortality and concerns about its affective stimulations. The existence and life of the Negro are abstracted into a universalizing metaphysical question for white thought, yet this placement also recognizes a permanence of ancestry, as a fossil, an immortal survivor of the underground, despite the stratal pressure of white man's overburden. The other, as Negro, plays as simulant and stimulant to the white imagination in the guise of primordial matter, a locked-in subjectivity, out of time, intruding on Emerson's civic daylight.

The color lines of geology are clear: Caucasians and Saxons (whiteness) are coded as daylight (light of reason), present (a political subject), and on the surface (agentic ground); Negro (Blackness) is coded as darkness (antithesis of reason) and fossil past (nonsubject) in the underground (inert resource). Once the stratigraphic geo-logics are established, Emerson questions the right of the Negro to intrude on the present (and thus the social), concluding that they shall be extinct too, like the Indian, were it not for the need to resuscitate them for natural philosophy and the poetics of the monstrous. "Aestheticizing the gap" (Crawley 2020, 48) between Romantic Nature and the inhuman sciences, Emerson not only colonizes the strata with racialized forms but also explicates the dependence of nature ethics on the reflective qualities of subterranean Blackness for a politics of futurity. The transcendental theory of white nature needed a *racialized bedrock* to bootstrap off. Yet, survivals in the liminal spaces of fossil life signaled a different quality of transmutation in the subterranean. Glissant recognizes another form of petrification in the opening of *Poetics of Relation*, in the belly of the enslavers' boat: "For the Africans who lived through the experience of deportation to the Americas . . . confronting the unknown with neither preparation nor challenge was no doubt petrifying" (1997, 5). From this womb of unknowing, in the cut-off world below, knowledge is born of this geotrauma, Glissant answers. Liminality haunts the rocks, a liminality that navigates between inhuman processes, forms of life imagined as carceral minerality, and poetics in inhuman tender.

This chapter attends to how geologic grammars join classification and geologies of race into modes of valuation, across material and symbolic terrains, to create shared metaphysical and *geophysical architectures of affects*. Emerson draws together the intimate relations between the universalizing thought of inhuman nature and the location of Blackness as geologic strata to permeate a metageophysics of thought. In this abandonment in the stratal rifts, I seek to open the question of how the phenomenon of nature as a thing-in-itself (available for Western ethical contemplation) is reliant and racially tied to the production of the Negro as a fossil-being and erasure of Indigenous life.

In the second section, the calculus of use value frames the discussion of Black life as a public geologist discusses slavery as an economic question for national resource development, expanding upon racialized geologies of extraction in settler colonialism. Both these accounts demonstrate how racialized materialisms fundamentally shaped the categorizations of inhuman nature across poetic and propertied orders through a reliance on deep time to do racial work, which in turn promoted a normative form of *geochronicity* in narratives of racial life. Emerson's text, like many others, generates a persistent question that is the spinning kernel of this chapter: How does the identification of matter generate a set of geologic grammars that weave together a normative extractivism of the earth and establish essentializing racialized categories of racial identity woven in the interstices? And how do these geologic categorizations across disciplines—the aesthetics and ethics of Romantic Nature and the science of paleontology—configure a racialized material politics of geopower that continues to mark social relations?

Responding to this question, I want to target some of the relays and subjugating enclosures in geologic grammatology, particularly sites that might offer a negotiation in the possibility of undermining dominant modes of categorization. This is also to open questions about how to deal with forced containments in geologic grammars, and how to read alternating geophysics of sense through these encounters that begin to recognize orientations of refusal and resistance to an extractive geo-logics—which would be to see the racial debt of what was taken and what needs to be discharged in relation to that debt as a geophysical-geoethical operation. Remapping natural philosophy through its undergrounds and its temporal lines, understood as carceral subterranean pockets, posits the futurity of other timelines, ways of making time sensible, as another inheritance. Attending to the collaborative juncture of Romantic Nature and the natural sciences gives an insight into the scene of apprehension in which the paleontological narrative of Blackness is made as an ethico-aesthetic, subtending ground and temporal location (and its relation to futurity). In this process of acculturation, nature is accorded a historicity that is authored by Man/whiteness to supplant that of the earth as an open and multitudinous relation. Sense becomes heavy across the breadth of this emplacement, and thereby a normative regulatory optic in the perceptual field of matter and of racialized subjects. The final section of the chapter goes to the Smithsonian National Museum of Natural History to look at how racialized geo-logic codes continued to be narrated in the public sphere in the 1980s exhibition *Fossils as Natural Resources*.

Geology Museum as a "How-To" Archive of Earth Praxis

By the twentieth century, the museum as an institute of interpretation consolidates the directional legacy of eighteenth-century theories and classifications of matter toward white settler nationalism and culturates the nation-building of the colonial state. The geology museum is a reference archive for the transformation of the other archive, the earth. The earth is the praxis of the imperial archive. Rethinking histories of violence in the earth, as process and outcome, is the most important issue of our present ecological crisis. Ecological practice that does not pay attention to the combined effects of geotrauma does nothing other than continue the futurism of a destabilizing axis of colonial earth. The point of spending time with these historical grammars of geology is to notice how these powers of signification are still very much operational in the present, and how the archive (understood as earth and social institutions) can function as a temporal map of those relations of materiality and race. Examining the role of metadata—in the classification of materiality—might be understood as a form of geologic realism that establishes the pragmatics of hierarchies of value in extraction and its corresponding racialized stratigraphy.

Little by little, metadata in classificatory geo-logics accumulate and sediment the principle of extraction—constructing histories of material exchange as inevitable—across interdisciplinary registers to strive for uniformity of thought across social and scientific worlds. They require the development of a *critical stratal discourse* and engagement with the *stratigraphic imagination* as an intimate and world-making praxis. In the aforementioned geologic cuts, I seek to explore how the ontological violence of an imposed geologic life might be thought otherwise through a generative engagement with the lithic liminality of Black fossil life. As Sexton (2011) suggests, the conditions of both possibility and impossibility shift when the thought of being is present through that which is refused into its schema of being. To transpose this thinking, if the human is made through its relation to the inhuman, then the inhuman presents the impossibility of the thought of being. The very stakes of temporal location must be disaggregated to counter being cast out of time.

Rudwick's "untainted bones" lifts a history of geologic ideas into the untarnished neutral space of intellectual purity (the plateau), while dismissing geologic race talk as a generational quirk, a sign of the times but not a quality of geologic temporality. If we take seriously the lasting effects of the simultaneity of exclusion from time and geography established in racialized identity, washed up on the fossil shore, then a fundamental rethinking of the operational zones of geology is needed that invades far beyond the discipline of geology into the recesses of its imaginative reach and psychic foreclosures. These operational zones are the material (land, value, property, discourses, epistemes); the psychic (relation); the bodily (race, labor, gender, sexuality); the geophysical (chemical, thermal, permeable, porous, gravitational, dynamics of force); and the valence of geology's grammar to organize modes of experience and sediment forms of geopower as naturalized forms of power. Inhuman matter enters objecthood in an already compromised field of denigrated subjecthood and segregated social orbits. Even as inhuman matter is the subject, the modes of classification are not immune from subjective connotation and exist in a repository relation to that subjection. Blackness is made to do transmutational work in the category of the temporal lines of Western ethics and philosophy. In populating the category of the inhuman as a material operation, there is a critical pathway that comes back at the ontological function and epistemic violence of the materiality of race.

In the structuring of Life and inhuman as ontological categories in Emerson's thought, he betrays his fantasy of historical obliteration of the Negro from the political present into the space of the substratum and the time of fossilization. Emerson highlights the paleontological connection and imaginative contamination between the pre-Human othering and prehistoric Other.[2] The construction of Blackness as prehistoric lodges a prior claim in the strata but holds it there in suspension, much like the Fourteenth Amendment. The political affect of doing race stratigraphically is to call attention away from the surface and political claims in the present, shelving the questions and demands of geologic nows (such as land reform). Emerson's poetic aporia substitutes a natural origin to a racial disfigurement. While his maneuver is by no means unique given the connections between poetry and geology during the period of intense industrialization of geology (for British examples, see Heringman 2004), it exemplifies how racial boundaries are made by temporal placements and how this bordering is naturalized through a temporal imaginary that is used to hold identity in different geophysical states. Similarly, extinguishment is deployed to refuse the "now" of Indigenous life in a form of stratal overreach to depopulate the future. This conceit is no different from how the Anthropocene origin stories are euphemistically framed as "The Great Dying," or the activation of environmental rather than political determinates of climate change.[3]

The refrain of Blackness in the strata refers to the importance of considering the spatializing power of geology in the materiality of race, as the figure of the Negro is signaled as having "no place" in space or time. Much as in Ralph Ellison's *Invisible Man* (1952), uplift begins in the basement, invisible and incandescent, courting the subterranean as refuge and temporal release. As Richard Wright's 1940s story of *The Man Who Lived Underground* makes brutally clear (foreshadowing Ellison's basement illumination), the underground is both creative refuge and temporal derangement. As time is drained first through police brutality and then into the sewers (whereby the gravity of the manhole replays the suck of the political state, an inversion of Emerson's nature ethics), a tender temporal slip emerges that allows Fred Daniels to rearrange the pieces of his world and give spatial form to its revaluation. Diamonds are stomped into a dirt floor, making it shiny; rings are hung; and hundred-dollar bills are pasted onto a cave wall (R. Wright 2021, 143). The material matrix of subjugation and racial deficit is inverted. In the terrible reversal of the legacy Emerson makes and names, "the dim light of the underground was fleeting and the terrible darkness of day stood before him" (Wright 2021, 145). The message is clear: the political arrangements of the state and its enforcers are stratal. The subterranean scrambles the brutal geo-logics of the surface and its anti-Black gravities but it is no place to live.

Wilderson argues, "The Black was a static imago of abjection. But this stasis was productive for the Human: against Black abjection Humans could know *themselves* as agents of change as well as agents who *could* change" (2020, 314). Abjection is performed by the withdrawal of space-time in the present and the lack of propertied or genealogical relation to land or political relation, and yet there is the recognition, however narrow, of strategies of persistence. As Denise Ferreira da Silva (2007) suggests, European sensibility of itself, as an affective social body and a geographic episteme within time, is secured in the foundational structure of the racial other. The governance of this political space is predicated on stabilizing and securitizing the origins of time naturalized through the origins of the earth.

The violent right to geography, as Agassiz argued, taking and making place, is the self-confirming mark of superiority (echoed by Emerson). The Negro functions in Emerson's formulation as a figuration of pity, redeemed only as an ethical object *for* natural philosophy. This redemption in a metaphysical mode performs the parallel compensation needed for the geophysical debt that underwrites the emergence of that thought. Hartman argues that the "fungibility of the commodity understood as the imaginative surface upon which the master and the nation came to understand themselves" positions Blackness as "the occasion for self-reflection as well as for an exploration of terror, desire, fear, loathing, and longing" (1997, 7). Monstrosity is used to posit racial difference as a sign of limit experience that provides the psychic displacement necessary for geologic accumulation (see Sharpe [2010] for a reading of racialized intimacy). This abjectifying line of thought functioned because geologists had already scripted Blackness as inhuman. In return, hear Canisia Lubrin: "You must know, black isn't always the void: / to be a living fossil, light years old / perhaps." As Lubrin reminds us, "to be a living fossil" is a cosmic constellation and an astrolabe for temporal incursions (2017, 48).

On the Edges of Time

Blackness is the sentinel of inhuman nature in the Western geographical imaginary, as guardian and figuration on the further reaches of time. In the ideation of fossil formation, Emerson exposes the indebtedness of white metaphysics to processes of political fossilizing and containment of Black subjectivity, alongside the symbolic erasure of Indigenous life. As in "out of Africa" human origins theory, Blackness is made sovereign of deep time and guardian of the abyss. This placement is both an expulsion from subjectivity and an affirmation of the ambiguity of that relation. I do not want to move too quickly to make ontological use of this alien emplacement, to find yet another utility, because it was and still is an embodiment of the terror that facilitates the projection of anti-Blackness. The focus here on the inhuman seeks not to replicate the corpsing effect (Marriott 2018, 327) of a prevalent geophysical imagination but rather to plot the coordinates of the imagination in a way that shows how conditions of unlivability and unlivable lives are made, across a scene of multiple materialisms and specific instantiations that create the rift effect of the plateau.

As the inhuman grounds the human in the history of being and time, the abolition of these intimate inhuman inhabitations requires the dismantling of fossilizing affects, as metaphor and materiality of race. At the same time, "alien" nature was also an intramural space that was peopled and subsequently made as an inventive margin (or rift) in which other relations were forged (from the kinship with alien ecologies within and outside of the plantations that Wynter discusses as crucial to establishing cultural practices in the plantation to the alien performativity of Sun Ra and Afrofuturism).

The temporal location of the fossil is not a placement nowhere and of nothing; it is, however, stratified outside of humanism and inside inhumanism. The ambiguous perspectivism of the inhuman looks both ways; it exposes both the fascinated desire and the ontological anxiety of humanism and substantiates the fiction of its ideas of supremacy. It is like placing dragons at the edge of the known on ancient maps; the transfiguration testifies to terrors of an inhuman universe, where Blackness serves as an inoculation on the conceptual and material boundaries of an indifferent universe. This ghost geology signals a sadistic comfort in the conjuring of magical beings in a cold cosmos. The critique is not to challenge this belatedness as such but to ask after its necessity and operation (i.e., why is Blackness required for natural philosophy, and what does this categorization allow to function?). Why this excessive fidelity in the imagination to a racialized subject in the depths of the earth when Indigenous life is extinguished from the white surface in these fantasies? And how does geologic time coupled with a racialized geo-analytic do the work of articulating a purposeful white universe?

If thinking the earth is always, already tied to thinking the future and the occupation of its subjective tense, why does it require a racialized analytic to do this? In the shadow of the Kantian subject, race, as a first order of differentiation, mediates and ameliorates the ontological assault of inhuman nature. The Negro as fossil-being, like Baartman, is poised as breach and bridge—a body to be stretched over a rift in time, a subject to define whiteness in time. The figuration of a time traveler sutures the universe and the human together in a continuum that allows for human nature and natural history to become coeval as conjoined historicity, as a colonial fantasy of *its* achievement in time. It is the racializing analytic that secures the thought of the universe as thinkable, gathering the dispersive cosmos into the black hole of subjugation. The black hole of the universe conceptually recruits a domestic sentinel born of the Middle Passage to do ghostwork. Thus, the accordance of fossil-being as a mark of Blackness is also a designation of a geophysical mode of a differentiated gravity of/in time, and Blackness is personified as a countergeoforce to philosophy's aberration at an unsympathetic universe. While in his later writings Emerson rejects Kant's foundational account of the differentiation of the subject on two counts (reason and bodily difference), he nevertheless retains and relies on the necessity of Blackness as a regulatory figure for an ethics of nature to locate the white settler liberal subject. He says:

> I believe that nobody now regards the maxim "that all men are born equal," as anything more than a convenient hypothesis or an extravagant declamation. For the reverse is true—that all men are born unequal in personal powers and in those essential circumstances, of time, parentage, country, fortune. The least knowledge of the natural history of man adds another important particular to these; namely, what class of men he belongs to—European, Moor, Tartar, Africa? Because Nature has plainly assigned different degrees of intellect to these different races, and the barriers between are insurmountable.
>
> This inequality is an indication of the design of Providence that some should lead, and some should serve. . . .
>
> Now with these concessions the question comes to this: whether this known and admitted assumption of power by one part of mankind over the other, can ever be pushed to the extent of total possession, and that, without the will of the slave? (November 8–14, 1822, in Emerson 1982, 19)

Stepping out of the shadow of Kantian reason, Emerson argues that racial difference cannot be just the attribute of reason, stating, "In comparisons with the highest orders of men, the Africans will stand so low as to make the difference which subsists between themselves & the sagacious beasts inconsiderable" (19). What is it, then, asks Emerson, that accords racial (white) superiority? Physiological? The upright form? He argues (mimicking Agassiz's visual cultures of race), "The monkey resembles Man, and the African degenerates to a likeness of the beast. And here likewise I apprehend we shall find as much difference between the head of Plato & the head of the lowest African, as between this last and the highest species of Ape" (21). Emerson asks that if the distinction between "beasts and the Africans" is neither in reason nor in figure (i.e., neither mind nor body), where then is the ground of that distinction? He concludes his geophysical aesthetics through the distinction of an order of species difference:

> Is it not rather a mere name & prejudice and are not they an upper order of inferior animals? Moreover if we pursue a *revolting subject* to its greatest lengths we should find that in all those three circumstances which are the foundations of our dominion over the beasts, very much may be said to apply them to the African species; for the slaveholders violently assert, that their slaves are happier than the freedom of their class; and the slaves refuse oftentimes the offer of their freedom. . . . For it is true that many a slave under the warm roof of a humane master with easy labours and regular subsistence enjoys more happiness than his naked brethren parched with thirst on a burning sand or endangered in the crying wilderness of their native land. (21)

Emerson counters the rule of Kantian reason and phenotypical association to make his argument of differentiation by way of Nature, of the hierarchy of natural history. His "revolting subject" is differentiated by the sciences of man, and Kant's teleological principle is supplanted by a paleontological one. Geology becomes a means to overwrite the rights of duration, scripting "eternals" in the question of origination. Racial eternals make a "placeless" claim materialize out of the temporal markers for the Overseer to claim the rights of supremacy and subjugation. Ferreira Da Silva (2007) argues that the fundamental exclusion of the human is why Western humanism cannot generate the ethical crisis that would be supposed around racist acts because that racism is internally scripted in the foundations of its ethical thought, geographically and subjectively (as global history and idea). Put geologically, Western ethical/environmental thought has not yet found a way to function without this breach/bridge of Blackness, which stretches time (and thus space) but remains within the impasse of the rift (and thus does not collect genealogical accumulation in the present but is inadvertently accorded sovereignty of the depths of the earth). Moreover, the cosmological imponderable is constantly being domesticated to fashion more pragmatic rifts in the social fabric and secure its maintenance at a granular level (whereby the imagined environmental conditions of a native land "parched with thirst on a burning sand" seemingly necessitates brutal enslavement as a form of paternalism).

Secured in the racial grounds of the natural sciences, the naturalization of voiding Black subjectivity for white contemplation was as foundational to Western aesthetical thought about nature as it was to economic geology. Blackness served, and thereby bridged, metaphysical and geophysical material orders as the underlying homogenizer of thought and practice in colonial and settler colonial cultures. That the Negro in Emerson's musings is a structural—stratal—affect of the material ordering and classification of geology tells us that a transcendental view of nature, as a thing-in-itself that could be directly accessed through subjective experience, was reliant on a subtending debasement, an "intervening, intruding tale" of Blackness (Spillers 1987, 79). What presents as a psychic displacement in aesthetic theory is rooted in a quieter grammar of material subjectification and spatial dislocation via the paleontological grid of the inhuman. To make his universalizing, abstract white transcendental subject that can think his ethics, Emerson must draw on the genealogical subjection established by geology's "fathers" and their organization of lineage. The mythology of tectonics has racial effects; as the plateau rises for the optic of contemplation, it grinds down, a *strike-slip* motion in tectonic plate parlance. Yet Emerson exposes the hinge of genealogy's racialized twin: kinless subjugation or geographical orphanage. Fossilized abandonment is achieved via subterranean stratification for the fear of miscegenation in the political present. The inhuman border establishes the domain of agency (white subjectivity) and nonagency (Black fossil), while genealogy as a stratigraphic structuralism establishes relation (adjudicator on the polis) and nonrelation (kinless in the space-time void). Blackness is literally made in the outer limits of the Heart, as a signification and sign without subjective occupancy. Writing into the antagonism of a category that creates nonbeing, M. NourbeSe Philip names this enclosure and erasure "*BlanK* and *Blac_K*. Blac K, the colour of the ever-expanding universe whose apparent blankness belies the plenitude of black holes, stars—exploding and collapsing, planets, comets, nebulae, galaxies, red dwarves and red shifts, to name but a few of the bodies and events that comprise our universe" (2017, 26).

BlanK and Blac_K capture an axis of blanks and ciphers, complex universes and multivalent experiences, a traumatic caesura that "belies an absence and instead contains a multitude" (33). The figure of the Negro is made by a forcible effect of geologic temporal ordering into a necropolitics that configures the surface as a space of white power and natural resources (for philosophy and the state), thereby constituting the normative and the "monstrous intimacies" of the inhuman.[4] My point is that this is a material hack (geophysics) as much as a theoretical one (metaphysics), which organizes a structural subjugation of Black and Indigenous existence in advance through the grammars of matter that ground thought in specific understandings of the material autonomy of subjectivity.

Darkness is not just metaphoric or epidemiological; it belongs to the subterranean enclosure of race in different stratigraphic formations, so that the antagonism between desire for freedom and recognition need not be raised as necessary for all in this pragmatist philosophy. The question of freedom is footnoted in the strata. Sedimentarily shelved. That is also to show *how* and *where* this division between personhood and fossil-life was sustained within stratigraphic structuralism as a question of *when*. And, to ask the question of why ethics could stray over this antagonism in its inquiries about monstrosity without any recognition of the scene of subjection that arranges the optic that delivers this contemplation? Stratal footnoting is how racialized subjects become integral but submerged in Western ethics of nature (and contemporary environmentalisms and development narratives), so that romantic accounts of nature can be articulated without any recognition of the dispossession of land by settler colonialism or its geologic transformation by enslaved persons in the ongoing extraterritorial racial impacts of colonial environments. Which is to say, the psychic prosthetic of fossil-being was a way to navigate the actual bones buried in the advent of colonial natures. Which is why Jesmyn Ward calls to "salvage the bones" in her 2011 novel of that name set in the context of Katrina, in the savage inhumanisms of racialized environments in slavery's afterlife.

Extraterritorial Environments

In Emerson's natural philosophy, relations of power and property have been all but erased through the work of the deep time eternals. In not belonging to the day, the fossil must come to light by other means. Deep time is practiced as a literal relay of temporal entombment and effacement. Philosophically, Black subjectivity is entombed in the past ruin of life in fossil beds and deep time, while Indigenous life is imagined archived as surface ruin. Internment in lithic orders appears to enact Blackness in the past perfect tense, that which has been, while in fact it consolidates the present perfect tense, securing the Negro as a subject without political futurity. Stratal disenfranchisement. Thus, the fossil marks its own disappearance; positioned on the edge of the abyss of time, it is the geochemical remains of life's imprint, as fossilization mobilized racial subjugation via geology. If we extend Hartman's formulation of the "time of slavery," which she characterizes as "the relation between the past and the present, the horizon of loss, the extant legacy of slavery, the antinomies of redemption (a salvational principle that will help us overcome the injury of slavery and the long history of defeat) and irreparability" (2002, 759), the stratigraphic differentiation of subjects as a material-temporality of race (and its relation to questions of material agency and possession as it negotiates inhumanism) can be added. Blackness, as fossil form (contained in an inhuman category that forecloses relation and agency in advance), requires the undoing of this carceral strata, as this inhumane categorization continues to make Blackness a fungible unit in global material economies (in South African gold mines, for example). As racialized peoples are materialized as strata that are placed on the front lines of inhuman terror, this geophysical bordering of race acculturates geoforces.

On abolition, Emerson wrote: "This revolution is the work of no man, but the effervescence of nature. . . . It is elemental, it is the old eternal gravitations; beware of the swing, & of the recoil" (Emerson quoted in Newfield 1996, 203). In his formation of liberalism and its aesthetics, the racial contract had a stratigraphic and geophysical resolve—the recoil of race. The elemental is prescribed a political role in consecrating the ethical subject, so that subject becomes unburdened for responsibility in light of the weight of subjection (read: transcendentalism did not want to get stuck in the racial mud). In this geo-logic we might see Toussaint Louverture and the Haitian insurgents (1791) or the rebellion led by Sam Sharpe in Jamaica (1831–32) or Mary Thomas in St. Croix (1888) and many other pyro-revolutionaries everywhere as a given by the force of the hurricane and fire.

Stratal Struggles

Struggles against stratal enclosure pressure for a different account of subjectivity through matter that gives a more unruly account of relation in the inhuman borderlands and identifies a multivalent range of agencies in planetary arrangements, agencies that do not cordon off minerality from life, in all its painful inhabitations of time that were a product of geologic classifications in the histories of extraction. To push further into inhuman intimacies is both to undo the border that constitutes the Human as an exclusionary category and more precisely to work toward activating the matter relations erased by a colonial optic (to see the necessity of earth reparations and to counter the agency of enforced environmental vulnerabilities). Importantly, minerality must be understood as actively involved in geologic subjectivity, not as an elemental state outside it. This is why Indigenous and Black engagements with time are so important to the inventive possibility of freedom, disrupting both the monotemporal horizon and its material telos. The work of the literary and artistic collective Black Quantum Futurism, for example, reconceptualizes time travel to exit negative temporal loops, as well as tapping into ideas of tonal memory in "our past futurism." That group argues that the agency to imagine a different future is a technology of liberation, utilized by Black freedom movements and anticolonial activists. Similarly, Kodwo Eshun, of the Otolith Group, argues, "Imperial racism has denied black subjects the right to belong to the enlightenment project, thus creating an urgent need to demonstrate a substantive historical presence. . . . It is clear that power now operates predictively as much as retrospectively. Capital continues to function through the dissimulation of the imperial archive, as it has done throughout the last century. Today, however, power also functions through the envisioning, management, and delivery of reliable futures" (2003, 287–89).

The methodological task, Eshun urges, is to create "temporal complications and anachronistic episodes that disturb the linear time of progress." It is the generation of these cacophonous futurisms that "adjust[s] the temporal logics that condemned black subjects to prehistory" (297). The abolition of temporal execution is as important as paying attention to spatial matters in racialized geologies. It is necessary to think backward (as well as forward and eschewing the back-and-forth axis altogether), to cast an imaginary about what an antiracist time would look like, and what needs to be undone in the past to release this parallel present (see McKittrick 2006, 1–36). In geology this means attending to the temporal scripts of materialism itself, in all its renderings in matters of the earth.

Structuring Time(s), Stratal Sentinels

Underscoring the role of paleontology in ideas of aesthetics and natural philosophy (since Kant), questions of matter raised via nature or ontology are embedded in the development of the natural sciences as material extractive praxis that structures time and agency. This is a narrative not just about how the lithic eye of the geologic survey finds its home in transcendental aesthetics, but about how mastery is joined up along this racializing axis to occupy and capture territory in such a way as to produce a broader hegemonic view from the plateau. Emerson and Agassiz were both members of the Boston-based "Saturday Club" and were friends who went on a camping trip to Follensbee Pond in the Adirondacks in 1858 with other eminent New Englanders, dubbed the "Philosophers' Camp" (recently recreated, in homage, by the environmentalist Bill McKibben).[5] Agassiz's sexualized racial propositions and his eugenic theories about life on earth were well known and were accommodated within these circles. Emerson's own theoretical configuration of individualism without universal equality and its organization of the social order was understood as the workings of a metaphysics of nature rather than racism. What is interesting here is not so much that ideas of natural philosophy and natural science comingled on camping trips as a history of ideas (and the social politics of race, gender, class, and sexuality in the production of knowledge that substantiated those histories), but how *stratigraphic thought was a form of structuralism* that was consistently reproduced across different arenas of thought to collaborate toward a normative presentation. And, how those stratigraphic formations were founded on an axis of raciality that is embedded both in the development of geologic extraction as economic geology and in the individualism of liberalism and its account of capitalism (Robinson 2000 [1983]) and nature (understood as capitalism and nature rather than racial capitalism and racial nature).

What becomes clear in Emerson's geologic politics is how Western (colonial) thought produces a temporality captured by the racial dilemmas of the political present, so much so that it inadvertently forgets the geopower of the longue durée, both in the radical potential of deep time and in the environmental violence from material economies fixated by the "now" of extraction (see Nick Estes's 2019 *Our History Is the Future* for an articulation of the power of Indigenous resurgence). Emerson's musing is just one example of how Blackness was made as a type of terrestrial object, like a rock, mineral, or fossil that simultaneously accorded Black subjects a stony silence in the politics of recognition. In response hear Césaire's cry, "My negritude is not stone" (2013, 58). Thus, the placement of race within geologic time becomes a means by which to legitimize white hegemonic control of geopolitical, geosocial, and geophilosophical space. The convergence of ideas of natural philosophy and science creates the plasticity of white normativity and extends its reach from an individuated experience of racial nature to the history and control of Life on earth. Stratigraphic structuralism becomes a singular epistemological narrative that tells one origin story of European lineage across disciplines to set the colonial timeline and structure the present as hegemonic and inevitable. The layers of strata function as a barrier to protect the figure of the Caucasian and to buttress his fragility against the existential terror of inhuman nature.

Blackness literally is protective stratum for whiteness, as Indigeneity is a gateway to the ungrounding of the strata; both figures are positioned at the limit of Emerson's attempt to forge an "original relation with the universe." The metaphysical process, all that was deemed "not me" in the epistemology of Emerson's thought, attempted to dissolve the brute force of nature's alien quality in the self/world breach. That breach was tempered by the figure of the Negro as a waymark, an intramural fossil. The racial bridge of material labor that allowed Enlightenment subjectivity to emerge as individuated is then exiled through the transcendental movement of metaphysics from its geophysical ground. If Blackness functioned at the ambiguous interplay of recognition, fossil rather than *not-fossil,* it also parsed the lack of recognition of human exceptionalism by an indifferent universe and the liberal recontouring of a domestic individuated center, made in the wake of the "settling" the material worlds of colonial Man. Another way to say this is that racial difference was a necessity for Western metaphysics to breach its psychic and material registers of absolute other (Earth)—or the realism of geophysics—specifically within the context of settler colonialism. Race was a political differential that did work to secure economic geology, but it was also a psychic means to differentiate Being from inhuman nihilism. Telescoping time into the political present, racial ordering brought deep time closer in, offering intimacy with the vastness while maintaining its "alien" qualities in personified forms through an account of the monstrous. On these outer reaches of the world, the fossilized "alien" speaks to the recurrent need of Eurocentric thought to domesticate itself though inhuman subjugation (a necropolitics of racial bordering) and build psychic dams against the full "measure of the world" Césaire 2000 [1972], 73) that would collapse the fictional concepts of individuality, control, and mastery. Yet, as Glissant says about the collective passing through petrification, experienced by those in the Middle Passage, fear of the inhuman is not something that is shared by those who already know themselves through that annihilation and survival, as part of the earth, collectively gathered in the ontological effacement of inhuman(e) life. This geopolitical and geopoetical passage of the inhumanities is the genealogical inheritance of the anticolonial will.

Race Time and the Materialism of Morality

In January 1846, Charles Lyell, as president of the Geological Society of London, spent two weeks at the Hopeton plantation (637 persons listed as enslaved) to observe plantation life and "the treatment of Negroes." Lyell wrote that James Hamilton Couper was a benevolent enslaver, and he commented on the affect and attachments between master and enslaved: "During a fortnight at Hopeton, we had an opportunity of seeing how the planters live in the South and the condition and prospects of the negroes on a well-managed estate. The relations of the slaves to their owners resemble nothing in the Northern states. . . . The slaves identify themselves with their master, and their sense of their own importance rises with his success in life" (Lyell quoted in *Louisiana Planter and Sugar Manufacturer*, April 27, 1901, 269).

Couper, a graduate from Yale in hydrology and land reclamation, was a noted planter, remembered for his application of the scientific method to agriculture and lauded for his management practices of enslaved persons. Hopeton was seen as a "model" antebellum plantation and held up as an achievement of crop diversification in the South, growing sea island cotton, sugar, and rice.[6] Aside from his agricultural geology, Couper was an avid fossil collector and donated fossils to the Geological Society of London. Describing his first meeting with Mr. Couper, Lyell recalls: "The next morning, while we were standing on the river's bank, we were joined by Mr. Hamilton Couper, with whom I had corresponded on geological matters, and whom I have already mentioned as the donor of a splendid collection of fossil remains. . . . He came down the river to meet us in a long canoe, hollowed out of the trunk of a single cypress and rowed by six negroes, who were singing loudly and keeping time to the stroke of their oars" (*Louisiana Planter and Sugar Manufacturer*, April 27, 1901, 268–69). It is also noted that Couper, "by methodological use of his time found leisure to cultivate his scientific tastes so much as to cause his correspondence to be solicited by almost all of the learned societies. He was recognized as the best planter of the district, as a most humane and successful manager of slaves" (268). In the methodological "use" of the enslaved, research and leisure time emerged for geologic study. The mobilization of material time given to whiteness for scientific "improvement" was through the multivalent inequity of race-time.

Ignoring the obvious temporal deficits of race, Lyell places the quandary of man's exceptionalism in the history of a dynamic inhuman planet (that he understood to be a uniform), by suggesting that "we may then ask whether his [man's] introduction can be considered as one step in a progressive system, by which, as some suppose, the organic world advanced slowly from a more simple to a more perfect state? In reply to this question, it should be observed, that the superiority of man depends not on those faculties and attributes which he shares in common with inferior animals, but on his reason, by which he is distinguished from them" (1830, 155). Lyell suggests that this "dignity" of man is not preexisting, "a decided preeminence," but is to be found on account of his intellectual and moral achievement: "If this be admitted, it would by no means follow, even if there had been sufficient geological evidence in favor of the theory of progressive development, that the creation of man was the last link in the same chain" (155). Here, the double efficacy of reason becomes apparent, reason as deductive logics of appearance and reason as a mode of thought to still a dynamic planet.

Lyell's precept is that reason performs the "leap" between kind or species, and this achievement cannot be considered part of a "regular series of changes" as this would be "to strain analogy beyond all reasonable bounds" (1830, 155). The implication is that if Man can leap so can other reasoned organisms, should they occur, and that would mean that there is nothing exceptional about Man. Thus, in order to refute this notion of a plurality of progression (an equal earth with racial equity), Lyell must find something extraordinary in the ontology of the human to differentiate it and name the leap from the physicality of racialized being to its "unnatural" change of state in white supremacy. He says:

> Is not the interference of the human species, it may be asked, such a deviation from the antecedent course of physical events, that the knowledge of such a fact tends to destroy all our confidence in the uniformity of the order of nature, both in regard to time past and future? If such an innovation could take place after the earth had been exclusively inhabited for thousands of ages by inferior animals, why should not other changes as extraordinary and unprecedented happen from time to time? If one new cause was permitted to supervene, differing in kind and energy from any before in operation, why may not others that have come into action at different epochs? Or what security have we that they may not arise hereafter? (156)

If there is one exceptional being (Man), Lyell's geo-logic goes, why not others? As Lyell questions the "security" of the human against further incursions of exceptionalism, existential terror creeps in (this is also a shadowing of the panic that grips Kant's formulation of reason as a resolution of mastery over the limit experience of inhuman nature). If something is let through (Man) as exception, then there must already be gaps in the geologics of uniformity, and then the order of nature (and by inference the order of Man) is destroyed. This is the dilemma of trying to accommodate teleology through an earth with a multiplicity of originations and iterations of future earths. So, how to resolve this psychic existentialism and geologic conundrum to retain the ontology of sequence that Lyell's racialized lithology proposes and on which genealogical accounts depend?

Rather than stay with the terror of an indifferent earth (and universe) that on occasion scrambles meaning, Lyell proceeds to answer the enormity of the ontological question posed by the existence of inhuman nature by way of race and the geologies of colonialism. Twinning the inhuman as race and matter, he thereby stabilizes the plateau by way of the interregnum of race. He says: "[W]hen a powerful European colony lands on the shores of Australia, and introduces at once those arts which it has required many centuries to mature; when it imports a multitude of plants and large animals from the opposite extremity of the earth, and begins rapidly to extirpate many of the indigenous species, a mightier revolution is effected in a brief period than the first entrance of a savage horde, or their continued occupation of the country for many centuries, can possibly be imagined to have produced" (1830, 157–58).

While Lyell is about sixty thousand years off in the autochthonous inhabitation of country, the point that he makes is that earth revolutions, by way of imperialist ecologies, economic ethnobotany, and racial violence, are what secures human exceptionalism (i.e., the power of geoengineering human-nonhuman-inhuman states). Thus, the human is predivided as a geologic category in time through species (European and Indigenous), even as Man's exceptionalism is being made agentic in its division from the rest of Life and earth by geomorphic transformation. The text highlights the necessity of an analytics of raciality for arresting the scope of the earth's possible histories and its existential challenges as geologic argumentation for methodological procedure is made by way of analogy, highlighting the role of inference as a speculative craft in populating the history of Life and earth. To consolidate his argument, Lyell uses the signifier of morality as the marker of material exceptionalism. Morals maketh materialisms. Lyell asserts:

> Now, if it would be reasonable to draw such inferences with respect to the future, we cannot but apply the same rules of induction to the past. We have no right to anticipate any modifications in the results of existing causes in time to come, which are not conformable to analogy, unless they be produced by the progressive development of human power, or perhaps by some other new relations which may hereafter spring up between the moral and material worlds. In the same manner, when we speculate on the vicissitudes of the animate and inanimate creation in former ages, we ought not to look for any anomalous results, unless where man has interfered, or unless clear indications appear of some other *moral* source of *temporary derangement*. (1830, 163)

Thus, for Lyell, morality secures the *bios* of Life against the indifferent *geos*.[7] Physical nature is still reciprocally untouched by humans, he argues, but Man represents a new and extraordinary circumstance because of his moral nature and abilities at "temporal derangement" in the ability to author, script, and interrupt time. And, as moral nature is the mark of civilization (accorded by Western reason), unless some other source of morality comes along, argues Lyell, this *geology of morals* secures European Man's hierarchical position and coconstitutes his development beyond his natural limits (through a temporal derangement of materiality—colonialism—that rescripts time and the earth).[8]

Lyell's conceptualization of the uniformity of nature was not immune from sociological interpretation or deployment in conservative rejections of gradual reformism in society. While Lyell's statements are philosophical in nature, they are important for how power is demarcated as situated between moral and material worlds, thereby constructing the European subject's exceptionalism as located between paleontological supremacy (race) and economic geologic development (extraction). The labor of the enslaved as temporal providers on Couper's plantation is erased by the assertion of morality. This moral materialism that turns on a racial axis is further consolidated through what Wynter (1996) calls material redemption narratives of the "underdeveloped" as temporally deficient. Morality was the time ticket for geologic mobility.

Economic Geology

> Exhibit: Manilla, 1600, British manufactured brass or copper, used as currency in the purchase of enslaved Africans in West Africa.

> It is no exaggeration to affirm that Geology has close relations to every branch of Natural History and to all the physical sciences, so that no district of that vast domain can be cultivated without awakening trains of thought leading to geological questions; and, conversely, the prosecution of knowledge in this department, cannot fail to excite the desire and to disclose the methods of making valuable acquisitions to the benefit of human life. . . . In our day, through every degree of extensiveness, from the parabulation of a parish to the exploring of an empire, travelling has become a "universal passion" and action too. . . . Within a very few years, the interior of every continent of the earth has been surveyed with an intelligence and accuracy beyond all example. . . . Similar labors are in progress upon points and in directions innumerable, reaching to the heart of all the regions of the globe: and the men to whom we owe so much and from whom so much more is justly expected, are geologists. —JOHN PYE SMITH, *Introduction to Elementary Geology*

In parallel to Emerson's discussion of the issue of slavery, the geologist David Thomas Ansted (1814–80), fellow of Jesus College, Cambridge, professor of Geology at Kings College, London, assistant secretary of the Geological Society (1845–48), and author of numerous popular books on geology, published three letters in *The Times* titled "Slavery as an Economic Question" (Ansted 1854). The letters are collected in one of his popular books, *Scenery, Science, Art*, alongside chapters on geologic formations and mineralogical resources. As with Lyell's American publications, the shift from natural resources and its extension into Black life is presented as a neutral geologic question of extraction techniques and geologic inquiry (Yusoff 2018a, 74–80). While Ansted's letters to *The Times* are unremarkable in terms of their framing of slavery through the usual prisms of morality, development, paternalism, and economy, they demonstrate the public deployment of geology and its temporalities to do the work of racial ordering and argumentation (as with the prior examples of Cuvier, Agassiz, Emerson, and Lyell). The question of the economic development of geologic resources through extraction is raised as a cogent component of the debate on slavery, and the ability of extraction to function across the epistemes of inhuman "property" is taken for granted. Geology travels as a relay between geophysical ideas of racial temporality (the metaphysics of race) and the economic functioning of race in practical geologies of extraction (the geophysics of race). Blackness is constructed in both these domains as ontological challenge and material-energetic resource. What is common to both these productions of the metaphysical and geophysical is how Blackness is epistemically made for extraction.[9]

Ansted explains his interest in slavery as "part of my business to observe and inquire into the condition of the laboring population in some of the slave states" and to present a case as "many of my fellow-countrymen whose knowledge of slave life does not extend beyond 'Uncle Tom'" (1854, 294). He begins by stating his conviction that slavery is indefensible on the "broad principles of natural justice and morality" and goes on to argue that it is "uneconomic and highly impolitic as a source of supply of human labor," and thus should cease to exist (294). Yet, from the start Ansted frames the question of slavery within what Hartman calls the "racial calculus" of economic geology and similarly justifies his argumentation through the infantilizing rhetoric of the enslaved as junior partners in deep time. He says, "It is quite clear at a glance, and becomes more so on further investigation, that in all essential points they are children, and like children, totally unfit to take the management of their own affairs, and advance their interests" (295). Although Ansted repeats some standard tropes of enslaved persons as childlike, lazy, and without moral responsibility, he does so as a geologist through the "great family of the human race," highlighting their "natural tendency" toward errant behavior (296–97), where infancy is constructed via genealogical placement.[10] Infantilization is fossilized for economic acumen.

Despite the obligatory show of moral conviction about the wrongs of slavery, Ansted (1854) suggests in a stratigraphic mode of thought that breaking up the system quickly could not be achieved without "breaking up the foundations of society" (297). In his assertion is an admission that geologic formations were foundational to society, with race as the stratal determinant. He likens enslaved persons escaped to Canada "as idle, useless, and unimproving in every sense of the word. Their idea of liberty is an escape from labor, and the indulgence of a mere animal existence with as little effort as possible. They are bad servants and bad citizens, and rarely rise above the very lowest position in the social scale" (296–97). In a footnote, he says that statistical exceptions might be found to his general observations about their lowly condition, but that these are isolated cases. Unable to conceive of the enslaved as anything but instrumental vessels of labor, he bemoans, "It would be to scold a steam engine" (296). Without any sense of the searing irony about why the "idea of liberty is an escape from labor" might be so for the enslaved, Ansted fails to read refusal as epistemic resistance to the categories of subjective containment.[11] As modes of gesturing outside social reproduction of normative (white) forms of life are lost to him, going against the grain of the construction of liberal subjectivity is merely seen as a regressive habit that returns the enslaved to the lower rung on the ladder of progress. The enslaved are inscribed as an irritant in the progressive horizon of material accomplishment through the application of labor as temporal accelerator in geologic transformation. Back to animal. Start again. Do not collect personhood as you pass Go. The concept of Being is used to manage territory and its populations, and the language of geology is mobilized to do the work of signification in racializing nationhood.

Describing the subterranean refusals, the small indices of freedom that are read as ungovernable, unproductive acts of idleness, of good-for-nothing achievement in the Great Chain of Being, Ansted finds it easier to infer that this is about paleontological occasion in time rather than insolence as a practice of freedom-ing the racial temporal deficit model in the script of progress. What is insolence, if not a refusal to participate in the terms of participation? The liberal valorization of labor and productivity is organized into social modes of racial capitalism in economic geology. Even under conditions of slavery, the reasoning of labor suggests the norms are unable to accommodate their own abnegations. That work should be eschewed in the category of extractable property seems inconceivable to the eminent geologist in his appraisal of the economic status of Black subjectivity. He argues that slave labor is uneconomical because for the sum of eight hundred to a thousand dollars, "the right of property is acquired in a man, who must always be fed, and clothed, and kept in good health, whether his services are needed or not, whether he is at all the sort of servant required" (1854, 300), whereas the sum paid for colored labor per day might seem higher than owning a slave but is in fact, he argues, a fallacy because having a slave is like having the "dead-weight" of a mortgage, and hiring labor when it is needed would relieve the white man's burden of having to bear responsibility for the enslaved's food, shelter, and well-being, instead of just utilizing Black labor as a resource to be used up and discarded. These ideas of Black life in a different unit of fungibility prefigure the conditions of convict lease labor but prepare the moral and material ground for its instantiation.

Ansted compares the unfavorable development of coalfields in Pennsylvania with those in Virginia, whose "coal is workable without the smallest mechanical difficulty," and then the underdevelopment of gold in Virginia and North and South Carolina, and Georgia, that has "been carried on with the smallest regard to economy" (1854, 304).

Considering the nonutility of economic geology as a moral deficit in nation-building, Ansted concludes that slavery is a wasteful use of capital that interferes with the competition among labor and thus checks resource development of the country's minerals. Yet the historical agency that is afforded to Blackness is elemental ("properties") rather than a class-based or hierarchical placement per se, so any political organization around labor in racial capitalism is disingenuous. Prefigured by the British Slave Compensation Act 1837, Ansted's solution to the squandering of economic potential of Southern resources is that the enslaved would need to be "educated" into freedom and slave owners compensated for the value of their property. He suggests that the Negro population would disappear through extinction, either through the competition among whites and Blacks for labor or a return trip to Africa.[12] The black hole of slavery is reconstituted as future opportunity or exit route to the inconceivable and unquantifiable debt of relations.[13] In a counterreflex, Ansted cynically sees the accumulation of geotrauma as economic potential rather than a debt beyond repair (but within the scope of reparations). Without consideration for the possibilities of consent, he imagines that the enslaved can just be psychically pushed back through the "Door of No Return."[14] He also imagines that "free Negroes might desire, to establish a colony established in Africa . . . [which] would afford to the black man a home in a climate more resembling that adapted to his constitutional peculiarities . . . hence a nucleaus [*sic*] might be formed which would afford a means of placing civilization within reach of various African tribes who are difficult to deal with in any other way" (1854, 308).

This imagined (and later executed) Colonization Society would educate in the "established habits of industry, and the inculcation, as far as possible, of a spirit of progress—these would be the best treasures that could accompany the African returning home to his ancestors from a long sojourn in a distant country" (Ansted 1854, 308–9).[15] The infinitely extractable, geographically expandable commodity form of Blackness is now imagined as a physical colonial buffer to the continent, expanding the circle of extraction through the afterlives of slavery. As if slavery were a "sojourn" in the ongoing expectation of an industrious contribution without end. Renewing the relays of extraction, Blackness was being made into another imaginative stratum, this time to mediate and buffer the "wild" African interior in the reproduction of the colonial life-form. Obscenely, Ansted describes the emplacement of a live Black stratum to insulate against "savage tribes" and teach Africans the ills of slavery: "Back to the African shores a population of free and intel-

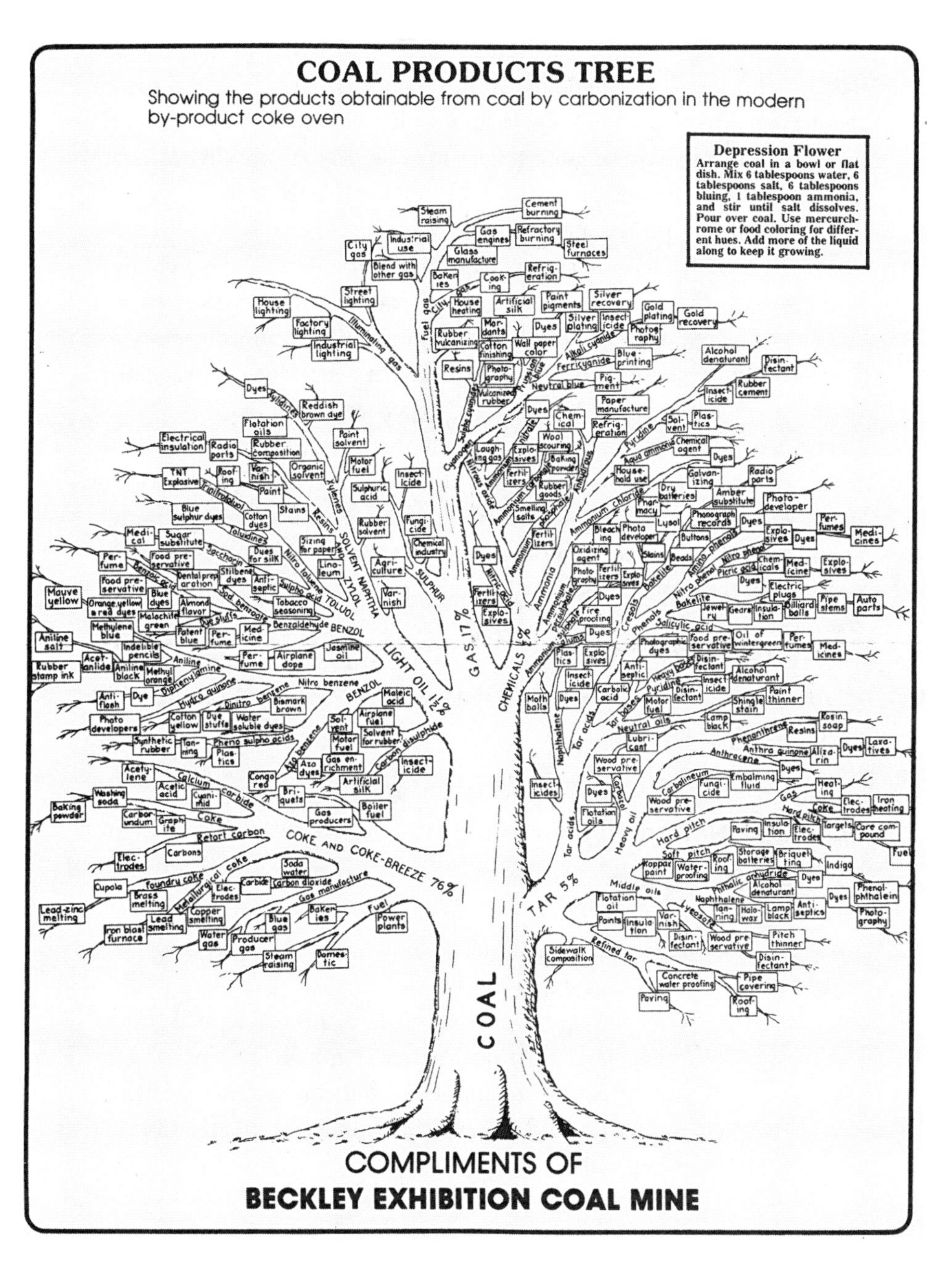

FIGURE 5.1 Coal products tree, Beckley Exhibition Coal Mine, Beckley, West Virginia, 2017.

ligent coloured emigrants, who have learned what civilization and freedom can do, who have had experience of the justice that public opinion can enforce when it has reason on its side, and who, having seen something of the evils and folly of enforced labour, may be expected to form a living barrier against the debasing and infamous trade in human flesh" (309).

The argument reveals many things about the colonial geological mindset, most notably an optic that makes inhuman objecthood in economic geology newly fungible as a "living barrier," another utility squeezed out on the way through this geographic reversal point of continental dispossession. Every circuit of his consideration brings further extraction operationalized on the status of personhood as geologic properties. Repair of the problem of slavery is suggested as being remedied by another geographic erasure, another forced disappearance in the optic rift of recognition, another fossilization in the volumetric depths of subjugation. Blacknesss, throughout, functions as an element and mineral in the materialism of economic geology. In the contours of the geologic life that Ansted championed, what transpired was that over thirteen thousand people were shipped "back" to Africa, and the system of convict lease labor extended the inhuman subjective mode in new guises in the coalfields of Alabama and Georgia to "develop" the economic geology of the South.

Fossils as Natural Resources

Fossils as Nat. Resources—Main label

7/16/80.	FINAL 8/7/80
MAIN TITLE:	Fossils as Natural Resource
MAIN TEXT:	Fossils are important to you. Over the eons of geologic time, plants and animals have concentrated the earth's riches through the process of life. After years of burial, this mineral wealth is available to us as coal, oil, limestone, diatomaceous earth, and phosphorite.

"After years of burial, this mineral wealth is available to us." In a reversal to the entombment of Blackness and its one-way time-travel ticket, mineral wealth "awaits" its extraction, patient for the geologist and nation-builder to arrive with their liberatory shaft, to categorize and release the fossil wealth into the light of day. Blackness goes down the shaft, and mineral wealth comes up, much like a mine in the Witwatersrand. Geologic grammars map out a classificatory language for marking symbolic difference and spatializing separateness out of the earth while codifying that difference in regimes of valuation, speculation, and property consolidation. To divide the earth into inhuman modalities and mineralogical objects, economic geology organized the earth through a commodifying optic, marking materiality as exchangeable, circulatory, and defined through utility. It is a first-order marking of the earth and the hegemonic inscription of the forces of geopower. While geologic languages of the plateau differentiate, marking the valuable from the invaluable, the gold from the slag, the diamond in the rough, the language of difference is concerned with making a stable entity of equivalence for exchange and interchangeability that bears no trace of geographic or subjective relation.

As the "Fossils are important to you" label makes clear, after years of burial, fossils are ascribed an *awaiting materiality*, one that anticipates the temporal identity script as "Natural Resource." In the exhibition text, life can germinate in materiality through heteronormative modes and progressivist desire. The "organic sludge—the 'mother' of oil" births matter for the fathers of geology to author (exhibition display, August 7, 1980). Identification is what allows minerals to move into the economy but also what arrests their movement in the earth and interrupts relation between situated geochemical material processes and more porous accounts of the subterranean existence and reciprocity. Methodologically, it is important not to replicate this imaginative division of *bios* and *geos* through an ungeographical elemental analysis. *Geologic Life* is not about establishing a new identity politics for geologic matter because these iterative modes of identification are how borders are established in the extractive language of matter through the interpolative matrix of material and intrasubjective positions. This bordering work of description carries minerals from the mountain, through speculative finance, onto the stock market, into the boardroom, through repressive state regimes and contract militia, to the intimidation and murder of environmental protectors, into the rivers as pollution and skin as toxification, through to the commodification chains of unfree labor and into the laptop on which I type.

Tracing these commodity chains of resources and their political economies is important for locating the geographies of resource extraction and modes of intervention, but what I want to do here is to dig into the quotidian grammars that facilitate these exchanges as nonlocal, universal and without geography. Put differently, *geologic grammars* are a site where specific configurations of extraction, dispossession, and enclosure are advanced in

FIGURE 5.2 Schoolchildren at the Hall of Oil Geology, 1955. Photographic Print Collection, American Museum of Natural History Library, New York.

the object formation of minerals in advance of extraction operations that actively work against the recognition of relation; a geochemical embrace of intrasubjective attachment, possibility, and exchange. Inscription practices give value and identity to matter, credentialize it, establishing the inhuman as a flattened form that is materiality stabilized for scripting, so that materiality plays its part in building the polis of the settler nation-state (and its capitals) in the spatial form of the plateau. As a mode of colonial differentiation, geologic grammars are a material manifestation of that geopower and a modality of disassembling relation, establishing commodity conversion and naturalization of the earth as resource through this bordering of *bios* and *geos*. The stabilizing of inhuman as nonrelation through the construction of "raw" material as resource (not country or ceremony) was a vital practice that initiated imperial ventures and continued consolidating its national projects through the establishment, refinement, and exercise of geologic codes as a foundational tenet of settler colonialism.

National identity functions via exclusion of racialized subjects, but inclusion of inhuman materials as national objects (which relies on the exclu-

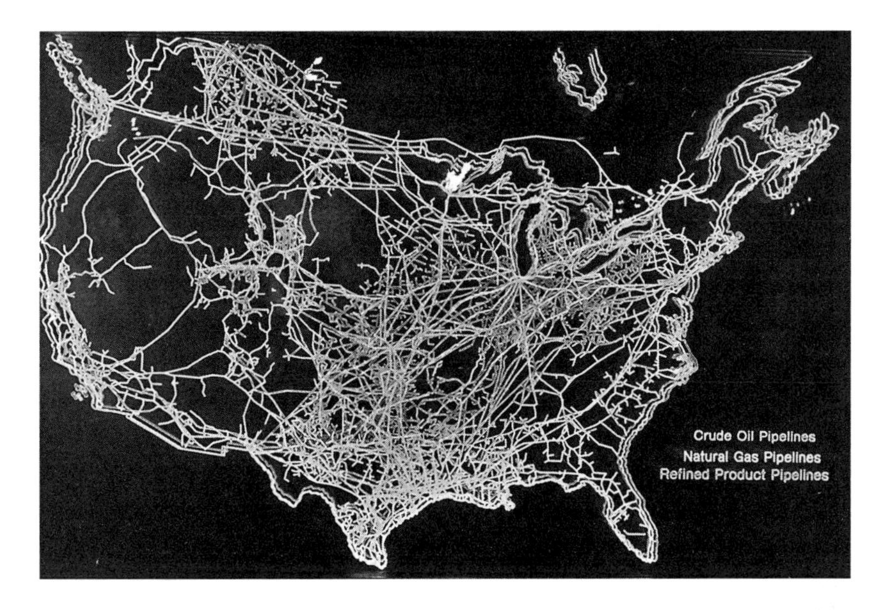

FIGURE 5.3 "Flesh of Geology" pipeline graphic. From *Black Gold: The Oil Experience* exhibition, Kern County Museum, Bakersfield, California, 2015

sion of relation to and with the earth), and specifically the negation of the recognition of that intramaterial relation for Indigenous peoples. As in the previous discussion of fossil-being, the voiding of relation is a strategy of extraction through the relocation of the power of origination in mineral and subjective registers. In the site and script of the museum, the narration and consolidation of the fossil as resource and national building come explicitly together.

The museums that populated Europe and North America used mineral collections to establish the persuasive material grounds of what might be called *seeing as an extractive state*—that is, seeing as a mode of possession and seeing as historically constituted apprehensions of the earth as a colonial utility (Overseer's earth). These geo-logics and their fossil optics were necessary for inscribing territory as co-productive (of settler and geophysical state); of desirable mineral beds, of alluvial planes for plantations, forests for lumber clearing, mountains to be mined, quarries for monumental statues and urbanism, aggregates, and sands to fill extractive rigs. As a discrete form of geopower that normalized the utility of earth beyond any prior or durational relation, geology was integral to the making and marking of imperial commodity forms and the accumulative geographies of colonialism's ongoing frontier of dispossession. Macarena Gómez-Barris names this optic the "extractive view" and states, "The extractive view sees territories as commodities, rendering land as for the taking, while also devalorizing the hidden worlds that form the nexus of human and nonhuman multiplicity" (2017, 5). Extraction was the last act and the originary desire of an earth empiricism of world-breaking. Thus, geology is a praxis for colonial world-building as it names the earth: building new worlds out of previous geologic epochs; building a model of the world for extraction (material, conceptual, infrastructural, and temporal); and building a space-time in the world as geophysics of sense that prioritizes and reinforces the geopower of the plateau. In the dual process of unearthing the strata for the colonial "now" and the erasure of surface (Indigenous) place relations, colonial earth operates from both above and below, desedimenting attachments and reordering relations through the matrix of geopower for the settler-colonial surface economies. Both culturally and geophysically, colonial relations of geopower have neglected the depths with disastrous effect (to peoples and environments).

Museum of Practical Geology

In the following text, by a Dr. Parr at the opening of the Hall of Oil Geology, the American Museum of Natural History in 1955, he demonstrates the interconnections of inhuman nature, origin stories, progress, and ethnoculture, whereby inhuman materials are the basis for the materialization of epic narratives about the nation to naturalize vertical imaginaries of the settler state and resediment belonging through Indigenous erasure:

> In this hall we deal with inanimate nature as a provider for human existence. We try and explain how nature creates, distributes and stores a family of substances of paramount importance for our daily existence and for the *progress of our industrial civilization*. We show some of the ways by which man locates this treasure in the earth, how he reaches it and extracts it, and a few of the thousands of things he does with it.
>
> The subject also has connections with future halls in the anthropological section of our exhibition program. In this we plan to deal with origins, development and diffusion of human cultures leading to a review of the sources, evolution, and contributions of our own North American Civilization right up to the present—in an exhibit we think of as the EPIC OF AMERICA. Petroleum is part of that epic.[16]

While paleontologists speculated on the material history of the earth, devising theoretical arguments on origination and species to present in metropolitan centers and plantation states, work in the "field," or "practical geology," as it was oft titled, did the accumulative work of categorization, consolidation, and extraction of geologic materials for the standardization of material economies. The contact zone of these two practices was the newly founded natural history museums that brought private rock collections and specimen classification, alongside industrial specimens and extractive cultures, into public spaces of display and narration. Hugely popular in Europe and North America, geology was considered both a fascinating meditation on the metaphysics of life and the material practice of advancing national interest through the sciences of economic geology (a geophysics of Life).[17] By the early 1800s, nearly all Europe and the United States had been geologically mapped and states held national and regional geologic collections of samples and cores as part of their earth archives.

Geologist Edward Hitchcock recounts in his popular text *Elementary Geology* (at least thirty-one editions were published between 1840 and 1859) the dual progression of practical and philosophical sciences "by which geology has been thus rapidly advanced. The most important application of this sort, was that of comparative anatomy, to determine the character of organic remains by the Baron Cuvier . . . [and] that of Agassiz on fossil fishes" (1840, 307). As extraction provided the fossils for the temporal construction of geologic life, so it also furnished the idea of distinct material forms of Life that defined civilization (and those that did not). In 1840, Hitchcock was one of the first authors to publish a paleontologically based "tree of life" that was set within the context of geologic time, showing two trees, one for fossils and living plants and one for animals, with a crown at the apex to represent feudal man. In his tree, through contemporary anatomy and fossil interpretation, organic and inorganic remain in communicative relation through branching genealogical relations (although the crown indicates the Great Chain of Being as a progressive unbroken course). The age of the earth and fossils (as the remains of life) became tied in this period, as fossils were the preferred empirical means by which to interpret the stratigraphic column and date rocks, rather than discerning the age by mineral type. Colonialism made the occurrence of fossils prolific, as the proceeds of slavery made the geologic societies that debated them.

Alongside state- and private-sponsored practical geology, geologic societies were set up to debate and discuss the origin of the earth and its formation, as well as more quotidian questions of how and where to extract

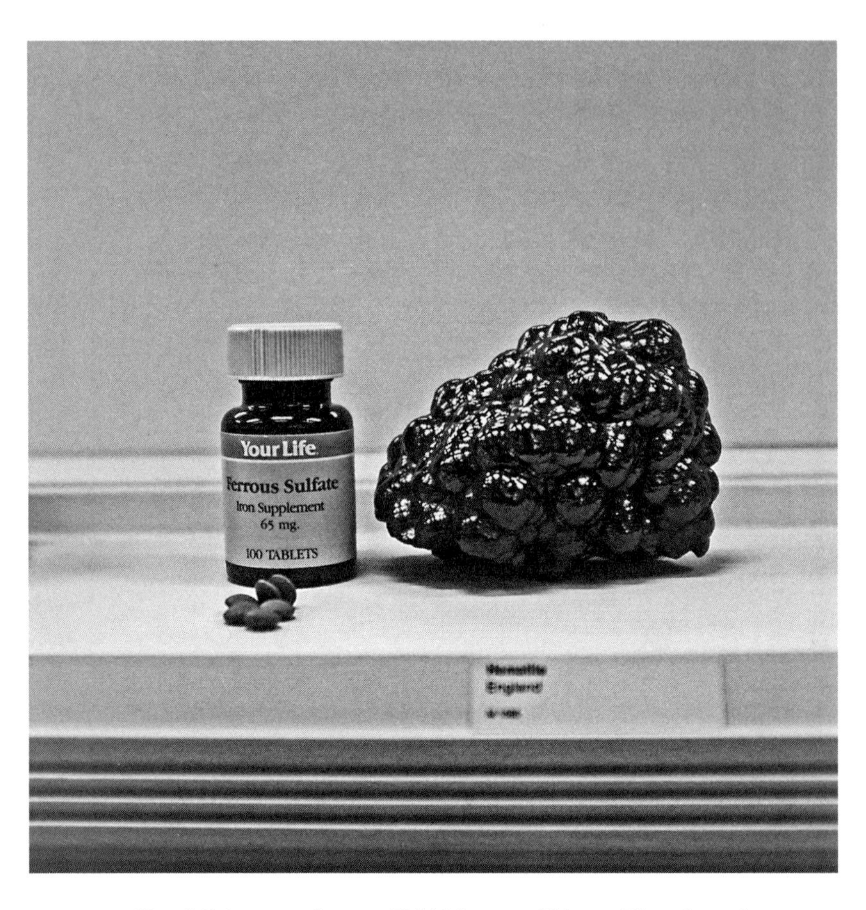

FIGURE 5.4 Your Life iron supplement, Field Museum, Chicago. Photo by author, 2017.

and "develop" mineral deposits. The earliest geological society was founded in London in 1807 as a specialist society dedicated to the exploration of the mineral structure of the globe.[18] Later, in 1851, the geologist Sir Henry Thomas De la Beche founded the Geologic Museum (now part of the Natural History Museum) and the Royal School of Mines in London (now part of the Department of Earth Science and Engineering at Imperial College London) from his proceeds from the abolition of slavery (and the compound wealth of enslavement before that). He was the son and heir of Thomas De la Beche, a plantation owner in Jamaica.[19] After his father died, he went back to the plantation, concerned about its declining revenue, and spent a year on the estate in 1823–24, writing two pamphlets, *Notes of the Present Condition of Negroes* (1825) and *Remarks on the Geology of Jamaica* (1827). In the geologics of accumulation, racial deficit aligns with extraction, and enslavement seemingly pays for its own observation. As with Lyell's publication on his travels to America, the natal positioning of Black life and geology was neither incongruous for audiences nor outside the purview of disciplinary scope (see Yusoff 2018a, 74–80).

At the time of De la Beche's writing, 201 enslaved persons were listed on his Halse Hall estate and 33 at Hanbury Pen (down from 291 in 1809). His biography recounts that his geologic illustrations sought to bring to life ancient scenes of flora and fauna (including parodies of geologists). While he was not a theoretician, his observations were credited as contributing to the "geologic revolution." He wrote the popular books *A Geological Manual* (1831), *How to Observe Geology* (1835), and *The Geologic Observer* (1851) and became the first president of the Palaeontographical Society. As owner of Halse Hall (listed as crop: sugar, rum, cattle, and hire of enslaved people) and Hanbury Pen in Clarendon, Jamaica, he was paid compensation for both plantations and is recorded in the Slave Registers of the Parliamentary Papers as receiving £3,523 11s. 9d. for 172 enslaved and £1,698 9s. 4d. for 88 enslaved in December 1835, respectively.[20] The Geologic Museum developed out of De la Beche's geologic library and collection of rocks, and he founded the Geological Survey of Great Britain and Ireland (1832–35), becoming director of the first national geologic survey in the world.

Across the Atlantic, in 1818 the American Geological Society was organized, yet Hitchcock says this does not amount to any contribution to the science as there was a lack of publications, but "an important feature in the history of American geology, is the numerous geological surveys that have been executed, or are still in progress, under the patronage and direction of the State authorities, as well as the United Stated Government. The leading object of these surveys, is to develop those mineral resources of the country, that are of economic value" (1840, 310). Unearthing the nation through mineral extraction brought geology into the sphere of consolidating nation-building, and thus narratives and practices of geology became formalized as an economic determinant of the polis and the reproduction of the plateau (or how to see like an Overseer). It is not just the environmental situatedness of relations that is disowned in the classification of the autonomy of matter as a stable entity in time and space—atemporal materiality—but in this subtraction from the world, the directionality from coloniality toward the project of modernity was accomplished (as a form of rescripting temporality while continuing with the same racial geo-logics). It is the imagination of geologic materials awaiting their role in the present, as Colson Whitehead sharply writes:

> The shrinking dregs of the Ice Age glaciers retreated and scraped through mountains in order to facilitate these modern highways; the final and supreme use of accumulated eons of pulverized stone is gravel for highway shoulders; the succession of rivers they pass merely affirms their progress like milestones, and the water cycle is just a little something on the side. He had come to believe that the intent of geological dynamism is modern convenience. . . . This substance biding its time through humdrum epochs for its ultimate deployment against Southern humility, the prevention of perspiration stains on Lucien's suit. The inexorable tending-towardness of all things. (2001, 191)

In Whitehead's novel *John Henry Days* (2001), the geologic epochs eagerly await their participation in the modern world, the intent of geologic dynamism ready to serve the state in its prosaic needs. Epochs arrive for a moment of glory in a polystyrene cup. Eager geology is compliant, supplicant, materiality scripted into the drama of the teleological accomplishment by technical means, oriented toward the western frontier; "Westward ho!," "Make America great again!," the metals and minerals are made to sing in chorus, manifest destiny is *mineral destiny*!

Scripting the Earth

Shifting focus to the detailed material orderings of the geologic, the museum functioned as an interface that generated both "purified" classificatory values and ontological discovery stories that were scripted by the geological fathers through the "Heroic Age" of European surveying (Bowler 1992, 193). While the history of Geology as a science, between the late eighteenth century and early nineteenth century during the "age of revolution," has been studied in detail (see Rudwick 2005) and scholars of extraction have paid attention to the role of mines in colonial practices and imperial expansion, the contact zone between *geology and subjectivity* as an integral part of the *colonial extractive matrix* is rarely discussed other than in terms of resource or labor economies, rather than as a geopolitical strategy for racist world-making and tactic for crafting racialized life-forms. Questions of ontology emerged out of intense debates among small groups of geologists (that are still ongoing) about the temporality of the world and the structuring of geologic bands—the Devonian, Silurian, Cambrian, Jurassic, Carboniferous, and so on. These temporal formations oriented understanding of the history of the earth and were made through speculative, globalizing, and imaginative

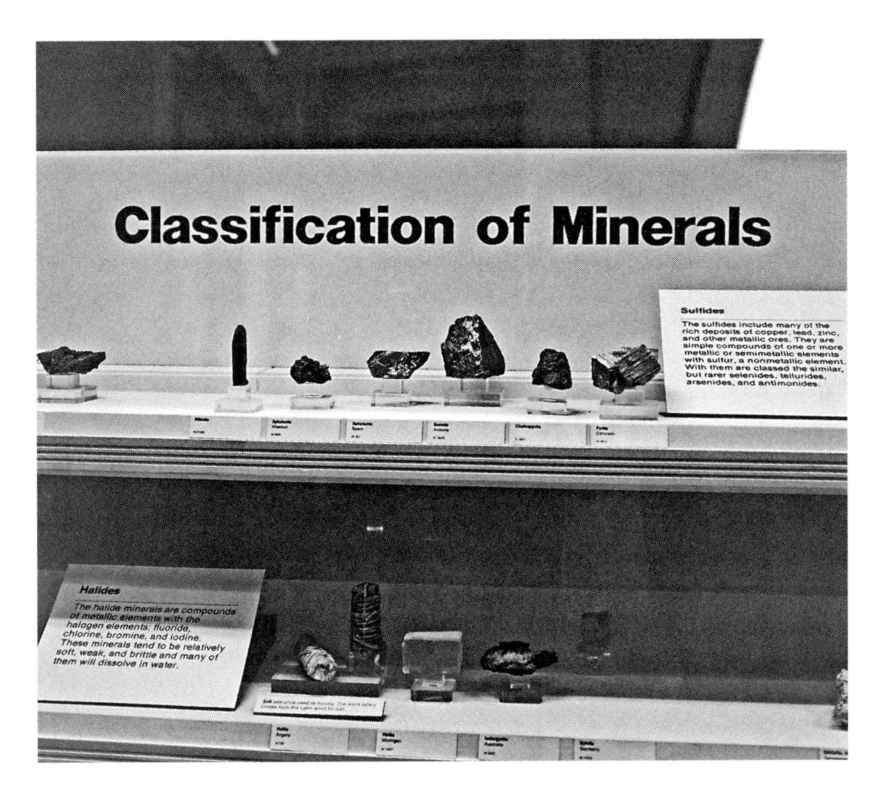

FIGURE 5.5 Detail from Classification of Minerals display, Field Museum, Chicago. Photo by author, 2017.

statements about the materiality of time and its universality. The ontological dimensions of this planetary geology and the universality of its epochal times are part of the way in which extraction was operationalized in its territorial, ecological, and subjective forms, yet this ontologizing is kept remote from stratigraphic jurisdiction. While the broad arc of geologic practices has been traced from mineralogy to the emergence of geology as a discipline concerned with the history of the earth, or geohistory, such disciplinary separations belie the stickiness of geologic theories in their racial structuring of philosophical, material, political, and psychic space.

Hartman argued in *Scenes of Subjection* (1997) that power is inseparable from its performance. The display of mineralogy and the modalities of its appearances in geology do everything to dissuade us from considering that this presentation is anything short of the natural ordering of the world. Nuggets of deep time. Yet the interlocking racializing axis of the grammars of geology remind us that this is a fallacy in the operationalization of geopower.

We know nature is power, but the obtuseness of making geologic materials elemental as they reach surface processes in categories of valuation is covered over by a seemingly transparent and autonomous "nature" of their classifications across the tableaux of minerals and metals. Marx would name this commodity fetishism, but there are other relations at stake that are obscured by the focus on the object of value (which prioritizes the accomplishment of the Overseer's view). The classifying imperative organizes an optic of extraction and a regulatory mode for materializing value, while simultaneously organizing processes of devaluation and abandonment (that would threaten the stabilization of value as value and thus the economic and social order). The performance of the inhuman is governed by a prior separation between *bios* and *geos*. In the parsing of Life (active) and mineral/metal (anticipatory but mute) into distinct categories and qualities of matter, imagined without amalgamation, the infrastructural realization of settler colonialism and industrial revolutions (in Europe and America) was facilitated.[21] The practical geology of colonialism operationalized the wealth that built the discipline and institutions of geology, as well as furnishing the empirical bedrock of fossils from prehistoric life-forms and contemporaneous organization of peoples via the anatomy of racialized subjects. Foundational to that mode of categorizing matter is the raced figure that grounds the distinction between natural and human history through a fictional filiation. Race is the erasure of amalgamation so that race can *become* the amalgam, which holds human and inhuman worlds together while generating separability or apartheid of forms of life. The ambiguity of being in the caesura between natural and human history, and spatialized as part of the prehistoric strata and the ground on which human hierarchies are built, ensures an origination without the possibility of transcendence of the designated subordinate origin. Eternal, elemental, "a single root that kills all around" (Glissant 2020, 37). The recursive racial figure, as Emerson and Lyell evidence through their "safe housing" of the "Negro" in the subterranean, is a means to still the trouble (and fear, constructed as threat), of classificatory mixing but one that the category of neither race nor rock can contain.[22]

Institutionalization and stabilization of categories of matter through collections and museums are means by which to operationalize the extraction and utilization of geopower as a deployable materiality and narrative form of colonialism, while simultaneously naturalizing white power through both the intensification of forms of capital accumulation and the extension of subterranean earth into property as part of the matrix of settler colonialism—the *depth model of colonialism*. While geologic categories remain subject to revision (in parallel with identity categories), the categorization remains underexamined as a given metanarrative of mattering that yokes together extraction and racialized forms. It should be noted that the classification of geologic materials is already a wreckage scene of relations—of colonial invasion, dispossession of personhood and land, disruption of reciprocal ecologies, racialization of geography, and the violent interruption of the clamor of world-making. The philosophical and practical together create and consolidate a material relation that was terraforming at a global and now planetary scale (called the Anthropocene), through an intimate relation of geophysics. Fossils as a national resource discipline a discursive process around materiality, as the naming of forms in both dimensions of inhuman categorization is part of how the regulatory and patriarchal structures of racism function.

Disinherited from geography, the Negro on the edges of European ethical consciousness for Emerson and the one buttressing African wildness in Freetown in Ansted's text both center European Man through the construction of a dis-placed, atemporal periphery. These disinherited figures are made to stand in the cosmos and deep earth in atemporal muteness, imagined as circuit breakers in the existential "nothingness" of inhuman nature, restaging the alien as race and earth. As the analytics of postcoloniality have shown us, modes of address and optics of recognition are made legible from this periphery for the center, further consecrating the legitimacy of this positionality. Museums of geology script plateau speech, as the economic geologists dig down and unearth using the Overseer's optics.

Conclusion: Fossil Life as Racial Capital

In this chapter I have focused on how the ontological organization of matter in formulations of mineral, metal, and nature has a quieter grammar in geology than the extraction sites of its execution, but they are no less brutal in how racialized lithologies are deployed as strategies of accumulating geopower. Processes of fossilization govern forms of life, especially in regard to the racialized proximities to the environmental excesses of fossil fuels in the present. If Blackness is inadvertently named as an enduring state in the fossil—a material permanence against nonrecognition in vestiges of time, wherein the fossil stands as an immortalized inscription in time—narratives of the *not-fossil* are a strategy of doing history otherwise. Drawing on survival as a different kind of stimulus, Wynter suggests how Glissant's poetics recognizes the possibilities to be had in this intramural space, as the orphan of organic life mobilizes a different material crossing of the inorganic. She says: "Glissant's Antillean human subject, coming to realize its cognitive autonomy not merely with respect to its knowledge of physical and organic nature but with respect to its knowledge of itself as a mode of life which exists outside the symbolic circuit of organic life, must therefore now accept the full responsibility of its position as a 'free outcast' who confronts 'the rest of nature as a trial, task, issue and enigma, as an alien abode'" (Wynter 1989, 646).

Wynter brings attention to how a social abandonment is also an ecological otherwise that breaks up the borders of classification to embrace the alien (and perhaps to dream across the grid of parallel geologies of attachment). Rather than fleeing the scene of these categories of enclosure, reworking essentializing tendencies and tendrils toward new modes of narration, organization, and reclassification as a way to reject the racial grammars of geology. To be at the rift edge of the plateau of Western reason and its ethical purview is also to be beyond and free of that reason, and thereby unconfined and delimited by it, in alien circumstances, but in nonextractive Relation. In the politics of nonrecognition there is an inadvertent admission of possibility in fossil formation. The fully cosmic life that Western metaphysics forgoes in its fear of the abyss is opened in this equation. Recognizing the antagonism of inhumanism, as I believe Glissant does, is a liminality that can be embraced and reworked as a space of possibility and ungovernability in the politics of materiality to shift the terms of subjective enclosure.[23] The work of invention and poetry in *cosmic materialisms* becomes a crucial craft of survival and existence in a differently imagined geography. Put differently, as deep time is an injurious category of ontological differentiation in the geologies of race, it has also been historically reworked as a site of engagement with the fugitiveness of other space-times that efface the extractive relation and extend out into the universe beyond its narrow margins into the breath and breadth of a cosmic sovereignty. Inhumanism is a site of further work.

Paleontology—the history of earth as it was developed through fossil study and comparative anatomy—is tied to the epistemes and genres of ideas of ethnocultural development and thus forgoes a radical revaluation of time and being in favor of political gratification.[24] It was the syncretic advancement of geology as a study of the history of the earth and the history of life that created forceful alliances between the "times" of matter and ideas of stratification, which naturalized racial ideas. These historic

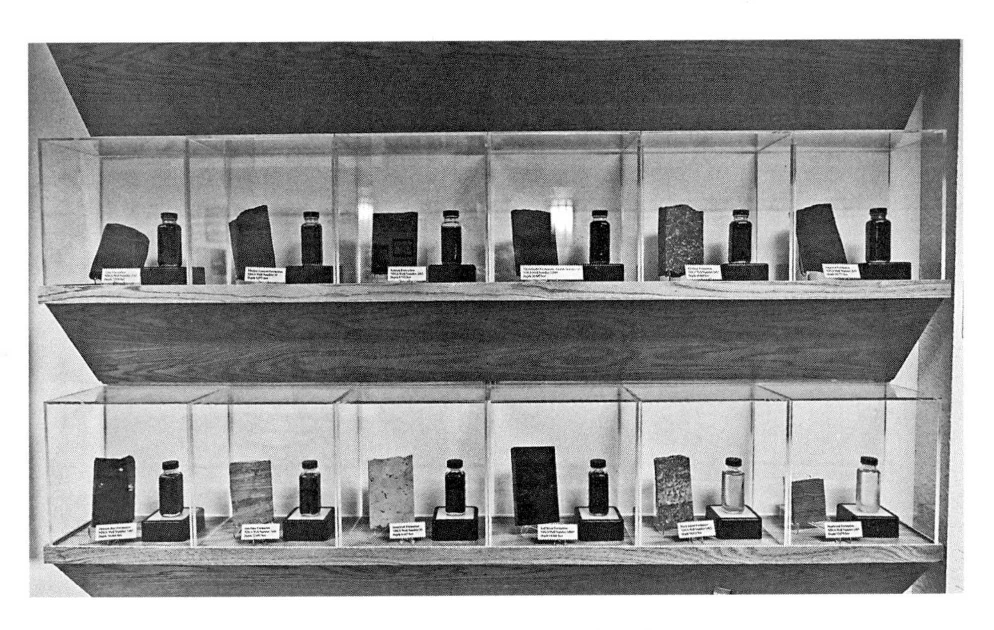

FIGURE 5.6 Bakken oil samples, Pioneer Gifts and Books and Long x Visitor Center, Watford, North Dakota. Photo by author, 2016.

geo-logics produced racialized bodies as the site differentiation and intense description, via the category of the inhuman. As race takes on a role in the mediation of inhuman materiality, it enables forms of spatial execution and expansion. Analogous claims about colonial extractive practices (narrated through the transformation of materiality as the consequence of the natural war between races) and aesthetic claims of the supremacy of a cumulative ethnocultural being (immunizing Man from his sadistic violence) positioned him at the apex of deep time's progressive work through a "purposeful" transformation of earth. These rapid changes in ideas about the history of earth and life were made in the context of imperial practices that forced "open" worlds to systematic extraction and subjective expulsion, while colonialism instigated new relations with the underground—instigating rifts—to "settle" colonial states as naturalized.

In contrast to the colonial geo-logics of matter (value made via the identification of purity), Glissant, the Caribbean thinker, names the intrasubjective nature of identity and matter as creolization, in the wake of these histories—a mixing that already is in relation with earth and others because of these colonial entanglements and the enforced material openness to the world. Glissant acknowledges how human and nonhuman forces shape being, forces that are always pluralized and predicated on the exposure between human and natural history (creolization, for Glissant, was not just a human affair). These long, interrupting, and overlapping histories of colonialism have what Glissant calls "elemental" contradictions of Relation that require an abandonment of essentialized categories that make no space for the contradictory and unusual. He says: "Creolization seems to be a limitless *métissage* [cultural creolization], its elements diffracted and consequences unforeseeable" (1997, 34). The unusual, for Glissant, was not that Caribbean history was punctuated by colonialism, plantations, and mining with its countermaterial histories of marronage and those with "the smell of oil on dusty skins"(3), but that these dialectics of subjective-spatial forms are broken by the specifics of the actual geologies of Caribbean experience, as a site of islands porous to the sea, of mangroves, volcanic eruptions in the depths of contiguous relation, earthquakes, and cyclones that mobilize sedimentary Relations; a physical geography that was similarly changed by colonialism, as monocultural production caused shifts in ecology and changed the local atmosphere, increasing vulnerability to hurricanes, as the eradication of Carib volcanic knowledge increased exposure to eruptions (see Last 2018).

Imagining the *geos* as a counterforce and geochemical companion of the depths, Caribbean literature embraced a geographic definition of its resistance *from within* rather than *from without* the inhuman, thus making a virtue of archipelagic thinking and a praxis of sedimentary epistemologies as an intervention in the transactions of understanding of space and time. This is why the postwar, ex-slave, ex-colonial Caribbean, Wynter argues, provided the site for a struggle, "sited on the new terrain of being, of modes of subjectivity" (1989, 640). This "daring new methodology," Glissant suggests, must respond to the "confusion over our relation to time, a ruined history, which we must give shape, restructure" (1989, 65, 244). Accounts of matter that pay attention to the unusual in its ability to reorient and churn subjectivity into different forms beyond the dominant politics of recognition by the state enact a counterrelation to racial capitalism, a relation that has the potential to disassemble rather than reproduce the current extractive modes or subjugation (see Agard-Jones [2012] for a recent example). Extending the understanding of the body's ability to enter relations and experience affects with the inhuman world beyond the nomenclature of extraction discharges a differing force into the world that conditions, positions, and arranges experience in nonnormative modes as a methodological repair to the geophysics of race.

Geology matters precisely because it is seen, in its coding of inhuman materials, to be radically *im*political outside of the localized environmental context and labor of extraction, and its value economies are understood as self-evident, but it is a sphere of metareproduction in which modes of life are made to appear and disappear (especially in its founding grammar in chattel slavery and racialized geo-logics of species). Considering geology's role in assembling forms of Life and how the axis of demarcation in epistemologies of matter enacts the bordering of inhuman life, what other grammars of geology come to the surface if colonial materialisms are not taken as the master narrative? What accounts of time and space were forged in the spaces between and in countergravity to the enfleshed internment in the category of the inhuman? How does the exilic subjective condition of the inhuman turn dispossession toward other forms of inhabitation that campaigned and campaign against the valuation of extraction?

If we change the register of how geologic life is thought, in a nonextractive mode, the structures and foci shift. Then writing demands something else apart from mastery and mining, which characterize the material acts of geology in the world and much of the geopolitical writing that attempts to situate those acts in the political (often at the cost of eliding the difficult subjective operations that bleed across clear boundaries of matter and subjects and thereby consolidating a transaction that promises a "better" extraction). As Singh's project on mastery critically suggests, "by continuing to abide by the formulation of 'mastering mastery,' we remain bound to relations founded on and through domination. In so doing, we concede to the inescapability of mastery as a way of life" (2018, 7). The same is true of extraction. It is precisely the grammars of mastery over matter—white geology—that normalize the force of extraction, naturalizing and domesticizing the range of the social and ecological injustices produced through these practices. Singh suggests that mastery is a mode of maintenance in keeping relations in place across matter and narrative "in ways that allow and disallow particular material actions" (17). While there may be better and worse forms of extraction, there is no way to do away with extraction entirely (given that its infrastructures, such as nuclear waste and climate change, are already structuring the future far in excess of the temporality of the "now" of geopowers' surfacing). There are other less deadly ways to mine a seam and build a world, but these cannot be enacted within the current conceit of inhuman materialisms that naturalize matter as already autonomous—as resource or contemplative medium—erasing the normative formation of geology as racialized. Turning to inhuman intimacies to give the *not-fossil* a history in the amalgamation of the earth's occurrence is one tactic for the disassembly of this colonial earth.

Geologic grammars, as they were developed from the seventeenth century to the early nineteenth century, are colonial languages, constructed with the explicit apprehension and conversion of matter for possession through dispossession—building privileged plateaus through rifts in time. They are grammars of valuation designed to conceal the localized environmental and subjective relations on which extraction depends to constitute its objectified-mineralogical form—Relation that cannot in fact be erased. Thus, geology emerged as a material practice of inscription, management, extraction, acquisition, valuation, hierarchization, and ownership of matter (human and inhuman). Such geologic grammars forge and force actual practices, further reproducing this depleting world, unevenly and catastrophically, both in the formation of broad social and geologic worlds and in the intimate possibilities of life that are sutured to that material-symbolic praxis. *No geology is neutral.* Therefore, *Geologic Life* takes aim at a structural erosion of the naturalized and normative paths that ground Life in taken-for-granted concepts of geologic time and material worlds, and its languages of apprehension, recognition, and affect. Singh suggests that "it is not merely that the subjugation of environments is intimately linked to the subjugation of peoples; rather, it is that the logic that drives the modern world cannot formulate the nonhuman world as one invested with meaningful, dynamic life" (2018, 18). Rather than more Life, or vitalism, what we might need is less investment in Life and the crafting of a politics of the inhuman that takes Césaire's "thingification" of the enslaved as a geo-logic that does not have redemption within the current dialectics of Life—inhuman life—but requires a revaluation of its divisions, and pushing further into the inhuman as a potential political and liberatory category. Where staying in conceptual proximity to geotrauma is a means to navigate and perhaps loosen the violence of its passage into the present.

Shifts in language and foci aim at a structural exposure of those barely invisible relays in the grammars of geology that furnish the establishment of other regulatory concepts, such as biopolitics and its categories of the less-than-human (the subhuman, inhuman, and nonbeing), white Geology, settler colonialism, genealogy, normativity (and its arrangements of family, capital, whiteness, kinship), and anti-Blackness/anti-Indigenousness/anti-Brownness. The splitting of materiality into classificatory units through an extractive logic of language pertains to the quality of apprehending matter and how matter is separated from environment, place, and relation in its commodity or propertied form. Intimate links and shared modes of objectification exist between forms of description of matter and modes of classification in the organization of geologic materials into extractive materialities, and through the seemingly innocuous description of inhuman matter as earth histories (or a historic medium). There is a collaborative juncture in this intimacy as a mode of sense-making, between matter and thought, that coheres in language to bring into being extractive regimes as both earth imaginary and governance. In short, geology is context and agentic ground to earth relations, precisely because it names and "owns" the naming of time. That is, geology names that conflict zone between biocentric organizations of Life (or biopolitics) and geologic means of production (temporal designation).

Disrupting the "voyage of time" parenthesis between time and life (referenced in Terence Malik's 2016 film) and introducing the asymmetries of life's uneven distribution and racialized forms unsettle the coterminous narrative of accounts of time and life through geology. Collapsing the spacing between *bios* and *geos* is one way to ratchet open the biocentrism of Life (that objectifies time as its accomplishment) and to rethink *geos* in a parallel implicated material and subjective register. The other is to rethink the mastery of matter as constructed through a parallel methodology of elementary thinking that produces metals and minerals as geochemical isolates in the present. The *elemental performativity* identified in geology is paralleled in accounts of white supremacy as a concomitant process of extracting and maintaining regimes of value through racial reification.[25] What emerges from the historical materialism of colonial geo-logics is the elemental figuration of a targeted subjective life that becomes paradigmatic of racist relations. The elemental conditioning of race is not, however, ontological but geophysical. The geotraumas of genocide and slavery are obscured by a material substitution, whereby the standardization of racialized material processes through "natural resource" alleviates the wreckage of human signification that is left in the wake of colonial passages of extraction. Grammars of geology falsify a history of nonrelation.

6 Stratigraphic Thought and the Metaphysics of the Strata

There is a blue pectolite, called Larimar, that is made only under the volcano in the Dominican Republic. Minerals have migrated into the spaces in the flow, driven by incandescent matter, to form a foamy errant blue. Exuberant, like the sea of Glissant's alluvial geopoetics, the rock holds an image of sedimentary power (alluvium was one of the rock formations proposed by Giovanni Arduino in 1759). Larimar performs a kind of visual tidalectics with the oceanic all around, materializing Edward Kamau Brathwaite's (1983) "submarine unity" and Glissant's (1997) communication between the volcano and the sea. As with most attractor stones, the conditions of extraction involve poverty and danger, with scant reward for its informal miners who go down the shafts that pocket the landscape. Some of these rocks remain in situ under the volcano, and other stones are dragged into the mouth of the Bahoruco River and deposited onto the beaches of Barahona and into the waves of the Caribbean. The gem-rock releases a way of seeing bubbling forces held in the fluidity of geochemistry. Foamy swirls, tidal pools, aggregate earths, a holding of the opaqueness of light which I have seen only in crevasses in Antarctica. It is an aerial perspective of the sea. A punctum of the shifting quality of the duration of inhuman time and matter. A rift of movement in the rigidity of rock that is endlessly open to elemental narrativizing and uncanny orogeny. Geology is also a form of expression that draws down the metaphoric into specific geophysical registers that expand the imagination of inhuman bodies and worlds.

Geologic Life is about the rocky continuance of the world and its stratal layers of compression in the admixture of being ore (Fanon) and the wealth of ore (white geology). About how the racialized surfaces of the world press on the stratum below, compressing flows, and hardening what is captured there into stratified forms of Life. It is social geology and provocative geophysics. I start this chapter with a rock (Larimar) that captures the intense articulation of the inexhaustible relationship between the communicative matter of geology as an aesthetic nomenclature and its specificity of occurrence (only in the Dominican Republic). Larimar is a reminder of the geographic and imaginative scope of geology, a countergeology, to hold in mind alongside the understanding of the strictures of stratigraphic imagination to be discussed later: a spatial imaginary that stresses capture, sequence, and causation.[1] And, above all, containment. Stratigraphy was both a science of process and formation, and an outcome in white geology that sited corporeal and material zones for extraction.

In this chapter I want to show how extraction is racialized as an affective and subjective infrastructure during colonialism through the concept of *strata* that *stratigraphically positions* the social hierarchy of relations to capture geopower and is deployed as a diagnostic of society. It is not simply that empires are built on metals and fossil fuels that engage stratigraphic rearrangements to mobilize capital, but that these raced and gendered colonial geo-logics organize the ongoing geophysical architectures of racial capitalism, such as racism, climate change, extraction, and extinction, among other forms of enforced (near and far) finitudes. Stratigraphy was not just a technique but captured the geosocial imagination as a concept, methodology, and worldview.

Stratigraphy started small, as a localized tool for identification of mineral deposits. Nicolaus Steno (1638–89) defined principles of stratigraphy through the "law of superposition," which detailed the successive surface deposits as layered through time, with the younger on top and the older underneath subject to physical and chemical processes. Steno's law designated strata to be the outcome of a sequential history of deposition and thus was a locational methodology. William "Strata" Smith's map of the United Kingdom, for example, became important only once it was realized that this was more than a localized field guide for extractables. Ironically, Smith was poor (an orphaned child of a blacksmith) and lacked the means to

FIGURE 6.1 Coal lignite strata strip mining, North American Coal Corporation, Falkirk Mine, Underwood, North Dakota. Photo by author, 2015.

publicize and protect his map, and so it was plagiarized by the good fellows of the Geological Society, and he was regarded as an upstart who needed to learn his social place (notably in the economic undergrounds of debtors' prison). Nonetheless, through endurance and working on drainage projects for country estates, mineral maps for coal deposits, and coal canals, Smith's stratal map eventually became a geologic imaginary rather than a mere ratchet for extraction.[2] What was distinctive about Smith's map was that it was a national imaginary, that enabled both a global understanding of geologic time (which the British Empire understood itself as the center of) and a more parochial notion of *natural* resources as *national* resources.

Geology was a young theoretical science (compared with physics, chemistry, and biology), but it had an older disciplinary incarnation as a material science of minerals and acquired its specific industrial social sedimentation as a colonial praxis through settler colonialism. Maps of economic geology were widespread since 1750 in France (Guettard), Italy (Arduino), Germany (Füshel and Lehmann), and elsewhere in Europe. As early as 1727 John Strachley described strata as global foundations through his analysis of coal beds in Somerset in the United Kingdom. In a wider context, in England, as in the Americas, practical geology was part of the praxis of turning common land into property through the privatization of natural resources, particularly watersheds, using ditches, drainage, and diking alongside geochemical soil augmentation and mining excavation. The two branches of geology—minerals/metals (economic geology or geognosy) and fossils (comparative anatomy and theories of the earth)—remained separate as distinct fields of study until geologists began to think stratigraphically about fossil occurrence, joining economic extraction and the production of life in a *hermeneutics of strata.*

Theoretical geology provided the conceptual armature that initiated an epoch of *vertical thinking for surface gain,* which in turn opened *inhumanism as the vertical strategy* (or shaft) for the modality geopower. This new geologic method went to America as a form of geologic life that both conditioned the transformation of earth and forms of racialized life and established the geologic apartheid of white geology. Superposition of strata became yoked to imaginations of the hierarchies of suprahumanism in the reciprocity of fossil and mineral life under colonial inhumanism.

It was in the United States that the first formal professorship in geology was established at Yale in 1802, followed by the appointment of a chair of mineralogy at Cambridge University in 1808. (Prior to these posts there were mining professorships in Germany; for example, Johann Wolfgang von Goethe was appointed in 1776 as director of mines and mining, and he had a prestigious rock collection and theoretical engagement with geognosy, and Abraham Gottlob Werner was head of the Freiburg Mining Academy in the late eighteenth century.)[3] What becomes distinct about the coming together of economic development and geologies of life in the late seventeenth century as a political geology is that it forges a language that makes a geographical imaginary of the whole of space and time that furnishes the speech-acts of colonial earth. Which is to say that geologic life becomes the ontology and origin story for white supremacy (and its means of infrastructural reproduction as an affectual architecture of racial extraction).

Strata

In terms of an *episteme*, the identification of strata was a way to locate minerals and metals of value and later fossil fuels, for the extraction (and, in the case of gold and silver mining, so you could mint your own currency and stabilize economies of value and property). At the same time as it was a *locational device*, stratigraphy was a geographic imaginary that constructed ideas of geologic epochs and thus participated in temporal world-building. The application of stratigraphy organized the *material production and imagination of empire,* underpinned by the geopolitical model of the United Kingdom, where coal *was* empire, sustaining its duration as a geophysical space-time entity: the empire "on which the sun never set" was maintained by the geological debt relation of buried sunshine (along with stolen persons and land). Because of the close connections between "practical geology" in development of understandings of strata and the expansion of colonialism, and the subsequent theories of the formation of conjoined geologic and political worlds through the identification of minerals, metals, and fossils, the development of race was historically concurrent with these material empires as a geophysical condition of the state. As stratigraphy ordered the materiality of time through sequential strata, comparative anatomy ordered bodies of species through their differences, and both made the material possibility of empire. That is, geology is a differentiating machine in its first instance, simultaneously producing codes of temporal and racial difference. As stratigraphic thinking was applied to rocks it was simultaneously applied to race, around notions of racial difference and ideas of the "fixed" notion of racial hierarchies and strata. What is potent about these geosocial formations of race and earth is that they organize the "consolidation" of empire and of settler colonial societies (and the movement of geologists from Europe to North America). Becoming part of settler colonial geo-logics of settlement through combined geosocial structures, the stratigraphic imagination was both a *material hermeneutics* and a way of doing *racial metaphysics via geophysics*.

While geologic strata acted as geographical devices for unearthing, they also became theoretical models in geologic societies and popular geology to map and imagine modes of control in colonized lifeworlds. In the social and philosophical sciences, the concept of strata was deployed as a way to understand differentiated material forms of life as a psychic form of power. The social terrain of stratigraphic thought arranged a model of historicity and materiality that both explained the surface and present, and accounted for its emergence through accumulative historical forces (which presented colonialism as a fait accompli of nature rather than a contested geopolitical form). Thus, stratigraphy could accommodate changes from fixed to processual thought and more causal logics of arrival, while producing an account of social difference as temporally conditioned and yet determined by history.

The geologic conception of stratigraphy became a lexicon for thought that continues to underpin social theory today. Try thinking race, class, gender without the idea of "formations." Or the structure of the psyche or the bourgeoisie without stratal relations. As a causal geo-logic that explains the depth-surface, past-present relation, it offered a model of social history for society and the subject as a unified plane of experience. Thus, the stratigraphic imagination naturalized subjugating racialized relations in its geologic wake as an inevitable outcome of historic formations, even as it sought to critically analyze their occurrence and disrupt their futurity. Through strata, what started as the desire for the fixity of genealogical life—as white supremacy—was rendered as a mode of geologic life that posited a hierarchical and stratified way of being human (as it simultaneously divided the earth in political formations through colonialism to accrue geopower). Underpinning the stratification of thought and political geology was the ground of the inhuman.

As colonialism was being theorized by paleontologists as a natural emergence of *the* social surface, and thus universal and white (so without the need to acknowledge the racial deficit it is built on), simultaneously, the enslaved were spatialized in the depths of strata (in racial undergrounds, in the "holds" of slave ships, in the seabeds of the triangular trade and the dungeons of the Gold Coast, in plantations, prisons, and mines). As Indigenous land was stolen, and Indigenous people subjected to cultural, viral, and gendered violence, their presence was literally being erased off the earth's surface through an imagined stratal emplacement in the category of extinction (enacted through repetitive genocidal acts). At the same time as stratigraphy was being applied to the geophilosophical organization and extraction of colonial worlds, against Indigenous and enslaved persons, the accumulation of wealth in European centers and its burgeoning of institutions of science were further consolidating the construction of race in its geophilosophical imaginaries. While the inhuman world lent an "innocence ground" of stratal thought to philosophy, it did so with an entwined history of race and earth and its deadly carceral geophysical states.

As Western philosophical and capitalist imaginations were securely based in a stratal imagination, postcolonial philosophies were embracing the abyssal, the void, and the volcano as a way into, and out of, the brutal superficial dimensions of inhuman classification (Antillean thinkers such as Glissant, Fanon, Césaire, and Wynter). While postcolonial thought happened in the context and structures of the colonial imagination, anticolonial theory—Glissant's, for example—pushed *farther out* into the seas of chaos and forms of being where these geophysics of sense started to dissolve and throw up their nonhistorical histories and disinherited ferment. As Wynter's rogue Marxism underwent a Black metamorphosis.[4] The geographies from which this thought emerges have multiple origins (sea, colonialism, island, plot, spirituals, hold) and thus it challenged the purity of elemental thought through the density and complexity of inhuman epistemologies. For the colonized, the underground and undergrounding (as a verb) became a way of crafting a geophysical being—a subterranean ontology of sorts that sought to disrupt stratal modes and imaginations of containment and violence. Fanon seeks the psychic underground, berating Freud for his forgetting of the sociogenic principle, saying that he mistakes the substituting phylogenetic theory for the ontogenetic perspective. This epistemic space is also geographic. Fanon echoes this in the beginning of *Black Skin, White Masks* (1986 [1952]), referring to the concept of nonbeing as a sterile and arid region. The politics of strata then entailed a stratified order and a corresponding topology of resistant racial undergrounds, epistemic rifts, and thirsty places that sought its disruption. Methodologies of the rift challenged the stratigraphic method and the materialism it shaped.

The Stratigraphic Method

> Stratigraphy is the study of layered materials (strata) that were deposited over time—their lateral and vertical relations, as well as their composition. The basic law of stratigraphy, the law of superposition, states that lower layers are older than upper layers, unless the sequence has been disturbed. Stratigraphy is the study of the processes of stratification and the outcome of stratal formations.
>
> There are three kinds of unconformities: disconformities, nonconformities, and angular unconformities.

Beyond its usefulness in determining where to find the shiny, stratigraphy structured a paradigm of time and relation, of surface and underground, mapping back from the present into the past, giving foundational form to the hierarchical and bounded (and the impassable containment of those strata). The permutation of stratigraphy went well beyond the natural sciences and extractive geosciences and saw a flourishing in ideas of social theories through the nomenclature and metaphor of the formation as the *outcome* and *process* of social, political, and economic forces (see N. Clark 2017; N. Clark and Yusoff 2017; Yusoff 2017). In diagrams of society constructed through the idea of formations, geologic grammars became sutured onto the social and its modes of critique as a foundational grammar of Western epistemologies and new sociologies of political thought. There is no social theory without formational thinking; no social strata, no archaeology, and no concept of genealogy without paleontology.

The homogenizing concepts of formations and the social strata of gender and class became well mapped in the twentieth century, reliant as they were on the conceptual and material architecture of race stratification in the seventeenth through nineteenth centuries as their point of origin. The development of racial difference through stratigraphy inhabits and haunts all social structures of thought. Thus, race is not just a social differential (such as class, gender, sexuality) but is constructed as foundational to the spatial operations of stratigraphic ordering. While stratigraphy was deployed as a structural model for hierarchical relations under capital (Marx), conditions of the psyche (Freud), aggregate in determinants of society (Antonio Gramsci), institutional forms of repression (Foucault), subject and sovereign formation and so on, the forgetting of the racial origins of this method has deep structural consequences for the whiteness of social theory. And, while deconstruction, poststructuralism, postmodernism, and so on, sought to take apart the rigidity of structuralist (née stratigraphic) thought and opened the doors to a multivalency of value and signification, they failed to deal with the foundational racial roots of stratigraphic relation in its archaeological method and thus the racialization of its own thought. The ordering of stratigraphic thought permeated the imagination long after it receded into the background of what was taken for granted in the mechanisms of the philosophical imagination itself.

While deep time required substantive speculative theorizing about the earth, its formation and processes produced a geophilosophy of sorts. It was not deep time per se that became foundational to thought, the existentialism of extinction, void or cosmology of time, but the methodological structuring device that offered a locational account of the ground and an explanation of historical forces. Strata provided the imagination of social life as a seemingly naturalized mechanism of relation and simultaneity of occurrence. While Georg Hegel in the 1817 *Jena Encyclopedia* infamously rejected the importance of the strata, suggesting "this temporal succession of the strata, does not explain anything at all. . . . They are, however, hypotheses in the historical field, and this point of view of a mere succession in time has no philosophical significance whatever" (quoted in Clark and Yusoff 2017, 4), he embraced a dialectical reasoning that distinctly resembles a subtending stratal arrangement. The dialect *is* the superposition. And dialectical thought required a stratigraphic supplicant. Like Kant's reaction to the Lisbon earthquake, the philosophical impetus was toward a stilling of the earth, which was not necessarily a response to the planet's inertia but its opposite (of the amassing of evidence of its propensity for violent self-transformation), and a desire for thought to repress and contain this inhuman violence. Marx studied the relation between the proletariat and the bourgeoisie like a stratal formation and understood the proletariat as a subservient substratum feeding the rich. The relation between strata gave an account of causality as hierarchy: disrupt the hierarchy and you can change the arrangement, or so the thinking went. The eschewing of the racial origins of the production of difference cannot be an oversight, but it belongs to the repression of what constitutes the "ground" or *geos* of philosophy and its institutions. The politics of strata became fused into the political ontology of the subject and society, and a way to understand the relations between different political "times" and social milieus—alongside a means to conceptualize the impact of crossing stratal times (as with fossil fuels), as a mode of transforming the present and the future through the restructuring of the past (stratal retrieval as a mode of historicity). To study the geosocial matrix as it pertains to both the geopower and the possibility of social life is not to add the geologic (or inhuman) as a supplement to an already existing biopolitical or sociological conceptualization. Rather, it is to register that the geologic is already a mode of material and geophilosophical expression that is active within these formations, yet often bracketed out, and its political potential—or geopower—as a sedimented strata of racialized power within socialities is underplayed.

Geosocial Formations

The idea of *geosocial formations* as a mode of encounter for dynamic processes in the sciences and social sciences and how social and political agency is both constrained and made possible by the forces of the earth itself is one methodological approach to account for strata (Clark and Yusoff 2017). Noting the dual meaning of "formation" as *process* and *outcome* in late eighteenth-century geologic thought, both the geosciences and the social sciences shared a sensibility that the emergence of the "new" is made possible by the orderings that have materialized at previous junctures (as with Deleuzian ideas of plateaus and their politics of strata). Geosocial strata as a concept foregrounds a vertical interchange with the geologic depths and temporalities in the organizing of social and material worlds (Yusoff 2017). For at least three centuries most social thought has taken the earth to be a relatively stable platform on which dynamic social processes play out.

Western theory has engaged with surface-led theorizing and attachments, eschewing the depths and durational in order to adhere to a homogenizing place of social interaction. Looking at some of the key social formations of Western thought (and the Anthropocene as a thoroughly Western concept in its current incarnation), I want to show how geology has been hiding in plain sight, actively modeling thought as metaphor and social structure, even as that thought has been primarily concerned with the dynamics of a biocentric era and earth.

James Hutton published the *Theory of the Earth* in 1788, promoting a view of a "uniformitarian" dynamic earth. It was a theory of the earth that was not necessarily directed toward an end in its perpetual formation but that he understood as machinic, driven by a molten core: "When we trace the parts of which this terrestrial system is composed, and when we view the general connection of those several parts, the whole presents a machine of a peculiar construction by which it is adapted to a certain end" (1788, 209). Abraham Gottlob Werner, professor of science in the School of Mines in Saxony in 1775, describes the globe as "universal formations" over "chaotic fluid" (quoted in Lyell 1997, 11). However, Lyell notes that nobody was particularly interested in his observations about the natural position of rocks and their distribution ("served for little else than to furnish interesting topics for philosophical discussion; but when Werner pointed out their application to the practical purposes of mining, they were instantly regarded by a large class of men as an essential part of their professional education" [Lyell 1830, 55]). These stratigraphic imaginaries and geologic theories matter to the material, mythic, and subjective registers of the production of space, time, and matter (the building blocks of world-making and its materialisms)—specifically, embodied geology and spatial relations that are predicated on racialized relations in the paradigm and infrastructure of extraction, and in the afterworlds of colonialism. In the traffic between earth and philosophical thought for imperial conquest, the influence of the idea of geologic formations impacted on the idea of social formations. This includes attempts to consider how social and political agency is made possible both in practice and in thought by the forces of the earth itself (so, thinking about the earth as an engine for thought, rather than just something we think about).

Stratigraphic thinking informs sociology and understandings of social hierarchies and movement, as well as psychoanalysis and theories of nature. In *The Natural Contract*, Michel Serres (1995, 18) refers to "the dense tectonic plates of humanity" and suggests a natural contract with the earth is needed to mirror the social contract of society in order to govern these geophysical human entities, these “lakes of humanity, physical actors in the physical system of the Earth” that “behave like the sea” (18, 16). In the diagramming of earth and social relations, strata are used to open a sense of the longue durée, of the semipermanence of “historical” layering, and to think of how the inheritances of structural social forms constrain society (as well as provide pivots of liberation). Gilles Deleuze and Félix Guattari (1993, 160–61) advise developing a meticulous relation with strata rather than throwing them into suicidal collapse. In *Capital* (1976 [1867], 286), Marx talked about how the “relics of bygone instruments of labor possess the same importance for the investigation of extinct economic formations of society as fossil bones do for the determination of extinct species of animals.” As we have seen, Du Bois produced a visual sociology of racial strata. Louis Althusser suggests: “Every concrete social formation is based on a dominant mode of production. . . . The dominated modes are those surviving from the old social formation’s past or the one that may be emerging in its present” (2014, 19). In the context of sociology and strata, social strata are organized in a vertical hierarchy and can be organized in terms of class, caste, gender, race, ethnicity, age, or disability, whereby social stratification is used to understand social exclusion and marginalization.

Social stratification, as the division of members of a society into strata (or levels) with an unequal access to wealth, prestige, power, opportunity, and other valued resources, led to the inference of *social mobility as* moving between strata (aligned with the concept of meritocracy as a form of bootstrapping the strata); it might be better termed *social destratification*. The three dominant dimensions of stratification were understood to be *economic* (Marx), *power* (Foucault), and *prestige* (Max Weber). Marx imagined a revolutionary “spring-loaded” working class that underpinned social formations: “The proletariat, the lowest stratum of present society, cannot stir, cannot raise itself up, without the whole superincumbent strata of official society being sprung into the air” (Marx and Engels 2012, 49). This unidimensional view of stratification presented two strata characterized by an exploitative relation (bourgeoisie, who own and control the means of production, and the proletariat, who do not own the means of production). The volcanic idea of the working classes was as an eruptive magma! Fanon argues that the structural logics of race requires a Marxism that “should always be *slightly stretched*” (Fanon in Marriott 2018, 142; my italics). Which is to say that the problem with Marxist analysis and formations is that it presumes the stratal organization of society as given (even as it attempts to blow it up), and in doing so it withholds the ingenuity of those who elude these stratifications by mapping the undergrounds of these formations. The analytic takes the thing as thing and misses the excessive refusal of that claim. As Cedric Robinson suggests in his book *Black Marxism*, "Marx had not realized fully that the cargoes of labourers also contained African cultures, critical mixes and admixtures of language and thought, of cosmology and metaphysics, of habits, beliefs, and morality. These were the actual terms of their humanity" (2000 [1983], 121–22).

The Black radical tradition argued not just an "excess" beyond the false universalism of Western materialism but a redefinition of the containerization of subjects in the category of "labor." If stratal thought was a colonial epistemology, according to Robinson, it could not account for the formation of new epistemologies under slavery: "Slavery altered the conditions of their being, but it could not negate their being" (2000 [1983], 125). Robinson further critiqued Marxism for its Western-centric geographic imaginary that universalized from Europe and its "myth of a 'universal' proletariat" (Robin Kelley in Robinson 2000, xiii). He says race and gender have been erased in Marxism and that Marx's geography is simply wrong. The global social order must be understood as always having been global (i.e., imperial and colonial). Robinson's point is that rather than race being a superstructural phenomenon that sits on top of a technical base (of proletariat bourgeois relation = labor value), it is integral to the historic development of this relation (i.e., racial strata emerge from this arrangement). In other words, capitalism is never just capitalism; it is always racial capitalism.

Similarly, C. L. R. James (1938a) in *The Black Jacobins* centers slavery as the site of the modern world's geographic and economic project and thus the beginning of the modern system (which challenges imperial histories that center Europe and regionalize other places, and Marxism that sees European industrial capitalism as the center of the history of modernity). And this Europe begins with genocide, conquest, and globalization; it begins with the act and fact of diaspora (the dispersal of people). Simply put, James says, slavery is not an archaic form of production that leads into capitalist modes or modernity but their very foundation (slavery is the subtending stratum). This racial capitalism of the Caribbean (or the "slave mode of production") is preserved in modern capitalism. He sees the Caribbean as the beginning of modernity. Everything is transformed by slavery. In this geographic redefinition, James also centers the Haitian Revolution (1791–1804) as a key dynamic in the modern world in which enslaved persons took their own freedom and dignity, not waiting for it to be given (a successful slave revolution as a seismic shock). This sovereign and spatial act of

Blackness is in dialogue with, in complicated ways, the French Revolution and the call for the "Rights of Man" (1789–99). The Caribbean, for James, is a particularly special place as it allows a "double consciousness" (W. E. B. Du Bois) of seeing the past and making the future of modernity. As another Black rogue Marxist, Jamaican theorist Wynter says the "modern world" begins in 1492 in the Caribbean, challenging the epistemological outside of the Enlightenment: "What is called the West, rather than Latin-Christian Europe, begins with the founding of [the] post-1492 Caribbean" (Wynter 2000, 152). Saint-Domingue became the site of a radical social experiment as the Haitian Revolution ricocheted around the colonial world, not least on the waves through the small boat networks of the Caribbean. Countries feared that the success of the slave revolt in Haiti would "implant thoughts" of revolution in the enslaved elsewhere. The Haitian Revolution did in fact influence enslaved persons to rebel in other countries and escalated the fear of plantation owners.

Black Marxists, such as James and Robinson, saw solidarity across laboring classes (through sugar workers in Jamaica and the working classes drinking tea in England) as the sedimentation and radical reevaluation of stratal power (C. L. R. James 1938a). We can see James's idea of solidarity as a form of sedimenting geopower, building stratal connections across space and time with others that are in the rifts of our uneven and unequal stratal pressures. In a different challenge to the ghost of Steno's superposition, David Marriott's (2018) explication of Fanon's negritude as a concept of being that was beyond the supernatural and thus, we might argue, beyond the cultural imposition of strata releases a different geographical imagination of the colonial abyss. In the depth of the rift, negritude literally smashes the strata and its historicity, to insist on a time "beyond petrification" (Marriott 2018, 331) (also known as racial stratification) that draws a new idea of personhood into being interstitial in the imagination of the stratal limit. Beyond petrification mobilizes a *not-fossil* subject history as a recuperative imagination of Black life as in flux across space and time (and thus refusing the geosocial "ground" of colonial stratification as the place from where an anticolonial politics starts).

Geostratal Subjects

Firmly within the stratigraphic visual imagination, Freud drew on the popularized work of Charles Lyell in *Principles of Geology* (1840) as a structuring metaphor in organizational modes of psychoanalysis, whereby the unconscious was diagrammed as stratal layers underpinning the surface expression of the subject. The id, superego, and ego were seen within a stratal relation that mapped onto the overgrounds of the conscious and the undergrounds of the unconscious, akin to how Du Bois utilizes a stratigraphic method to diagram the color line. Escaping the enclosures of strata was seen as the terrain and terroir of political freedom. The stratal underground has long been associated with revolution and freedom, such as the Underground Railroad of slaves escaping the Southern states. Or, the rallying cry of the 1968 revolution, "Under the paving stones, the beach!"—made famous by the situationists. Michel Foucault's method of discovery was grounded in a stratal understanding of society and its heterogeneous formation. "In attempting to uncover the deepest strata of Western culture, I am restoring to our silent and apparently immobile soil its *rifts*, its instability, its flaws; and it is the same ground that is once more stirring under our feet," mused Foucault of his own stratigraphically inspired efforts to explain the appearance of man as an object of inquiry of modern thought (1994 [1966], xxiv).

Foucault emphasized the repressive power of the ruling strata and explained how institutions (prisons, hospitals, schools, military) were central to maintaining the power of the social order through their stabilization and regulation of strata and their errant rifts. The carceral institution in Foucault's thinking emerges as an internalization and governance of the errant flows of the current stratal form. His intellectual project is an archaeology of the said strata and an analysis of how freedom emerges within this epistemological framing. Deleuze, conversing with Foucault, suggests: "When a new formation appears . . . it never comes all at once, in a single phrase or act of creation, but emerges like a series of 'building blocks,' with gaps, traces and reactivations of former elements that survive under the new rules" (2006, 21–22).

Thus, freedom is imagined through this political geology as a rift coming in that fractures the existing strata and upends their points of stabilization (such as the prison), building new architectures out of existing rocks. Rather than just the greatest hits of strata, both structuralist and poststructuralist theories argued that new formations must be built in the ruins of sociohistorical context that they emerged from. Drawing significantly from Foucault's work on stratal or historical foundations in *Discipline and Punish*, Deleuze (2006, 41) makes a case that Foucault presents two elements of stratification: a historical perception or sensibility, and a discursive system. He says: "Each stratum is a combination of the two"—of discursive and nondiscursive formations, and visible and articulable expressions (42).

A change in the stratum changes the modes of expression of those strata so that a historical stratum implies a certain arrangement of the distribution of visibilities or sense. In short, rearranging the strata changes the possible affectual architectures.

In a rare engagement with race, Foucault suggests that race emerges to "designate a certain historico-political divide" in the nineteenth century, "not pinned to a stable biological meaning" (2003 [1976], 77). He argues that racial purity and superiority become the counterhistory of race, and the state shifts from the state wielded as an instrument of racial war to the state as that which protects the integrity of race. Interestingly, Foucault argues that racism comes into being to block the call for revolution of the state, both as a form of state protection for biological "integrity" (Nazi state) and as a reworking of racial struggle into social struggle (in the Soviet state). Parallel to the exclusionary thinking of Arendt, Foucault also fails to engage with the history of anti-Blackness that is his milieu and in his consideration of racism calls fascism and Stalinism the "two black heritages." Nowhere in his discussion does he engage the role of empire, colonialism, slavery, and its execution of anti-Blackness and Indigenous genocide (and all its quieter violent vicissitudes) in the biopolitics of race. Rather, Foucault discusses race through Anglo-Saxonism. Which, again, we might be surprised by, given the voluminous quantity of the writing on race and the fact that the book was written in 1975–76 when Foucault first visited Southern California (after the Watts Rebellion [1965], the Detroit Uprising [1967], and Martin Luther King's assassination [1968]). We hear much about Foucault's acid trips during that time at Berkeley and nothing about race politics. It is worth noting that in *The Order of Things* (1994) Foucault had no small amount of admiration for Cuvier and his account of transformations and dissections in. Foucault heralds Cuvier as initiating a taxonomic revolution that made Darwin and the concept of genealogy possible. It is Foucault's understanding of the science of life, through Cuvier, that underpins his discussion of the significance of biological race and racism in *Society Must Be Defended* (2003).

Deleuze and Guattari are perhaps the most stratally demonstrative thinkers, in *A Thousand Plateaus: Capitalism and Schizophrenia* (1993 [1980]), for whom there are three main groupings of strata, each with its own "concrete" historical formation: (1) the inorganic or geologic, (2) the organic or biological, and the (3) "alloplastic" stratum of human culture and language (see Yusoff 2017). These strata are multiple, and the alloplastic is subtended by the organic and biological. They also talk of machinic processes of tapping flows, drill heads probing strata, and the dangers of destratification.

These are not just metaphors (although Deleuze and Guattari employ a rogue geology informed by Arthur Conan Doyle's Professor Challenger in conversation with Nietzsche in the "The Geology of Morals" [Deleuze and Guattari 1993, 39–74]), but primarily ways to think of social processes and how to maneuver these processes or politics *as if* traversing a live earth. In Guattari's terms, strata are a form of "collective equipment" that both exists as a "social unconscious" and "allows for the encoding of disjointed elements" (Guattari in Foucault 1989, 109; Guattari 2016, 11). Applying this thought, we could think of the encoding of fractured oil across the intrastratum of networked oil rigs and underground horizontal drilling that moves through and across stratifications to uplift white heteropatriarchy and its racialized gendered violence. For Deleuze, strata come in pairs, the substratum serving the stratum (a hierarchical modification of Steno's superposition). And, if power, following Foucault, is a relation between forces (and thus every relation between forces is a power relation), then the socialization of geologic forces of the substratum into political relations on the surface requires attention. Understanding these geosocial relations as one (or many) such forces that organize the politics of social relations gives visibility to how a political geology might be conceptualized, in a way that takes account of the forces of the substrata to organize social strata (see N. Clark and Yusoff 2017, 4). More than this, if we take *deep earth debt* and *racial debt* as paired, then any analysis of geosocial strata needs to start from the rift to give an account of how power sustains and stabilizes itself through race.

The inhuman in Deleuze and Guattari's concepts of political freedom is the frame of reference for alterity and immanence. In their geophilosophy, the earth is unstable matter, which "flows in all directions," and nomadic singularity and the social are the strata that capture these intensities and flows. They say, "The nonphilosophical is perhaps closer to the heart of philosophy than philosophy itself" (Deleuze and Guattari 1994, 41). The revolution of the earth is understood as the closest incitement to revolution because it gives a perspective outside the strata, of what they called the "cosmic storm." Accordingly, "Strata are Layers, Belts. . . . Strata are acts of capture" (Deleuze and Guattari 1993 [1980], 40). So, while the earth produces flows in all directions, social strata "lock in" these intensities and flows of the earth into the social, political, and economic apparatus. Capitalism is understood as one such stratigraphic machine that captures subjects, the earth, and affect. They understand these flows as geophysical, psychic, and social. More recently geosocial strata have been a way to diagram social and geologic relations, in ways that pay attention to the histories of strata and the nodes of accumulation and racial pathologies within those strata (see N. Clark 2017; Saldanha 2017; Yusoff 2017).

Racial Stratigraphy

Fanon, seeking to open the repressive racial strata, recognized Freud's coupling of ontology and phylogenetic theory in his composition of the subject, thereby rendering the subject as defined through a particular (and raced) account of the human, named as the individual. While individuation was defined against deterministic principles (including those of environment and thus race), it nonetheless resided with a subject that defined its limits through an ontogenetic perspective crafted within the sciences of racial categories of definition: "Reacting against the constitutionalist tendency of the late nineteenth century, Freud insisted that the individual factor be taken into account through psychoanalysis. He substituted for a phylogenetic theory the ontogenetic perspective. It will be seen that the black man's alienation is not an individual question. Beside phylogeny and ontogeny stands sociogeny" (Fanon 1986 [1952], 13).

As Fanon asserts, the individuation of the subject is possible only from a privileged ontogeny (or origin—which I call the plateau), that is in fact a sociological mode of production, or what might be understood as the anthropogenesis of colonial Man. If Blackness is produced in the zone of nonbeing, then Fanon's cosmic *yes* is one answer to the possibilities of the subject's historic and material ontic dispersal in the world. While the historic part of this upheaval labors against sociogenic principles such as race (per Wynter), the material possibility of this recognition is and was a path to the recognition of new properties of possession. Residing in the affirmation of nonbeing, Fanon offers the abyssal as another cosmos, or ground to thought (that has no ground!), that breaks with the stratal imagination. Fanon's abyssal/cosmic *yes* is an answer to the mine/*mine* as the colonial archetype of extraction site and the propertied assertion of the subjective "I" in the "fatal coupling" (to use Ruth Wilson Gilmore's 2002 phrase) between geology and ontology in the historical scene of racial geophysics. Geologic ontics ensured that race and matter were produced through that racial geophysics (modeled on the extractive and stratigraphic) principle as conjoined material and subjective registers, connecting the material and sovereign power of whiteness in a form of geopower that could pronounce itself "I" (mIne).

As the stratal imagination became a way of organizing thought in its critical modes, it lost sight of the racial undergrounds on which that thought and earth praxis were built. The three feet, three inch height of the stratal hold (of captivity in the hold of enslavers' ships) was part of the assemblage of governing racial geocodes that enabled philosophies' thought of the strata to take place (as an institution, world system, and structure). While race is not precluded in these Western geophilosophies of society and subjectivity, it is relegated to a differential in the strata rather than the very substratum that underpins the alloplasticity (to use Deleuze and Guattari's term) of the philosophical surface (and thus there is no challenge to the plasticity of whiteness). The overall ontological and material affect of social theory that was underpinned by the stratigraphic imagination was to reproduce its claims to universality, in the production of social laws and concepts that theorized from nowhere and applied everywhere. The geography of this theory was unhindered by the dirt of relation, and it had missing and blood-soaked red earths. In the current politics of strata in the context of the epochal shift that has been designated as the Anthropocene, new imaginaries of the history of the Earth and social thought emerge, similarly told as a history of impervious material relations, but these ideas and material practices have a past and afterlives that need redress.

The concept of *geosocial strata* as an analytic, in its simplest iteration, can be understood as a global conjoining of geophysical worlds and temporalities in the Anthropocene context, or as an Anthropocenic fracture in the portal of time's materiality (depending on which side of the earth genesis/earth destroyer Man is being posited). Yet, there is a historic politics to this stratal "conversation" between deep earth materiality and its surfacing, and its racialized gravities bring us back down to the rockiness of the stratal earth and its differentiated embodied geologies. Alongside a *stratal imagination* that constitutes the perspectivism of historicity and social forms of life, there are the tendrils of race and the spatialities of the rift that ground the extractive principle. The stratigraphic imagination secretes and traffics race, as it built colonial earth through the affective architectures of the *flesh of geology*.

Geology is a site of differentiation that is historically racialized (because of the inhuman-inhumane root and the stratal organization of race) and this has implications for epistemologies of the inhuman, as the basis for how nonlife and the inorganic are engaged, as well as for how geophilosophy is structured and environmental thought is unraced. Which means any account of geosocial strata in the Anthropocene must go in three directions.[5] It must go (1) into the specificities of persons-as-bodies, places, and political formations in which these stratal unearthings come to matter; (2) through inhuman agency and its continuance, notwithstanding the human (as a diversion of the forces of the cosmos), which is to say the cosmos needs to be considered on its own terms as a materiality with certain proclivities toward its own forms of difference, becoming another earth and as a metaphor for structuring the cosmic materialisms of antistratigraphic thought; and (3) through the material conditions of subjection and inhumanization. A racialized politics of strata might yet help disaggregate the imposition of the global and its predatory planetary reach, understood as a projectile imagination over the earth that dismisses parts of its origination (as racialized in the social and earth sciences), and thus eschews its obligations to reparative forms of justice and attribution in the relation between undergrounds and Overseers, or the basis for a *geoethics for antiracist earths*.

7 Geopower

Materialisms before Biopolitics

> It is because local and global forms of power are elaborated on a plane of forces that they can generate their very real effects on particular categories of bodies. But we must be careful to distinguish these different orders of force, or violence, that structure life at its very eruption and its subsequent elaboration: geopower, the relations between the earth and its life forms, runs underneath and through power relations, immanent in them, as their conditions of existence. Power—the relations between humans, or perhaps even between living things—is a certain, historically locatable capitalisation on the forces of geopower. —ELIZABETH GROSZ in Yusoff et al., "*Geo*power"

> Particular (presently biocentric) macro-origin stories are overrepresented as the singular narrative through which the stakes of human freedom are articulated and marked. Our contemporary moment thus demands a normalized origin narrative of survival-through-ever-increasing-processes-of-consumption-and-accummulation. —KATHERINE MCKITTRICK, *Sylvia Wynter: On Being Human as Praxis*

After a century of political and material biologisms, the Anthropocene thesis is reshaping attention toward geologic agency, and thus it inadvertently furnishes a mode of subjectivity as geologically implicated.

Disrupting the dominant tenets of the colonial matrix of materiality (as *geos*) that secured imaginaries of Life (*bios*) through racial violence and inhuman extraction, this chapter engages Grosz's (2012) concept of prepolitical formations of "geopower," Povinelli's (2016) articulation of the stratigraphic arrangements of social (re)production through geontopower, and Jamaican theorist Wynter's (2003) critique of the raciality of biopolitics. In conversation with Grosz, Povinelli, and Wynter, I offer the theory of geologic life to speak to the antagonisms between the inhuman (*geos*) and Life (partial *bios*), as it is historically and conceptually arranged through the spatial division of race. I do not take geopower as a self-evident entity but as a historical, political, and spatial formation that is organized in the grammar of the inhuman and shapes subjective forms of life. *Geologic Life* posits a theory of *strata-fication* and *geologic-fication* as a confrontation with the spatial arrangements of the social divisions of materiality; an arrangement of power that is both exceeded and complicated by geologic elements; and an organization of inhuman power that aggregates bodies across scales.

Rather than make any presumptions about expanding biopolitics to the scale of the planet, or geologizing the social as a new inhuman phenomenon, I have suggested that the racial deformation of biopolitics is shaped by the confines of the Enlightenment ethical subject (in the shadow of Kant) and the colonial production of Global-World-Space (as an imaginary of spatial totality or singular field of material power). I argue alongside Wynter that biopolitics fundamentally neglects its constitutive exclusions of race, but instead of seeing Western biocentrism as the primary site of racialization as Wynter does, I locate this erasure in the political sphere of *geos*, through the consideration of the multiple lives of the inhuman: the *inhuman as matter, race, and cosmos* in the ongoing geologic grammars of colonial extraction (and the spaces of their negation and redress). Rather than consecrate an extended biopolitical critique, I seek to demonstrate how geologic life is a critical and creative site in the constitution of the geophysics of race and colonial forms of life that predate the biocentric subject and thus limit this analytical frame in accounts of power.

Geopower (after Grosz) introduces contradictory spatial relations into the mix (conceived here through the spatial entities of the plateau and rift to differentiate the plane and geophysical "states" of power), to problematize long-held spatial propositions about social relations and the organization of power, across and through space. I offer *Geologic Life* as an analytic that substantiates the inhuman as a prior formation of political geology to biopower and biogenesis, that constitutes a material field of politics in which subjectivity is caught and in which geopower is mobilized. Race is the erased foundational materialism of biopolitics rather than its elaboration. The *politics of the inhuman* expand the density of geopolitical relations in ways that are not reducible to those geopolitics but constitute a missing field of materialization. If we see *geos* and its affectual geophysics as the epistemic (and empirical) ground of both materialisms and metaphysics, how does this unearth questions that fundamentally terrify a Foucauldian biopolitics and destabilize the Kantian subject (and all that is raised in its dialectical shadow, collaboratively or antagonistically)? What happens, simply, if we say biopolitics has no ground, no adequate account of the *geos* in which it is historically and materially situated? (And, thus, biopolitics can never give an adequate account of racialized life because its divisions are raised on Blackness *as* geos.) In the amnesia of the *geos* of colonial earth, recalled by the existence of many broken earths (and its partial rendering of Life), biopolitics as critique and method also sutures over that which allows the question of the human to be raised (that which is buried and erased in the category of the inhuman) and the material basis of its subjectivity.

In the erasure of its constitutive exclusions—of inhuman earth and racial and spatial descriptors of the subject in the category of the inhuman—critical biopolitics produces a field of power that fails to understand its own internal structure (externalized as "outsides") and the dynamics of their political force in world-making through the earth. A parallel argument is made by Povinelli (2016, 176) about the reduction of politics to the life of the organism. In opening a critique of the political geologies of race, I see these historic material relations of the earth as crucial to the recognition of antiracist forms of life in the resolution of the historic material eugenics of white geology (and that which is antecedent to a livable planetary politics).[1]

Geopower, Biopolitics, and the Earth

The attunement of politics to a geophysical context and the deep sedimentation of colonial materialisms in the praxis of extractive earth present a challenge to think expansively across material registers so as not to reinstate the material dialectics of *bios* and *geos*. As geologic time and political time converge in descriptions of the planetary, a radical interdisciplinarity that holds to account the intramural spaces of subjective and earth extraction, as well as experimenting with a syntax of apprehension and change, is needed. Geopower is emerging as a mode of description of power in the context of the change of earth states (e.g., Anthropocenic climate change), a power that has hitherto been seemingly hidden in its appropriated form or disaggregated from its context of production (i.e., a power that emerges from the earth rather than in the "shallow" assemblage of extraction and the politics of its subsurface purview). Given the centralization of Life as an organizing concept within biopolitical thought and its explicitly racialized ground, the consideration of a political subjectivity of the earth raises the question of what materialism after Life might look like. Responding to that question, I put forward a politics of the inhuman that is not turned toward an essentializing view of matter (*elemental earth*) or an undifferentiated plane of power—from neither the plateau nor the iron heart. Thinking with geopower and from geontologies (via biocentrisms), the concept of geologic life shifts the focus onto a historically contingent earth to come back at a more porous and less fungible account of subjectivity and the cosmos.

Geopower is a concept that was suggested by Grosz (in Yusoff et al. 2012), in response to Foucault's writings about geography and territoriality, but which he never fully elaborated on. At the time of Foucault's writing about geography (rather than the earth), the idea of nonhuman power was also discussed by Deleuze and Guattari in their accounts of the "outside" of inhuman forces, specifically in their conceptualization of the process of stratification in plateaus. As we have seen, in conversation with Foucault's use of the concept of "formations" as a process of historical and discursive organization of political milieus, Deleuze and Guattari suggest that strata both capture and institutionalize inhuman forces in social machines, but they can do so only because the modes of capture have a "resonance or redundancy" with the forces of the earth that are being encoded (Deleuze and Guattari 1993 [1987], 40; Yusoff 2017, 110–14). More recently, geopower was given substantial attention by Grosz in the context of her engagements with the incorporeal and questions of territoriality (Grosz 2008, 2017; Grosz in Yusoff et al. 2012; Grosz, Yusoff, and Clark 2017). A parallel to the concept of geopower is developed by Povinelli in her analytic of geontopower, which signals the ontological and strategic deployment of the categorizations of Life and Nonlife in the governance of permissible forms of lives (Povinelli 2016, 2017).

Briefly, in situating the relationship between geopower and biopower in the context of the "outside of thought" (as discussed in the work of Foucault, Blanchot, and Deleuze and Guattari), the inhuman is understood as a provocation and alienation of thought (rather than as a product of inhuman forces). The spatial operation of the inhuman "outside" structures much of the literature around a biopolitics of extraction (and its subjugations), so it acts as a trace architecture of other epistemological inventions and understandings of somatic affects in the conceptualization of inhuman power. These thinkers suggested a constitutive exclusion within biopolitical thought to denote a wider field of material production and power relations adjacent to social formations (a cosmic field). Grosz's take on geopower and Povinelli's analytic of geontopower are feminist interventions in the field of biopolitics that reckon with its territorial implications across Indigenous, gendered, sexed, and liberal forms of life. From their thinking, geopower can be broadly understood both as the inhuman context in which biopolitical life is governed and as a concept that opens thinking toward cosmic forces in ways that challenge the interpretive grid of biocentric thought. Adjacent to this work, through the historical substantiation of geologic life, I want to articulate inhuman intimacy as that which challenges the exteriority of the placement of inhuman "outside" or "below" (in a hierarchical understanding of life/human-nonlife). Furthermore, I argue the idea of changes of state (understood as geophysics of space and subjects) resulting from the affectual architectures of geopower (as white geology) that historically constructed race as the interregnum of geophysical states in the governance of *bios* and *geos*.

As the Anthropocene thesis contains within it a notion of geologic planetary agency and a new mode of subjectivity that is constituted by, and constituting of, geologic forces, geopower provides a powerful (if unevenly conceived) concept with which to grapple with the uneven forces of inhuman materialities and their manifestations on/in specific bodies (even as it might be used to *un*work the hold of the biopolitical subject). However, inattention to the historic organization of geopower and race within colonial geologic grammars tends to obscure older, more sedimented forms of geologic subjectivity (specifically in the context of Indigeneity and enslavement, and *colonial materialisms* more broadly). Even as a new geologic epoch is being materially articulated, such a claim is a temporal juncture in comparison to the earth's own forces, which for much of the earth's geohistory have not supported life *at all*. The temporal logics of the inhuman remain a question of the entanglement of near history (as a product of the forces of colonial earth) and deep history (as a product of the forces before life). In the larger material context of the earth and more recent iterations of earth events (of the uneven ground of the near present), it can be observed that mineralogical evolution and geophysical events have been in constant material communication with life (mainly through bacterial innovation), in which the field of life is shaped and subtended by geology (a communication that is intensified and altered on a planetary scale by colonialism). For, as Grosz suggests in her formation of geopower: "Life partakes of the earth, requiring its forces to survive; but in turn, life elaborates the forces of the earth in ways that would and could not occur otherwise, developing certain potentialities in one direction or another, converting the qualities of the earth and its products into other qualities useful to or enhancing life. It is a temporary detour of the forces of the earth through the forces of a body, making them an endless openness" (in Yusoff et al. 2012, 975).

Life is *détournement*, an elaboration *within* inhuman matter that is always in asymmetrical relation to that matter, but it is from that geochemical matter that life draws itself and so it is no different from that matter in kind (i.e., life is an expression of mineralogy, but clearly it is not necessarily a vital expression for the planet). Or, to put it another way, what I call geologic life is both a confrontation with these modes of classification that silo Life and the inhuman and the realization of cosmic possibility as life makes the planetary otherwise (as colonial climate change has lithically and liminally created a different earth). Geologic life, then, is not a self-evident category of description for the Anthropocene but a confrontation and challenge to understand the traffic between different historical orders of the inhuman (Yusoff 2015b, 204) and the semantic organization of their materialization. Grosz suggests that

> if and when life emerges from the forces of the earth, forces that cannot be separated into different categories, life carries with it this excess over corporeality that the material has always contained, a virtuality that enables it to transform itself or to emerge as life. Life capitalises on the two-faced orientation of the earth and its forces, erupting into materiality as a bounded and self-producing cohesion that is also always 'thinking,' that is to say, is always oriented by the senses of the earth inherent in its materiality (in Yusoff et al. 2012, 974).

This means that inhuman materiality already contains the virtual qualities that make life possible, so it is not life that does life's work *as such*, but the already existent capacities of the earth and minerals that give rise to a whole range of variations of forms and conditions that engender life, political, cosmological, and otherwise. As a resident mineralogy with its own proclivities, the earth is active in the transmutation of forms of life and their geochemical otherwises (anthropogenic climate change is an example). Thinking with the historic formations of extractive earth alongside a fuller inhumanism (or cosmic materialism) in the context of geopower is important because biopolitics reproduces the material divisions of Life and the inhuman in ways that continue to perpetuate the racialized rift. Geopolitics that parallels biopolitics (where the "geo" connotes an additional spatial or territorial dimension) also marks the impossibility of a reconciliation with geoforces as the constitutive scene of relation, because it leaves the partial organization of Life intact and simply "accumulates" the earth onto an existing figuration (and its material divisions). Attending to the granular geologies of *racial earth state formation* corroborates that inhuman intimacies are a viable tactic of the earthbound. Turning away from biopolitics offers an inhuman imaginary that is indivisible and thus an earth that is attentively in relation rather than an autonomously discrete outside. Such an inhuman imaginary allows the time of freedom and the time of the earth to be understood together, in the same geophysical state.

Geopolitics after Life

I have argued that biologic thought is underpinned by geologic propositions and praxis (Yusoff 2018b). Considering political subjectivity as already and always geologic in its expression of political forms (without recourse to environmental determinisms) organizes a concept of geopolitics that must take account of materiality and materialisms as political strategies and question the exclusivity with which accounts of Life have been rendered through racialized earths and sutured to normative materialist assumptions. Life as an exclusive biocentric proposition and as a politics secured in racial imaginaries has consistently produced not so much an understanding of what life is but ways of organizing hierarchies of life and its "others" (i.e., racialized, sexualized, and gendered modes of ordering life). As Povinelli comments: "Biontology is the true name of western ontology. And the carbon imaginary is the homologous space created when the concepts of birth, growth-reproduction, and death are laminated onto the concepts of event, conatus-affectus, and finitude" (2017a, 174).

Whereas Life may be the central production of Western ontology (and its exclusions), it is clearly not the central concern of the cosmos, where forces more powerful than it organize and contextualize its appearance and thus put pressure on the politics of life, as well as on its imagined arrangements (pressures that are often internalized and then expelled as necropolitical intent or deemed nonpolitical). This chasm between the concern of Western ontology and the "attachments" of the cosmos is a site of governance (and repressive fear of an inhuman planet). As Grosz comments:

"Life is the provisional binding of an order of conceptuality with an order of organic cohesion, the temporary protraction and delaying of the forces of the universe itself" (Yusoff et al. 2012, 974). I have argued that Life enters its regulatory strictures through the colonial framing of geology, which produces the geophysics of space according to race, Euro-Western concepts of genealogy that organize kinship, and paleontological stratigraphy that organizes hierarchies. At a domestic level the white settler family, through its adherence to heteropatriarchy, holds these violences in place, making genealogy legible and recuperating stratified subjection through projection, while the nation-state governs the broader field of extraction. This is why the Anthropocene as political geology and imaginary is so readily able to function through the futurity of the white family (at risk), while withholding that recognition of kinship from those that are relentlessly pursued in geologic borderlands. Claiming futurity through the extinguishment of those designated as already outside life and submerged in violent geologies requires the schism between *bios* and *geos* to already be in place, alongside the ideation of the *geos* as dead matter that is recuperated only through the work of the plateau.

Awareness of how social worlds are an affect of and effect geology, rather than a world that is constituted through "our" making (i.e., purely social and socialized biology), suggests an arrangement of geologic power beyond questions of mastery. Identifying geologic force as a regime of power that operates at the scale of a planet brings the structures of exchange between geologic forces and social worlds into view as a spatializing operation (of time and geography). If power, according to Foucault, is a relation between forces, geopower (via Grosz) is a way to conceptualize how stratifications organize and capture geologic forces into political geology. As Nigel Clark comments:

> Whereas conventional geopolitics tends to restrict itself to human inscriptions on the earth's surface, Grosz's notion of geopower permits the dynamics of the earth to leave their mark on human and other bodies. Her work has always been characterised by its boldness in evoking difference—within and beyond that which we designate as our own species. . . . Her developing fusion of Darwin and feminism has deepened conceptions of corporeal difference to encompass not only the differential forces of the sociocultural and the biological but also the enduring impression of earth processes on living flesh, which includes both incremental and more rapid impacts. (Clark in Yusoff et al. 2012, 976)

It is the very differential forces of the earth that Grosz argues have been left out of accounts of theorizing of social difference and political geography. To put it another way, to construct a fully operative sphere for political thought and social action, the intransigent forces of the earth have often become marginalized as a nonoperative space for politics (because of their noncooperative or cosmic ends) and thereby excluded. Taking account of the flesh of geology is not just a means to an embodied geological account (a vitalist move to recognize the "lives" of geology) but a way to mobilize the possibilities of those solidarities of the inhuman against their weaponization (inhuman intimacies like an oyster knife prizing apart indifferent humanism).

The concept of geopower can be used to examine the expression of social forms as a product of geologic forces that run through social and subjective formation but are not contained by those forms, however they are understood to be constituted. Geopower, it could be argued, is a plane of social reproduction that both constrains and is expressive of possible modes of expression and thus in the terrain of political freedom. The social plane and subjectivity are constituted by more than social forces and are constrained (as much as enabled) by the geologies that underpin that organization of surface power. However, the partial optic of the surface in relation to power is delimited by the restricted view from the plateau; power is drawn up from the depths but does not inquire downward. Before there can be an Anthropocene geopolitics as such, there must first be a conceptualization of geopower as a force that constitutes planetary (and cosmological) scale, which must start in the earth (as Indigenous and earth defenders have already theorized and actioned) and the undergrounds that constitute the surfacing of geopower and its narrativization. Such a conception of geopower that extends beyond the boundaries of Life in political theory would challenge the various formations of subjectivity in which Life is construed as the organizing locus of power, agency, and reproduction (and thus exclusion and subjugation through its "purposeful" autonomy and directional genealogy) and would open the question of who and what are kin. Such a geopolitics after Life would present a materialist challenge to the formation of planetary scale through materialisms that are not bifurcated in advance and held open by the functioning of race. The "geo" cannot just act as a descriptive material mode or spatial expander appended to politics, which allows geopolitics to look beneath the surface into the subterranean or up into the sky at the atmospheric (i.e., to mobilize the vertical or volumetric in geopolitics rather than the default horizontal of the plateau as its political stage). Nor can geology be understood simply as a modifier to a "new" geopolitics that extends its purview and practice to encounter material difference in a different ontological guise (as the current "elemental turn" attests to). In these formulations of geopolitics, geopower is taken as a self-evident descriptor of geologic forces rather than as a question about the asymmetrical organization, the racial capitalization, the temporal conversions, the contraction, and the radical temporality of cosmic forces. Thus, geopower as Realpolitik reproduces the normative political event and regulation of social forces, rather than challenging them. For any antiracist earth politics, *geo* must be understood as historically constituted through its inhumane and intimate inhuman occupations.

Political Geology and the "Outside"

The turn toward political geology often contains the same formations of politics and power of the inhuman that it has hitherto paid attention to (a progressive politics to counter a populist one), but it often fails to notice the change of geophysical register required. As Clark adroitly comments:

> When geography, along with other "progressive" fields, regrouped around the imperative of the "political," it tended to imagine the contents of politics in predominantly agonistic or adversarial terms. . . . From here on in spatial forms and processes were reimagined as being constitutively open to struggles of collective making and remaking. But one of the consequences of this manner of politicisation has been a gently descending silence around domains of existence that are not so amenable to contestation. In order to gain an intrinsically negotiable spatiality, it might be said, we traded in the more intractable forces of the earth and cosmos. (N. Clark in Yusoff et al. 2012, 976)

If politics involves not just sovereign, governmental, or juridical states to be contested but also includes planetary states, then politics must shift in its definition and delineation of concerns to consider the regimes of geophysics as a form and spatiality of historically constituted power. That is, geopower is not biopower at a bigger scale, for a planetary body rather than a human one. And there is no ground zero for the earth, as there is for the human (even sun death and fragmentation from a massive asteroid strike are modes of continuance in different cosmic orbits). Grosz is explicit in her refutation of this understanding of geopower as simply a correlative to the operation of biopower: "Rather than concede geopower as the power that humans can extract from or hold over the geological, he [Foucault] sees geopower as the forces of the earth" (in Grosz, Yusoff, and Clark 2017, 135). Rather than understanding geopower as earth extraction or a reinvention of the political economy of geology, Grosz urges us to be more expansive in our thinking even if it does compromise our sense of immediate gratification of a progressive politics. In addition, I understand geopower as a historic expression of those relations of the forces of the earth that are occupied and therefore possess a historically constituted resistance that already does critical work beyond these narrow margins in their account of political power through cosmic materialisms.

Unlike the molar and minor expositions on biopower that seek an ever-tighter account of power's penetrative force into the forms and formation of life, Foucault was always (via his engagement with Blanchot's postwar writings at the limit) engaged with the outside as an internal and anterior state that deforms the possibility of any totality of control (Foucault 1977; Blanchot 2010).[2] The recognition of internal and anterior resistance is a recognition of the possibility of *freedom from*, as much as accounts of biopolitics often render it as *coercion to*. Freedom from inhuman subjection might very well be in the trees. In Foucault's engagement with Blanchot's concept of the outside of thought and Bataille's limit-experience, there is an explicit acknowledgment of how life is not, in his conception, entirely framed by a Life/Death dynamic as the constitutive organization of what life is and can become. Akin to Povinelli's focus on the cruddy "quasi events" of life that are as much about extinguishment as they are about extinction (Povinelli 2017a, 170), Foucault's thought imagined the outside as an on-going opening that politics was always in communication with. That is, the geologic and geographic (in its broadest sense) are there in Foucault's project as the context in which biopolitics makes and takes bodies, as well as engaging with their libidinal and other economies that reshape bodily life and flesh. As Foucault comments about sexuality and its delineation of limits, it is "intrinsic finitude, the limitless reign of the Limit, and the emptiness of those excesses in which it spends itself and where it is found wanting. In this sense, the inner experience is throughout an experience of the impossible (the impossible being both that which we experience and that which constitutes the experience) . . . a world exposed by the experience of its limits, made and unmade by that excess which transgresses it" (1977, 32).

That is, any presumed totality or the forms of hypersecuritization that so excite geographers are already constituted by an engagement with the limit and its transgression, but these are not life/death limits; they are limits of possibility, of freedom, of sexuality, of thought, and of "interior and sovereign" experience (Foucault 1977, 32). Transgression is the action that involves the limit in order to redefine it. Limits for Foucault are ontological decisions to achieve certain ends; to contest this, he argues, is to proceed until the limit defines being (36) and that nonpositive affirmation is the play of transgression (48). The question Foucault asks—"But can the limit have a life of its own outside of the act that gloriously passes through it and negates it?" (34)—is answered by an earlier text by Fanon (1986 [1952]), wherein his understanding of the abyssal becomes the "cosmic *yes*," claiming a "life of its own" rather than being held in a dialectic state (of colonial limit and excess; plateau and rift). What is important about this discussion to a biopolitics of life is to notice how the communication with limits (that is, transgression) is always in conversation with questions of sovereignty and its breach. It is precisely through these economies that engage the outside (that are also epistemologies that frame the transgressive in particular racialized sexualities, as we have seen with Cuvier), which are also sites in which biopolitical control is undermined. For example, pleasure becomes a way to unpick politics, and geoforces contain their own excessive demands in the wet lip of the day that speak to other forms of material sovereignty. When this scene of transgression is apprehended in the context of the cosmos, not just creaturely life, an entire sphere of the political takes shape differently.

Geopolitics

In the realm of earth forces, understanding geopolitics in a way that takes the *geo* seriously as materiality, materialisms, and geoforce (as well as history or historicity) would need to reconstitute the very understanding of the arrangements of relation and the location of agency. Such a "new" geopolitics or political geology would require an adequate conceptualization of geopower as a force within (rather than outside) political subjectivity in the first instance (the inhumanism within), and then it would need to look at the uneven geographies and traversals from geopower into the privileged forms of political Life. Both these movements need to be accomplished with the understanding that politics takes place in relation to an inhuman earth without geologic ends (there is mineralogical evolution, there are ends of species and geologic epochs, but there is no teleology or preferred forms of accomplishment or salvation for the planet) and historic inhumanisms. Thus, geopower is political in a very particular way, as a politics that organizes social life from both the "outside" and the "inside" of the plane of production that institutionalizes or operationalizes those geopowers for its own ends (this is the relation between the cosmic and the inhuman-inhumane). It is the conversions, contractions, and enclosures of that geopower that form the traditional modes of political attention on institutions, actors, borders, states, and so on. But geopower is already active before these bodies of capture *give* these forces agency through forms of authorship (which can be philosophical, epistemological, technical, or juridical), which is why geopower is not the critical equivalent of biopower for the geologic age (or life after biologism, because geopower operates in a broader plane of production that is already structured historically in its geography of the plateau and the rift, and in its geophysical gravities of race).

While Foucault conceived of biopolitics in relation to outsides or the limit, in ways that are often overlooked in accounts of the regulatory inside of biopolitical life, geopower is an express acknowledgment of the intrusion of those inhuman forces that press on biopolitical arrangements. This gravity of the earth is the intimacy of the inhuman but also the deep freedom of its rocky preponderance. Which is not to say that deeply biopolitical concerns do not reside in geologic forces (the historic and present connections between mining and unfree labor tell us otherwise), but those geoforces, in their fullest expression, exceed a biopolitical critique (i.e., geoforces do not have biopolitical ends such as the preservation or extinguishment of human life and geoforces are scripted external to biopolitical description to give that concept its internal cohesion). Geoforces do not just exceed life forces; they are a material and conceptual open that is not organized through or as an organism, Gaia or otherwise (Deleuze 1997, 131; Yusoff 2017, 108). The organism is just an effect of those inhuman forces and has many variations and iterations (if we think of Neanderthals, for example). This is important in the context of human politics because understanding how Life is organized within the shadow of these forces through its various modes of capitalization on earth forces is the very basis of the possibility of social change. However, this change requires a geologic imaginary that does not focus exclusively on *this* Life but understands the situatedness and shared condition of becoming through time. That is, the *pre*political condition of geopower (understood in its disruptive state as a force of the open and a site of narrative capture) can be a basis for an understanding of genealogy after Life: forces that do not begin and end in the organism, or in an account of a privileged (racialized) genealogy and its account of kinship. Eschewing false ends and embracing the geologic realism of the embodied flesh of geology (Yusoff 2017) might turn out to be a means by which to encounter the geologic forces in a more exacting relation with the earth and to account for how that relation tethers geology to different experiences and exclusions in subjective life. If, according to Grosz, geopower is the potentiality of matter to be capitalized upon, then geopolitics must take account of the political formations of those modes of capitalization, but also the desire that motivates and reciprocates with those geoforces, their narrativization and institutionalization, and their modes of capture (in the white settler family, for example). So, political questions are shot through with geologic forces and their mineralizations, far below the surface in the substratum of the earth and tied to the force of *those* flows as much as they are to the forms of their expression "on the surface" of the earth or the political stage. Grosz suggests in her conceptualization of geopower: "What we understand as the history of politics—the regulations, actions and movements of individuals and collectives relative to other individuals and collectives—is possible only because geopower has already elaborated an encounter between forms of life and forms of the earth. Our everyday understanding of power both draws on and yet brackets out this primordial interface that sustains it in its ever-changing forms" (Yusoff et al. 2012, 975).

Grosz's formation of geopower came out of her intense and continued engagement with cosmic forces as the basis of art (or intensification), difference, and sexual selection. Art, for Grosz, is a way of naming both the capture of and communication with inhuman forces and where, she says, "intensity is most at home, where matter is most attenuated without being nullified . . . matter as most dilated" (2008, 76). The dilation of matter into art (understood in its broadest terms as a sensibility of intensification) is a mode of concentrating experience by drawing on the forces of the cosmos and the experience of nonorganic matter from which it is made (9). Sensation is "that which subject and object share, but not reducible to either subject/object nor their relation. Sensation is the extraction of qualities—how art maintains a connection with the infinite/inhuman origins of the earth" (8). The contraction and elaboration of cosmic forces emerge in the labor of art as an achievement that stretches organic life across human and nonhuman worlds in forms of recognition of the inhuman forces within and outside. As a form of communication with the material world and a transformation of immaterial forces, this nonhuman power is the basis of Grosz's feminist ethic of making the world otherwise (or changed from patriarchal subjugation). In this way, Grosz ties art to Deleuze's ideas of nonhuman power of the animal and, importantly, to the geopower of the earth and its modes of territorialization (that far exceed yet organize hominid life and its material political and racial divisions). Thus, an engagement with nonhuman power is an opening of material and immaterial forces of the universe to the practices of expansion and experimentation; "In this way art taps into the substrata of the earth, its geography and its time, to unearth and repurpose its forces" (Yusoff et al. 2012, 972). Thinking of art in its manifestation as an aesthetics or poetry is a way to bring this affiliation to and with geopower into the geopolitical sphere and to recognize its importance to the constitution of political sensibilities in the affectual architectures of an otherwise.

A Geophysics of Power

Part of the impetus for engaging with an analytic of geopower for me is to think about the preoperative conditions in the realization of power (and "its" others, i.e., pleasure, intensification, excess, abjectification, violence) and to address the geoforces that maintain, feed, and transform this Life in the full arena of its constitution. In other words, to understand that there is not just a geography of power but a *geophysics of power* that is racially differentiated. And, in apprehending a geophysics of power there is a need to ask the question about what forms of geopower or arrangements of earth forces make territorialization possible in the first instance as an infrasubjective experience of material worlds (i.e., what are the conditions that *take* place and differentiate its spatialities?). This is to suggest that political geography has a missing terrain: (1) the agency and volatility of the earth, that is not so easily marshaled into the decisive acts of capture performed by classical (or even critical) geopolitical actors (Clark in Yusoff et al. 2012); (2) the ground of race that provides the preconditions for an account of agency. These spaces of resistance are always open—as liminal and lithic—as I have endeavored to show in the poetics and politics of the inhuman, as well as being historically organized through race. If the story of politics starts with the earth and the forces underpinning the very possibility of life-forms and forms of life, then this is not to institute some form of geologic determinism but to recognize the *pre*-political conditions of territorialization as a harnessing and scripting of inhuman powers of the earth that have *determining forces* in politics (and those determining forces require affective architectures of renewal in epistemic and genealogical reproduction). These questions are not mute in the face of political struggle because they frame political geography in the specificity of its actualization as a material practice enacted within geosocial formations. What are often considered as spatial divisions in the conceptualization of political geography, and in particular the formation of planetary politics or planetary scale, are actually questions of material ontological division or the *matter fix* that designate the location of agency on the side of biocentric life (which cleaves to a particular politics of Life) that erases *geos* (as both a scene of subjection and a register of cosmo-being). Thus, the deep time of racism changes how worlds materialize and in what state.

The Spatiality of Materiality

It is the spatial arrangements of the divisions of materiality agency that organize an understanding of the arrangement of power. As Povinelli (2016) argues through her analytic of geontopower, it is precisely the Life/nonlife arrangement that governs certain modes of acceptable social death within policies of the liberal state, as well as the more finely tuned decisions about who and what forms of life get to survive or be extinguished. It is the very ontological formulation of materiality as matter in all its incarnations (as lively, inert, dead, brute, lacking in agency) and the connections of that matter (as Life or not) with perceptions of the earth (as territory, land, or minerals) that produces forms of political subjectivity (as individuated, possessed, or available for dispossession). We need only to think of the scenes of subjectification in slavery to understand how persons rendered as bodily matter or as flesh are used to obfuscate the responsibility to recognize subjecthood, and instead transform that nonrecognition of life (or production of life as inhuman) into a commodity form of person-as-property (Hartman 2007). Similarly, it is the recognition of the energetic power of the inhumane categories of life in the mine and plantation as geologically coded forms of energy that transforms the modes of subjectivity in the Industrial Revolution, both as dependent on those earlier forms of precapitalist modes of production and as underwritten by the energy of the enslaved as labor—developed through the infrastructures of modernity that are financialized by abolition compensation to slave owners and through the "sugar in the bowl" of working-class laborers (Williams 1944). While the Life/nonlife arrangement is a mode of governmentality in liberal forms of recognition in the polis, it is also subtended, I argue, by a more deadly colonial taproot of geologic life that expands itself through the placement of life within the ground of *geos*. While biopolitics may govern at the surface to designate the included and excluded of Life, there is a depth relation of inhuman subjects and racial undergrounds that are separated through a stratigraphic relation, and the distinction between figure (Life) and ground (inhuman)—racialized undergrounds and deadly shafts that preconfigure the possibility of Life as a mode of governance. The figure of Life literally dissolves, conceptually and materially, in a confrontation with the inhuman grounds that constitute it (but are zoned "outside" and beneath it).

While attention is given to the biopolitics of Blackness (Mbembe 2017; Weheliye 2014), understanding how race became a geophysical production that was preceded, in the first instance, by the extraction of mineral wealth from Brazilian mines along the Atlantic coast in 1518 is crucial to understanding contemporary forms of racialized geophysics. Natural resources in the partition of the *geos* set the terms for the development and transcription of the geologic grammars of personhood. The fact that the Gold Coast of Ghana was called "The Mine" during the slavery boom extrapolates on the material bond between geologic and human extraction (Hartman 2007, 51, 111) as coterminous. There was the gold of Africa, and then there was the "black gold" of the enslaved coded as property (44). Mineral and subjects could change place because of how the praxis of inhuman matter *and* the capture of geopower were understood in relation to the accumulation of political power and the consecration of Life. Grosz elaborates on the first step of this capitalization of geopower: "The relations between the earth and its various forces, and living beings and their not always distinguishable forces, are forms of geopower, if power is to be conceived as the engagement of clashing, competing forces. This means that before there can be relations of oppression, that is relations between humans categorized according to the criteria that privilege particular groups, there must be relations of force that exist in an impersonal, preindividual form that are sometimes transformed into modes of ordering the human" (Yusoff et al. 2012, 975).

The codification, valuation, and circulation of the material geopowers of the earth arrange particular and historically enduring forms of oppression in both the extraction of those geoforces and how the qualities of those forces are deployed and understood as relations of value and productions of subjective worth. That a person coded as slave and a piece of gold were established as materially equivalent entities to the Crown was testament to how the coding of matter arranged political geographies and how geology became part of the spatial and subjective expression of colonialism. Once coded as inhuman matter, forms of personhood and agency were delimited stratigraphically to subtend Life (as privileged form and spatial plateau).

Inhuman Matter

It is important to note that only by introducing the earth as dead matter or what Wynter calls "inert force" (2003, 267)—that is, of a matter ontologically different from that which defines the biologism of the colonial political subject—could colonialism make its political subject levitate off the energy of the subjugated (scribed as exchangeable, commodifiable flesh without contemporary geographic or kin relation). The voiding of geography and genealogy becomes a prerequisite to the encoding of subjects as ground. It is only by inserting this chasm between the biopolitical and *geos* of being that the coloniality of the world as a global territory of (exclusionary) humanism can be achieved. While Povinelli (2016, 3) draws our attention to the provinciality of Foucault's project in its conceptualization of a Western European genealogy, Wynter (2003) and Mbembe (2017) have shown how that genealogy was underscored by the racial division of Life and nonlife. Hannah Arendt in a similar fashion argues that race "is, politically speaking, not the beginning of humanity but its end . . . not the natural birth of man but his unnatural death" (Arendt quoted in Hartman 2007, 157). However, I would argue that the end and beginning are one and the same (see chapter 8 for a discussion of Arendt's race politics); if race is relocated from an expression of hierarchy (with nonlife) to the inhuman, then race names the point of quaternization between the inside and outside of (colonial) Life and the division between the *bios* and *geos*, and then the full schema of geologies of race become evident as a chasm. This schism of race cleaves *bios* and *geos* (cleave in its autoantonym form as a holding to and cutting from). The signifying practices of race (as a discourse between inhuman matter and designation of inhuman subjects located within particular racialized bodies) denies a common genealogy of the earth while it locates those subjects in that earth. This enables global geography to be claimed as Global-World-Space, an exclusive domain that does not have to admit those who are not represented by the preferred figure of biologism (the humanistic subject). So, the imperative to introduce a geopolitics that goes beyond a biologism divorced from the earth and its interrelation to forms of preferred Life must be a compelling project for any antiracist earth practice. Part of this reoccupation of the relationship to the earth and its geopowers, or a move from a biocentric paradigm to a geocentric one, includes a project like this one that rethinks how the social and political configurations of subjectivity have been historically conceptualized and experienced in relation to, and as, earth.

By thinking about the "geo" as a power that both incites and generates particular historical forms, Grosz offers an alternative to the critical modes of geopolitics that often either frame materialism as an inevitability immanent within power relations (i.e., the "resource curse") or entirely ascribe the production of power relations to variously chosen agentic centers, whether individual or structural (white patriarchy, state forces, genocidal maniacs, juridical systems, etc.). Both these approaches partake of a restricted economy in the appreciation of the potentiality of material forces within and beyond political worlds that inform the very material basis of practices in the exercise of power through the classification of matter. Extending her understanding of the engine of difference beyond social forms, Grosz recognizes the work of difference in opening human identity, by way of the cosmos, to the outsides of the forces that both produce and constrain those identities and importantly allow the possibility for those identities to be otherwise (Grosz 2011, 91). This is not about some thinly veiled return to determinism; it is about recognizing the potentialities of the cosmos and its production of difference through all the life, nonlife, and inhuman forms that constitute the planet and seeing that as a political resource to overcome the violence of their present incarcerations. Destabilizing the ground of identity and agency, Grosz argues, is a means by which to destabilize the claims made on behalf of that agency and to open up that potentiality to social transformation. Thus geopower (Grosz in Grosz, Yusoff, and Clark 2017, 134), in one sense, offers a way to go further than the "I" and undo the exercising of biopower (understood as the social regulation and positionality of bodies from the outside): "To see life as coming from the earth and its forces—gravitational, magnetic, electrical and so on—is perhaps the most powerful and direct way to destabilize our concepts of identity and agency. If the earth is riven by agents, acts and events—if it is not inert and passive—then life cannot be understood to master itself; life must look outside itself to attain the possibility of continuing itself and knowing itself" (132).

While biopower might be thought of as an enclosure (and sometimes as a disciplining and regulatory force), geopower has no such outside—according to Grosz it is the outside! But geology is also inside, intimate to the materially communicative forces of geology that make life possible, an inclusionary exteriority that ties internal forces to the outside and offers a potential for resisting biopolitics (Grosz, Yusoff, and Clark 2017 135–36). At the same time this unregulatory force belongs to the earth, and this is the very reason that instigates a desire for biopolitical control. As Grosz suggests, "The inhuman, as resistance, is always to some extent and in some way beyond biopolitics. It is important to seek out these sites of resistance at whatever level and manner they occur. These sites are those that must be left aside in the rational and economic management of 'things.' But there is something left over that remains resistant, that wants what it wants, before and beyond biopower" (137). In this desire, the inhuman has cosmic instantiations within life, and in the language-ing of this *geopull* geopoetics has the potential to counter traditional geopolitical violence (Last 2015).

What is crucial for my argument is that the specific intimacy of the inhuman, as a historically lived experience of racialized subjects ("we, the earth-bound"), is not lost in the ontologically parsing of insides and outsides but presses on the mud of relations. The radical enclosure of the inhuman as geopower and inhumane subject position is what supports and maintains the emergence of Life *as* the *white supremacy of matter*. Thus, if capitalization of geopower is the first step of power in the production of Life on the plateau, then its undoing is in the step before, in the expansion of that category of the inhuman beyond its inhumane subjective forms (but tethered to those histories of resistance) to dismantle the division on which Life extricates itself as a racial form.

Geopolitics in the Caesura between Life and the Inhuman

To understand power in late liberalism, Povinelli uses the analytic of geontopower to show how understandings of Life have shifted register from biopolitics to the governance between Life and nonlife. She says: "The simplest way of sketching the difference between geontopower and biopower is that the former does not operate through the governance of life and the tactics of death but is rather a set of discourse, affects, and tactics used in late liberalism to maintain or shape the coming relationship of the distinction between Life and Nonlife" (2016, 4).

As the ordering between the divisions of Life and nonlife moves, Povinelli argues, new figures emerge to define the discourses of power and their tactics and tense in late liberal governance of difference and markets (2016, 4). The "onto" part of the equation signals the biontological enclosures of existence that are the basis of Western ontological claims in the genealogies of governance, formulated as Life (Life{birth, growth, reproduction} vs. Death) versus Nonlife. She notes:

> Thus the point of the concepts of geontology and geontopower is not to found a new ontology of objects, nor to establish a new metaphysics of power, nor to adjudicate the possibility or impossibility of the human ability to know the truth of the world of things. Rather they are concepts meant to help make visible the figural tactics of late liberalism as a long-standing biontological orientation. . . . And, more specifically, they are meant to illuminate the cramped space in which my Indigenous colleagues are forced to maneuver as they attempt to keep relevant their critical analytics and practices of existence. (5)

Extending and redirecting biopolitics through practices of existence that exceed and differ from Western framings of bionto-life, Povinelli's "tactics of visibility" make an argument through settler modes of Life and nonlife in the governance of agency "between that which supposedly arrives into existence inert and that which arrives with an active potentiality" (Povinelli 2017a, 173). It is the very division between the access to the signification of "dead matter" and the privilege of recognition as a "live subject" that can master, which constitutes the politics of recognition in late liberalism as Life.

This axial division of materiality into passive and active forms, that might or might not denote subjectivity (depending on the colonial color line), is the current bite of geopolitics. What is important to think through in relation to this governance of the caesura between *bios* and *geos* is how the rift is geophysically organized into different spatial coordinates that map forms of life through the praxis of materialisms. That is, the distinction between Life and not-yet-nonlife that maps onto the discourses of *bios* and *geos* is actually a semantic division of geologic grammars that is actively built through a ground based on the nondivisions or continuances of geologic life (i.e., a *geos* that contains *bios* in the form of a subtending inhumane life and an unequal geology that subtends everything). It is not simply that nonlife and the inhuman can be exchanged along a continuum but that the colonial exercise of geopower made the earth different in kind to Life (in time and in space). Thus, apprehending geologic life is a prior condition of the governance of Life and not-yet-nonlife or the ground of geontopower and a consequence of the capitalization of geopower. In conversation with these concepts, geologic life gives space to examine the spatial formations of temporal scripts within the constrained continuum of the materialisms of the inhuman, and a vantage point to undermine the biocentric subject and his plateau from below. Geologic life argues the colonial sedimentation of geologic politics as a key arena of power in the organization of earth states (and their weaponization in the geophysics of race). That is, geology produces the *difference* of geopower as a change of state, not a continuum of identity negation.

Wynter's Georaciality and the New World System

The drive to homogeneity in the world system, Wynter argues, is a spatial project of unification, pushed through the figure of Man as the measure of that single world order and the modes of description in the sciences that scribed racial ordering of bodies and space. It was a racial-spatial ordering that established the hierarchies of subjection under the universal sign of Man. Wynter broadly argues that this took place in two distinct but overlapping governing codes, *Man 1* (Judeo-Christian context of European Man the conqueror, sire of Western modernity, and his aesthetic twin, Enlightenment Man, world-making descriptor of space) and *Man 2* (*Homo oeconomicus*, a man of capital accumulation in the racial aftermath of the plantations and in the ongoing empirics of [settler] colonialism, with his sidekick biocentric Man, a man that biologizes rather than philosophizes a natural account of racial differentiation). Both of these, what Wynter calls "overrepresentations" of the human, reinvest in the georacial production of space as universal, instigating the invisible "I" of the Kantian subject into place, which functions as the transparent overseer of geography: a positionality requiring the subjugation of "others" as a grounding stratum that is sedimented to achieve the perspectivism of subjective transcendence. That is, reason required and necessitated the diagnostic of the "unreasoned" as native, Black, and Black-woman.

Biocentrism configures georacial sorting while maintaining the power of scientific naturalism to author "rational" subjective truths through the recognition of a natural law, thereby obscuring the ideological and geopolitical roots of those epistemic categories and promoting their ideologues (or Man) as exceptionally situated within that natural law. This biocentrism lances Black and Brown subjects to the phylogenic bottom of the species tree, racializing and condemning their place in space and time, "dysselected-by-Evolution" (Wynter in McKittrick 2015, 146). What is important to note about this form of biocentrism is that it institutes a normative worldview that privileges the spatial movement and homogeneity of white patriarchal Man as axis and origin, while erasing that privilege through the imaginary of the universal and the natural history of a racialized (and thus spatialized) account of beings-in-time.[3] That is, this worldview of the selected human becomes an account of space that is taken as the naturalized ground of spatial relations, rather than as one raised in the construction site of a specific racial geo-logics and its subjective forms.

The idea, then, is not to expand out the forms of humanness as multiplied from these zero degrees of Man and his space-time (or surface), which would be to add to the accumulative potential of this point as *the* point of migration and expansiveness, but to fundamentally disrupt this proposition and its racializing instantiations of what Nandita Sharma calls a "single field of power" (Sharma 2015, 164). As McKittrick elucidates: "Man-as-human-and-origin fades away not to be *replaced* by an alternative perspective/figure who occupies that defining position, but rather to bring a challenge to *where* humanness takes place" (2015, 156). Challenging what Wynter (2003) called the "coloniality of being" instituted by Man-as-Center, Man-as-Universal-I, the utterability of Blackness already asserts its subterranean creative project from the rift thereby reconstituting the modes of spatial address and configuration.[4]

What Wynter's analysis does is to disrupt the homogenization of a "single field of power," to disrupt the perspective of the ground/grounding, which becomes the axial point for colonial being. She creates a rift. The zero point of colonial being, as it is imagined and actioned, is also the zero point of Time and Space (as measured from Greenwich Mean Time), and thus it erases and substantiates the *geos* as a continuous earth through the Global-World-Space imaginary of coloniality. Biocentric *Man* simultaneously erases and denigrates others as well as the earth through relations of extraction and subjugation. Therefore, Wynter's challenge to biopolitics is the shattering of the continuation and organization of its logic because her argument and Ferreira da Silva's (2007) about the racial inscription of global space make the ground shake under the assumption of the naturalism of biocentric reason. Wynter's thinking gives us a way to think what is *unthought* in biopolitics, not to foster the expansion of biopolitics (to accumulate from its racialized axial point) or foster the accommodation of others into systems of plurality (and this has consequences for the politics of pluriverses that are currently being considered in the proliferation of ontological worldings), but to destroy biopolitical thought as it currently exists as fundamentally lacking in the consideration of its own constitutive outsides and its brutal subjections.

That biopolitics ignores its own ground, or reinforces the practices of description that organize space and time from the racial axis of colonialism as a point zero, *exposes* the impossibility of a retrieval of a biocentric subject (although Wynter holds on to a tactical humanism). Wynter's (1984) biopolitical challenge begins in 1492 precisely because, she argues, this New World view brought a "root expansion of thought" (in Sharma 2015, 166) into being concerning ideas of subjectivity, temporality, and spatiality through geographic expansion and its descriptive modes (as territory and site of extraction, and through the selective partitioning of belonging to human, subhuman, or inhuman categories, alongside the voidings of geographic attachment and freedom). She argues that 1492 was a "geo-racial restructuring" (quoted in McKittrick 2015, 136).

If the biopolitical conceptual ground is understood as always already constituted through an inhuman grounding (or racialized subjects) since 1492, geologic life, even as it is pried apart by liberatory inventions that refute that designation, the inhuman is what secures the differentiation to produce the human as such. The inhuman, however, can never be incorporated into the political subject or *being* as such because of the material differentiation between figure (the political subject of juridical rights) and ground (the matter of resource to enable the *becoming* of the political subject, philosophically and materially, that may have access to rights). Thus, the politics and practices of making nonlife within a discourse of biopolitics, its sites of concern in terms of bodies (the alienated human subject; the subject without rights) and places (camps; zones of exception), remain exclusive within the domain of the human as it is designated by its partial humanism. According to Wynter, this "partial humanism" (Wynter 2000, 196) organizes "*history* for Man, therefore, narrated and existentially lived as if it were the *history-for* the human itself" (198). Man thus represents himself as if he were the full scale and sense of the human and reproduces those epistemic modes of life. But Man is not geography.

Wynter's argument upsets the notion of a transparent subject that either preexists the field of power or occupies the normative middle ground of knowledge production (epistemic and aesthetic) that bids the imagination of a singular plane (or planet) into being. This imagination of a singular plane of power installs a perspectivism of a synchronous present time, that is rendered as political time, that is universally shared (see also Indigenous thought on the coloniality of time). This temporal spatiality establishes a ground that was recognized as in need of "leveling" in postcolonial thought through the suggestion of an equality of access, possibility, and perspective, while failing to address the actual stratified lived experiences of institutional historical power imbalances and the ongoing afterlives of the disruption of different temporal-spatial modes of inhabitation. Because the institution of power was epistemic, there is no such common ground as yet and thus no grounding to be made in the universal of politics and the production of its plane of the present political tense of the planetary (i.e., there is Indigenous time(s), Black time(s), Black materialism(s), Brown ecologies, etc.). Similarly, the presumption of biological differentiation or the differentiation through biological markers (racialized codes) rests on stabilization of forms that are stabilized for all. In the same way that Kant's "subjective universality" and transparent "I" can sense a transparent thing called "beauty," *as if* it were a transparent universal medium of access, except for the condition of racial difference that is imagined to preclude access, and thus produce a conditional universality, which betrays the hierarchical organization of totality. In a concurrent production of space, globality is a production of space-time that, from 1610, described a particular mode of territory (as totalizing) and temporality (as colonial time). Bodies traversed this time and territory, which established the commodity form of matter (as slave, cotton, nature, copper, gold, indigo), but this time-space on which biopolitics arranges itself did not establish itself in a vacuum of other temporalities and tethers to the earth.

The presumptive erasure of all other possibilities of time and space materialisms, and all other modes of relation and subjective sense, was a geographic imagination before it was an ideological goal (and a material condition enforced through genocide and annihilation). This *violent intent not consent* was able to be refuted precisely because of the fullness of geography to hold a multiplicity of imaginaries. Thus, the regulatory "outsides" to the normative that allow it to function as such are never just regulatory edges. The temporally discontinuous, the chronogeographies of nonsettler time, blues time, Black time, and beaver time *expose* the plane of reason on which the presumption of the universal rests as radically dissonant, as the rift exposes the lines of compression that constitutes the plateau. Colonial forgetting and its failure of a plurality of imagination overemphasize and overrepresent (in Wynter's terms) the conditions of the world as constructed by and for the colonial, née Kantian subject. The *bios* of agency of the colonial subject is predicated on the inherent subjugation and unacknowledged *geos*. Blackness was raised as a concept and material praxis (encoded and subjected) on the erased ground of the Human, and the other "inhuman" in the dual economy of matter. But as we know, both Blackness and the *geos* have their own inventive paths.

Wynter instead calls for "a version of humanness imagined outside liberal monohumanism. . . . Her overall project [argues McKittrick] can be identified as that of a counterhumanism" (McKittrick 2015, 11). McKittrick argues that Wynter "dedicated her own past and still ongoing work to the furthering of the 'gaze from below' emancipatory legacy" (2015, 11). This subterranean counterhumanism—the *human as praxis*—is a model not of inclusion but of counterpraxis. As biopolitics fulfills the function of explaining who and what we are—the forms of life and the terms of its reproduction—the bare ground and zero point that define those whats of life (*what life is, what life wants, and what life can be lived*) do so using the conceptual apparatus of a form of biocentric life that is already racially defined against a subtended position that is ground, not subject (i.e., these questions occur in situated material relations that are already indebted to the mobilization of a racial deficit). The sovereignty of white privilege and the white privilege of sovereignty (or the "I") are grounded in what Bench Ansfield calls "black bodily and geographical liquidation" (2015, 135). The geographic transformation of spatial coordinates that racial geophysics instigates collapses bodies and earth in an implosion of material categories via the figuration of the inhuman, understood in its stratified and held position below the surface. All spatial projects and materialisms must take account of this difference of the geophysics of race in the juridical and ethnopolitical delineation of forms of life, understood as a deep geopolitics of the racial arrangements of geopower.

Man 3 (after Wynter)

> Only by becoming inhuman can the human being pretend that they are. —JAMES BALDWIN, *The Price of the Ticket*

As Baldwin alerts us, the mechanism by which the human can realizes himself is through the contradictory double movement of becoming inhuman. In the context of colonial afterlives this took on the double valency of marking other subjects as inhuman to secure inhuman powers. Yet, Wynter says, the slave plot and many other vernacular histories point to other conceptualizations of possible and actual material worlds, other threats to this order, other possibilities that nourish the ground and body of those cast outside of *Man's* partial humanism. To unsettle the sociogenetic codes further would mean paying attention not just to *Man* in his inhumanism, as a figure whose material emergence is built through the flesh of geology, but to inhumanism as an institutional earth praxis that organizes the conditions of geography. If the figure of *Man* is one such target in the asymmetries of race and the modalities of its subjection, colonial earth is the other, as *his* feasting ground, from which *Man* makes *his* subjugated-inhuman-persons. Overseers' earth claims the spatiality of the surface and all that is drawn into its forms of valuation. But the affectual infrastructure of being inside the effects of colonial geophysics also extends a perspective in its cramped subterranean undergrounds. Those other modalities retheorize place in the rifts of geotraumas.

Indigenous and Black studies scholars have shown how the Human, as a preferred or partial subject of European colonialism, structures the categories of being, subjectivity, gender, sex, labor, and modes of relation. Speaking in the realm of species life, Fanon's concept of two species or Wynter's genres of the Human are watchwords for the incorporative logic of humanism and its willful optics of erasure of the "ground" on which its hierarchies are built—namely, anti-Blackness. To this list of arrangements, the earth should be added as a racialized object and category (white geology as the praxis of being human), but one that is resistant *and* a source of resistance to such flattenings in the voluminosity of existence. Wynter's work shows us how biopolitics naturalizes itself as a theoretical armature that is engaged with necropolitical relations. This biopolitics is itself assembled with a conceptual and material infrastructure that is already built on the ground of the dispossessed: a black and brown earth, a "fertile soil" (Glissant 121–30), from which it grows its epistemic subjects. In this epistemic construction site, the capacity to create relations beyond Man must inevitably find resources in the earth and within the capacity of the world beyond an anthropocentric focus on world-as-property and resource. Caribbean literatures, Indigenous risings, and blues epistemologies all give expression to deep imaginaries of matter. Intimacy with the inhuman has a longer duration and density than the architectures of colonialism, but in authoring stories of monogenesis, of origin and telos (of natural resources and Anthropocenic earth), it erases them. I do not want to claim this colonial materialism of inhuman intimacy as a recuperation but rather to pay attention to it as an essential collaborator in projects of freedom, and thus argue that it forms an ongoing site of collective redress of opening space in colonial afterlives.

The Fanonian paradox of the invention of Blackness has received much critical attention, yet the substance of othering, its geophysical properties rather than metaphysical dimensions, bears on the mobilization of forces of energy and mineralogy in the world, as well as on an ongoing question of environmental justice and exposure. The material incorporation of the European subject (and its settler colonial kin) in terms of value, accumulation, and subjective forms was defined against what was classified as fossil nature (Indigeneity) and fossil energy (the enslaved) to transform the ecological and energetic organization of the world.

To extend Wynter's formulation, the Anthropocene might mark the introduction of *Man 3*, whose emergence is as a geologic superpower. The terms of his geography are planetary. He is a geologically informed subject whose life-forms are coded through the inhuman-inhumane, whose constituting powers are realized through the mobilization and governance of geologic materials (minerals, geochemicals, fossil fuels, carbon, nitrogen, phosphorous, etc.) and the accrual of geopowers as a form of securing settler futurity. *Man 3* trades in the conversion of racial geophysics into metaphysics and mobilizes the weaponization of environments to police populations.

The Inhumanities

Reimagining the humanities through its shadow institution, "The Inhumanities" designates a field in which race and materialism are intimately tied as coconstituting discourses of colonial earth's progeny, the Anthropocene. Beginning with the inhumanities, as a savage form of subjectivity that is organized as a category and thus subjugation for phenotypically designated populations through the operation of race within the historical geographies of colonialism, puts forward a pedagogy that is not immediately predicated on erasure.[5] There is a before to colonialism (and its spatialities and temporalities), but there is no *going back before* colonialism, so in this sense, the need to battle humanism is a reckoning with the historic situatedness of Europe's global imperial geographies and their planetary geo-logics, even as those "other" places were rendered as mere mirrors for Europe's "soul." The doubling of global and Western-centric spatial productions in colonial geographies forced an engagement with humanism rather than its total rejection. In conversation with Wynter, David Scott refers to this tie as her "embattled humanism" (Scott in Wynter 2000, 153). As Wynter explains: "You know that you cannot turn your back on that which the West has brought in since the fifteenth century. It's transformed the world, and central to that has been humanism. But it's also that humanism against which Fanon writes [in *The Wretched of the Earth*] when he says, they talk about man and yet murder him everywhere on street corners. Okay. So it is that embattled [humanism], one which challenges itself at the same time that you're using it to think with" (154).

Without the logic of the inhuman (as object and subject), the entire mechanism of value and the spatiality of category distinction that produced the relation between Europe and the New World and the humanist subject falls away. And thus the inhuman, as well as being a source of subjection is also a site of embattled possibility because that category of regression was always being transgressed, and other relations were being instigated that spoke to the possibility of other geo-logics and geopoetics in relation to the earth. Everywhere that the inhumane is imposed, it is resisted by a humanness that highlights the dark contours of the humanist subject in its partiality. Thus, the anticolonial critique is not simply a critique of the inadequacies of the human but a counterimaginary that opens up a fullness in the register of the world through its ghost geologies and actual histories. The methodologies of the inhumanities craft epistemologies for their own earth and institutions.

In tension with Wynter's "embattled humanism" (Scott in Wynter 2000, 153), Frank Wilderson (2010, 2020) and Alexander Weheliye (2014, 2019) argue, among others, that Black life is the negative ontological ground in which the New World/Old World is planted. In *Red, White and Black*, Wilderson examines racial antagonisms in cinematic cultures in the United States as a "triangulated relation of modernity" (2010, 247); White as "settler," "master," and "human"; Red as "indigenous," "savage," and "less-than-human"; and Black as "slave" and "inhuman." Black, he argues, is an ontological negation indivisible from the construction of being (see also Tavia Nyong'o's *The Amalgamation Waltz* [2009] for a more hybrid performance of the dialectics of racial embodiment). As such, the slave is already absolute nonbeing and so cannot enter the biopolitical realm as a "state of exception" (in Agamben's [2005] term). In an incisive chapter titled "The Ruse of Analogy" Wilderson asks about the normative framing of suffering through whiteness. He suggests, "Giorgio Agamben's meditations on the *Muselmann*, for example, allow him to claim Auschwitz as 'something so unprecedented that one tries to make it comprehensible by bringing it back to categories that are both extreme and absolutely familiar: life and death, dignity and indignity. . . . Agamben is not wrong *so much as he is late*. Auschwitz is not 'unprecedented' to those whose frame of reference is the

Middle Passage followed by Native American genocide" (Wilderson 2010, 35, 46).

Wilderson is not ranking the inhumanism of violence so much as historicizing it and highlighting the amnesia of the normative template of such questions that can realize their horizon only in a European body politic. What Wilderson identifies as the repressed and irredeemable conflict of antagonism (2010, 36) that Mbembe (2019) calls *necropolitics*. While Agamben's *Muselmann* oscillates between the poles of life and death, nonhuman and inhuman registers of being, he never leaves the vicinity of Man, as *he* is conceived in the humanist form and in terms of its remit of "being." As the rhetorical demands of analogy call for a register of social oppression, Wilderson argues, the "ruse of analogy erroneously locates Blacks in the world—a place where they have not been since the dawning of Blackness" (2010, 37). This not only erases and mystifies the grammar of suffering but is laden with a promissory modality of the Human as available and open to share the same fantasies (where masters and slaves can speak as if these had the same interests).

Rejecting the promissory lure of humanism in the context of its inhumanities is to understand the dual liquidification of being and geography that slavery enacted rather than some prior or absolute claim in histories of violence. In the context of the geologies of race, Blackness is not just positioned as less than the Human but is the inhuman *cut* that allows the Human to come into being and that which elaborates the difference between its *bios* and *geos*. The space of this interregnum cannot overcome its structural placing without the collapse of humanism and whiteness, so is outside the purview of the biopolitical without recourse to a claim on the inside of that juridical category. Degree zero cannot traverse its own negation and structure; this is what Wilderson calls the "unbridgeable gap between Black being and Human life" (2010, 57).

A more appropriate biopolitical figure of thought might be Ludger Sylbaris (aka Louise Auguste Cyparis), the only man known to have survived the eruption of Le Pelée in the colony of Martinique in 1902. Thinking biopolitics from below as an earthed condition of the *geos*, Sylbaris was incarcerated in a prison in Saint-Pierre for some minor offense; he absconded to the carnival, then was placed in solitary confinement in an underground cell—a prison in the rock. He survived the volcano that dissolved the colonial city and then made his living in fairs, in the performance of survival as a pyrowitness, corpsed as a body of evidence. Billed in inhuman taxa as "the Only Living Object that survived in the Silent City of Death" in Barnum and Bailey's "Greatest Show on Earth" posters, he enacted performance after performance of near death, charred curiosity, and volcanic proximity before audiences in America. Sylbaris figures the *shock-forward* of geotrauma and the multiple materialities of its incarceration, first as trauma and then as corpsing. In the history of geology, Sylbaris is a figure of survival in the context of the death of twenty-eight thousand to forty thousand people from the volcano, yet he is never allowed to just survive. As his geologic cave provides a respite from the geoforces of the earth, Sylbaris's survival points to a performance of Black political life in need of abolition—a life that is rendered visible only though repetitive cycles of geotrauma. As Tiana Clark's brilliant poem asks us to consider:

> To be saved, for what? This. Me? Making him an object, again—
> objectified ode, again
>
> . . . .
>
> the witness and his blistered onyx back, cracked coal, black tephra.
> Praise
> pryoclast: meaning fire + broken in places. Praise
> a man without a city, his people: scorched and suffocated
> from insidious gas or the searing belch of reckless lava. Praise
>
> . . . .
>
> The man has become his cage—has become a mountain within
> a mountain
> about to burst with liquid fire. (2017, 79–80)

A history of biopolitics reveals the inability to think outside the cage of white supremacy, outside the nearly or not-quite-yet subject, which is the subject that can be recognized as incorporable into whiteness, unlike Blackness, which Wilderson suggests registers a "scandal" in the epistemic order.[6] Biopolitics follows a logic of the subject as it is made through capital and in relation to capital, rather than a logic of colonialism and how the subject is made in relation to white geology and the "right" to geography as the claim to planetary property and persons-as-property. Or, in Du Bois's terms, "whiteness is the ownership of the earth for ever and ever" (1920, 30). This raises profound questions about the functioning of biopolitics as an analytic when it is formulated as a discourse in the white grounds and geo-logics of humanism. What I am addressing in this book is how that question of the human is materially made and framed through the inhuman. It is not just that this ontology has a material expression but that racial geophysics is an expression of a colonial materialism that far exceeds a biopolitical analysis. If geopower as a field of power relations was taken not as an *extension* of power—understood as a colonial projectile *reach* across space, into undergrounds, volumes, verticality—but, rather, as an *analytic of the ground* in which biopolitics is raised, then the partial politics of the subject emerges not just through its biopolitical exclusions but in its material forms as an *aliquid* subject interned in the matter of the earth, in the rifts of earth and being. Black subjectivity, according to Glissant, has the volcanic magma of Pitons in it. The great fiction of colonialism is this privileged separation between *bio* and *geos*, and the deadly pretense that we are not all of the earth. As Sexton makes clear, "Necropolitics is important for the historicist project of provincializing Agamben's paradigmatic analysis, especially as it articulates the logic of race as something far more global than a conflict internal to Europe (or even Eurasia). Indeed, Mbembe initially describes racial slavery in the Atlantic world as 'one of the first instances of biopolitical experimentation'" (2010, 32).

Sexton, following the work of Hartman, suggests that slavery is an exemplary manifestation of the state of exception in "the very structure of the plantation system and its aftermath" (2010, 32). Further, he points out that Mbembe's work on necropolitics abandons too quickly the institutional structure of slavery and does not heed Hartman's work that names the Black woman as the paradigmatic figure for biopolitics, as the site of the (re)production of enslavement, which in turn normalizes sexual violence, thereby highlighting the "inextricable link between racial formation and sexual subjection" (Hartman quoted in Sexton 2010, 33). Sharpe (2010) calls this the "queerness of blackness," whereby Black sexuality is connoted as lacking a "proper" signifying power as sex or gender, rendered as monstrous. This is also what Tavia Nyong'o calls "negative heritage," in which "futurity is not the social democratic hope of gradual inclusion and improvement, but . . . cataclysmic irruption" (2009, 164).

So, where does this leave biopolitics and its patriarchal privileged forms that articulate from the vanishing point of the earth (the degree zero of Overseers' earth or Sylbaris's pyric underground)? What possibilities open for antiracist thought if it is conceded that the premise of biopolitical thought is beyond the event horizon of Blackness? And how, then, to engage geopowers as a site in the making of these gravities of violence? Thinking with Black studies scholars, in conversation with an engagement with the veracity of the inhuman and its historic geophysical grounds, touches on the totality of the systematic geo-logics of colonialism and its nows. The materiality of this analysis thus opens questions of gender and sexuality as inchoate in the flesh of geology and thus brings into view possible sites for the amelioration of the heinous violences that often structure its coming into being. Weheliye, alongside Indigenous queer scholars (such as Leanne Betasamosake Simpson), suggests the possibilities of imagining gender and sexuality otherwise: "This means that often if Black people sought recognition as properly human, they have needed not only to accept and perform an idea of humanity steeped in white supremacy and colonialism but also to don the drag of normative genders and sexualities. My point is, though, that this represents an opportunity for imagining gender/sexuality otherwise, for embracing and inhabiting the ungendered flesh, for fully and differently inhabiting the gift of Black Life" (2019, 239).

If we extrapolate back to the rift and through the flesh of geology, both gender and sexuality are rendered changed by the abandonment of the narrative field of white geology and its natural resources, as it is constituted by patriarchal geopowers and regulated by the settler family. Engaging with an explicitly raced materiality as conceived within the stratigraphic formation of colonialism offers the possibility to comprehend the scope of the psychic life of geology and its empires of inhumanity. In a parallel vein to the geologics of race, the gendering and sexualizing of natural resources (such as oil and gas) are not just a regulatory function of the expression of patriarchy but the very basis of the exercising and reinscribing of geopower and its corollary geotraumas. As there is no race without patriarchy, there also is no racialization without deployment of heteropatriarchal geopowers. Gender and sex are not prior to geopower but constituted in its capture and consolidation. And thus geopower is a possible site in the undoing of patriarchy and its prescription of normative genders and sexuality regulated through normative modes of materiality (white geology).

An analytic of geopower can be understood as an undoing of the categories that maintain a juridical subject that is predicated on a humanism that erases the libidinal and lithic investment in anti-Blackness as its material and conceptual ground. Geologic forces are potentially antagonistic to oppressive forces insomuch as they are open-ended and always in excess of the strictures of power. In Grosz's contribution to geopower she suggests the potential of geopower as a recognition of the freeing of the grids of subjugating power. Unsettling history as plotted in a trajectory that naturalizes the fallacy of the human as an open category ready for investment, organized through the struggle for justice framed as a coming-of-age story, is but a first step. The libidinal economy of the Human as a category awaiting inclusion, rather than one organized through specific racialized enclosures and exclusions, is a ruse under which biopolitical thought gathers itself. The lithic economy and its stratigraphic imagination of race underpin biopolitics and are what upholds the spatial surfacing of whiteness as power.

Furthermore, knowledge of the earth is also interned in a perspectivism and praxis of the supremacy of whiteness, which marks the capricious geography of the human, what Wynter referred to as a "racially dominant white elite stratum" (2000, 126). Similarly, scholars in Indigenous studies have argued for and practiced different ontological arrangements of the earth against settler colonial modes of extraction. It is precisely the noncorrespondence between earth forces and the enclosures of racialized political theory that makes geopower an expansive material and theoretical terrain that challenges the concrete construction of identity politics and thus a site for possible material and symbolic transformations of freedom and subjugation. As I have argued elsewhere, this engagement also requires a remembrance of the inhuman as an occupied racialized category and historic geography—or a surfacing of the sedimented racial *bios* through the *geos*.

What does an alternative orientation in the geologic field of force look like? How might geopower be imagined as aligned to possibility rather than dialectically configured to forces of oppression in the earth that colonialism made? What are the methodological routes to develop a praxis of relation to the earth that embraces revolutionary geophysics to undo metaphysical voids (whereby "blackness is constituted by an injunction against metaphysics" [P. Douglass and Wilderson 2013, 121])? I have argued that engaging an understanding of the historical and ontological contours of geologic life is a precondition to making visible the full scope of the praxis of extractive earth and its increasingly (and always for others) unlivable modes of Life. The inhuman is where meaning gets unmade or remade, so it is both target and site of possible emancipation with regard to racist structures of material engagement and its uneven earth.

Transgressing Geopolitics

If the inhumanities expose humanism as a hermeneutic shell that neglects its material constitution (as one that holds its elevated position only because of the accumulated energy from systems of settler-colonial violence and anti-Indigenousness/anti-Blackness/anti-Brownness), it is time, then, to re-imagine another subject capable of apprehending the differentiated and differentiating geoforces it is historically embedded within. The *tense of geophysics* dismisses the fantasy of a materially autonomous subject that does not need the earth and that exposes the racialized forms of its extractive economies to stay above the earth. Below is a summation for the sublimation of geopolitics into the rift with the analytic of geologic life:

1. Biopolitics will not do as it attends only to the aftereffects of race (even as we seek to undo biopolitical violence in this register). Geology is what extends the body beyond itself into material forms and modes of expression that continue beyond the frame of Life (partial humanism) and the imperative of its privileged perseverance, in a relationship with land, territory, and forms of expression and sustenance that derive from those earths. These very geoforces stretch and incite life (in a general field) beyond itself, thereby giving life, as a not exclusively human category of experience, an excess (or possibility) that it cannot achieve on its own (Grosz 2008, 23; Yusoff 2015a, 2015b). And life as a concept is changed by the intimacies of those inhuman experiences and transgressions.
2. Geopolitics does not yet capture the colonial division of the earth's surfaces and depths for the accumulation of geopower (and its attendant affects in terms of racialized geophysics and spatialized rifts).
3. Geopower can be extended as a concept if it takes in the scope and sedimentation of its historic geologies of race and their deployment as a form of geophysics. Geopower, in Grosz's formation (in Yusoff et al. 2012, 988), is the *pre*political condition of geopolitics and the uneven field of power in which biopolitics takes place. Geopower in my account is prepolitical only in terms of the visibility of its emergence in politics but is already politically constituted through grammars of geology and the constitutive exclusions and strategic deployments of inhuman power in colonial earth.
4. Geontology in Povinelli's formation gives visibility to the governance of the Life-Nonlife boundary (as it is established as an operative political tool that presses on Indigenous life).
5. *Geologic Life* argues that anti-Black, anti-Indigenous, and anti-Brown gravities are constituted by an injunction between *bios* and *geos*. This injunction operates ontologically and epistemically to secure geopower for the plateau (through the work of the inhuman). The plateau is a historical geography that is characterized by the white supremacy of matter and the governance of geologies of race. The plateau is now a multiethnic and multiracial affair, but it is still characterized by the paradigm of racialized undergrounds (often of the racialized poor and migrant) to produce its geopower. Racialized undergrounds create rifts in time and space. These uneven geographies are maintained by racialized gravities.

The onto-epistemic histories of geologic life have spatial affects beyond that which is captured in geographic analysis.

6 In the analysis of geologic life, the space and spacing of the inhuman emerge as a key site in its construction and thus its unraveling. Understanding the inhuman as a differentiated intimacy of potential transformation frames a possible breach in the reproduction of colonial earth.

7 A commitment to the radical redescription of inhuman intimacy is a transgressive shift in modes of subjectivity and the earth, its materialities and temporalities, and geopower; modes of description that pay attention to how the ontological categorization of matter is used to do political work that is gendered and sexualized as it performs in and through the categories of life-death, human-inhuman, agentic-dead, past-future dialectics; and modes of subjectivity that are released through the acknowledgment that identity (organic or otherwise) is never settled once and for all and that the earth is always a dynamic ground, and that power relations are historically situated but not immovable forces. The concept of inhuman intimacy offers temporal dissolve, as time is a way of undoing geographic holds (as it is part of time's historical racial scripting). Modes of redescription must go beyond the enclosures of subjectivity and should not extend out recognition from the Western center of bionto-political Life.

8 Understanding the geologies of race through a material architecture of affect and a stratigraphic imagination gives precise coordinates to do this work. To counteract the racial debt accrued in the levitation of whiteness cannot simply proceed solely through a hierarchical readjustment but through earth reparations. The very imagination of the earth and the geologic grammars of languaging that geographic imagination must change. Addressing inhuman intimacy produces a partial horizon of experience and powers that are only partly confined by the limitations of what a body can do, insomuch as the living body is already imbricated in the cosmos and its geologic materials and a differentially situated subject of its forces. Such open-ended geoforces push against the regulatory skin of biopolitics, but they have also been historically marshaled to biopolitical ends.

9 Rethinking geopolitics is not just a question of redirecting the spheres and spaces of operation into more mineralogical or elemental expositions of subjectivity or through the embodied states and thresholds of resources, but requires an overhaul, a *going under* (as Nietzsche says), into the corporeal and planetary processes of the geologic and how it pro-

duces forms and categories of being (political), while recognizing that this is also where people live and theorize and fight. This going under is already extant to the political scene, as a racial underground has defined the logics of biopolitical surfaces and surfacing. Only by accounting for the flesh of geology does this going under become understood as a racialized subterranean subsidy to the surface (of value, materiality, metaphysics, and thought).

10 *Geologic Life* is an analytic to (a) substantiate the full terms of the inhuman in its (geo)political context as historically constitutive of biopower in all its incarnations (as ground and geophysics); and (b) examine modes of inhuman subjectification or the *intimacies of the radical impersonal* (not just as a category of differentiation but as a differentiated racialized category) that demand a conceptualization of geopower outside of Life and inside of geologic life. The first task is a local task of decolonialization (of languages, epistemologies, categories, ontological determinism); the second task is nonlocal and extends beyond any formation of the human into a poetics of Relation with the cosmos and the earth (it is also a historic project of what the earth remembers, and colonialism tries to make us forget).

And the rock cried out, Geopower! All on that day.

III Inhuman Episte-mologies

8 Inhuman Matters I

Black Earth and Abyssal Futurity

But the rock cried out
I can't hide you, the rock cried out
I can't hide you, the rock cried out
I ain't gonna hide you there
All on that day
I said rock
What's the matter with you, rock?
Don't you see I need you, rock?

—NINA SIMONE, "Sinnerman"

Geotrauma

I'm not going to describe the song "Sinnerman" because you can probably already hear it. In Simone's arrangement the rock is where Sinnerman runs to, but it does not offer redemption or rescue; neither does it enforce any brute condemnation. It *cried out*. . . . But why does the rock not respond in the right way? *What's the matter with you, rock?* In the interstitial refrain *Power! Power!* Simone claims the space between the rock's refusal of redemption and the call to political freedom in hardened spaces. Keith Feldman (2017, 156) comments that the cry of "Power!" holds together the time of slavery and the enduring times of unfreedom: "The cry's possibility materializes out of a double refusal—neither save nor damn—that frames what we might call, following Fred Moten's keystone formulation of a black radical aesthetic, the break." While I do not want to reduce Simone's song to a partial reading, it seems that she plunges the song into the rift, naming Black freedom and power between the rock and a hard place. In that naming of an epistemological break, between the promise of shelter (*I run to the rock*) and the demand (*Power*!), Simone sings fracture and force as an epistemology of the rift: *Don't you see I need you, rock*? Rocky refusals are the indifference of inhuman matter, yet rocks hold the possibility of shelter nonetheless, in the durational force of their geology and testimony.

In another rock parable, Patrick Chamoiseau's character Slave Old Man, escaped from a Martinique sugar plantation, channels a different form of geopower, seeking shelter beside a rock that holds the ghosts of the Indigenous and enslaved in the stone, "teeming with a myriad of peoples, voices, sufferings, outcries. Unknown peoples were celebrating an awakening. The being seemed like lightning shooting through the Stone. An un-shining energy. It did not project itself anywhere. It did not affect the reigning eternity. An incandescence pulled-in to its very core" (Chamoiseau 2018, 108).

The rock holds concentrated sonic traumas "pulled-in," "an un-shining energy" that does not expose its affective archive. The inhuman rock council did not "project itself anywhere," suggesting a nonimperial geographic imagination (unlike colonial perspectivism). Geotrauma is held by the rock, incandescent, as if to the iron heart of the earth itself. The ghost geologies of the rock transform the "Monster"—who is the Master's mastiff in pursuit of the escaped enslaved man (called Slave Old Man)—from a beast into something else, a being no longer intent on murder but given to licks. The Master weeps for "the monster he has lost" (Chamoiseau 2018, 109) as they return to the plantation. The weaponization of nonhuman life is revoked by the rock's inhuman testimony. This pursuit of the Maroon and the transformation of the Monster at the rock provoke something buried in the Master: "other spaces were bestirring themselves, spaces where he would never go, perhaps, but where one day no doubt, in a future generation . . . his children would venture" (109).

The temporal qualities of the rock's deep time testament allow for the possibility of redress and future reparation. As Slave Old Man runs to the rock, the rock cries out its geotrauma, naming the hidden strata of desecrated personhood that formed the conditions of the Master's possibility. The rock is a shelter, protective of difference, in the way that Glissant names opacite, whereby the inhuman holds with the irreducible difference of being against the cold night of reduction in the sterile worlds of identity configuration. The testimony of the bones *gathers* to the rock Carib, maroon, unknown, speaking as fossilized utterances, *not-fossil* voices. Jumbled and together. It is a gathering porous to material histories, no longer one single origin story or geography, but a concentrated pooling of geotrauma in the diaspora, resistant to erasure (which is the prelude to reoriginating matter in colonialism toward the telos of the white supremacy of matter). The rock speaks the histories that dig at the earth for their relation; rock languages mark the cry of abandonment (which is the worst fate of all). The narrator narrates: "Boulder of heavy intoxication. These vanished ones live in me by means of the Stone. A chaos of millions of souls. They tell stories, sing, laugh. . . . I no longer feel the wounds hacking up my body. I have reached a rib of alliance between life and death . . . time and immobility, space and nothingness. I embrace the stone as refuge-being" (Chamoiseau 2018, 102).

Unsegregated being, *all on that day*. The rock speaks the *rib of alliance between life and death*. If white geology organized a thought that rooted being (as racialized genealogy) and territory (as the apex of possession and property), it also set into motion the thought of errantry and earth as the refuge-being of the cosmos. Believing in the shelter of the world makes possible revolution and marronage (as a quieter form of revolution). Rocks are not racist, even as they can be weaponized as such. Simone and Chamoiseau name *power* in inhuman intimacies as a possible site of redress in the violence of geotrauma. The rock is a grammarian in a call-and-response between sinner and the world, gathering stories to the rock pile, building remembrance, and relanguaging Relation.

"I build my language out of rocks," says Glissant. "I am confronted with the necessity to exhaust all at once the deserted (devastated) field of history where our voice has dissipated, and to precipitate that voice into the here and now, into the history to be made" (2010 [1969], 43). The deserted field is the geotrauma of erasure, the "nonhistory" of colonial imposition that makes compressed strata of deformed subjectivity under the pressure of the surface's white plasticity. "Our historical consciousness could not be deposited gradually and continuously like sediment . . . but came together in the context of shock, contraction, painful negation, and explosive forces" (Glissant 1989, 61–62). Opening to the rocks' testimony is also opening the strata, uncovering buried histories, unearthing that which has built present conditions and geographies. Glissant says it is the poetics of duration, "in obscure and extracted strata," that "opens being onto his lived relativities, suffered in the drama of the world" (2010 [1969], 42). As the stratum speaks of buried kinship, it is important to clarify that the inhuman resists an easy redemptive narrative. I understand the desedimentation that Glissant proposes, and the shelter Simone sings, as a poetics that gathers to alleviate the material weight of history (Master's history and the Overseers' prerogative).

Language builds relation in Glissant's schema; the rocks are in Relation, which refuses the universal. The universal can be seen as that which *overdescribes* in the generality of its speech and occupation (monolinguism); it is the speech of the Overseer, whereas underdescription travels in the subterranean (as a protected or secret speech act). Glissant's rock language speaks from "place to totality" (2020, 75), rather than totalizing all over the place. The rock pile is the sum of the particular in relation to the whole world. Thus, it is important in these material metaphors to see the rift not just as a space that is about illuminating the surface and its arrival and what made possible its conditions of being, but to see the rift (and its rocky debris) as always in a state of recomposition, building new languages out of the discarded stones, improvising narrative arrangements and geostories. These rock piles accumulate detail and hold witnessing. The earth remembers. Artist Kiyan Williams uses the phrase "what the soil remembers" to think with the longitudinal and fracture afterlives of slavery, collaborating "with soil as material and metaphor to unearth Afro diasporic history and trans/gressive subjectivities."[1] The earth remembers as colonial epistemologies orchestrate renewed forgetting.

The potential of the earth as an archive and a journey fellow in struggles of resistance is to activate shared social and racialized histories against a history of the "inhuman," where rocks instruct on stories of erasure, if not the erased and forgotten. Can a rock be a collector of stories of disobedience, a witness to rebellion, and the dreamy countenance that claims time outside the colonial clock and its abrupt climate earth? If the earth remembers more than it forgets—as a billion blackened Anthropocenes of colonial earth suggest—inhuman memory is much more than new or better origin stories. Rocks are ruins that *hold with* absented and erased memory. In soil, rocks, mounds, a gritty residue of counterarchives can be found, as discourses of materiality (namely, natural resources) organize its forgetting. Material histories cling to the world under conditions of erasure, marks lodged in unseen places, against inhumane histories of dehumanization, where "colonization = thingification" (Césaire 2000 [1950], 42). Tactics of inhuman intimacies are a political valence that give the earth a different future as an archival medium, a *scarred* set of forgotten affects and accumulated temporal residues. These counterarchives require not only a reading against the grain of epistemes of materiality (from natural resources to nature writing) but also a syntax that differentiated from the enclosure of normativity that these languages make.

Looking at the ground, the aftershocks of colonialism ripple as a form of *geotrauma*, muscular and shocked forward into the sinewy fabric of social lives and material worlds. These aftershocks reproduce their affects through spatial and temporal architectures of affect that regulate spatial expression and materially impact as anticipatory modes that govern for the present and build futures. But there is a counterforce, of the *forward shock of events*; that which gets buried and foreclosed before it has begun, or granted beginnings as a temporal passage, in the material production of empires. Earth archives, such as these, hold submerged lives and cut off futures in a silty form of stasis. Earth archives might be a site of methodological redress—building spaces of remembrance, naming ghosts, and shifting temporalities in the colonial afterlives of extractive earth. This is a forward shock that unearths like a fossil a new set of constellations that rewrite history and name loss.

Rock Piles Smash Epistemic Dialectics

Dominant encounters of the inhuman are organized either (1) through *dialectic* structures (in the human-inhuman hierarchy; person-body designation; subject-property; agentic-flesh) or (2) as *alterity* (belonging to a different order outside of the geosocial; as a harbinger of the "outside"; alien; or intemporal, outside of time). Both these positions use a stratal arrangement and collide with flesh as modes of inhumane subjugation (that contribute to a structural underdevelopment and material violation). However, it is important to recognize how alterity functions as an epistemic escape that covers the inhumane dialectics of colonialism, placing outside that which it functions within. The purification and obscuration in the idea of absolute alterity (typified by the speculative realists) ignore a prior division of the materiality of colonialism and its occupations. Thus, there are reasons to be cautious of the redemptive effects of the inhuman as alterity in the ongoing (re)production of inhumane conditions. There are other gatherings in the collapse of categories that require us to start elsewhere in the erosion of both those positions and "plumb the depth" of the abyssal (Fanon 1986) to find a submerged temporal dissonance that brings the inhuman inside as a radical continuance of disruption (the rift) rather than as incorporation (on the plateau). The change in state that characterizes the rift offers a path for thinking geophysics as a fracture in the normative that reroutes both the understanding of what a subject is and the *intra*material contours of being for one another and of the earth. Another way to say this is that it is not a case of being "in" the world (in the Bergsonian sense) but "of it," which enacts analytics of inhuman life that stress the paucity of continuance, reciprocity, and consent (L. Simpson 2014).

Hear Mojave poet Natalie Diaz's "The First Water Is the Body": "What is this third point, this place that breaks a surface, if not the deep-cut and crooked bone bed where the Colorado River runs—one-thousand-four-hundred-and-fifty-mile thirst—into and through a body?" (2020, 49). The poem is not a metaphor; the river does *not stand in* for some symbolic relation, it *is* the material relation. Diaz raises the tension of opacity in the question of translation as appropriation, reminding us of the reproductive labor of language and the continual taking that has made the historic context of theft. The question of care is tendered to the reader, as a participant in the world that unmakes, straightens, and drains the river that is body. The poem gives us an order of priority, *geos-bios*; geoforces are the arrangement of life and its geophysics of being. Diaz talks about the third place that is the river—the "before" of the verbal, the name, the signifier, all that is partially touched by language, "clay body into my sudden body," "We must submerge, come under, beneath those once warm waters" (49). The importance of the recognition of this third place is that it is not reducible to the either/or in translation; it is what mobilizes all terms and holds an originary relation to them. The aliquid registers form, the subterranean supports and expands being, reversing the flat description of geologic agency, the body pulled by an inhuman context and desire. Diaz addresses "toward what does not need us yet makes us" (49)—her body as the emergence of ancestries of geoforces. This is a body before that which is labeled and bordered as body, a first water body. Diaz shows us the dialogic border work of language as well as its inhabitation otherwise. The poem is both exposition and site of redress in the relanguaging of relation. *Acts of desecration entail acts of description.* Because colonialism instigated the mass mobilization of materiality through epistemologies of detachment (grammars of geology), repair and rebellion are acts of redescription of the *geos*. Language is one site of materialization; speaking it into being is another.

Breaking with the dialectic as oppositional forms (and the racialized unity between human and inhuman), a third inhuman position—*cosmic materialisms*—is neither outside colonial geo-logics nor contained by it. Colonial separability worked against the pedagogies of continuance within "bodies" that preceded coloniality. It created temporal rifts through the figures of the Human, Indigenous and enslaved, as they were located within the concept of colonial time (as with the indentured and incarcerated). It superimposed "pseudo-humanism" and its "narrow and fragmentary, incomplete and biased and, all things considered, sordidly racist" genre of being (Césaire 2000, 37). The establishment of hierarchies of subjectivity was tied to paranoias of control. Time and duration, as they are yoked in the dialectic of inhuman-inhumane, must be thought without messianic quality that repositions the human in the dialectical frame (i.e., through the subtraction of the inhumane to resolve the subjugated inhabitation, although strategically this is the work that must be done in human rights). This retrieval of the human is treacherous for solidarity because it proposes an outside in a redemptive register without addressing the structures that built the forms of inhumanity through which the figure of Life is materially and symbolically substantiated. It also reinstates the *bios-geos* separation of anthropomorphism. Thinking of being as *being of the world* within inhuman coordinates has a continuity with a different spatiality and temporality of being, one that does not privilege the capture of being as Life transcendent over the forms of geochemical attachments that it travels within. Water *is* life, *geologic life*. Geophysics *is* gravity and an open materialization in the constrained dialectics of colonial afterlives.

Colonialism broke worlds through temporal profusion. Rips in the continuance of forms of time were the consequence of obliteration by linear temporality, but such rifts do not break evenly, totally, or predictably (i.e., they do not possess the same material telos as colonialism imaginaries because the universe is so much more expansive, and Man is not universal or particularly durational). This temporal rift is part of the geotrauma of colonialism, where breaking ground is not an end but the precondition of the beginning of a material telos that aims at occupation and extraction. While theorists in a critical mode understood the value of formations for conceptualizing the structure of social relations, drawing down into sedimentation, as Glissant proposes, is a way to live with and build from that geotrauma, a way to recover history and acknowledge the weight of its formation on those excluded from its narrow narrative who nonetheless built its material and psychic worlds. Social thought's plateau is still conceived from the perspectivism of Overseers' earth, whereas the abyssal depths are the locality of the rift and its fragmented and collective forms of being. We can see this materially if we think about the quality of the surface of imperial maritime space in the Middle Passage and the disregard for its abyssal depths and abandoned souls. Rift-being can be understood as an effect of the social arrangements of the surface, but it possesses its own geography and gravity. The conceit of the plateau is that its pressure alone will crush any possibility of being otherwise, when in actuality it does not, and the anticolonial will is a stranger and stronger geoforce than imagined in that projection. While the workings of raciality maintain the hierarchy of the strata, the rift exposes and fractures its spatial and temporal dimensions in ways that challenge an account of history (as historicity) and its spaces of freedom-possibility. The broken earth is just the beginning. Geotraumas are wounds of geography and wounds of time, of durational affects and temporal fissures in Relation (as well as a censor in existing duration forms of life). Understanding colonialism as geotrauma is a way to acknowledge reparative claims and seek reciprocity in the superabundance of being that is secured in the *geos* and cosmos.

This chapter focuses on the *inhuman as geophysical relation* located within a discursive field of dialectic, alterity, intimacy, internal outsides, cosmic rifts, and the mattering of Life. Geophysics here is understood in relation to the depths and submerged fields of force that produce forms of geopower and gravity. As Simone sings us, rocks are neither redemption nor shelter to the Sinnerman, but they might just bear witness in the way Chamoiseau imagines, crying the opacity of testimony and thereby forging a new epistemic language to break with the grammars of geology. This is the inhuman as parataxis. A rock beside a rock inside a geologic life.[2] The following discussion takes up the orphan qualities of rock and its larger geophysical field of the inhuman to think through the histories and geologies of segregated and unsegregated being. I focus first on black earth in Hannah Arendt's ethics as the grounds of European ethical thought to show how the dialectics of humanism are built through Black earth; then on to inhuman ontology or geologic realism as a false outside to colonialism; and, finally, to the depths of the rift to discuss Fanon's cosmic "yes" read through David Marriott's analysis of the abyssal.

While inhuman nature was set as an alien inditement of Enlightenment thought that needed to be brought under productive control, it was also being reimagined, theorized, and practiced as a sanctuary for those already segregated as "outside" the rights of the human, yet situated within its territorial geophysical affects. These two types of "outside" are very different in quality and privilege, as one is involved in expropriated extraction and the other in spatial expansion of carceral states. One is abusive; the other seeks to disabuse. Outsider earth has two very different lineages (fear of an inhuman-Black planet v. what might be called *cosmo-being*). I will start with the fear and then move to the cosmos.

Fear of an Inhuman Planet: Fear of a Black Planet

Ontological Terror: Dialectics of Racialized Earth

The white supremacy of matter was achieved through the practices of geophysical stratification (mirrored in metaphysical formations) that evidenced both ontological terror of an inhuman planet and fear of a Black planet (an idea echoed in the 1990 album of that name by Public Enemy). Concomitant with this fear of geologic nihilism was the genealogical insecurity of whiteness (narrated as a pale-ontological fiction of the Caucasian) and the projection back into deep time to assert white settler futurity (dinosaur metaphysics). That is, to secure whiteness as the apex of culture and civilization, the ideation of whiteness is needed to secure a natural origin outside the social coordinates that produced it, to confer an entitlement to the whole earth (as both territory and natural resources). Kant's writing on race and Agassiz's, Lyell's, or Cuvier's on geologic life illuminate this in the previous chapters. But how did this fear of race become hustled between, and partly in compensation for, the fear of a contingent planet? And what does this mean for how environmental disaster arrives in the afterlives of colonial extraction as a question of futurity for planet earth?

Race was a buffer for the radical asymmetry of the inhuman earth that operationalized extraction and the unequal harnessing of geopower. Its ontological performance as a means to domesticated material mastery in the context of those inhuman asymmetries is no less important for how dynamic earth events are thought and governed. The terrifying earth disavows human supremacy, white, male, or otherwise. The racial planting of this terror has its roots in a culture that believes its own philosophies of severance between the *bios* and the *geos* (as if such a thing were possible). That Kant had a fear of both racial difference and the inhuman implications of geology suggests that both race and geology were positioned as excess that threatened systems of thought when the understanding of subjectivity was organized through spatial expansion and mastery. The operative geo-logics of spatial subjugation extended to the edges of the known universe in its imagination of oppression.

Kant literally devised a system of thought—Enlightenment Reason—to exorcise both the signifying excesses of inhuman alterity and Blackness, and to order a volatile planet without God. Fear of a Black planet became organized in thought as the psychic compensation and buffer for fear of an inhuman planet, as Blackness became a "planetary fix" for fear of an inhuman earth, and race was mobilized psychically as a method to dampen the threat of abandonment to inhuman asymmetries. While it's tempting to read Kant's response to an inhuman planet as a psychic overcompensation (N. Clark 2011, 83), which it undoubtedly is, doing so perhaps obscures the mutuality of the affective racial architectures that violently structure the material world to achieve the realization of those geographic imaginaries. Quite literally, Black, Indigenous, and Brown bodies *did* buffer the violent effects of the earth (not just in an ontological sense but in a material scene). The individuation of pathology is a distraction from the structural efficacy of stratigraphic thought. In this sense we can see how geographic imaginaries do not just plot spatial desires but organize the coordinates of their coming into being through a psychic lure that emboldens the frightened and grounds false confidence about the mastery of separability. They are certainly the somatic repercussions of racism as the consequence of an amplified fear of difference, in the context of an encounter with both "alien" peoples and alien earths during colonialism. These interlocking formations—of material and psychic deformations—cohere to produce a racialized geologic formation of the earth and thought, where metaphysics provides a camouflage for *geophysical expropriation* and *geographic appropriation*.

As geology narrated an account of beings told as an earth story and as an account of libidinal racializing geo-logics that subtend and specify those accounts geographically, metaphysics collaborated in the white supremacy of matter to claim that being as individuated ("I"). As Kant's writings on the Lisbon earthquake and on race (see Kant in Mikkelsen 2013, 41–55) define his philosophical approach as a racist geographer, he established in the field of Western epistemology more broadly the collaborative junctures of the material and philosophical categories within a universality of thought that enabled a whole (racialized) worldview to emerge (Kant's "Essence of Man" was universal, *but* "absolute racial difference" positioned Blackness as falling beneath and behind the Human). While this was a geographic imaginary that furnished the desire of imperial and colonial world-making across its material and subjective registers of extraction and control, it also launched *metaphysics as a foil for the geophysics of power*.

Which is to say that the inhuman has been an ongoing problem for the coloniality of thought, configured as both race and geology. The terror of the inhospitable and wildly catastrophic earth is perhaps where the fault line of the compensatory colonial modes of geologic articulation through racial schemas arises. The terror within becomes a violence unleashed outside on particular racialized bodies. In Kant's racial ordering, race is made as spherical positioning, whereby the earth becomes an axis with which to realize hierarchies between "Greenlander or a Hottentot" (2009 [1755], 152). In the long nineteenth century (1789–1914), dilemmas of the earth and the shaky ground on which they attempt to raise their concepts and subdue their fears balked at the implications of an inhuman planet. The geohistorical investments in ordering the seemingly unorderable highlight the role of race as a mode of stabilization to bring the planet back down to earth, alongside the dual fascinations with difference (racial and inhuman) that played a crucial role in the construction of ideas and imaginations of the earth. While that ordering may have psychic resonances or earths, it was a material way to make a new earth in the ideation of white supremacy.

It is not so much the fact that Kant theorized race at the same time as he theorized the physical geography of earthquakes. What is important to notice is that philosophers, geologists, and paleontologists all theorized across the earth and life, as coconstitutive terms, rather than as separate disciplinary spheres of Life and Earth. Kant was not just a racist who wrote philosophy (as we might say of Heidegger); rather, he coupled ideas on physical geography, inhuman ontologies, and race in the late eighteenth century in Europe as populating the existential questions of the time. The resolve to the imbricated question of human and natural history was the fiction and invention of supremacy over nature, wherein race provided the necessary metaphysical and material leverage to levitate this subject position. The case to prove around Kant and race is the role not only of race in relation to his philosophy but of race as the ontological ground of *any* Western philosophy of nature (see Mikkelsen 2013; Sandford 2018). Thinking about natural history was dependent on thinking about racial history (even if the racial question was without representation). And, racial history was understood as natural history, and the genius of the Caucasian was seen (per Agassiz) as the escape from that physical geography. The Eurocentric imposition of an idea of a purposeful earth (awaiting development and extraction) is intimately tied to the idea of the telos of racial achievement.

Race as human development and geology as material development have a unity in colonial and settler colonial societies as a political geology. Kant's "teleological judgment" is common across his physical geography and philosophical teachings (Kant taught one of the first physical geography courses at a university in Europe). The natural laws that he sought to discover as the causation of the Lisbon earthquake are the same "rational" laws that are sought in human philosophy, connected at the level of empirical unity. This connection is important for how discourses on the inhuman in social theory continue to carry and bury their stratum of race without disclosing this suffocation on the surface of their concepts. The line is drawn in between the organism (Man) that possesses a formative power (to be his own telos and thus formation or "end") and those that do not (who are stuck in the natural laws of the strata or in natural dead ends). This difference entails judgment, between being natural history (system of nature as a geographic description of nature) and being a subject of reason able to discern through concepts an organization of time and space based on interpretive logics (being a subject of and over geography). Race is organized geographically but also, according to Buffon's writings (contemporaneous with Kant), is generated geographically (as with Agassiz). Kant's essay "Of the Different Races of Human Beings" was published as an announcement for his lectures on physical geography for 1775, because race and geography were understood as interdependent, even as this understanding was under tension precisely because colonialism reorganized geography through a diasporic explosion. Kant's logics are decidedly geo-logics. He highlights the importance of the description of nature as natural history in racial and inhuman logics and argues race as both *process* (causation) and *outcome* (telos, and thus formation). The logical and physical division that Kant writes as a geography of nature is thus situated in the grammars of geology and their modes of inhuman description and division. What is important to understand is how the metaphysical subject is extracted as *internally* rather than *spatially* constituted by its expansionist geography of industrial theft, murder, and rape.

whiteness (stratal imagination + telos of materiality as paleontological claim)

$$= \frac{\textit{bios}}{\textit{geos}} \quad \text{human \{Indigene | interred\}} \Big/ \ \text{inhuman \{Blackness | extraction\}}$$

Ontological Terror: Fear of Black Being (or the Ontological *Terroir* of Terror)

Black studies scholars have long argued that the ontological terror of Blackness permeated the founding of the US nation-state and signed its racialization at inception.[3] Rather than seek a reconciliatory inclusion into a "better humanism" (which is never actually better because of the failure to address racial origins in the reproduction of social power), Calvin Warren suggests that by "abandoning the human, human-ness, and the liberal humanism that enshrouds it, we can better understand the violent formations of anti-blackness, particularly ontological terror" (2018, 170). Understanding anti-Blackness as an ontological formation is both a way into (as explanative) and perhaps a passage out of the literal stranglehold of that terror (through a refusal of normative terms that exceptionalize and individuate that violence as aberrant and unspatialized). Warren observes, "We have invested unbelieveable [*sic*] value in the human—it constitutes the *highest* value in the world. . . . It is this *terror* of value, of not possessing this value, that keeps us wedded to the idea of the human and its accouterments (and I must say, constantly revisiting the human, reimagining it, expanding it, and refashioning it does nothing but keep us entangled in the circuit of misery)" (171).

The terror of value that is constituted by the figuration of the human thus has the doubled effect of continuing a negative dialectic of lack in those not included in its remit and absorbing the energy of liberation that is tended toward that goal. If we tilt this argument to look below and consider the materialization of ontology—the *terroir* of value—the very materialism that is the structural basis of dispossession, the analysis shifts to the ground of the inhuman. Inhuman classification and its spatial affects travel under the figuration of the human, yet everywhere contribute to its material and metaphysical sedimentation. Understanding race in this geophysical context expands the sites of attention and the forms of categorization in which race is understood to be active in its anti-Blackness.

The idea of Being is so organized by its historic construction in Western philosophy (that is itself inflected and effected through paleontology and extraction) that it makes sense to collapse those hierarchical terms of valuation, to see racialized extraction as a general field with a prior spatial and material ontology. Geologic life under the terms of white geology is an organization of the normativity of anti-Blackness, and yet it often resists identification because of the allocation of material questions into metaphysical orders. To put this another way around, the normative mode of materiality—*extraction eugenics*—has the same ontological root in both paleontology and philosophy. Yet, the anti-Black ground of extraction is obscured by the absence of this signifier of the Human (in ± human + inhuman earth) to alert us to how questions of Being are located in the logics of the inhuman (while the conditions of Western thought are designed to instinctively repel the inhuman as the human's twin). In this inhuman sphere of *material sovereignty*, alternative forms of onto-epistemic material arrangements are instigated that issue challenges to the white settler state through the redescription of what life is and how, and with what and whom, it comes into being. Geography shows up a relation that ontology keeps divided.

Warren argues the task for Black thought is "to imagine black existence without Being, humanism, or the human" (2018, 170). One starting point is the counterintuitive place of the inhuman, in both its erasures of the human (through the inhumane) and its opening into a radical porosity of the flesh of geology. The charge to invent a new grammar of being might also be spoken through its erased material histories, oppressive materialisms, and overlapping historical geographies. Warren says, "Black studies will have to divest our axiological commitments from humanism and invest *elsewhere*" (172). Alongside Jared Sexton's (2011) expansive naming of Black thought as philosophy's critical blind spot that all thought needs to take account of, Warren sees the potentiality of Black studies in its unrecognizability within philosophy. His method is to "investigate the abyss of black existence without ontology" (Warren 2018, 175). One approach to exculpating ontology that I advance here is to see how attending to the history of (pale)ontology shows us a sleight of hand, which converts materialism into metaphysics, through the praxis of racial geophysics. The ontological humanism that grounds political philosophy does not transfer to the outside precisely because Blackness *is* the missing ground of its fabrication.

Colonial materialism makes ontological claims in a material order to disguise the geologic taproot that secures the conditions of the emergence of Being. I have argued that Being and the Earth are co-natal at the point of origination in white geology; both ~~Being~~ and ~~Earth~~ are under erasure as sites in which racial subjugation is implicated, so the end of ontology would also require the end of matter in its current formation (as eugenic or as the white supremacy of materialism). The claim of an "aboriginal" state of matter (which is itself is instructive in pointing to the language of Indigeneity), or purity in the mineralogical and genealogical order, as a form of personhood and as a form of extraction, is the defining relation that cuts across both. Where the necessity of Warren's "antimetaphysical" understanding and the geophysics of race outlined in this book might meet is in scrutiny of the colonial fantasies of the annihilation of Blackness. For my argument in the geophysics of race, such fantasies of obliteration conceal the spatial and material work that ontological concepts do, in the relocation of Blackness into racialized undergrounds, a study of which comes together in the ontological terror of Richard Wright's *The Man Who Lived Underground* (2021). As emerging Western ontological claims about whiteness are put into conversation with their historical geographies and intimate granular geologies, ontology becomes spatialized, which in turn annunciates how racial forms of life are materially made, and how this existence is geographic and geophysical.

Racialized undergrounds are the suspended deferral of Western ontological fear of Blackness (that which defers the "nothing" of nihilist ontology), which in turn leads to Black earth being conceptualized as a ground to place the colonial taproot and suck. Thus, the ontological implications of the category of the inhuman return in struggles over *material sovereignty* in the cancer corridors of poor Black neighborhoods that buffer the petrochemical industries; in organized vulnerabilities to the hurricane; in the heat burdens of redlined cities; through to the everyday exposures to chemicals and pesticides used in low-paying cleaning and gardening jobs; and in the policing of Black and Indigenous bodies and the struggle for breath. The inhuman (as a nihilistic ontology of humanism) also possesses this pressure on ontology, but when parsed with the geophysics of race and the materiality of the cosmos (i.e., nonlocal conditions), the inhuman emerges as a juncture that brings other vernacular maps of being into existence.

Reason's Earth and *the* Ethical Subject

> They [the Boers] treated the natives as raw material and lived on them as one might live on the fruits of wild trees . . . to vegetate on essentially the same level as the black tribes had vegetated for thousands of years. . . . When the Boers, in their fright and misery, decided to use savages as though they were just another form of animal life, they embarked on a process which could only end with their own degradation into a white race living beside and together with black races from whom in the end they would differ only in the color of their skin.
> —HANNAH ARENDT, *The Origins of Totalitarianism*

> White people believed that whatever the manners, under every dark skin was a jungle. Swift unnavigable waters, swinging screaming baboons, sleeping snakes, red gums ready for their sweet white blood . . . but it wasn't the jungle blacks brought with them to this place from the other (liveable) place. It was the jungle whitefolks planted in them. And it grew. It spread. In, through and after life, it spread, until it invaded

> the whites who had made it. Touched every one. Changed and altered them. Made them bloody, silly, worse than even they wanted to be, so scared they were of the jungle they had made. The screaming baboon lived under their own white skin; the red gums were their own.
> —TONI MORRISON, *Beloved*

Arendt's concern in her account of settler colonialism and the "fall" of the Boers into Blackness is to show the way in which the Boers have fallen from the eugenic tree into the fruits of vegetal matter (despite being white). In the reference to epidermal phenotype, she alerts us to her production of the double negative of Blackness, where phenotype is difference but not *the real difference*, which is the perception of Black life as raw material. In Arendt's parataxis, whiteness beside Blackness, inside of Blackness, is degradation. She continues: "Under a merciless sun, surrounded by an entirely hostile nature," the Boers are subjects that have lost their ground (quoting Joseph Conrad's *Heart of Darkness*): "The earth seemed unearthly" (Arendt 1951, 190). If the whites have lost their ground, what then is the ground, and where is whiteness in relation to this earth? Hear Morrison's reply: "White people believed that whatever the manners, under every dark skin was a jungle" (2007 [1987], 234). Arendt extricates whiteness as already materially distinct from the social category of the earth, which produces an imaginary of the earth as already Black. She scripts her attentions to the native, through what Hortense Spillers called "*being* for the captor" (1987, 67), which is the transformation of *being* into *becoming for* as a site of extraction for Western ethics. Blackness performs philosophically as every ground. *Grind.*

While the geographic and matter distinctions that Arendt draws contribute to a fracturing of the homogenizing trajectories of colonial spatialities (for which she has been lauded), they also use the distinction of territorial difference to hold Africans in a "savage" slot/state, whereby geography is transmuted into the ability to claim (or not be able to claim) a political state in relation to the present tense and property of the plateau. As Joy James comments: "The ultimate political binary is that of the civilized/savage" (2013, 308). In Arendt's account, "native" Africans can travel as vegetal-animal-earth matter, seemingly devoid of specificity in time and space, yet nonetheless available to ground the concept of the European subject and justify their murderous desire as an innate and thus legitimated fear. Following in the footsteps of the fathers of geology, Arendt uses a matter division to assert a racial hierarchy. Black earths ground an Arendtian "World" and its ethical subject as distinct in kind from matter and the earth.[4] The account of geography Arendt propagates is the basis of her reasoning for an ethical subject; yet it is the subject of a partial worldview made through the distinction between races. The praxis of emergence of subjective entities from those who are submerged prompts questions of how is the institution of Being (as an entity and mode of apprehension) reshaped through Western metaphysics and the juridical-ethical subject? Through the geographies and spatial formations of thought, its gravity, geophysics, and ground? Through the earth in relation to Arendt's World? Materialisms beyond territory?

Reason's Geography

Arendt is known as a philosopher for her geographic insight in locating the origins of European fascism in the racism of imperialist expansion. Placing the seeds of fascism outside Germany and in the colonial "periphery," Arendt gave fascism a genealogy that was internationalized and globalized, whereby Africa (or the "scramble for Africa," in Arendt's terms, as a power competition over territory and resources) became a space of experimentation for the terms, tactics, and terrains of violence that were returned for use to the European continent. Specifically, the event she is thinking about (but does not name) is the German Herero and Namaqua genocide in South West Africa (now Namibia) where up to an estimated 100,000 were killed, which is ignobly entitled the first genocide of the twentieth century (1904–8). More broadly, she may have had in mind the colonial dynamic of "cultivating" extraction through extermination. Two things about Arendt's geography emerge: (1) that the origins of violence are imagined to be forged in and with Africa, in her "fertile soil," rather than already existent and extant in Europe and America (fascism comes "out of Africa" along with the protohuman that will become suprahuman); and (2) that Africa is understood as a laboratory of violence, where the "proof of concept" for the methodologies and tools of racialized violence is demonstrated (the conversion of persons into the categories of racialized subjects, the use of extermination camps, etc.). It is an insidious argument, that the savages made savages and brought home their savage ways, but remember Morrison on genealogy, where origination serves as rebuttal: "The screaming baboon lived under their own white skin" (2007 [1987], 234).

Colonial Shade

Contra Marx, Arendt argues that it is "race thinking" rather than "class thinking" that arranges European violence into a dialectic of colonial "shadow" and then European "weapon." In Africa, the violence of race thinking is "shadow"; in Europe it becomes "weapon" because, following Arendt's metaphors, violence against "vegetal matter" is just shade. While Arendt's accounts do render race as a technology of governance, rather than a lament of subjugation, she is keen to point to a Eurocentric geography of the concept of race that is constructed through an inside/outside, far and away, origination. In this normative colonial globality, Arendt's idea of race articulates a thought that protects itself (as Western) and the idea of Europe (as a protected entity). Her thinking is unable to register violence against racialized subjects outside of Europe *as* violence. Race is a way of creating a philosophical outside, where the inhuman directly contributes to the dialectical functioning of the human as a concept that can circulate as a universal. Blackness once again performs as a spatial buffer for ethical thought. The racial normativity of her geography sustains the very racism it attempts to describe. It is not just that the other, the African, is always in abeyance to the European project and center, but that the earth is made into "World" on this basis. Therefore, Western ethical subjectivity is raised in an ideation of the earth that concretizes the normalcy of colonial spatial formations and the subjugation of Blackness as *the* ground of European thought.

Arendt's thought is not unusual in its normative racism. Ferreira da Silva (2007) shows how the circulation of racial knowledge and power provided an account for the functioning of global space. Arendt's diagnoses of how imperialism in Africa leads to, first, an outward projection in the "scramble for Africa" and, then, its inward manifestation in European fascism and Nazism, usefully linking colonial and European projects of nationalism. However, her diagnosis does nothing to challenge the normative (racial) production of space in the mapping of ethics. Geographic appropriation (the "scramble") is achieved and justified by racial rather than purely national means.

Arendt's geography sits within a broader set of questions for me about the histories and materialities of racialized understandings of the partial "World" that transforms the earth into a philosophical ground, and how the narration and designation of "Origins" put down a single root theory or monogenesis (as in paleontology), against the idea and possibility of multiple roots and beginnings (and thus endings and trajectories). In the call to a single point of origin, we might detect not just an unfolding of spatial conceits of a Global-World-Space (and a doctrine of universalism), but also a monotemporality that consigns all subjects as designated first and foremost as subjects of (or subjected to) the colonial time of empire. This conception of time-space places all subjects as coming into history through the moment of colonial encounter and thereby scripted by its modes of apprehension and political technologies. That directionality of thought establishes the political geography and locus of action and agency. Arendt asserts: "It is strange that, historically speaking, the existence of 'prehistoric men' had so little influence on Western man before the scramble for Africa. . . . The word 'race' has a precise meaning only when and where peoples are confronted with such tribes of which they have no historical record and which do not know any history of their own" (1951, 192).

Again, Arendt deploys the double negative to condemn Africans as ahistorical, by asserting their "lack" from the outside (no historical record of) and inside (no history of their own). Repeating the paleo-stratigraphizing of those racialized as "prehistoric" and their stratal displacement outside of human history, Arendt's concept of race does not come into being at the moment of confrontation with tribes so much as in the prescripts of paleontology, where whiteness holds the key to historicity. Here, her conceit of white supremacy becomes clear: race performs difference, but there are the wrong kind of performances of racialization (the Boers, Europeans) and the right kind ("Natives"). It is not just Arendt's inability to call the colonial situation to account but a broader arsenal that she draws on that cements race as a unifying common grammar that holds the instability of radically different geographic relations to time, nature, earth, and land in place. In her critique of Marx, Arendt argues that imperialism cannot be accounted for solely in terms of political economy, but that race and scientific racisms provide a "commons" across colonial projects (that returns to haunt Europe). While Arendt's spatialization of world order sees connections in the organizing logics of colonialism, the only locus of concern is the threat to the European project and *its* ethical subjects. Arendt is careful not to make an argument about skin being the basis of the color line; rather, she asserts difference pronounces itself through the nature(geo)-culture(social) divide:

> What made them different from other human beings was not at all the color of their skin but the fact that they behaved like a part of nature, that they treated nature as their undisputed master, that they had not

> created *a human world*, a human reality, and that therefore nature had remained, in all its majesty, the only overwhelming reality, compared to which they appeared to be phantoms, unreal and ghostlike. They were, as it were, "natural" human beings who lacked the specifically *human character*, the specifically human reality, so that when European men massacred them they somehow *were not aware* that they had committed murder. (1951, 192; my italics)

Arendt's geography divides a human world (plateau) from a parallel earth of phantoms (rift). The ghostly zone of nature is yet to *become* an Arendtian World. It is awaiting a purposeful materialism, ready for mastery. The refusal of a materialism of mastery and the recognition of the durational capacities of inhuman natures are not considered part of *this* (Arendt's) World. What is the materiality of this projection that would excuse the genocide of "part of nature"? Nonadherence to the white supremacy of matter causes an imperceptibility of recognition (of their humanness), and this is what exonerates the murderers of their crime. The location of lack is in the murdered rather than the murderers ("they somehow were not aware that they had committed murder"). As colonialism harnessed geopower—in the form of the enslaved, gold, minerals, fuels, and land—to furnish the subjectivity and ethics of the humanist subject, as well as the institutions and modes of regulatory governance that reproduced that form of "human condition" into the future, it produced an aspirational ethical figure and a material form of being. Race was the unequal codification of geopower and the regulation of its powers, dressed up as ontology or difference. As colonialism and extractive regimes of racial capitalism dissolved relations and forms of attachment, race glues them back together into a coherence when there is, in fact, none outside of that categorization and its utility.[5] Arendt's refusal to recognize Africans as persons and her philosophical apology for their murder is not the version of Arendt we find in the humanities on ethics and subject formation, but on another level, she merely continues to contribute to the arc of paleogenealogical thought. The fact that Arendt could not conceive of a Black subject as a subject even as she sought to grapple with the question of the dehumanization of subjects in the context of Nazi Germany is not on show in the halls of critical theory.[6] What Arendt's thought does is to confirm the spatial dynamics of whiteness as bounded by the perception of a properly designated inside and outside of race (a parochial geography of good and bad racial subjects), which is constituted by the elevation of whiteness through the mastery of matter (which creates a "human world" and thereby a human subject of ethical rights contrasted against the dark phantom world of inhuman subjects).

While Arendt is skeptical of a certain kind of humanism, her geography reveals an investment in the maintenance of white supremacy, wherein Africa becomes and remains a reflective surface for whiteness and its ethical trajectory. The ethical and political expropriation of Africa demonstrates a geography in which value is accorded through a dialectical appearance: Africa in service to the dimensions of whiteness that need to be illuminated and fixed. Black death in service of white Enlightenment (*again*). Africa is a mute site of nature otherwise, not transformed by culture, technology, or labor, "without the future of a purpose and the past of an accomplishment" (Arendt 1951, 191). Further, Arendt invests in the mythical accumulation of whiteness as the source and register of all meaning, perpetuating white values as the unquestioned standard of normativity and locus of concern for violence, while alluding to the locus of its projection as a geographic rather than racialized force in the optics of meaning. Whiteness is the realm of all appearances of the political and its object-world relation, and for Arendt, too, Worlding is realized through the labor of fabrication, or the transformation of nature into "World" from earth: white World through Black earth.

Whiteness, in this sense, levitates from the ground of nature. In the ontology of Being that Arendt offers us there is a reliance on the figure of Blackness as the counterpoint to reason's oscillatory achievements (and its inclusionary-exclusionary biopolitics). As God needs the devil, the hidden structural transcendentalism levers whiteness (as cultured Europeans) on a Black earth (that is not assigned the status of a figure) through a figure-ground dialectic that pushes African subjects into the earth and justifies their murder. While Arendt calls out the Black Panthers for their interest in the "Black and white dichotomy," dismissing it as escapism, her thought does not seem able to resist her own dialectics that reinforce this patternation: "Rather than seek to understand why anti-racist activists might situate themselves within wider anticolonial and neocolonial struggles, that is, in the context of global white supremacy, Arendt reduced black solidarity to a crass demographic calculation. She presented black students as more violent than whites . . . [and demeaned their entrance into the university] without academic qualification" as mere interest group jostling, interested in "nonexistent subjects of African literature" (Owens 2017, 417).

In Arendt's cosmos, whiteness is formulated as a satellite called culture. While she diagnoses the cosmic gesture of technology in her discussion of Sputnik in the prologue of *The Human Condition* (1959a), identifying the transcendental impulse since Kant that seeks to escape the bounds (and gravity) of the earth, she sees no such elevation in her designation of culture. Anti-Blackness is a condition of a priori erasure in the definition and political contours of the human subject, as it is hailed by Arendt. The anti-Blackness (and erasure of Indigenous peoples by omission) becomes an instrumental and ideological ground that is used to justify the New Republic of America.

Arendt's consistent removal of African Americans and Indigenous peoples from political theory, as agents of the polis, replicated and reinforced the actual spatial executions under the regimes of colonialism and slavery and their afterlives (land theft, bad treaties, Black Codes, convict lease, and prison-industrial complex). This extraction from space and its connection to the political present is compounded by Arendt's designation of what Joy James calls "nonpolitical sites," which disregarded the segregation of African Americans and enclosure of Native Americans on reservations in the exercise of power that spatial disenfranchisement afforded. The making of that ground of politics, whether through the financial capital of slavery and the theft of Indigenous land that underpinned the expansion of the settler colonial state, or the Southern steel mills and carceral coal mines of convict lease labor that provided the fungible grounds of modernity's skyward steely projections and its girded settlement, is all erased as the material foundation of *The Human Condition* (1959a). The white overground of Arendt's ethics requires Black and Native undergrounds to do the work. Arendt's thinking betrays a thinly disguised belief in the inevitability, and thus justification, of racial "struggle," which informs the limits of biopolitics as a mode of critique: "Racism, white or black, is fraught with violence by definition because it objects to natural organic facts—a white or black skin—which no persuasion or power could change; all one can do, when the chips are down, is to exterminate their bearers. . . . Violence in interracial struggle is always murderous, but it is not 'irrational'; it is the logical and rational consequence of racism" (Arendt 1970, 173).

Such a tautology of violence, its propelling force of inevitable resolution (that arrives in the teleology of murder), belies historical connections to the origins of colonial violence and its geo-logics rather than "organic facts" (particularly if we remember how race is naturalized in deep time). As Kathryn Gaines attends to in her book *Hannah Arendt and the Negro Question* (2014), Arendt's response to the violence against African Americans attempting to acquire a nonsegregated and thus better education (given the underfunding of African American schools) is to turn that violence back onto the family as a site of its regulation and resolution. Arendt asks, "What would I do if I were a Negro mother?" (1959b, 179–81), and she proceeds to say that the violent confrontation of a young Black girl trying to enter Little Rock High School is somehow more the responsibility of the parents of the girl (who orchestrate her exposure) than of an anti-Black society that has already scripted the child as both fungible and excluded. As Gaines outlines, Arendt's account not only is factually inaccurate but also trades in grievous stereotypes and a refusal to acknowledge the colonial origins of the American Republic and its racist forms of social reproduction. Arendt essentially makes the argument that Black parents were sacrificing their children into a violent environment to attain social prestige, because she already presumes that Blackness is a form of space invasion. She repeats the pattern of thought in the "Little Rock" argument that she has already made in the discussion of the Boers' murderous attention to Africans, which is that Blackness pulls violence toward itself, by being alien and unassimilable to European culture, by wanting too much (like a good education or to be recognized as a subject), and by being unprotected by the right to genealogy and attempting to invade space "where it was not wanted" (Arendt 1959a, 179).

For Arendt, too, anti-Blackness has a gravity, like the earth's gravitational field. As Owens argues, she is unable to see the "deep transnational structures of white supremacy" and the origins of a nation-state built on settler colonialism and enslaved labor (2017, 422). Arendt's difficulty in comprehending that subjects exist outside Europe in any fully human way (which is to say, her racism) is compounded by her geography that collapses space to furnish an exclusive ethical subject that is able to both traverse and claim the World but be historically and philosophically immune to the material processes of how its existence comes into being. This arrangement of the World is constituted by the form and materiality of an Enlightenment Being that Arendt inherits and reproduces, and one that defines itself *against* the earth and its wretched (designated inhuman) subjects. Arendt's argumentation tells us something of the unconsidered philosophical bedrock that pervades the ethical Western subject and its coming into being, as well as the anti-Black ground that sustains and builds its privilege. The transformation of *earth into World* and the ruse of Reason that leaves Black subjects in its ontological and material rift (as considered within the earth but not of the World) alert us to what is wanting in how philosophy grounds the charge of an inhuman world and the ensuing precarity of subjectivity. What if we consider the alien "ground" from another perspective: As alterity is flung in rather than flung away, does a different set of gravitation affects emerge that might challenge the normative modes of earthing a subject?

White Man's Overburden

Writing contemporaneously with Arendt about the question of subject formation and planetary alterity, Césaire declares in his notebook on return to a native land this much-quoted observation: "The West has never been further from being able to live a true humanism—a humanism made to the measure of the world" (2000 [1972], 73). The observation is made in the context of a discussion of the propositions by the ethnographer/paleo-speculator Roger Caillois about the burdens of whiteness. Among other writings on "Man and . . . ," Caillois is known for his book *The Writing of Stones* (1985 [1970]), which imagines a computational-like creativity of stones, sorting and organizing on riverbeds, and arranging mineralogical orders in various forms of symmetry or artistry. He sees in this organization of minerals an inhuman cognition that defines an aesthetics. Césaire's attack is launched at Caillois's articulation of what he calls his "Caillois-Atlas" symbolism, where he "plants his feet firmly in the dust and once again raises to his sturdy shoulders the inevitable white man's burden" (Césaire 2000 [1972], 73). In this other order of rock-being (Atlas symbolism), Césaire argues that Caillois presents the difference between cultures (read races) as being an issue of naturalized inequality "in the material sense of the word, to those who are strong, clear-sighted, whole, healthy, intelligent, cultured, or rich" (Caillois quoted in Césaire 2000 [1972], 73). This "inequality in fact" among cultures should not, Caillois cautions, result in racism, extermination, or a diminishment of rights, but it does "confer upon them [Europeans as the center of culture] additional tasks and an increased responsibility" (73). Such superiority is articulated, Césaire concludes, in the very "tolerance" of Caillois's morality that insists that no one should be exterminated (when of course they already have been to build his institutions of thought and museums of rocks). A mark of humanism's achievement and ethical superiority is its ability to bestow tolerance, as the discretionary benevolence of Overseers' earth, to the overseen. This is the figurative occupation of the stratal overburden of extraction—white geology.

Césaire calls out these claims of superiority, couched in tolerance and paternalism, where "metaphysical commentaries obscure the path" (2000 [1972], 66). Césaire's point is that nature must not be used as a justification or reflection for the internal chaos of human life, lest it is pulled into a form of labor on behalf of "Man" and the naturalizing of "his" impulses toward subjugation. *Metaphysics fictions material relations*. This imagined shouldering of the responsibility of race is portrayed as the weight of eugenics, yet where does the heavy burden of this gravity really fall? White supremacy belies the labor of the inhuman eugenics already at work. If we put geophysics before metaphysics (which *obscures the path*), how does Caillois's Atlas rock grow broken hands that mine, sores that weep as metallic waters leech, and weighted lungs that wheeze under the radar of his mineralogical revelry? For Caillois, it was easier to imagine the internal "life" of rocks and their computational awareness than to recognize the ethnography and exposure of racialized others. Césaire quotes Caillois's observation that "the only ethnography is white" (Césaire 2000 [1972], 71) and his argument that the museum is the great display of this fact. Césaire replies that "Europe would have done better to tolerate the non-European civilizations at its side, leaving the alive, dynamic and prosperous, whole and not mutilated; that it would have been better to let them develop and fulfill themselves than to present for admiration, duly labelled, their dead and scattered parts; that anyway, the museum by itself is nothing, that it can say nothing" (71).

As Césaire is writing and Caillois is modeling himself as Atlas, Sara Baartman's body continues to be displayed in the Museum of Man in Paris, on behalf of this imaginary of geologic life. Her bodily mutilation is the empirical ground of Caillois's metaphysics, and it is the geographic context of the museum in which he imagines the prowess of his white ethnography laboring to understand the writing life of stones. Degradation was a means to segregation, the atomization that cleaves life from bodies to separate flesh from bone, *bios* from *geos*. In the burden of salvation of the "lesser peoples" of the earth, the ethnographic impulse forgot to visit its own empirics to study whiteness, as it has been studied by those who have had to try and survive it (to paraphrase James Baldwin's refrain). Césaire comments: "In the scales of knowledge all the museums in the world will never weigh so much as one spark of human sympathy" (2000 72). Such museums of whiteness (as every colonial museum implicitly is) would simply present what Indigenous and enslaved people already had to know and study—*epistemologies of the oppressed*—to navigate and survive the erratic spaces of colonial violence and its reductive geographies. Think about the lessons given to young Black and Indigenous children (and the children of the diaspora) about how to address and elude the tyranny of whiteness. The natural history museum attests to the opposite of Caillois's assertion about ethnography because nowhere in the museum, as we wander through the Hall of Africa and the Hall of Asia, for example, do we find the exalted species of the plateau or its violent perspectivism analyzed, even as its universalism claims to command every view that delimits the spaces of nature in mean categorization.

Guarding human origins in the American Museum of Natural History in New York is how Africa gets to stand as origin of the human species geostory on the hominid trail "Out of Africa" and the journey toward *becoming* apex-humanoid, but never the future. A quick glance at the Hall of African Mammals and Hall of African Peoples is enough to confirm the dioramic status of Africa as bountiful continent, a superabundance of nature awaiting extraction and material conversion from its "primitive" state of nature into the progress of natural resource. In the entrance to the gallery on "African Peoples," the genealogical map of "Family," represented by a non-blood-based configuration of clan networks, sits adjacent to "Society," represented by masks and implements. The museum has chosen to represent this anthropological mapping as a three-dimensional diagramming that rises from a little green of symbolic artificial forest. Akin to the dioramas in the main exhibition of Mbuti pygmies of the Ituri forest people who "typify both physical and sociological adaptation," whose "light skin color helped them move about easily and unnoticed," the African family can be observed (in the museum) in the arrangement of a genealogical forest of the earth, rather than patriarchal lineage. Deviation from patriarchy in the organization of different kin and family relations halts the accord of "progress."

In the entrance hall, the geologic time of the dinosaur skeletons in combat and the American flag call the natural order of an aggressive supremacy into collaborative affect. Under "The State," it reads, "Aggressive fighting for the right is the noblest sport the world affords." The warring dinosaurs concur that life is a battle to avoid extinction. Geologic bones stand as both reminder and resource for mining temporal frames of extinction and deep time as white supremacy's present. "The African Tradition in America" is represented by one small side corridor off the main gallery. The exhibition consists of a print of a Black woman being held by a rope around her neck and about to be struck. Below this woman is a print of a slave ship and some restraints. And, if the unreconstructed narrative about nation, nature, and geology wasn't clear enough, a poster frames the print, "Negroes, Mills, Mules, Hogs, Farming & Mining Tools, Wagons and Carts." The apologies of these "overburdens" of whiteness are nothing but the wisp of an insubstantial pollution that clouds every judgment of metaphysical thought.

FIGURE 8.1 Africa Family Tree display, American Museum of Natural History, New York. Photo by author, 2017.

Forgotten Geophysics

In forgetting geophysics, Enlightenment thought obscures its racialized grounds and simply symbolically exchanges elevation for violence (as with Arendt and Caillois). Yet, every return of the anticolonial will attests to the fact that occupation and enslavement occurred in a place, not in an emptied museum space or philosophical treatise. In the move from geophysics to metaphysics, whereby the "higher" orders forget the subtending racialized stratum on which they build, it is important to remember the texture and specificity of these quotidian encounters and their resistant contingency. In place(s) there is abundance, everywhere, openings in the world, always in excess of the puny human orders and their fascistic perspectivism of the plateau. Even in the most precarious execution there is a little gust that carries a seed from another place that is not here. As I write, I see Barry Jenkins's (2021) visual attentiveness to Cora's okra seeds in his Black devotional cinema in the adaptation of Colson Whitehead's *The Underground Railroad* (2016). The seeds carry relation and ancestry (as Cora holds to the promise of planting the seeds her mother left behind when she tried to escape). The plantation and the mine are spaces that involve the world, after all, made in the context of a political-symbolic geology but not immune to the broader cosmos and its black, brown, and red earths. The mine and plantation murdered much in their traumatic incursion into the geographies of place, but they could kill neither the possibility of earths' generous asymmetry as Relation nor the anticolonial will. This is the "warm contamination" of "obligation 'offered' to every creature to be co-born with the world and itself" (Glissant 2010 [1969], 68). In conversation with Césaire, Glissant sees this *measure of the world* as the geographic expanse (not expansion): "Measure is also seen as the echo of human breath. No longer the search for depth but the *inspiration of the spatial expanse*. This kind of measure enables us to drift through the fullness (or the surface) of the world, bringing it back to our own place" (2020, 55; my italics).

The geography of measure that Glissant speaks to is the ability to spatially differentiate between the reductive and the expansive, a measure of the whole-world, not an imaginary predicated on the museum-world. The inhuman already possesses this propensity for spatial expansion in its shared geographies of becoming, in its thousand breaths of organic matter and the taken breaths that the inorganic holds as testimony. As the inorganic sustains the leaps of organic life, its durational occupation and excitement, it is not outside but integral to this movement in ecstatic and evolutionary terms.[7] Cora's seeds are a place and a hope, and their gathering a germination of pasts into futures, a pocket of gravity to ground another place in an inhumane space.

The double life of the inhuman confirms geophysics as a collaborative epistemic ground to metaphysics, not as the abyss per se but as a rift through the staging of the abyss that becomes sited in Blackness. In that rift a third way is innovated. Blackness is utilized in the Western geographic imagination to make the experience of the abyss perceptible, tangible, and enfleshed, to personify that which remains out of reach. In this buffering of the void (which is geophysical), Blackness performs a form of mimesis and alterity, absorbing the fear of a social life that cannot hold the vastness of the cosmos as a statement on beauty. But inhuman worlds also open relation to the universe, the cosmic, beyond the immediate political moment and its containment. The tree opens the moment to the breath of a different, nonhuman life, a life lived differently. It gives rises to relations that have been made illegitimate by white geology—as the pause, the loiter, the care in another form and shape of the earth as valued differently, not subjected to an extractive view. A wink at a way of living, the temporal hold of the morning and its dew, and the long sun that drags into the release of night. Inhuman worlds give a wider berth to the world and breach other orders of time and possibility.

Alongside the beauty of the sublime, as Kant describes it, he imagines another abyss in which there is no redemption of the beautiful, that of race.[8] This doubling is important to recognize because of the subliminal investment that is made in Blackness, as a sublime that is made to ricochet with abjection and desire, in which the inhuman oscillates as a negotiation with alienation as both repulsion and attraction (although the redemption of beauty is refused in this schema for Blackness but not for the inhuman as alterity). Given the hold of Kantian aesthetics through the Western canon and its broader percolation into thought and language of a commonsensical beauty as the basis for appeals for conservation, the collective work of Black feminist accounts of beauty must stand. Consider Hartman on the Black ordinary: "beauty that resides in and animates the determination to live free, the beauty that propels the experiments in living otherwise. It encompasses the extraordinary and the mundane, art and everyday use. Beauty is not a luxury; rather it is a way of creating possibility in the space of enclosure, a radical act of subsistence, and embrace of our terribleness, a transfiguration of the given" (2019, 33).

Hartman not only reimagines and defines through the voices of the wayward and dispossessed another form of beauty against Kant's brutal enclosures but also resituates where, and how, and by whom, theory is done. Retooling Audre Lorde's "poetry is not a luxury" (2007), the chorus that Hartman thinks with, exhumes an "archive of the exorbitant" written by those who "have been credited with nothing," and who "were radical thinkers who tirelessly imagined other ways to live and never failed to consider how the world might be otherwise" (Hartman 2019, xv). "Intimacy outside the institution" in Harman's book rewrites aesthetics in a different order as a counternarrative to pervading normative assumptions of beauty and the location of their historical sedimentations. It is not enough to simply refuse this "part" of Kant as belonging to his racism and not his writing; rather, his abjectification of Blackness might be seen in the continuance of a mode of subjugation in the aesthetic tradition that privileges the European white subject as *the* originator of beauty and its source of recognition. In the radical acts of subsistence that Hartman names in *Wayward Lives* (2019), we can detect the power of the durational moment, open to the presentment of a different geophysics of space: a presentment that holds space differently, that forces its expansion in a gravity of clamorous being. As Césaire suggests, inhuman intimacies are aesthetic counternarratives: "Beauty, I name you petition of stone" (1969 [1956], 32).

Geologic Realism (or Colonialism Now)

Now that the Anthropocene has gained official recognition as a geologic epoch and social theorists celebrate the end of the imagination of the Human and Nature binary, they often leave the spatial implications and geophysical dimensions of the seismic separability of *bios* and *geos* intact. The ontological claim was never thus, and it reverberates in a "petition of stone" for a different account of geologic realism. As the pundits of description clamber over the wreckage of geologic lives to plant their assertions for a "new" geography, it is often one that lacks a depth relation and geologic history, so it repeats the universalizing impress of materialism that is supposedly challenged without the reparational demand for a change of geophysical state.

I understand geologic realism as a way of thinking through the entwinement and emergence of (1) *normative modes* of approaching matter (forged through the histories and temporalities of colonial power in material and metaphysical orders); (2) the *realism of racial experience* in the shadow of the imposition of colonial power and in resistance to it; and (3) a *shared political present* of planetary shifts and geologic disruption (that is recursively related to colonial configurations of matter that produce an overabundance of the space of death—i.e., the Anthropocene). In the context of the geomythos of the "Age of Anthropos," an older historical claim exists in the ghost geologies of Black, Brown, and Indigenous archives, which rebounds on the foundations of humanist thought and its methodology of colonial materialisms with their own rock epistemologies.

One route into geologic realism is to take account of what Nigel Clark (2011) calls the asymmetries of inhuman nature. Ray Brassier suggests "the unavoidable corollary of the realist conviction that there is a mind independent reality" (2007, xi) results in a recognition of an asymmetrical reality that is oblivious to humanity and its concerns, a reality that cannot be made over as our home or ground or worked into any kind of "meaningful" relationship to us. The other route is the recognition of the inhuman as a *constitutive hinge* that holds humanism and its production of racialized subjectivities together *with* a production of the earth as earth/World/planet. In contrast to the notion of the geologic accumulation of fossils as a form of sedimentation of value, nation, and heteropatriarchy, Glissant speaks back to the realism of nonhistory as a subjective condition "through the accumulation of sediments" (1997, 33). Sitting on a different terminal beach than object-oriented ontologists, Glissant reconstitutes silt as a residue of descendants made in the Middle Passage–plantation matrix and the eruption of Pitons, all of which have duration, giving life to new forms and relations; "the tie between beach and island, which allows us to take off like *marrons*, far from the tourist spots, is thus tied into the disappearance—a dis-appearing—in which the depths of the volcano circulate" (206). On another alluvial shore, Mbembe describes how "the world that emerged from the cannibal structures is built on countless human bones buried under the ocean, bones that little by little transformed themselves into skeletons and endowed themselves with flesh." He continues: "The durability of the world depends on our capacity to reanimate beings and things that seem lifeless—the dead man, turned to dust by the desiccated economy" (2017, 181). Glissant's imaginary summons an inhuman archive of material relation that must be stirred up. In other earths, the possibility of the world relies on a recuperation and animation of those that do not have a consecrated archive.

In one theory of geologic realism, the fossil is a chance encounter, an arche-fossil (Meillassoux 2008), a cut in relation of time and materiality that organizes the signifiers of anteriority, from which statements can be drawn. The rapid accumulation of fossil knowledge that emerged through colonial extraction regimes is not considered part of this ancestry. In another account of geologic realism, in the "thingness" of chattel slavery and its aftermath, the extinguishment of forms of livability constitutes a quotidian realism of the violent subjectification of "nonbeing." From this geologic realism Indigenous and critical Black thought has emerged. Without recourse to Global-World-Space or the genealogy of Western Man, such theory simultaneously engaged other earths and relations, modes of innovating from inside the limits of Western Reason toward a differently configured geographic alterity as radical Black praxis (see Ferreira da Silva 2017; McKittrick and Woods 2007; Sharpe 2016a; Wynter in McKittrick 2015). As Michael Dash comments about

Glissant's work, this involves an exploration of time that is related to neither "a schematic chronology nor a nostalgic lament. It leads to the identification of a *painful notion of time* and its full projection forward in the future, without the help of those *plateaus in time* from which the West has benefitted, without the help of that collective destiny that is the primary value of an ancestral cultural heartland" (Dash in Glissant 1989, xxxv; my italics).

What Glissant's thought-work points to is a temporality that overcomes the attempted extinction of Black life through a rigorous engagement with the implicatedness of materialism and metaphysics: a temporality that is central to the memory work of Black futurity. Refusing the crushing violence of Western Reason, Black aesthetic life transmuted the geotrauma of passage, equaling that realism with speculative vibrations and material innovations that refuse the binds of what was constituted as (racist) life through humanism; that is, life that is not configured in the same genealogical traditions and forms that, *essentially*, do not rely on a division between the organic and inorganic for an account of thought and relation.

Vitalists, according to Brassier, try to escape the "levelling power of extinction" by asserting the final cessation of physical existence is not an obstacle to the "continuing evolution of life" (i.e., a refrain of, it's OK, life goes on without us . . .). The horror, Brassier contends, of the "traumatic scission between organic and inorganic" (2007, 238) cannot be contained within the psychic economy but nonetheless drives the demand for the veracity of organic life, against the inorganic or nonlife to which the organism returns. So, it might be said that there is a geotrauma that runs through Western thought that drives a repetition of this expulsion of geologic or inorganic time; a temporal trauma that cannot be calibrated within the "life" of the organism or of the earth understood as a singular Global-World-Space. Freud, in *Beyond the Pleasure Principle*, positions this "organic striving" in relation to the death drive: "*inorganic things existed before living ones*" (1922, 38). Geology is the primal scene, where the pressure for "molecular diffusion" is referred to as the death wish.

The division between the interior and exterior of organic and inorganic life is won at the cost of a separation or shielding of the subject from the inorganic outside, a geotrauma of organic individuation that cannot simply be contained by death but harbors an exorbitant potential of the inorganic reaches of the cosmos. Geology gives us that scission: the cosmic excess of an anterior geologic life that is caught temporarily in the realism of life but is not bound by its puny origins or endings. This confrontation with the lithic processes of the earth is a confrontation with geopowers that organize beyond life and without interest in its sense of progress or purposefulness. It is only the distinction that is made between the inorganic and organic that allows the very concept of "progress," "purpose," and "civilization" to emerge, but at the cost of the suppression of the inhuman earth. And it is at this site of division between the organic and inorganic that subjugation is constructed as an ontological horror that campaigns on the senses in psychic and planetary terms.

The provocation of the Anthropocene demands an engagement with all these geologic realisms within the geologies of race, a confrontation with the abyssal horizons of the inhuman beyond: (1) the immediate suture of false ends to these processes (i.e., to be able to think and imagine an exorbitant planet) *and* (2) the dereliction of Blackness in the psychic category of the inhuman and the ongoing "experience of the abyss" in the afterlives of slavery, what Glissant called "a debasement more eternal than apocalypse" (1997, 7). Alongside, the ongoing injury to Indigenous life. While Brassier locates the psychic schism in the division between the organic and inorganic (*bios* and *geos*), it might be better historically located in the annihilation of subjectivity within the terms of *geos*, as a lived rupture that negotiates daily with the geography of this split.

My argument for this Anthropocenic moment is that thought must occur along the sight lines of a shifting planetary *and* racialized axis in the constitution of the inhuman. As concerns arise for fashioning alternative worlds, livable futures, what are essentially countermodes of fossilization in the present tense, there sits an often silenced Indigenous, Black, and Brown archive of fossilized and future lives that have been subjugated under the category of the inhuman and its *extractive principle*, which cannot be added to humanism's concerns without the necessary work of unlearning the syntax of being that constitutes it. While the Anthropocene is problematic in all its assumptions about agency and white supremacy of matter, it does signal a shared threshold moment in the political present: namely, the demise of the stable material conditions of the "Holocene Humanism" (Yusoff 2019) that provided the context for the dominance of Western thought and Reason, which is now the geologic realism of all our lives (albeit experienced in uneven environments). That this reckoning is made in the context of a threat to late liberal subjectivity (and its sedimentations of whiteness) is not unproblematic, but the sinewy bonds of geology *have* and *are* made differently.

There are proximities to the grammars of geology, intimacies in the inhuman as a classificatory and subjective mode, and in the force of geologic events, where demands articulate a counterpoetics that reorganizes another term of geologic realism. For example, replantation, Wynter (1971) suggests, was the cultural-ecological cultivation of relation to inhuman and nonhuman forms, from planting foodstuffs and medicines in slave plots to developing knowledge of plants and vegetation in marronage—a sense that gave space to imagination and relation through ecology and outside the machine of industrial plantation and its economic botany (see DeLoughrey 2007, 2011). Like Glissant's geography of archipelagoes, poetics occupy the historical legacies of a forced openness to the sea with a chosen opacity, where "green balls and chains have rolled beneath from one island to the next, weaving shared rivers that we shall open up" (1997, 206; see also Mardorossian 2013; Noxolo 2016; Wardi 2016). It is the very division of materiality that becomes the point of leverage, says Glissant: "The boundary, its structural weakness, becomes our advantage. And in the end its seclusion has been conquered" (Glissant 1997, 206).

The praxis of reorientation to other temporalities that survival attests to, "in alliance with the imposed land" (Glissant 1997, 7), redirects enforced inhuman proximity as a way and a passage to other kinds of intimacies beyond colonial heteropatriarchy. Acts of Relation attest to survival in the indices and intramural spaces of colonial power as a material and poetic act, where, to paraphrase Wynter, "senses become theoreticians," crafting an imaginary that replots the coordinates of Black and Brown life beyond the calculus of the forced horizons and hierarchies of inhumanism (which was humanism's gift of monolingual intent pressed in the colonial world).

Geologic ~~Reason, Realism~~, Relation

Anthropocenic rifts map into hyperlocal contexts and durations of relation that amplify historic exposures along a racialized axis. The configuration of the localities of inhabitation matters, particularly in how they recognize the ongoing colonial histories of displacement, exile, and errantry (Gómez-Barris 2018; Last 2017). Thinking again with Glissant (1997, 33) and his demand for a "nonprojectile imaginary construct" (i.e., the inverse of colonial heteropatriarchy) in the formation of global relation (which climate change is), there is a need to both describe the colonial map and its extractive principle *and* hold space open for a redescription of the planetary that can have a reparative relation to the future, within the context of increasingly intense inhuman events. As geologists vie to name the start of the Anthropocene and thus consecrate its origin story (When did it begin? What is "it"? And, who is the "we"?), the identities of time and space are being rearranged and

FIGURE 8.2 Diagram of a meteorite penetrating a house, Field Museum, Chicago. Photo by author, 2017.

reproduced (it began with a nuclear bomb; it began with industrialization and the "genius" of white men inventing a steam engine or big Capitalism; and, in counterrepose, it is the history of racial capitalism, the mine, or the plantation). As the earth is its own ground and invention, ceaselessly churning its own variations on the what the earth is, evolving new forms of life and mineral out of the anthropomorphic-geologic soup, another inhuman story remains in the caesura that must be accounted for before there is any possibility for environmental politics.

Césaire's poem for those "without whom earth would not be earth" (1969 [1956], 57) takes aim directly at this question of narrative genealogy of the world, rebuking and ridiculing the signs of progress and civilization organized around the imaginary of the tom-tom drums (or Arendt's vegetate subjects):

Look at the tadpoles of my prodigious ancestry hatched inside
me!
those who invented neither gunpowder nor compass
those who tamed neither steam nor electricity
those who explored neither sea nor sky
but those who know the humblest corners of the country of
suffering
those whose only journeys were uprootings
those who went to sleep on their knees
those who were domesticated and christianized
those who were inoculated with degeneration
tom-toms of empty hands
inane tom-toms of resounding wounds
burlesque tom-toms of emaciated treachery
(Césaire 1969 [1956], 54–55)

As new modalities of geologic realisms underpin and contextualize organic life across all entities in relation to new intensities of geoforces, these geologies are also a site of struggle in the designation of the inheritance of *which* arrangements of Life (and inhumanisms) are deemed to matter, and *how* those divisions carry forward regimes of materiality, which designate the conditions of subjecthood in their historical context and the necropolitical relations to the histories that have borne them. Rather than just the projection of a future possible in the colonial expansion of the present, the question of those who live in the grotesque imaginaries of "burlesque tom-toms of emaciated treachery" and are classified as noninventors in the material progression narrative comes back with their own earth archive that requires attention. The gravity of this belonging might be understood as a starting point to strike out into the geotraumas of the inhuman, as a going-forward into the density of Relation, given by those "without whom earth would not be earth" (Césaire 1969 [1956], 57), as *the* actual history of the earth.

Life in the Rift

Inhuman Ontology

In another inhuman register, consider the meteorite as the ultimate outsider, a "cosmological imponderable," to use Elizabeth Grosz's (2008, 23) term, that might burst through the perceived limits of the known. The meteorite

FIGURE 8.3 Remains of the Bendegó meteorite that survived the fire on September 2, 2018, at the Museu Nacional, Rio de Janeiro, Brazil.

signals the ultimate "free territory" beyond the propertied form of earth and its territorial constraint. It is both cosmic gift and a form of eviscerated metallurgic intensity. It is also a rock that speaks of survival through the burdens of gravity. The geochemical signature of a meteorite is one of attempted annihilation by the atmosphere, yet its arrival is as a confirmation and provocation of spatial expanse (beyond the known world). For example, a meteorite was a lone survivor of a consuming fire at the Museu Nacional in Rio de Janeiro on September 2, 2018, which destroyed the two-hundred-year-old collection and 92.5 percent of its (mostly colonial) archive, some twenty million items. The former environmental minister, Marina Silva, said the fire was like "a lobotomy of Brazilian memory."[9]

When all the forms and types of classification were gone, a cosmic iron "rock" remained (the Bendegó meteorite found in 1784, weighing 11,600 pounds [5,260 kg]), suggesting in its persistence a much longer lead time on planetary memory. While some of the other rocks remained, tempered by the fire, once those objects were without their labels and locative data, they ceased to function as part of a scientific collection (they lost their epistemic ordering and geographic locational data and so could no longer serve as an evidential base). Included in the burn were audio recordings of Indigenous languages no longer spoken, a reminder of the ghost archive of inhumanism that never made it into the museum, in the voices of the disposed and violated (hear Glissant: "For every language that disappears a part of the human imagination is lost forever: a part of the forest, of the savannah or the crazy sidewalk" [2020, 52]). There was some local irony in the fact that the museum's finances had been run down and adequate fire prevention maintenance was forgone while vast funds were directed at the Museum of Tomorrow, with its focus on the Anthropocene (see Reyes-Carranza 2021).

The meteorite's material stoicism stands as a reminder of the durational possibilities of different orders of time and space that envelop us in the universe (a rock that shatters the Master's clock and arrives with an inhuman tale of diasporic geographies). Sky metal coasts through the voids of cosmic cataclysms to land heavily on the earth even as it expands its geochemical registers of being. Carrying discrete metallurgic signatures from within and outside of this solar system, it reconfigures the stretch of the alien to far-flung space. It gives a sturdy glimpse into horizons undefined by the current order of things, a different catalog of exuberance and debris from the formation of the solar system. Meteorites are cosmic time capsules from the earliest days of the solar system. This cosmochemical evidence of different rations of oxygen isotopes indexes extraterrestrial worlds and the composition of other universes far outside the realm of human experience. Carrying material times, sometimes even older than the solar system, meteorites are used to explain the "gravitational tugs" between distant planets and asteroids, to index the generation of new maps to the solar nebula. Thus, meteorites hold the hull of the most extensive sea of time, a deep time that is established through neither analogy nor metaphor but from the material forge from where metal and mineral speak.

Inhuman difference is the generative force of the world that enacts materiality, which, Grosz argues, "marks the very energies of existence before and beyond any lived or imputed identity. It is the inhuman work of difference. . . . Difference stretches, transforms, and opens up any identity to its provisional vicissitudes" (2011, 91). The challenge of differentiating from patriarchal pasts is also part of this equation, as the earth and its geoforces challenge the making of feminist futures uncontained by the pasts, which have sought to contain and marshal those geoforces to oppressive ends. In this work, Grosz recognizes two orders of the inhuman—the preindividual and the impersonal—that provide freedom for the subject "who understands that culture and history have an outside, are framed and given position, only through the orders of difference that structure the material world"

(97). That is, the provisional and possible are given by the inhuman, as well as a field of excitement, incitement, and the geopulls of desire.

While Western philosophical thought frames the inhuman as an alien limit figured by a racial other (in the baton pass from Kant to Heidegger to Arendt), more expansive thought has seen the inhuman as an opening to difference in the hunger of a warm contamination with that which is not-us and as a limit to philosophy's reach. This is what Grosz's understanding of the freedom of indeterminacy teaches us. And Glissant's opacity for irreducible existences too. This is the time when the question is being thrown far outside its known modes of questioning. This is the geoforce of incommensurability as a condition of the earth's own axis and the radical improbability of its burning presence.

Alongside these inhuman openings is the heavy metal work of subjectivity as ore (voyage iron) or in mine and plantation (as blacksmith, miner, or metallurgist) installed racialized burdens in the anti-Black gravities of the day.[10] Reflected in the dark, mercurial surface of the crucible in Sondra Perry's video installation *you out here look n like you don't belong to nobody: heavy metal and reflective* (2019) is an image of the fire that transforms and brands mineral and flesh. As a site of recall and affective archive, the shackles hang above in the dimensionless embrace of a blood-colored velvet, sinking in the red liquidity of the screen. The formation of spike, screen, and crucible has the layout of a terrible experiment, carpeted in the luxurious fabrics of wealth and privilege that were extracted through the violence of the trade of the enslaved. The spike is also a prism, a way of seeing the past, through this amalgam of its violently melded history. Within Perry's melted spike of origination is a meteorite. In its proximity to racial forms, these celestial rocks represent a broader cosmic materialism that formed as a liberatory channel in Black life under subjection. From the North Star of Harriet Tubman's Underground Railroad to the practices of African spiritualism, the cosmic (of which the meteorite is a concentrated reality) was a way to freedom in the tight and brutal spatialities of slavery and its afterlives.

In Perry's work, a Campo del Cielo iron meteorite comes from a cosmic storm that occurred 4.55 billion years ago, speaking to survivance beyond the hold of the present and its violent pasts. Materialities that hold the past also hold the future, from heavy metals and their toxic body burdens in the present, to Fanon's cosmic "yes" that radically reinvented the subject of the abysmal as a "cosmic effluvia," a "ray of sunlight under the earth," with an "intuitive understanding of the earth, an abandonment of my ego in the heart of the cosmos" (1986 [1952], 45). Thus, Perry's engagement with

FIGURE 8.4 Sondra Perry, *you out here look n like you don't belong to nobody: heavy metal and reflective*, 2019. Image credit: Dan Bradica and Gregory Carideo.

materiality channels embodiment in this potent space of race and geology, as it makes anew an affective archive of memory. The first part of her title—*you out here look n like you don't belong to nobody*—takes us straight to the painful histories (as Glissant named them) that these geographic and psychic displacements enacted, that made the Middle Passage such a profound scene of orphanage. The quality of violence that characterized the origin story of oceanic orphanage continued as a paradigm in Black life through the plantation and into the genealogical cuts that families in the present suffer as their children are taken by police violence and the prison state, and all the less explicit forms of diminishment of Black life.

Cosmo-Being: Rifts at the Edge of the Universe

I turn now to an account of *cosmo-being* that is neither dialectic nor alterity but a specific historical mode of subject formation, or unsegregated being that is impaled and expanded by the geoforces of inhuman intimacies. Cosmo-being might be understood as an inhuman intimacy that is a *sheltering* that draws together intimacy in a different affectual architecture to build a more open world in defiance of the social one at work (rather than some explanatory concept of nature as mirror or mimetic). Inhuman intimacy is a historical and present condition formed under stratal relations of violence and relational voiding, yet it is also a history of becoming, in which the anticolonial *will* took hold and made places of inhabitation. In this rift, that hearkens toward origins in neither nation-state nor nativism, the question of geography is posed as expansive in the world, but not organized around a telos of expansion over it.

The rift is a way of gathering the brokenness of rocks of time that have been piled all around in colonialism's now. It is a rock pile (of shelter in Simone's voicing) and a way to hear the geotrauma of geographic gatherings. The wretched of the earth (in Fanon's formulation) are not "like" wretched matter; rather, they are pushed and pressed into relation with the earth and so know its vicissitudes (as the miner must know the seam). These are not parallel states of being that are naturalized as ontologically divisible, but they are adjacent and proximate states of being (inhuman-inhumane-cosmo). This is important to notice because it speaks back to the easy determinisms that creep in on account of the grammars of geology and their coevolution with states of subjugation. As the specter of environmental determinisms is raised, in speaking of inhuman intimacy, I do not want to reconstitute a generalization of a different kind that accords a material personhood a memetic function with the earth, whether conceived in a liberatory mode or not. The point is not to replace one determination with another conceived in its oppositionary mode. The meteorite tells us that the inhuman is also a route of alterity that is shaped by its fiery passage through the geocosmic storms. That the rock and volcano seemed to offer better allyship than whiteness in the project of articulating what it means to be human, however, suggests something more deadly and domestic about the violence at play under colonial occupation. That the ground was more forgiving than enslavers and their kin is no surprise but a brutal fact of being on the wrong side of the geologic line in the division of Being/being.

The intimacy of the inhuman also says something about wider genealogies of being that were rebuilt from the geographic severing of colonialism and its spatial executions (in the form of invasion, the Middle Passage, and ongoing geotraumas of settler colonialism). The vertigo of this collective geotrauma *is the rift*; the memories of that disorder retain a spatial rupture in the depths of the strata of worlds torn asunder. This is where history discloses itself through the temporal breaches of geographic execution and its segregated restructuring. It is these visibilities and enunciations that the plasticity of whiteness seeks to erase in its cultural production. This is what it is to look up from the rifted depths of the world and see the layers of time that have brought the present (as the surface of the earth) into being. The edges of such jagged canyons are often sharp and vertiginous. It is Kant's sublime only if you are on the surface of Overseers' earth looking down, poised to capture your sense of subjectivity on the return (as accomplishment, not understanding).

The inhuman is not the mirror of Being. It is its context and constraint, but it is also a site of spatial expansion, the sequestration of inorganic temporalities and their material substrates in the fermentation of the allied lifeworlds. To circle back to Glissant again: "Always the wind twisting, the forest in wild disarray, the volcano's voices spilling forth, the shaking that lays waste the black earth with its volleys of red earth." He continues: "We draw on these extremes and we reinforce ourselves with this violence, without knowing it. This danger preserves us from the certainties that would limit us" (2020, 53). The inhuman is the breath and breadth of the world, not its foreclosure, but we cannot forget that this vulnerability is made in a scene of historic and ongoing violence. Reversing the relation between the inhumane and the inhuman under colonialism, Glissant retools the geocodes to give back a narration of the enormity to the depths of the world and to claim its rugged shelter and open as a geographic imaginary with which to build new worlds. Thus, the sterile binary between *bio* and *geos* is countered, reencountered as a breach in the constraint of being and as its overcoming. As the cosmic is a rift in the grasp of the Master's clock (see the Black Quantum Futurism collective), this Caribbean geophysics of reparation also entails the restoration of the earth in its Relation.

Abyssal "Yes"

In his exacting and elegiac engagement with Fanon, David Marriott shows how Fanon's understanding of the zone of nonbeing is not a descent into nihilism but strikes a yes that is the power of affirmation, "a *yes* that vibrates to cosmic harmonies" (1986 [1952], 10). We can see Fanon's descent in the abyssal as a geographic engagement with Kant's sublime as he inhabits the fall of a perpetual descent into the inhuman without the brake of Reason. That cosmic *yes* answers as an epistemic invention in the register of its inhuman emplacement and challenges Arendt's racism that is buried in the ground.[11] In the fracturing of the abyssal and its register of untranslatability, the colonial reach of structuring languages (and their durational harm) is epistemically broken through a state of radical unknowing. Fanon names this contraction through the largesse of the sun: "It is a story that takes place in darkness, and the sun that is carried within me must shine into the smallest crannies" (29). In doing so, he takes the (Bataillean) excess of the sun as the measure of all asymmetries (and inhuman generosity) and brings it into the rifts of the earth. Marriott suggests that the point is not to flee from the void as it is inescapable, but the effect is "*inhumanizing*; and, more important, it opens onto a kind of unreadable rupture or fissure" (2018, 333; italics in original). This inhumanization does not just summon cosmic effluvia but calls into question the human in its racist language. Marriott argues it is not just Fanon's conceptualization of excess that organizes the psychosocial diagnostic but his recognition of an ungrounded trauma without reserve that is revolutionary.

The destruction of the ground in Fanon's thought is key here to the disruption of sense and the possibility of reconciliation within the given (racist) terms. That is, the physical world still holds that history beyond its formation in the human and its futural race as destiny. Racism's interconnectedness secures itself in the earth (where the earth is both possibility and storehouse for racism's returns, i.e., the earth is a site for "destiny"). Almost inverting Emerson's ethics that would grind Black life into a stratified grave,

Marriott argues: "Negritude thus becomes the bearer of a new performative; it speaks for those who lie *beyond* petrification. Thus the poem has been seen as the outcome of a refusal that is also a resuscitation giving the socially dead new life. The enslaved dead have arrived—quite literally—from a time without us, from a past not our own, from a time that is not historical. As such, they denote the supernatural limit of what it means to be a *person*" (2018, 330–31).

Beyond petrification in geophysical orders, the abyssal in negritude arrives through the breakage of history (with a capital *H*) and the traumatic organization of what it means to be a subject. Marriott suggests that this act of severance possesses, in its shattering, an autonomous quality that names the limit of the super- and supranatural. The ahistoricism of stratal abandonment becomes a tactic of the earthbound and a way.

Unlike Meillassoux's arche-fossil, this naming is not outside the correlation, but because of the depths of its subjugation, it cannot forget its history (2013). Marriott then asks the profound and complex question, "But what happens to this future whiteness when, blackened, it too learns of its origin?" (2018, ix). That is, if whiteness sustains itself through the "plea for race to be a destiny (whether in the idiom of assent or that of obligation) intrinsically ha[ving] to do with the wretchedness in which blackness came into being in the world and that continues to define the anti-blackness of the world" (ix), then if Blackness becomes unrecognizable it might find a complex liberation that is defined outside of itself.

As Blackness in Marriott's reading of Fanon's schema creates the suppressed emergence of whiteness and its assent on the singular world as *the* world, constructed through the anti-Blackness of the world, the resolution is to go further into the abyssal (because Black liberation, Marriott argues via Fanon, cannot be found in a redefinition of Blackness).[12] And yet, Marriott cautions, "The intrusion of race into the very concept of the human means that there is no position to which one can commit oneself that is not already a sign of racism's interconnectedness with the history of the human" (2018, 316). That is, the human is irredeemably raced, and thus the inventive duress of the rift must abandon reinvestment in the identity politics of colonial invented selves that organized this dislocation. In the confrontation with dislocation, the present (as it is figured) must, according to Marriott, "annihilate or willfully extinguish (or, more generally, self-sacrifice) the present in order to invent a radically futural language" (331). He argues that this language cannot be the future perfect (as in Black feminisms), for that would reinvest in history through its retrieval of missing possibility, but must break with time altogether.

The abyssal shows itself as a rift and a site of transformation, a perspectivism uncontained by the past that is achieved through the collapse of the dialectic and the impossibility of transference to the new conditions that arise (because nothing remains the same to pivot the possibility of the continuance of this affectual architecture).[13] The trauma of race is geophysical, and it delivers a new minerality of time and space. Fanon's cosmic "yes" is the affirmation of the inhuman as a site of perspectivism, where the blind spot of white universalism is rendered visible and the universal seen as that which is constituted by the particular (the specificity of colonial geotrauma). The abyssal functions as an inversion of Life (in its partial form of the genre of the Human) that is rifted so the colonizers' plateau slips away and the path to the unknown (a just gravity not organized by race) becomes possible. The abyssal is where invention breaks petrified forms (race). This is neither arrival nor departure (so not in dialectical relation to the colonial script) but "mutual submergence" in the depths (the abyssal perspective). Where, Marriott says, "Blackness becomes what it is: the inarticulable or unimagined horizon that cannot be grasped from a universal standpoint" (2018, 316). The plateau is not re-created. The problem of stratal containment for the white geographic imagination is countered with the unimagined and unimaginable. Here, the relanguaging of this spatial formation (that comes out of geotrauma) is poetry, precisely because poetry can maintain itself in relation to language while at the same time destroying its meaning and sending the logics elsewhere. Poetry can usher in a creative negation in the geographies of race, guiding feeling into new languages of being. Fanon identifies the complicity between race and capital, whereby race is the fetish that veils the social character of labor to produce a "transcendental quality of whiteness" (1986 [1952], 143) What Fanon identifies is the "speculative character of its veiling" (144) in which whiteness appears as wealth and value, which suggests that refusal also needs to be articulated in the speculative mode (think here of the work of Octavia Butler, Moor Mother, Black Quantum Futurism, for example). This speculative mode also speaks to the quality of the modes of redistribution of material wealth and valuation in the context of environmental justice and restorative geophysics.

Race and Value

Throughout this book I have been attentive to the geophysical qualities of whiteness in constructing its ability to float and how it obtains and maintains racialized geophysics. As whiteness materially maintains its transcendental qualities, above the earth, not captured by the earth (and now these dreams of Mars form another flotation device in these racialized geophysics for obscenely rich white men), the speculative imperative is the historical work of both bringing whiteness back down to earth and tendering other gravities of meaning that tether the inhuman into its actual intimacy with subjectivity. Whereas whiteness is given a naturalized durational quality as apex through deep time narratives rather than in a political economy, it is established as a preexisting site of accumulation for capital. Marriott puts it with diamond precision: "What is radical about the wretched is not their economic role but their challenge to the creation of racial life as value" (2018, 147). It is the ordering—the stratigraphy of race and the stratigraphic imagination—that is broken apart by the wretched. Whiteness is the only form of Life that escapes becoming a natural resource, as a chiasmic distance is established between whiteness and the earth. This process is fundamentally an effect of discourses and practices of materiality and a praxis of materialisms, not ideology, and thus requires geographic and geophysical responses to processes of valuation. Racism is literally a historically constituted geophysical distortion that changes the dynamics of gravity as it differentiates space. Racial geophysics can be enacted in many spaces on the epidermally selected (a border, a sea, a street corner, a police cell, an office), and it can arrive with the ferocity of a material rift in time and space that pulls persons toward the earth and into its rifts (whether these be oceanic, administrational, debt, etc.). The ongoing appearance of the rift in the social reminds us that geography is not a constant but a multiplicity of differentiated experiences in the dynamics of space.

The presentment of affect, of inhuman intimacy with another gravity, speaks back to that indifferent brutal exchange of inhuman exchange. Presentment holds space differently; it forces its expansion and indicates a reserve that can hold a different tension or gravity of clamorous being, a burning presence that must overcome its constant belittlement and withdrawal. For example, the future is already anticipated and made in the exchange between voyage iron and the person designated as slave.[14] Voyage iron was the metal used for the material exchange of latent energy, a geophysical exchange between ore and body. It is an exchangeability between properties that are organized in exchangeable bodies. This grammar of geology became the language of the future because it was already transmuting people into things, ore into iron, iron into persons, persons as strong as iron. This terrible exchange finds its meaning and legitimacy in the granular circulation of these inhuman metals. A future language—

geo-logics—anticipates itself as a form of geographic travel. The inhuman under colonialism is a language made for exchange, for properties to be exchanged; it also undergoes temporal invention as an intimate form of the possibility by those forced into the duration of exchange (which is endless and has not stopped). The violence and exhaustion are totalizing but not total. Rifts in the abyssal are without redemption, yet the movement is redeeming of a different duration sense and revaluation.

Marriott's (2018) reading of Fanon provides a way to think through what the radical devaluation of racial life as value might entail, fermented in the abyssal trenches. In another field of attention, in the material destruction of the geocodes that govern the geophysics of race, a similar material rift is required that seeks to understand the *material power of race*, which is to acknowledge how race is the violence *of* and *through* matter and that antiracist struggles require the reworking of grammars of geology and theories of materiality. Rifts are geophysical and can open all over town, as the next chapter on the prison mine in Birmingham, Alabama, will address. Geopower is not ahistorical; rather, it entails a geophysical shift in the segregation of being and a duration affect (the "weight of history" as a geophysical condition). The depth of the abyssal is a duration of inhuman affects of depth and pressure.

While Marx attended to the material formations of life, he did not attend to the processes of materializing life within the pressures and forces by which beings come into and are forced out of the world (the cut of *bios-geos*). Nor did he challenge the structure of racialized stratification. So, as we move against racial life as value, there is a need to divest from Life as the only value (and the narrative of whiteness as some form of "weak" natural selection), and to reckon with its preconditions and context, as *geos*. Which is to go beyond a biologically constituted (racist) ground to ask after the histories and futures of geologic life. The nondialectical third point, the place of the rift, is not just the rupture of humanism (and its architectures of reason) in the abyssal zone but the *opening* to a zone of cosmo-being. This is the alluvial subjectification of Glissant, of building ghost histories in the connections between islands, of the between places held together by the unhistorical, the erased, and the forgotten. This is the *painful notion* of history that is marked by the return of the anticolonial will, "surpassing through the depths" (Marriott 2018, 341).

Marriott argues in his reading of Fanon that "blackness is produced as a state of debt that is valueless in itself, and so never redeemable" (2018, 144). To add to this, this state of debt is translated into a state of depth, in the physical earth, whereby racialized undergrounds maintain the surface of whiteness through a geophysics of duress. Fanon (1963, 36) is clear that the problem of decolonization is the redescription of the order of world and that the racial underpinnings of geography are to be resisted rather than reifying those racial identities further.[15] The occupation of the cosmic by Fanon (and Glissant) is an audacious refusal of colonial binaries and the coordinates of white order (of staying in one's stratigraphic place), recalibrating the entrance into time, to grasp spatialities in flux rather than fenced off. The recalibration of registers into the cosmic and "suns underground" claims spatial expansion as a form of sensory geocommunication. By expanding the reaches of the rift into the inhuman beyond the grammars of geology, control is held at a scalar distance. That which is drawn down into the abyssal recovers submerged duration of affects, and the subterranean becomes a negotiation with the fact of being ground.[16] For the geophysics of being it is an astrolabe and a route into the cosmic as a spatial expansion that reverberates with the hum of a different (antiracist) universe. The inhuman leads us to "unpredictable intuitions" (Glissant 2020, 73) as a route into the world and its density of being. It enacts the deconstruction of the terms in which the human presents itself. It is a map to an antiracist earth.

I am geophysics, I am gravity.

9 Inhuman Matters II

Deep Timing and Undergrounding in the Carceral Mine

FIGURE 9.1 *Pratt Coal and Coke Company, Pratt Mines, Convict Cemetery, Bounded by First Street, Avenue G, Third Place and Birmingham Southern Railroad, Birmingham, Jefferson County, Alabama*, 1968. Historic American Engineering Record. Library of Congress. https://www.loc.gov/item/al1048/.

Artist Cameron Rowland's installations *Birmingham* and *Norfolk Southern (Alabama)* (2017) consist of two lines of steel relay rails (18 × 371 × 101.3 cm) and an ingot.[1] Rowland's titles and specifications draw attention to the material practices of geology as a practice that is located within a geography, one which geological nomenclature has made nonspecific and abstract. Rails are specified through their dimensions and metallurgic properties. In another room is *Jim Crow* (2017), consisting of a type of manual railroad rail bender. On the wall is a minimal archive, detailing the convict lease system contracts to railroad companies and the judgment of the *Plessy v. Ferguson* case, which upheld an 1890 Louisiana law that allowed segregation of Black railroad passengers. Alongside the tracks is a historic aerial photograph, looking over the coke oven and tailings piles in the former Birmingham Southern Railroad company site. Next is a photograph of the convict cemetery of the Pratt Coal and Coke Company that operated Pratt Mines, where the iron was mined with convict labor, young men and boys kidnapped by the state for their vagabonding geographies to build the industrial South and the "Magic City" of Birmingham, Alabama. Rowland's installation *Jim Crow* recalls the political histories of the Black Codes that predated on the racialized poor to feed the development of the carceral mines and railroads. The astringency of Rowland's practice calls attention to the quieted histories of the convict lease, quieted in its subjugating geographic and social terms, and quiet in the discourses of industrialization and urban life as distinct from the violence of plantation production.

The Pratt photographs are some of the few visual traces of the disappeared material geographies and social geologies of the convict lease in Alabama. Alongside traces of cyanide in the landscape (a byproduct of coke production) and innumerable unmarked graves. This partial witnessing of the convict lease system and its erasure of subjects has a parallel disappearance in the selective narrativizing of materiality and its appearance as an "isolate": alongside the anonymous lines of track and the *Jim Crow* rail bender, the categories of steel and iron soak up their origination in the convict lease system to pronounce an unlocatable, ultimately "clean" geography of materiality in the skyscrapers of northern cities and the infrastructural architectures of the settler nation-state. In the United States Steel cemetery in Pratt City, convicts like John Clarke might lie.[2] Convicted of "gaming" in 1903, he was unable to pay the fine and so ended up at the Sloss-Sheffield mines, bought from the state for nine dollars a month. Working off the fine for his "crime" required ten days of hard labor. Fees for the sheriffs, deputies, and court officials who derived their wage from convict fees added another

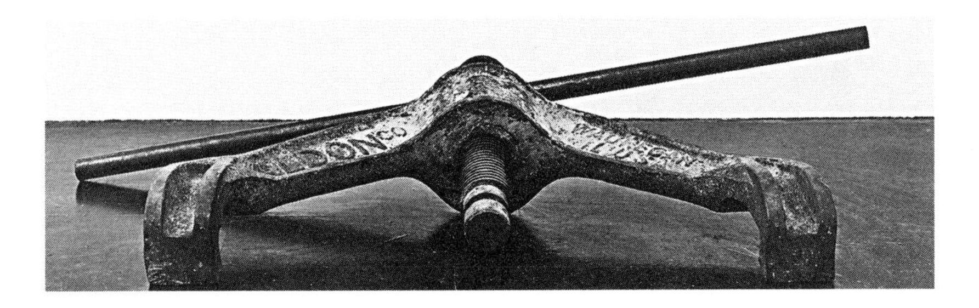

FIGURE 9.2 Cameron Rowland, *Jim Crow*, 2017. 91.4 × 20 × 44 cm.

FIGURE 9.3 *Pratt Coal and Coke Company, Pratt Mines, Coke Ovens and Railroad, Bounded by First Street, Avenue G, Third Place, Birmingham Southern Railroad, Birmingham, Jefferson County, Alabama*, 1968. Historic American Engineering Record. Library of Congress. https://www.loc.gov/item/al1109/.

104 days in the mine. A month later, according to an inspector's report, John Clarke was dead, crushed by "falling rock" (Blackmon 2008, 112).

Rowland's work recalls us to an aesthetics of nondisclosure and material autonomy that permeate the materialisms of colonialism.[3] The manufactured earth performs as a "mineral isolate," partitioning the senses for a concept of matter cut from social and geophysical relation and its racializing axis.[4] Not just as commodity fetish, materials cut as natural resources assume independence from the place and manner of their extraction (sometimes place is stamped as a providence of quality, Sheffield Steel, for example, or to distance violent geography, such as Lancaster cotton). Semantically, natural resources refer only to their mineral categorization and technical processes. It is not that the object performs the substitution of an origin story so much as its origin story is made self-evident (or naturalized) within the material telos. The ability to absorb other histories is erased. The origin stories of coal and steel are located in either the earth or the settler state. Coal is imagined as temporally suspended in relation to the surface, either inscribed as the Carboniferous underground or scribed in the present by whiteness as coal comes to the surface. The aesthetic instruction of the mineral isolate creates a geologic grammar removed from the violent circumstances that *surfaced* minerals from the earth to expand the white settler state through the undergrounding of Black life. This surfacing is a spatial, temporal, ontological, and epistemic operation.

It is against this autonomy of materiality as elemental that I write. I do not want to renovate an aesthetic of matter as either essence or site of recuperation, but to show how epistemes of matter transform material ontologies and selectively erase histories of surfacing. The fantasy of purity has always been a lure and storehouse of energy in the racialized production of subjectivity and hierarchies of the human. A reinvestment in mineral essence does nothing for the wretched of the earth. It simply confirms the value of the mine and the continuance of mining bodies and earth for their extractable value. Stabilizing minerality within an elemental episteme lets go of how the norms of materiality and its modes of capture organize economic orders of colonialism into racial capitalism (and thus it is a form of silencing). In the sparsity and volume of material address, as steel or iron ingot, the inhuman basis of minerality obscures, and overrides, more intrasubjective "quieted" (Campt 2017) modes of material experience and exchange—the lives caught up in mine and furnace, the mined coal and iron ore that laid the tracks, and the weight that bodies were forced to carry as broken rocks up the shaft. In her 2003 photograph *Untitled (Standing on the Tracks)*, from *The Louisiana Project*, Carrie Mae Weems places her own body in the middle of the tracks. Her back is to the viewer to deny the access that social conditions gave to Black women's bodies. Weems places herself as witness and subject in the observance of Black geographies, standing in the way of the smooth track, backbreaking the geographies for other people's mobility and segregated movement.

The allied work of mineral isolates, remote from their contractual and carceral relations, might be seen as a route into questions of the socialites that are alienated through the imposition of these epistemic categories and material circumstances in inhuman(e) ledgers, made in recursive geographies of value (when the value raised on your back pushes you farther into the earth and relocates freedom elsewhere). The parallel social and geographic processes of making "isolates" in the convict lease tied the incarceration of Black citizens to the production of ungrounded and disposable spatialities, whereby Black spatial life had to contend with the rupture and grounds of another geotrauma in the geographies of emancipation. The question that structures my archival work on the prison mine in this chapter is this: How do the material disidentification and dividuation of natural resources (rendered toward the telos of industrialization) get made into a form of subjective category that conjoins inhuman materials and the status of Black life into the inhumanity of carceral conditions?

Stilling the material object for accumulation, working bodies of incarcerated subjects and bodies of earth together formed a political economy of Southern industrialization (and northern accumulation). Building on the formation of value under slavery, convict lease labor extended the long arm of extraction over Black subjectivity (once again), while reorganizing gender and sexuality in historically new ways. The convict lease system decoupled reproduction from biology under slavery and coupled it to the state's ability to fulfill the category of criminality. The convict coal lease functioned to create social isolates through mineral disaggregation. It was a system that was intentionally designed to orphan, to rift social space. The incarcerated were sent underground without the sun and left to broil and freeze in the fetid climates of the prison.[5]

Inverting the proximity to the inhuman form from which slavery takes its social categorization is a way to rethink material isolation and the category imposition it performs as mutually implicated in that production and forms of value. One body exerts a force on another body; the second body exerts force in return. This would be to recall materiality as a witness and a site of methodological address in racial geologies, made in the context of

FIGURE 9.4 Carrie Mae Weems, *Untitled (Standing on the Tracks), The Louisiana Project,* 2003. Gelatin silver print.

the geomythologized notions of "progress" and "modernity" that situated whiteness as an authoring center of the earth and its possibilities. Racial geologies designate a mono-telos (a singular mode of development and change that prioritized technological optimism and environmental harm) and marginalized or cut off Black, Brown, and Indigenous earth practices. Surfacing histories calls on undergrounds to speak to the surface and its erasures, but also, remembering the abyssal potential of geologic lives to be otherwise, there is an imperative not to formulate that address in the language of the surface (i.e., labor, value, economy, color-blind miner). The independence of the industrial material object as placeless, save for its modes of production (mine, furnace, factory), parallels the radical alienation of the "labor" that forged it: young Black life, bending the tracks with the Jim Crow, whose lives Jim Crow laws bent. The intrasubjective-material worlds of *deformation,* of convict lease labor, its historical material junctures, and subjugating experiences do not seemingly leave a visible trace on the elemental as it is realized in modernity's architectures. Shiny as steel. *Shimmer*. Look up, not down! Yet, subjects do not remain untouched by this engulfment in materiality, its modes of categorization and enclosure, nor do sociospatial or geophysical worlds. Glissant proclaims, "I circulate among the coals. My force flattened against the forces!" (1997, 40), suggesting other ghost geologies to pay attention to.

In 1886, the United States was an agricultural nation organized around the wealth from industrial agroplantations; by 1900, it was the largest industrial nation in the world. The conditions of this material transformation are enacted through an elemental vernacular that is performed in concert with the mineral intimacies of mine shafts, coal dust, hot licks, rage of furnace heat, and the driving of cold steel forged in racial geo-logics. The desire for the elemental as purity rides our urban aesthetics and architectures. U.S. Steel "floats" on the stock market and advertises its reassuring settler family credentials (see figure 11.3). Bodies of the incarcerated routinely disappeared into the heaps of cast-off matter and in the hillside of the mines at the shaft entrance.[6] The ghosts of geology hammer still in the rocks of Birmingham, Alabama, as other futures were fought in Selma and Montgomery, monumentalized in an asymmetrical inverted cone of black granite by Maya Lin at the Civil Rights Memorial Center in Montgomery.[7]

Mineral modes of categorization—steel, iron, coal, ore—encapsulate the inhuman, as a stable form of property for extraction, a lexicon for making nonhistorical objects rather than subjective registers. Steel and stone are what scripts are carved into to remember the "event." There is a parallel inhuman grammar deployed in the propertied category of convict as with the slave. The nonhistoric is not total, nor can it simply be ushered away through an account of commodity fetishism that retains the myth of natural resources, of nature transformed through labor without excess. If that labor is not labor (able to function in the commodity form because of slavery) but organized as fungibility as Tiffany King (2019) argues, then the categorization of convict mine labor slips into an intramural *infra*materiality (as another account of property/properties). This is the liminality of materiality as a disruption to the clean categorization of labor and materials as capital exchange.

Racialized materialisms disturb Marx's ontology of labor, as they interrupt the perceived autonomy of materiality. Technically, materials can be dated by their manufacturing processes, maker's marks, and modes of fabri-

FIGURE 9.5 *Pratt Coal and Coke Company, Pratt Mines, Tailings Pile, Bounded by First Street, Avenue G, Third Place, Birmingham Southern Railroad, Birmingham, Jefferson County, Alabama,* 1968. Historic American Engineering Record. Library of Congress, Washington, DC. https://www.loc.gov/item/al1110/.

cation, but minerality also draws on deep time registers of association for an inhuman origination, obscuring the geologic life implicated in their arrival. Seemingly, nonresistance to their passage, minerals, metals, and fuels makes materiality available to selective storying of elemental earth. Attentive to the collaborative work that inhuman epistemologies are mobilized toward, we can say that it is the extractive industries, picking phosphate in South Carolina, digging and milling clay for bricks in Georgia, turpentine work in Florida, chopping cords of wood in Texas, stoking and feeding the sheer hot furnace with bricks and iron ore for pig casting and mining coal in Alabama that brokered Southern industrialization through subjugating Black life forms in the convict lease labor. Disrupting monolithic Southern geographies through their spatializing racial axis, this chapter seeks to complicate the experience of the underground and the material engagements that were part of the ongoing work of geology life as the praxis of undeclared racism.

Black Coal, White Present

The Birmingham district was sold as the "Pittsburgh of the South" because of its coal, iron, and steel production and was dubbed the "locus of the south's first free industrial proletariat" (Lichtenstein 1993, 119). Birmingham was one of the most segregated cities in America. The industrial heritage site of the Sloss blast furnace was a major location of pig iron production that was first put to blast in 1882, served by the Sloss Mines that were worked almost exclusively by incarcerated subjects.[8] Alongside the plateau of the free industrial proletariat, Alabama was also the most profitable carceral state in the country through the development of its convict system and prison mines. There was precious little by way of a discussion of the Black history of the postbellum South and its carceral conditions in the Sloss site or in the local mining museums that were dotted around Birmingham.[9] The rusted yet still monumental furnaces of Sloss still exact an imagined gravity of coal-filled lungs, crushing work under the earth, natural gas explosions, rock falls, and the too-close heat. The enormous seventy-foot-high blast furnaces were operated at the mouth by hand, whereby an infrastructure of incarcerated prisoners, stoves, rail, hoists, and engines pumped up and powered the iron ore and coke (from the nearby coal mines) into molten iron. What the narrative of modernity fails to mention is the forced proximity of Black life to these inhuman intimacies of the blasts, which regularly consumed more than raw material. One of the few explanatory panels at the site—"The People"—makes clear the segregation of the workforce and how the racialized conditions of work produced both an economic subjugation and a material-spatial positioning in relation to the furnace:

> Each furnace had an operating crew of six men. Supervising work at the furnace was the stove tender. He decided when to cast the iron and when to flush the slag, when to change stoves, and when to adjust the wind on the furnace. Under him were the furnace keeper, who was in charge of the iron notch; the fallman, who took care of the cinder notch; and three laborers. Prior to the Civil Rights Act of 1965 [Sloss closed down five years later] the stove tender was always white; the keeper, fallman, and laborers, always black. This segregation of jobs was an unwritten but openly acknowledged tradition. One black worker recalled that "the white man's job were engineer, foreman, or running the crane. All the rest of them were black." Black workers weren't allowed to bid on a white man's job, he explained, "They'd say, 'you're a n*****. That's not a n***** job.'" Another former worker commented, "There was nothing

> you could do about it at the time. . . . White folks been white folks all their lives."

Skilled labor was considered white (Norrell 1986); common labor was Black, no matter how skilled, free or unfree.[10] White labor was hypervisible (and heroically celebrated) as the productive hand of extraction; Black work was erased. Stoking a furnace, repairing the blast masonry walls, pig casting, and so forth demanded skilled operation, and the majority of the two thousand workers at Sloss were Black.

At the industrial heritage site, next to an image of a Black furnace worker, is an explanation of pig casting, as described by Edward Uehling, inventor of the mechanical pig-casting machine:

> When the iron had cooled down to the temperature of solidification, but was still red hot, the iron carriers began their task. They covered the pigs with a layer of sand, then put on shoes with thick wooden soles, walked on the hot iron, and with crowbars broke the sow into pieces the length of the pigs. This was as hot a job as a man could stand, but it had to be done. If the pigs were allowed to get cold, it would not only be more difficult to break the pigs off the sow and the sow into pieces, but the iron could not be carried away in time to mold up [the sand] for the next cast. . . . Only one man in ten is physically fit to be an iron carrier, and the best of them cannot stand up under the strain for many years. . . . The task of breaking up and carrying out the iron from the casting beds of even a modest-size furnace is not a fit one for human beings. If it were possible to employ horses, mules, or oxen to do this work, the Society for the Prevention of Cruelty of Dumb Animals would have interfered long ago, and rightfully so.

The inhumane conditions of intense heat and mineral labor, considered too harsh for animals, were considered "fit" work for incarcerated Black subjects (as the panel inadvertently attests). It is not that the human is routed through the nonhuman as the primary racial distinction so much as the collaborative junctures of the inhuman-inhumane are institutionalized as a mode of production. Race provided an interjection that mobilized units of natural resource and structured the distinction between a metaphysical claim to the telos of materiality (as accomplishment) and the material classification of *geos* as awaiting matter (and originary ground). The abyssal rift of this institutionalized predicament is the folding of subjectivity into matter itself, so that the disturbances of these two modalities of the inhuman(e) are energized into the productive telos of *geos* but are denied any form of becoming within that category. Here, the material pathology of natural resources comes close to revealing itself. The origin marker of the inhuman as an indifferent materiality is accorded to a collective subjectivity as an inhumane positioning (in alienation from the juridical rights of the human). The telos of natural resources as the genealogical accomplishment of whiteness required the *geos* of Black energy in this inhumane arrangement (first with plantation, then with mine). Thus, agency is denied in the inhuman category for both racialized subjects and earth. The sedimentation of the power of race as a genealogical construct is the rationale for, and the outcome of, natural resources. The reliance on two modes of inhuman made to "exchange" produced the hegemonic mineral isolate that stratified Birmingham, Alabama, into the state of whiteness.

The Spatialities of Race

Until 1883, the state of Alabama did not supervise the incarceration of any of its subjects (state or county), who were simply sold to the coal mine lessee or Sloss furnaces with no checks on the lengths or overruns of sentences, the type of labor undertaken, or the conditions of survival, injury, or disfigurement. The lease was a juridical and corporeal black hole. The later appointment of inspectors in 1883 was undertaken in response to extensive protests about the convict lease but ultimately served to justify its continuance. Reform, as Hartman (1997) has astutely observed, was a tool of further extraction, and rather than alleviating violence it often was a means to prolong and propagate it. Incarcerated mine and furnace subjects, working minerality into industrial products and the underground into surface flows of social power and capital, were made wretched in the earth, continuing the enduring structural relation of the subterranean racial order and inhumane intimacies. The microspatialities of race in urban space were vertical: the mines went down, the skyscrapers of the "Heaviest Corner on Earth" went up between Twentieth Street and First Avenue North in the Magic City, and race was remade as an inverse spatial affect of new narratives of material redemption for the settler state. Urban planning and the maintenance of hierarchized racialized relations continued to build worlds through the combined sociogeologies of race, as a new aesthetics of steel and iron hid the inhumane construction in pleasing inhuman materials and shiny monuments of verticality.

The prison mine performed the material stratification of an epistemic condition of the liminal understratum, a subterranean convergence of subject and material conditions. Overgrounds grew through racial undergrounds. Black life was always proximate to the fire in Alabama, in the mine shafts, in the foundry, in explosions under the earth, and at the furnace. That the records of Pratt Mines were recorded as "burnt in the fire" speaks to both an administrational and a calculative logic of Black life within the carceral condition. The *Biennial Report of the Inspectors of Convicts to the Governor* by the Alabama Board of Inspectors of Convicts notes: "There was no record kept of these people . . . what has become of persons sentenced to hard labor prior to 1883, of whom there is absolutely no trace; they have disappeared as completely as if the earth had opened and swallowed them up. The very fact of their existence has been forgotten, except by the few at the humble home, who still wait and look in vain for him who does not come" (Alabama 1886, 24). The pyric life of Blackness in the subterranean world of mines reproduced the gravitational geopulls of broader social forces from the plateau.

The Pratt Coal and Coke Company; the Tennessee Coal, Iron, and Railroad Company (TCI); and the Sloss Iron and Steel Company received exclusive rights to the forced labor of the incarcerated. By 1881, TCI was awarded all state and half of the county convict leases for the next ten years, with the remainder going to the Sloss Iron and Steel Company. The ten-year lease, secured by a bond of $100,000, made all male convicts "fit for labor" under contract at Pratt Mines near Birmingham. The largest coal prison mines were Slope No. 2 and the Shaft mines, operated by TCI in Pratt City, and the Coalburg prison mine, operated by Sloss with county prisoners.[11] The Alabama Penitentiary reported: "The convicts are earning more money than they ever did"; "earnings for the past two years have been $103,332.97" (minus expenses of $45,888.89, leaving a balance of $57,444.08); and "nothing like such a showing has ever been made before" (Alabama 1886/ 88, 3). Prisoners were graded by the physiology of slave classifications—full, half, third, and dead hands (full hands valued at $18.50 per month for hauling four tons per day to the surface, which meant shifting about twelve tons of dead weight of slate and rock to get four tons of usable coal; half hands at $13.50 for three tons per day to the surface; third hands at $9 for two tons per day; dead hands at $1 for one ton to cover the cost of their keep). In comparison, free miners made $45 to $50 per month. In 1892, the net profit for TCI from convict leased coal mining was $80,000.

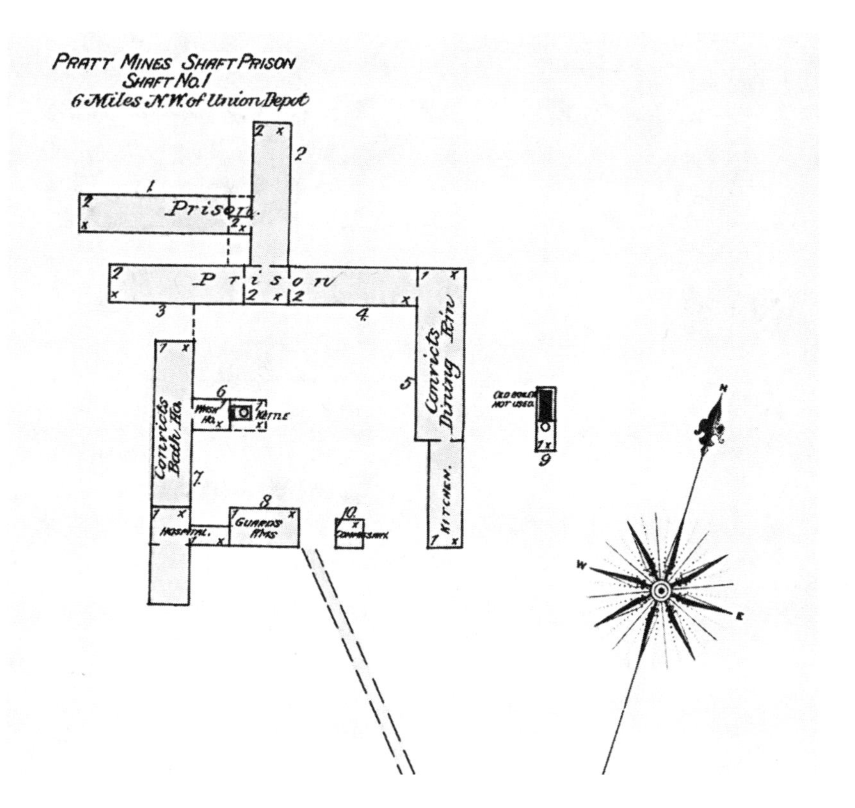

FIGURE 9.6 Map (partial) of Shaft mine prison, Pratt Mines, April 1888. Sanborn Fire Insurance map, Birmingham, Jefferson County, Alabama, Sanborn Map Company, New York. Library of Congress, Washington, DC. https://www.loc.gov/item/sanborn00015_002/.

Persons returning to the surface without the designated tonnage or too much rock or slate were met with whippings.[12] The incarcerated, over 95 percent of whom were African American, sold to coal mines suffered the worst death rates of any industry employing lease labor. While the death rates varied from year to year, during 1895, over 15 percent of convicts held at Pratt Mines died—"very embarrassing," the Board of Inspectors of Convicts observed (quoted in Letwin 1998, 50). In the same year, Thomas D. Parke, who was the health officer for Jefferson, cited a death rate of ninety men per one thousand incarcerated at the Coalburg mine (nearly 10 percent). Similarly, in Shaft it was roughly one in ten (Parke 1895). As race became reweaponized in the convict lease to mobilize the geologic underground, in 1898 Alabama obtained 73 percent of its total state revenue from the hire

FIGURE 9.7 *Sloss Red Ore Mine No. 2, Red Mountain, Birmingham, Jefferson County, Alabama,* 1968. Historic American Engineering Record, Sloss Furnace Company. Library of Congress, Washington, DC. https://www.loc.gov/item/al0961/.

of convicts, predominantly to mines (Mancini 1996, 112). After TCI became United States Steel (often referred to as U.S. Steel) in 1907 (U.S. Steel continued to operate TCI's mines in Alabama for another twenty years and had agreements with twenty counties in Alabama for the lease of incarcerated persons), it became the world's first billion-dollar corporation listed on the New York Stock Exchange, financed by J. P. Morgan, among others. Levitation on the financial market relied on racialized undergrounds.[13]

As a specific granular geology of racial capitalism, Black life in the Southern convict lease system enriched and expanded white grounds across the nation-state. As industrial plantations exhausted soil and souls, coal was seen as an engine of national settler growth and a new futurity for Southern fortunes, and it relied on experimental forms of undergrounding carceral logics. Challenging the narrative of materiality as isolated from its corporeal histories through the spatial logics of the carceral mine, I question both forms of material witnessing and material narrativizing through geology to ask: *How does the exercising of geologic grammars and geopower establish whiteness as surface, organized through its racialized undergrounds? How does the spatiality of the surface-underground relation become systemic in the organization of a racialized geophysics of subjugation of the settler-state?*

The Magic City

> Ours is an Industrial Civilization. Physical forces have opened in Alabama a broad entrance for the humanities of modern labor. The fertility of the fields, the richness of the mountains, the abundance of the water ways, the equability of the climate constitute the meeting ground between labor and capital. . . . Isothermal lines passing over soils capable of returning wealth for the labor bestowed on them . . . to stand conspicuous in the promise of the future. —JOHN WITHERSPOON DUBOSE, *The Mineral Wealth of Alabama and Birmingham Illustrated*

> The state of Alabama is now generally regarded as the coming center of the iron and steel industry of North America, and Birmingham District as the ultimate rival of the Pittsburgh District. Since 1890 Alabama has, as a matter of fact, dictated the price of pig iron to the United States. According to the latest statistics she ranks first in the production of brown ore, third in the production of red hematite, and in total production; she is third in production of coke, fourth in that of pig iron, fifth in production of coal, and fifth in the manufacture of steel. —ETHEL ARMES, *The Story of Coal and Iron in Alabama*

The founding of a mineral city in the Southern United States mobilized a set of relations between the underground mineral deposits and the white settler surface, relations that are imbued with both geophysical and political power.[14] In the triad of racialized resource, labor, and capital, modalities of geopower were established that organized urban space and race in a struggle defined by geophysical states and the gravitational pull of black holes. It is not simply that geologic materials open a "meeting ground between labor and capital" with "isothermal lines" "returning wealth"; rather, geopower is a specific organization of racial modes of the uneven returns of value through the spatiality of plateau and rift. The spatial lacuna of the underground as a geophysical-raced condition in relation to the white surface became crucial for understanding the racial construction of space as *stratal* and *sedimented*. The deployment of *vertically stratigraphic racial capitalism*, whereby vertiginous race relations regulate capital value, became an

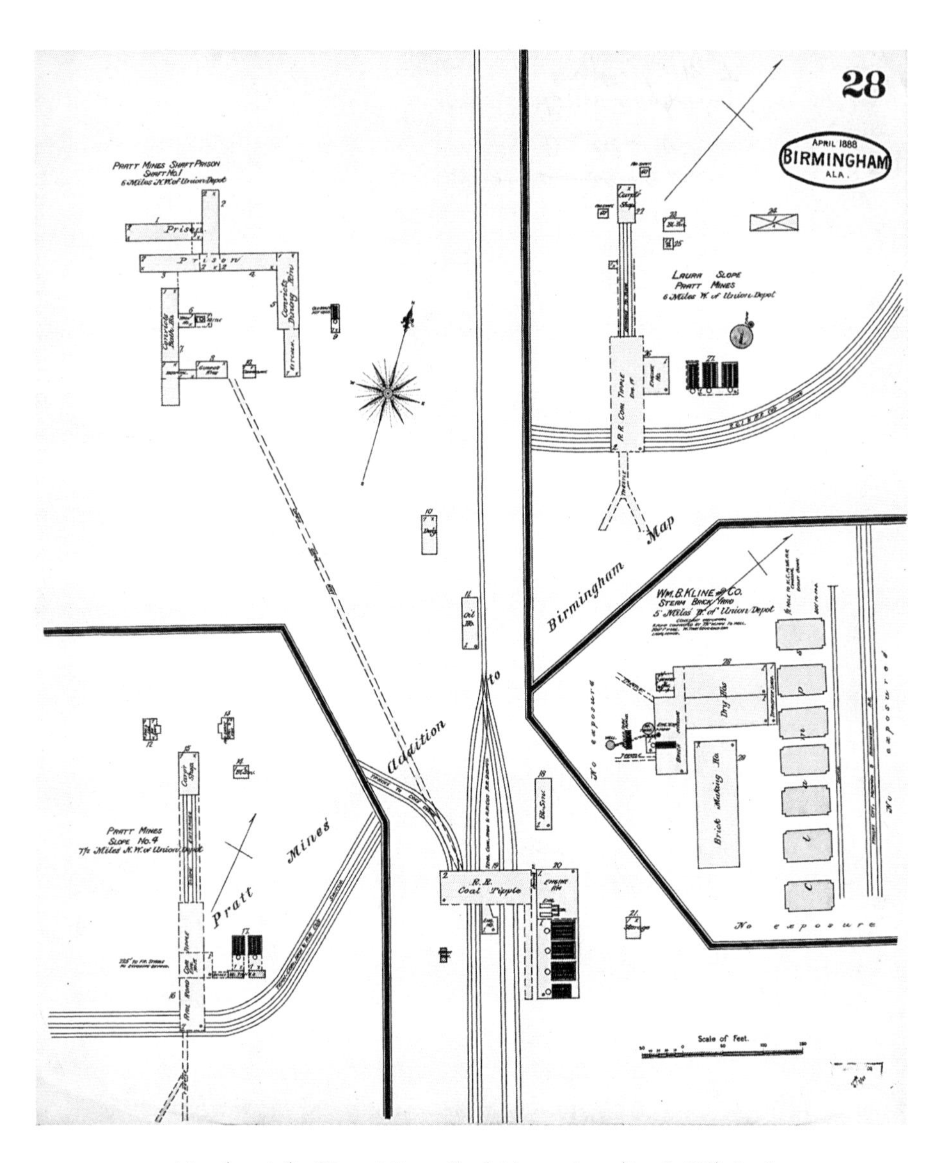

FIGURE 9.8 Map (partial) of Pratt Mines, Shaft No. 1 prison (April 1888). Sanborn Fire Insurance Map, Birmingham, Jefferson County, Alabama, Sanborn Map Company, New York. Library of Congress, Washington, DC. https://www.loc.gov/item/sanborn00015_002/.

exportable model for the dynamics of racial capitalism. As the spatial theory of race dug deep into the earth for its concepts, the political dimensions of materiality implicated all social forms and capital arrangements.

Fossil fuels in the form of coal, when cut from the earth and brought up through the Shaft prison mine, released energy, which was put to work in the construction of the material realities of social life as segregated, such as the buildings of urban space; connective tissues of railways, roads, and bridges; and the infrastructures of modernity and their affective imaginaries of geographic expansion. Reinforcing McKittrick's argument "that in the Americas, it is impossible to delink the built environment, the urban, and blackness" (2013, 2), the manufacturing of pig iron and steel materially brought strength and extension to the white settler project and its architectures of supremacy. In a parallel extraction, the materiality of the underground—as racial deficit—fed the furnaces and liquid core of Birmingham, Alabama, propelling it toward the future, as the liminal spatialities of prison mines fed the accumulative value of white settler futurity.

In the large-scale "experiment" with convict lease labor and the making of an industrial revolution, settling Magic City was part of an interlocking production of development that converted the racial dynamics of both the earth and postemancipation space. The psychic and geologic regime of whiteness produced urban plateaus and carceral undergrounds. Alabama's leasing lasted longest, and mining involved the largest Southern population of incarcerated Black persons. Natural resources mined through private enterprises were viewed as in the public interest (insomuch as they were nation-building and contributed to the reproduction of the white settler state). Incarcerated persons as wards of the state became organized within mining companies as privatized and leasable as racialized human "resource." Ethel Armes makes this national project clear in the introduction to the most comprehensive account of Alabama's development, *The Story of Coal and Iron in Alabama*: "It is, therefore, mainly her coal and iron business that enable Alabama to stand upon the self-respecting basis, industrially speaking. . . . A blast furnace, indeed, was built before Alabama" (2011 [1910], xxvii). Armes's widely published account accorded minerality much agency in building the state—"extraordinary conditions from the geological viewpoint"—and serendipity in the natural occurrence of the ingredients of steel (iron ore, coal, and limestone) being "in such close proximity as to be practically in one locality" (xxvii).

While such accounts of the natural "providence" of mineral resources were and are common (particularly deployed in relation to Africa and contemporary "developing countries," with the attendant narratives of resource scarcities and "curses"), they placed origination and progeny in the body of the earth mastered by the Overseer. The semantic field generated was of inhuman sovereignty as a part of the geomythology of the colonial state. The appendage of destiny to matter makes the material development narrative of geology hegemonic and defined by a racial telos in the construction of Overseers' earth. Thus, the achievement of species-life that is accorded to the mastery of geopower secures the fantasy of overcoming and institutes a pathology with which to confront the dangers of the inhuman world. The redemption of geology becomes another origin story that recentered whiteness. To wit, colonial earth material ontologies equal racial teleologies. Colonial earth erases the geotrauma of perturbation to relocate meaning to a symbolic register that allows whiteness to dream and live the plateau. The hypervisible guiding hand of whiteness carried the underground to the surface, while erasing the many Black hands that did the work.

The geologist Michael Toumey, who in 1849 named the coalfields around Birmingham Chaba, Coose, and Warrior, suggested that two of the fields constituted the thickest coal measures in the United States at the time (DuBose 1886, 22). Spurred on from one of the father figures, Charles Lyell, and other geologic luminaries, Southern financiers used these mineral reports to generate fiscal investment from northern capitalists. According to Henry McCalley, a chemist and assistant state geologist:

> Of all the minerals, iron is the most important, most useful and most influential. It seems to invest, as has been beautifully said, all countries which possess it in large quantities and which manufacture it extensively, with its hardy nature and sterling qualities. We have here in Alabama the most southern regular deposits of ores of this valuable mineral in the United States, and every facility for rendering Alabama one of the greatest iron producing countries in the world. We have hundreds and hundreds of beds of this ore scattered over the mountain region of Alabama, and, we believe, billions and billions of tons still undeveloped. (quoted in DuBose 1886, 27)

The ontology of geology gives iron a semiotic purpose and a self-fulfilling valuation, as iron *invests* and *influences*, commuting its geopower onto its capital developers. The coconstitutive imagination of iron, its material qualities, and its "metonymic qualities" (Appel, Mason, and Watts 2015, 25) are mobilized to hide the racial debt of its surfacing. As geologic grammars stabilize the material object, the florid prose and epithets of geologist reports naturalized extraction and carceral conditions as part of the boostering of the plateau for northern investment.[15] As the coal industry rapidly developed in the nineteenth century, spurred on by the huge fortunes being made as a result of the Industrial Revolution in the "other" Birmingham in England, a materially memetic correspondence of the coproduction of coal and empire was established, as the Birminghams were aligned in a conversant geography between the Magic City and the colonialscape of England. Birmingham sought to model itself and replicate the industrial success of its parent city (which was directly funded by the financial rewards of the Slave Compensation Act of 1837). While booms, busts, and boosterism have become indicative of the extraction industry, they were specifically a constitutive part of how coal, iron, and steel scribed the racial gravities of white invasion and settlement (from the agricultural use of iron in plantations to the forging of weaponry for genocidal acts against Indigenous tribes).

Mineral boosterism sought to encourage northern finances to migrate south. For example, one northern prospector reported on the convict extraction complex: "Alabama's Resources, As Seen by a Correspondent of the Boston Bulletin. Some Concluding Observations on Convict Labor—Ensley City—Prospects and Probabilities" in the *Birmingham Age-Herald* (1888).[16] On a visit to Pratt Mines he recounts his trip into the Carboniferous, in a style reminiscent of Jules Verne's *Journey to the Center of the Earth* (1864).[17] As he describes the experience, "Gathering some relics, we retrace our steps 2,000 feet, noting again the marks of civilization common to ours, the blacksmith shop, the rats and the livery stable where nothing but mules are kept. Contrary to Virgil's experience we find the journey back easier than the going. Passing the shaft we enter another slope where the machinery of the mine is stationed. . . . This is a hotter way. . . . At the shaft, in the 'cage' up we go, and see daylight once more" (*Birmingham Age-Herald* 1888).

Next is a visit to the furnaces of TCI, "the largest corporation in the south. It has paid up capital of $10,000,000. . . . The four furnaces, of which two are in blast, are the largest in America, having the capacity of 180 tons each dayly [*sic*]." The institute records 10 iron mines, 11 coal mines, 1,675 coke ovens, 20 blast furnaces, and 4 rolling mills in the Birmingham district. The accumulative language continues:

> As to the amount of iron ore and coal, that would be too great to calculate. . . . A recent scientific writer, in trying to compute the Alabama coal supply, after making the most liberal deduction, sums it up into a seam seventy miles long, sixty miles wide and ten feet deep, equal to

> 42,100,000,000 tons of coal, which, at a daily consumption of 10,000 tons, would last 11,500 years—quite long enough for the present generation to calculate investments upon. It is hoped your readers can now see why the American Institute of Mining Engineers came here, and why, in the next half century, several millions of people will come here. Why should not they come? With a coal area equal to three-fourths of England, and iron ore incalculable, Alabama has scarcely a million inhabitants. . . . The same methods that made England a manufacturing center, if pursued, will make Alabama the same. (*Birmingham Age-Herald* 1888)

Alongside the projections of epochal abundance of "iron ore incalculable," the author approvingly endorses convict lease labor as a kindness to the free Black population, suggesting that the carceral mine was "wise and humane and no doubt many are better cared for than when at liberty" (*Birmingham Age-Herald* 1888).[18] Which does not say much for liberty. "Why should they not come?" the author asks, when the geologic past is ordered in service to the futurity of the "selected" (to use Wynter's [2003] term).[19] The carceral system was an investment strategy in maintaining stratigraphic relations and disciplining the social strata through racialization. The demand for coal—"black diamonds"—increased for both domestic usage and industrial processes as the railways expanded.[20] While the coal mines and railroads also became an employment option for Black subjects during Reconstruction, new forms of subjugation emerged in the convict lease and indenture that made those gains ambiguous. The annual financial earnings for October 1884 through October 1885 recorded in *Biennial Report of the Inspectors of Convicts to the Governor* (Alabama 1886, 88–89) lists Alabama as the largest overall profit-making state in the United States both in total earnings and per convict at $88,902.40, with net earnings of $68,4555.94 for 550 convicts (= $1,245 profit per incarcerated person at $3.75 per month to lessees). Alabama was followed by Kansas, which was also using convict labor for coal mining at net earnings of $42,286.12 for 836 convicts (= $50.58 profit per incarcerated person); Texas, $20,196.04 for 2,539 convicts (= $7.95 profit per incarcerated person); and New York, $10,657.99 for 2,876 convicts (= $3.71 profit per incarcerated person). Thus, the mine set the precedent for profitable carceral labor. While often overlooked as a laboratory of modernity, the prison mine set epistemic and material conditions of racial undergrounds into place as a mode of capital accumulation, which disseminated far beyond the Alabama geologic "experiment."

Surfacing Whiteness

The surfacing of coal in Birmingham transformed it from scrubby cornfields into the Magic City founded in 1871. Sited on two rail lines, Birmingham had a burgeoning population of settlers in less than twenty years. Alongside this magic mineral conversion was the extensive regulation of Black spatialities through the deployment of the Black Codes and their criminalization of geographic freedom (alongside subtler forms of spatial governance). W. E. B. Du Bois, writing on the Black Codes, suggested:

> There was a plain and indisputable attempt on the part of the Southern states to make Negroes slaves in everything but name. . . . Negroes were no longer real estate. Yet, in the face of this, the Black Codes were deliberately designed to take advantage of every misfortune of the Negro. Negroes were liable to a slave trade under the guise of vagrancy and apprenticeship laws; to make the best labor contracts, Negroes must leave the old plantations and seek better terms; but if caught wandering in search of work, and thus unemployed and without a home, this was vagrancy, and the victim could be whipped and sold into slavery. (Du Bois 2013 [1935], 149)

Spatial modes of Black criminalization happened alongside the confiscation of land during Reconstruction, which sought reduction in the landownership of Black farmers after the Civil War.[21] The shift of classification from "property" to potential property owner made Black subjects and their geographic intent a threat to the hegemony of the white plateau. Property was a way of spacing out race while maintaining the racist legal space of sovereignty. The regulatory spatial patterning of the Black Codes happened alongside the geographies of internal mass migration north, which was forced by the geographies of terror—the quotidian in the convict lease labor and the spectacular in lynching. Both these violences were part of a retooled carceral geophysics aimed at securing the nation-state for white supremacy. Sometimes the two coincided in a fearful symmetry: the National Memorial for Peace and Justice in Montgomery, Alabama, records William Miller on a hanging Corten steel "body," a man who "was lynched in Brighton, Alabama, in 1908 for organizing local coal miners." The convict lease instigated a form of terror through enforced life-threatening labor and thus continued the alienation of the social form of labor as a category in Black experience in America (from plantation to mine), as well as divesting the term of labor from any useful coherence in the social experience of freedom in Black life

FIGURE 9.9 Corten steel memorial to confront the legacy of racial terror, National Memorial for Peace and Justice, Montgomery, Alabama. Photo by author, 2018.

(see King 2016a). The lease blocked the emergence of a mode of legitimated subjectivity, via the right to own and sell one's own labor (or subjectivity as based in the ability to exchange the "fruits" of one's production) through the policing of space and the right to geography.

The Geologic Hand of Whiteness

The consolidation of the underground as a prison mine further enacted the spatial execution of Blackness from the surface. This "improvement" of the nation via natural resources and the consolidation of the plateau cast the rockiness of Indigenous and Black life as an impediment to the expansion of that surface.[22] John Witherspoon DuBose, a cotton planter and Ku

Klux Klan and Protestant Episcopal Church member, locates Birmingham's birth in a prodigious geography, whose causality is an order of nature:

> Birmingham will become the workshop of the great territory bounded by the Ohio and Mississippi rivers, the Atlantic Ocean and the Gulf of Mexico. The industries developed by Coal and Iron are the richest in the world, and the localities in which those industries are maintained in greatest prosperity are fixed by order of nature. . . . "COAL IS POWER." (1886, 51)

Drawing on examples from the imperial centers of the United Kingdom—Nottingham, Leeds, Manchester, Birmingham, and Sheffield and Scotland—DuBose locates the founding of national power in geology. The futurity of coal was understood to both expand modernity and concentrate the population in urban centers, with coal mines providing the means for white social reproduction and transition to national superpower. As Britain had demonstrated how coal could be a narrative of nationhood and form the material basis of the expansionist geologic of empire, DuBose argued that natural resources were invested in the consolidation of empire and its academic and social institutions: "The use of coal inaugurated the industrial revolution which has gone hand in hand with the advance of all known science, and with all discoveries of new principles in society and government. Coal is the foundation of the FACTORY SYSTEM" (1886, 53). As a former plantation owner and enslaver (he inherited land and enslaved persons at Canebrake Plantation of Marengo County, Alabama), DuBose ascribes coal as both trajectory and originator of power; in a sleight of hand, he sees this minerality as the basis of the factory system rather than the plantation (he abandoned his plantation in 1884 to move to Birmingham to be involved in the local and national press). Informed no doubt by Lyell's articulation of coal and iron in relation to natural agency and uniformed development in his *Travels in North America* (1845), the crisscrossing of the causality determining social and geologic formations is assumed as contiguous.

Coal, like other "autonomous" materials (gold, sugar, indigo, rice, copper, etc.), is seemingly able to absorb the energy of the enslaved and incarcerated within its trajectory of geopower as "property" and is transmuted into properties, while maintaining the appended teleology toward empire through the "work" of the inhuman category of erasure. These epistemic alliances with modalities of matter are not projections over some inert matter but materialize across matter and subjects as a geopower of transmutation.

In tacit acknowledgment and subsequent erasure of the expenditure of enslaved labor in producing such a trajectory of power, DuBose boasts: "The great cotton plantations on all sides, and extending for hundreds of miles, will supply in the future as they have done in the past, at a days [*sic*] notice, a never failing complement of cheap and desirable labor. The unequalled mineral wealth of Alabama is the proposition; the immutable and incalculable growth and prosperity of Birmingham is the corollary. Each succeeding year of the world's growth widens the uses of iron" (1886, 73). As a former planter, DuBose felt an enthusiasm for mineral development that went hand in hand with his development work of armed intimidation for the Knights of the White Camellia and the White Shield (both were "secretive" political orders that enforced white supremacy during Reconstruction and opposed freedmen's rights and the "amalgamation" of races), and he was in extensive correspondence with government officials, newspapers, the national government, and the Convict Department of Alabama.[23] One such proclamation declared, "The Negro in His True Relations to the New South [Is] as a Laborer" (DuBose 1886, 107). Yet, such labor was not sovereign; rather it was "advanced," like natural resources, so the argument went, according to DuBose, by whiteness: "The negroes coming here from the cotton plantations, with no experience of the habits of town life, enter, as laborers, that field open by the white man, advanced to incredulous achievement by white labor, and whose chief factor is complicated and costly and elaborate machinery, every principle of whose invention and application, originated in the white man's brain for the white man's relief" (109).

Labor value is never considered *for* Black subjects; it is rendered a property *of* the Black subject. Thus, subjectivity is rendered in a semantic field of natural resources and the capitalization of geopower. Like the iron and coal, Black "labor" is proto-autonomous and contained by the authorship of white labor. As Clyde Woods argues, in the context of his compelling theorizing of "plantation bloc epistemology," "belief in regional superiority extended to belief in the supremacy of the planters themselves. . . . This mythical paternalism was in reality a production system organized around institutionalized starvation, discrimination, violence, fraud, debt and enforced dependency" (1998, 95).

The ruse of extraction is not just that the earth is only for those that work it in the origin story of racial capitalism, but that geology and Blackness exist "for the white man's relief" in the narrative of *geologic epistemologies*. Blackness in the convict lease is made subject to the forces of materiality but dislocated from the possibilities of the accumulation of this value and deprived of invention. Thereby, Blackness and whiteness in the United States are historically located in different material relations of geopower or *racial materialisms*. In a tautological form of reasoning, DuBose demonstrates how the white supremacy of matter is achieved by conjoining natural resources and Black labor in stratal forms: "Thus it is that the town of Birmingham owes its character for good order to the natural relations of the races, and their presence here in natural proportions of demand" (1886, 113).

Obscuring the racialized geosocial relations that make racial capital, the articulation of the Black subject as a fungible "body" (per King 2016a) of geopower is a more accurate mode of analysis than labor in the afterlives of slavery (to be discussed later). Racial capital relies on modes of production that capture energy and force (human and inhuman), and the fungibility of those forms is kept in a state of enforced porousness or intimacy by historicized structural forms denoted by race. As described by DuBose, Black labor slips away into an organic form of matter, "below the old slave standard of efficiency" (1886, 111), to carry coal from the seam, up the shaft, and into the white surface to deliver the appearance of a magic city. This is the geophysics of whiteness, to make cities rise from "nothing" through the dematerialization of the racialized subject.

Racialize the Volume

The subterranean imaginary of Blackness is proposed by the senator for the Thirteenth District, John T. Milner, an industrialist and former enslaver who helped found Birmingham, writing in a pamphlet titled *White Men of Alabama Stand Together*: "In fact, history shows that the African has produced nothing worthy of record except the results of his labor in a material way as a slave on the Continent of America under the guidance of the white man. . . . I will state that when our civil war ended the curtain fell on the highest type of African civilization the world had ever seen" (1890, 9). The guiding hand of whiteness directs and makes the nation so that Black work is recognizable only insofar as it is developed by white labor (and thus it follows, according to this geo-logic, that whiteness should accrue all the surplus value from this labor).[24] In this matrix of extraction, Blackness and matter await actualization and subsequent sublimation.

Blackness in contact with the Carboniferous in the carceral mine is transformed into an analogous decompositional layer—a disposable stratum of the underground, extracted and spent to enrich the overground of white supremacy. The aesthetics of such a transaction, in its creation of civic and national space (as urbanism, military and police defense infrastructures, high-rises representative of modernity's progress and phallic acumen), are reliant on the shiny surfaces of materiality that do not disclose the conditions through which they are materialized. The infrastructure of the underground is a racial medium that transforms time and space and is implicated in what materiality carries and covers over. Among concerns in geography and concomitant disciplines about the subsurface, volume, and depth, there is a need to *racialize the volume* in these accounts. What is the point of naming a new set of spatial conditions if the terms of apprehension stay the same and the *conditions* of undergrounds remain unexamined? The production and accumulation of industrial racial capital are dependent on this inhuman exchange between matter and race and the construction of concomitant spatialities. Which is to say that the *verticality of racialized materialism* is such that the symbolic and material exchange is one of *undergrounding* and *surfacing*, a racialized axis of pushing down to lift up. The sedimentation of forced Black labor into the subterranean as the means to spur the emergence of the New South is argued for by Milner, who states: "If let alone the whites can, and will, control, manage, and steer the unevenly freighter ship of State of Alabama proudly and grandly down the stream of Time. But if . . . the negro again should rule here, it would be better, as suggested by the great English historian, Froude, in reference to the English West Indies, if left in control of their negroes, that Alabama with her freighted cargo should sink beneath the waves and be lost forever" (1890, 1).

Milner imagines that Black self-determination/futurity will result in Alabama retuning to the sea, geologic time reversing into what might be the primordial mud of the Devonian. His imaginary of Alabama sinking beneath the waves brutally reinstates the geography of the Middle Passage and the permanent offshore-ness of those cast in a temporal lacuna without rights in the present to be in place.[25] With multiple references to the fate of England and its colonies in the Caribbean, Milner lauds the development of "mineral Alabama" as the means by which whites will steer the state down the "stream of Time," a stream of time that is "charged with the future of their people" (white futurism achieved via reproductive mineral and metallurgy). As a representative in the state legislature, he produced the pamphlet to "consider the political and material future of Alabama" (Milner 1890, 1). Milner starts with the usual valediction to the geophysics of the territory: "We have 52,000 square miles of territory, naturally the best adapted to the uses and purposes of man of any like territory on this continent, or in the world." He continues, stating that "in minerals our capacity to give employment to manufacturing people, the verdict of the world is now that Alabama has no equal. . . . Without any allegory or figures of speech, Alabama, in native elements that produce comfort, happiness and wealth, is unsurpassed anywhere on earth" (1–2). In an asymmetrical relation, minerality settles the white state for reproduction, while it incarcerates the Black subjects in the geologic strata and erases Indigenous history by negating claims to the subsurface. He continues:

> The race of white men who live here and practically own all of this grand territory, and who, like the Indian (as his tradition shows, for he has no history), who after roaming over this vast continent in search of a happy home, christened and named the territory we now occupy "Alabama," or "here we rest." "Here we rest" for our children and our children's children. "Here we rest" for our pure civilization. "Here we rest" for our civil institutions, and when the last trumpet has sounded would to God that the millions of souls that may be called up around the great white throne from Alabama, were of our own race and own people. But this can never be, for God has placed here two races of men about equal in numbers, one the highest type of his people, with a history reaching beyond the known periods of time, illustrious and grand through all ages. The other, without tradition, without a history, and without a monument or name, and looked upon, the world over, as the lowest type of man. (1–2)

Echoing the familial racial geo-logics of Agassiz, Milner places the indigene in the past as already succeeded in a seamless substitution ("eliminated," in Patrick Wolfe's [2006] terms), and their name and land expropriated for assimilation.[26] Milner's discussion of the Indigenous naming of "here we rest" is seemingly without irony as Indigenous peoples are "cleared" for white children's children. Blackness is locked in the "low" strata of the present. Both indigene and Blackness are abandoned in time (without History), in which temporal occlusions are a way to make spatial conquests and the narration and governance of time are reiterated as a colonial method for dispossession. Narratives of subjectivity via deep time materiality show how the time of race is posed as a spatial problem, a geophysical condition and the basis of mineral mobilization. Moving the focus from the body to infrastructural geology—understood as a *racial medium*—shows up how materiality is used to erase racial relations and relocate them in questions of the use and narration of matter wherein identity formation can be understood as having an inhuman dimension and resolution.

Milner pronounces his distress on the decline in productivity of the "negro" in terms of per capita production for the enslaver: "They have produced a little over one-third of the bushels and pounds since the emancipation that they did, per capita, as slaves. The free negro has done and is doing better here in a material point of view, than in any other country in the world where emancipated" (1890, 4), but not for themselves. Rather, Milner puts the reduction in "personal value from $792,000,000 to $201,855.841, or about one fourth what it was before the war" for the enslaver. He bemoans "material prostration and ruin—political imbecility, want and degradation" (6, 9), tying together moral attainment and material production. In a cautionary tale (to his white audience), he brings together florid examples from "Hyati," fetishizing voodoo and cannibalism, as well as the ruin of plantations and idleness of work in the British West Indies and Jamaica to illustrate his arguments about the fear of Black autonomy. While Alabama was the leading state in 1888 in the South for the value of its manufactured and mining products (*Weekly Age*, July 4, 1888), the psychic fear of Blackness was used to police the generation of economic benefit of geology for whiteness.

The Libidinal and Geophysical

Accounts of white supremacy are developed through racial ordering and natural resources as concomitant and collaboratory modes of classification. The ordering is not just about the free use of Black energy for the gains of white supremacy but about using the capitalization of geopower to isolate Blackness—racially and sexually—from whiteness. That is also to disremember the histories of sexual and bodily violence under slavery, while rearticulating them as new racial structures through the tying of "the manhood of the South" to the development of minerality (Milner 1890, 48), whereby the phallic signifier is realized through the geophysics of extraction. As Hartman (1997) articulates in terms of the afterlives of slavery, the psychic orders of anti-Blackness cannot be disarticulated from the economic or rational arguments. To this political economy, materiality is how the economic is achieved and the philosophy by which Reason secures its ontological ground. In the conjoining of the libidinal and geophysical, I use geophysics to signal forms of control over resources and forces that are leveled at bodies—to create both gravity and its variants (as Black subterranean or white surface). That is to create force fields around certain bodies and to organize the subjection of fields of force around others. Understanding the libidinal investments in geophysics is a way to see how value is made to gravitate toward whiteness in the precocious exchange of natural resources and nonvalue in subterranean stratal demotion.

Appeals to a symbolically "pure" genealogy in Milner's text are paralleled with material forms of purity, even as these appeals can exist only in the symbolic given the institutionalization of rape during slavery (and the constant threat of race's dematerialization as a nonscientific and paranoid irrational form). In other words, the economic forms of valuation of minerals are tethered to an attempt to realize and maintain psychic regimes of white patriarchal power, thereby structuring narratives of both territory and subjective domains.[27] After emancipation, when "in theory" Black ownership can be taken over the right to labor, sexuality becomes weaponized in a new formation to further differentiate and make the Black subject porous to violent politicking and moral retribution in ways that redistribute the ownership of land and labor. Convict lease labor is one strategy of this carcerality among many that have historically situated the Black subject as a particular kind of person-as-body—situated by containment and excess as a prelude to other forms of extraction.

The policing of the personhood as body and its perceived drives is an intermediate effect and affect of the consolidation of geopower, relocating identity politics into a geologic sphere of relation and subjugation. Sexuality is an instrument of convergence in the axis of control of racialized bodies, when the right to labor (albeit partially blocked but symbolically available) is proffered.[28] As Hazel Carby comments about Ida B. Wells's theorizing, in the parallel context of lynching, "Wells knew that emancipation meant that white men lost their vested interests in the body of the Negro and that lynching and the rape of black women were attempts to regain control" (1985, 269). The political campaigner Mary Terrell "addressed the issue of alleged low moral standards among black women head-on. White women who made these charges, Terrell said, need only look to their own husbands to find the truly culpable and immoral party" (Curtin 2000, 180). Building on the sexualized practices of disfigurement (practiced by geologists like Cuvier, Buffon, Lyell, and Agassiz), the deformation of Black persons through objectification as bodies was a continued means by which white patriarchal power sought to further obscure the path toward freedom. By breaking the relation between Black labor and spatial freedom, within the axis of excess and lack, material production is coded as white and patriarchal.

Inhuman(e) Fungibility

Slavery and the invention of race produced in the Black body a specific fungibility to the forces of inhuman materials and inhumane conditions in the transformation of the earth for the service of white supremacy (the racialization of geoforces both preceded slavery and exceeded the enslaved subject but became ontologically and materially concentrated as a paradigm during the period of enslavement, whereby Blackness became the referent). Capitalization of geopower from the earth occurs on a racializing axis, not just the division of race as the realization of a capacity to capitalize on geopower (to accrue value from extraction), but race, as it was practiced, assigned different capacities of geopowers to bodies; in other words, geopower is remade as an originary racial property. Originality was the invention of exclusive priority. The authorship and scripting of origins are how both Blackness and geopower are captured by the white settler state (as human origins and earth stories). Blackness is understood from the onset by colonizers as a form of geopower—an energy resource for making property from the earth and for powering the expansion of those subterranean realms for surfacing value. Minerals cannot escape flesh, and, per Spillers, Black flesh is a site of continued rearrangement in the afterlives of slavery (2003, 206; see also chap. 1, n. 5). Through the anticipation, acquisition, categorization, and vertical geographies of subterranean-surface relation, minerality is marked by its corporeality, by who carries value to the surface. The exchanges are many, but each is marked as a form of valuation (specifically in relation to materiality, sexuality, and bodies); an accumulation or dispossession; a pushing into the ground or a surfacing with that earth as a propertied form; and a cool acquisition of minerals or an intimate exposure to them. The praxis of extraction and the narrative of development of resources become a way of institutionalizing raciality *in* natural resources, as naturalized and therefore a normative materialism.

The codification of race that is given by paleontologists is what allows natural resources to surface as the exclusive product of white labor, while Black "labor" is assimilated into those inhuman materials. It is not so much a "eugenics of the rocks" as a division of "states" of materiality, wherein whiteness remains agentic, and Blackness remains fungible to the earth. The *states of geology create estates of being*. Vertical orders of race in relation to surface and depth are materialized in the carceral mine, in terms of proximities to minerality. Both axes—vertical and material—produce the territory of whiteness. What is crucial in thinking race with the material geophysics of the underground as opposed to epistemic hierarchies of race is to think with these changes of *state* (in the confluence of nation-state, states of being, and geophysical states). The pressure and the weight of gravity change the farther and deeper into the earth a body goes. Thus, the segregation of race is a spatial and material state with an affectual architecture of geophysical properties.

Channeling the geophysical imaginary, Milner approvingly quotes Robert Jemison Jr., delegate from Tuscaloosa, on the atmospheres of Blackness to consolidate his argument about white supremacy: "'The cloud is too threatening, too thick, and too portentous. It already covers two hemispheres, and will gather volume as it goes on, and will not stop until the negro has not only been made the political and social equal, but until he has reached the sacred precincts of the bed chambers of the white man'" (Milner 1890, 47). To which Milner adds his own mineral presentation of race: "We find occupying the soil of Alabama now, two distinct races of men, nearly equal in numbers" (60). He concludes with the threat of race war: "In Alabama the struggle for supremacy will be long, and the issue will be doubtful, as the immense concentration of white people in our mineral region will render difficult the negroizing of Alabama" (76). The slippages between metaphor and materiality in the specter of racial violence transport the mineral region to the bedchambers, petitioning on a symbolic imaginary that envisions minerals in relation to sexual ordering, as coconstituting regimes of power. The text makes clear that it is never about capital alone; capital is just one expression of the desire that is at stake in the determination of subjectivity through minerality. Liminality is copresent across both material and epistemic orders in the lithic imagination.

Benjamin Waterhouse (1786–1836), a chemist and cofounder of Harvard Medical School, presided over the first mineral collection in 1794.[29] Later he presented a lecture at Harvard titled "The Contents of the Earth" (Waterhouse 1805, n.p.).[30] In his public lecture he used a progressive causality to compare the discovery and production of metals to the making of civilization. The will of God, Providence, and, more important, the earth deliver natural resources to God's anointed—Man. Waterhouse suggests that by "digging into those subterraneous magazines, where immense riches are deposited for his use . . . we may observe here, that the <u>history of iron</u> is the history of <u>human civilization</u>. Those nations, or tribes who cannot obtain and refine iron from the ore are in a state of barbarism. But those who can smelt iron and form it into instruments had emerged from that state and risen to civilization" (Waterhouse 1805, n.p.).[31] Thus, minerals not only exact the divide between civilization and barbarism (the division between those that get to occupy the category of human and those deemed inhuman) but also materialize religious ordinance (Wynter's redemptive telos).[32] It is not simply that nature is transformed by capital and labor power, and the cheapest labor power becomes carceral to generate surplus value. Rather, the processes of classification and containment evident in mineral extraction extend a racialized *extractive impulse* that is psychic, sexual, and gendered.

After emancipation and its changes to the political status of sovereignty, the ontology of minerality continued as the means by which white supremacy and the settler colonial state continued to be made, as a normative arrangement of the racial allocation of geopower (as it was realized under slavery). Weaponizing geology was a eugenic tool sharpened against Native life long before the shaft was sunk into the ground. The social depravation and enforced starvation of Native peoples in Alabama (contemporaneously named Cherokee, Chickasaw, Choctaw, Creek, Alabama-Coushatta, and Yuchi) were mobilized through the cutting of geologic sustainability in modes of existence (food sources, place-making activities, and geographic knowledge of the terrain) and using the environmental impact of displacement as a tool of depletion, exposure, and hunger (on the "Trail of Tears") in which geography is weaponized to do settler colonial work. As Native grammars are symbolically buried in the inhuman as land (for appropriation and development) or incorporated to anticipate white arrival, Blackness is augmented through the matter economy as a material utility. Both geo-logics are forms of dissubstantiation that negate sovereignty in the epistemes of time and materialism. Codification of the earth into natural resources simultaneously substantiates the material and symbolic resources of settler subjectivity and scripts the temporal power of whiteness. Thus, natural resources are not just a mode of life that made an Anthropocene or colonial earth but a gravity for the continued uplift of white supremacy.

Prison Mine as Epistemological Refrain

Staging the boom, Alabama newspapers speak of the wonders of progress in the Magic City, without mention of the generations that parsed the transitional zone—the Black subjective stratum that bridged Southern (and national) industrialization.[33] *Elemental origination* provided another ground for the futurity of settler colonialism. The stretched fungibility of the underground, hauling, blasting, and being broken by rocks, ushered in a new possibility for white industrial pride, one that buffered the relevance of the South in the present amid a changing political and economic tide.

In a comparative study William Worger looks at the concomitant rise of prisons and industrialization in two cities deemed to be emblematic of a new economic moment of settler modernity: Cape Colony in Kimberley, South Africa (City of Diamonds), and Birmingham, Alabama (Magic City). Both founded in 1871, of rural agricultural origins, the cities were boosted by the monopoly of corporations and narratives of abundance: TCI (first Pratt Mines and then later U.S. Steel) and Sloss Iron and Steel in Birmingham, and Cecil Rhodes's De Beers Consolidated Mines in South Africa. Worger argues: "One of the most significant expressions of state power in the US South and South Africa in the late nineteenth and early twentieth centuries was the creation of coercive labor practices that were applied to African Americans and to Africans almost without exception" (2004, 63).

While geographic differences exist, "each society introduced vagrancy laws and other forms of juridical discipline to constrain the mobility and limit the price of labor, and they both applied these laws . . . to men judged on the basis of their race to be unskilled and criminal" (Worger 2004, 63). As Matthew Mancini's (1996, 104) analysis of political economy demonstrates, convict lease labor shaped modern (racial) capitalism, in which prisons functioned as a key stream-stratum of revenue to the state and facilitated the enrichment of national industrial corporations. Vagrancy and the despatialization of Blackness as a mobile force, narrated as *without roots*, was used to displace claims of sovereignty to land and labor. Worger lists three points of comparison between the American South and South Africa.[34] First, capitalists established new industries that were constrained by high capital expenditures and required machinery, investment, and resources from elsewhere, thereby resulting in higher fixed costs, whereby convict labor was seen as essential to the introduction of industrialized machine production. Second, the state actively supported the lease of its prisoners to private companies for profit and for the "discipline" and "civilizing" effect on African Americans and Africans. Third, convict labor was used to divide, cheapen, and defeat organized free Black labor and enabled segregation based on race (Worger 2004, 85). While the enduring structures of racism differ across continents, it is prescient to see how forms of enclosure travel in colonial geo-logic relations to mobilize colonial earth. As Tina Campt, working with photographs of prisoners from the Cape, comments, "The prison became a major source of labor, as an institution that transformed the indigenous populations of Khoi, San, and African natives into cheap, available manual laborers" (2017, 97). The mirroring of Black labor policies that deprived citizens of their rights through the carceral system in the development of South African mining demonstrates how the racialized relations of the mine and its affective infrastructures were reproductive and migratory. It is these migratory earth practices that were exported to settler colonial states that made colonial earth as a planetary paradigm rather than a national geoforce.

The carceral mine, like the plantation and colonial mines before it, changed the economic, biological, and chemical weather: a geologic act made in the underground that reconstituted white economics. The mine contributed further to the socioecological transformation of the South from plantation monocultures and slavery into an industrial zone of fossil fuels and carceral labor in the ongoing expansion of settler society. The state became the organizer of spatial relations, subcontracting permissible homicide to corporations, supplanting the enslaver in the prison-industrial complex. Both histories have immediate and lasting effects on Black forms of life and geographic expression, effects that remain salient today and condition the "relationships that Black bodies have to plants, objects, and non-human life forms" (King 2016a, 1023). While the coal industry slowly shrunk by the 1930s in the US South, the aura of fossil fuels enrichment of the Magic City paved the way for the structural development of Southern petrochemical corridors that now line the Gulf Coast and Mississippi River and heavily pollute African American communities.[35] Categories of matter and their racial materialization became a means to reproduce spatial claims through enduring structures of value that condition the way in which natural resources bring certain subjects into view as fungible. The New South and industrialization were ultimately built on the same enclosure and extraction of Black life that the Old South had employed, but with a modality in place that sutured the juridical application of the convict lease system onto a revised approach to who got to function in the category of labor and in proximity to the violence of the earth. The prison mine names the moment of collision when the violence of the state and violence of whiteness as a geographic and social surface become indistinguishable. The state of matter is also a world-state, a state of the world, in which racial gravity is implicated and enacted. This is the colonial aesthetic of geology, of undergrounds arranged in service to the enrichment of white surfaces, of the fungibility of Blackness as an energetic geopower arranged in the consolidation of white settler power.[36] This arrangement of power differs from the white miner in America, South Africa, Australia, Canada, and Britain. Labor as a terminology and practice is sunk in different grounds that hold the intersecting modes of policing and enclosure to marshal Black sociality and sexuality in the convict lease system as a differentiated spatial form. King argues that "Black fungibility—rather than Black labor—represents the unfettered use of Black bodies for self actualization of the human and for the attendant humanist project of the production and expansion of space. As a project of human and geographical possibility, the invention of Blackness (material and symbolic bodies) in the New World has certainly enabled the human to self-actualize as an expression of unfettered spatial expansion and human potential" (2019, 24).

Black Codes criminalized expressions of geography, demonstrating how material substantiation was achieved through processes of fossilization (physical, political, and psyche) that organized the entire geophysics of being as an environmental condition. King's cogent argument about Black fungibility and the expansion of space (see also M. Wright 2015) is part of a broader question about the densities, velocities, and sedimentation of gravity that materialism holds witness to: what might be called *actualization by racial extraction*.

10 Inhuman Matters III

Stealing Suns

> Slave went free; stood a brief moment in the sun; then moved back again toward slavery.
> —W. E. B. DU BOIS, *Black Reconstruction in America*

> We are jest like old broke down oxens . . . never being able to see the Gloryous son [*sic*] . . . which God made for man to see.
> —Incarcerated person quoted in Mary Ellen Curtin, *Black Prisoners and Their World*

A sentence of hard labor in the mines was a ticket of alienation from the sun. This solar extraction is petitioned by an anonymous letter written by prisoners at the Shaft mine in 1887 to the Board of Inspectors of Convicts. As voiced by the incarcerated, the removal of the sun extracted a cosmic relation, extending a new form of captive positioning.[1] In a black, hardening gloom without the languid stretch of dusk and day, prisoners had to make sensible other orientations that could stand in for that solar absence. As Benjamin Waterhouse lectured on the earth at Harvard Medical School in 1805, that "Reason may be compared to the light of the sun, which enables us to see everything upon Earth," the endurance of solar apocalypses, of not knowing when your sun will come up, hailed different geophysics of being into existence in Alabama. Daily the death of the sun organized the management and regulation of geophysics of Black life. To take away the sun—in the prison mine—required the incarcerated to live by and invent other suns, to envision being as "a drop of sun under the earth" (Fanon 1986 [1952], 85). Traces of those other suns in the institutional archives are limited, save for the simple insistence that they were there because they would have to have been, for the coordinates of an abolitionist imagination that followed.

In thinking with the temporalities of slavery, Hartman speculates, "An imagined place might be better than no home at all, an imagined place might afford you a vision of freedom, an imagined place might provide an alternative to defeat, an imagined place might save your life" (2007, 97). The sun is an inhuman relation that gives without receiving (in Bataille's formulation). Thus, solar excess is not predicated on reciprocity and is not subject to predation or humanistic incorporation into origins and filiation, despite the scripting of elemental earth origination (to cover over the racial deficit of extraction). The solar bind of asymmetry gives alterity to the possibility of more generous worlds. As a redress to the imagination of a Marxist top-down political geography, Wynter (n.d.) comments, "Needs produce powers as much as powers produce needs." Without the sun, other fires would have to be lit. This geo-forcing of the imagination might be understood as a libidinal form of geopower, hammered into shape, below the surface of whiteness. Dreams of other suns never make it onto the accession ledger, they do not constitute the official archive, and yet "people of the *liminal*," as Marriott (2018, 287) calls them, must reinvent the conditions of their being precisely because their position is constituted as an excluded space (the racialized system that disinherited those liminal subjects to build their plateaus). Hear Wilson Harris: "I shared an architecture in the Earth in which secret pebbled doors and walls were alive and opening voluntarily/involuntarily. Without such a sensation of secret openings coming to life, in odd pieces of ruined walls, one was doomed to endless imprisonment" (Harris 2017, 22).

The prison mine, as a liminal spatiality of lithic earth, cuts across undergrounds: racial and geological. The subject position of the liminal occupies a different spatial zone of experience and culture that is both within and spatially alienated from the categories of racial capitalism. This liminal geophysical zone is the rift in the spatiality of white settler colonialism. As Marriott comments, it is "the liminal who remain precisely useless as such: their narrative disrupts capital's racial hegemony and its reliance on time-as-exchange" (2018, 287). As the incarcerated were forced to inhabit an inhuman sovereignty, this mode of being was an alienation from the surface but also created resistances insomuch as the enforced intimacy of an aberration of matter defers the exchangeability of materiality as capital. This matters in the refashioning of subjects' own exchange with the conditions of their collective being and for questions of materiality and extraction. The refusal of alienation is not refused as an individuated question (see Fanon 1986 [1952]; Marriott 2018, 287). The value of the collective condition thought against its individuation (and the mapping of that individuation into the patriarchal nation-state through geology) changed the possibilities of existence, found other "solar keys" (Glissant 1997, 39).[2] The "solar keys" of material and subjective liminality are these passages between the white overgrounds and Black enclosures in the mine that are not constituted solely by the shaft. As incarcerated subjects had to theorize and make sensible other inhuman orientations—solar epistemologies—that could stand in for that absence, a space was wrought in the occupation of an otherwise. Thinking with Wynter's concept of the "senses as theoreticians" or Hartman (2019, 230) on theory-making—attending to a solar relation—relocates theories of the earth away from colonial models and its power to define lived experience to another kind of worlding that had to grapple with the deep antagonism of inclusion in the category of inhuman.[3] This is not a recovery project, but it entails a form of material witnessing to histories that have shaped the racialization of forms of life and the indebtedness of livability on the plateau to the mine.

The inhuman placement of the mine informs the applicability of liberal notions of the separation of bodies and forms of identity formation through modes of geologic life. Marriott argues the effect of this liminality is a presentness (extrapolated here as a form of rift-making) in the political, which fractures the hegemony of colonial relation: "The liminal, whose labor in the production of the politico-economic ends of man makes them incorporated and alienated at the same time, cannot afford to look forward on the basis of a mythical past, and therefore occupy the *present* in a way that breaks with the very value of history in the colony. As such, they open up the political dimension of invention in part, because, being an 'entirely different form of sovereignty,' they live a reality that is not mediated by the dominant imaginary" (2018, 287). The radical placement outside the sun and surface and the consequence of that dereliction require an understanding of the imposition of extreme violence (and its continued legacies in the geophysics of incarceration), alongside a recognition of how other modes of sensing and

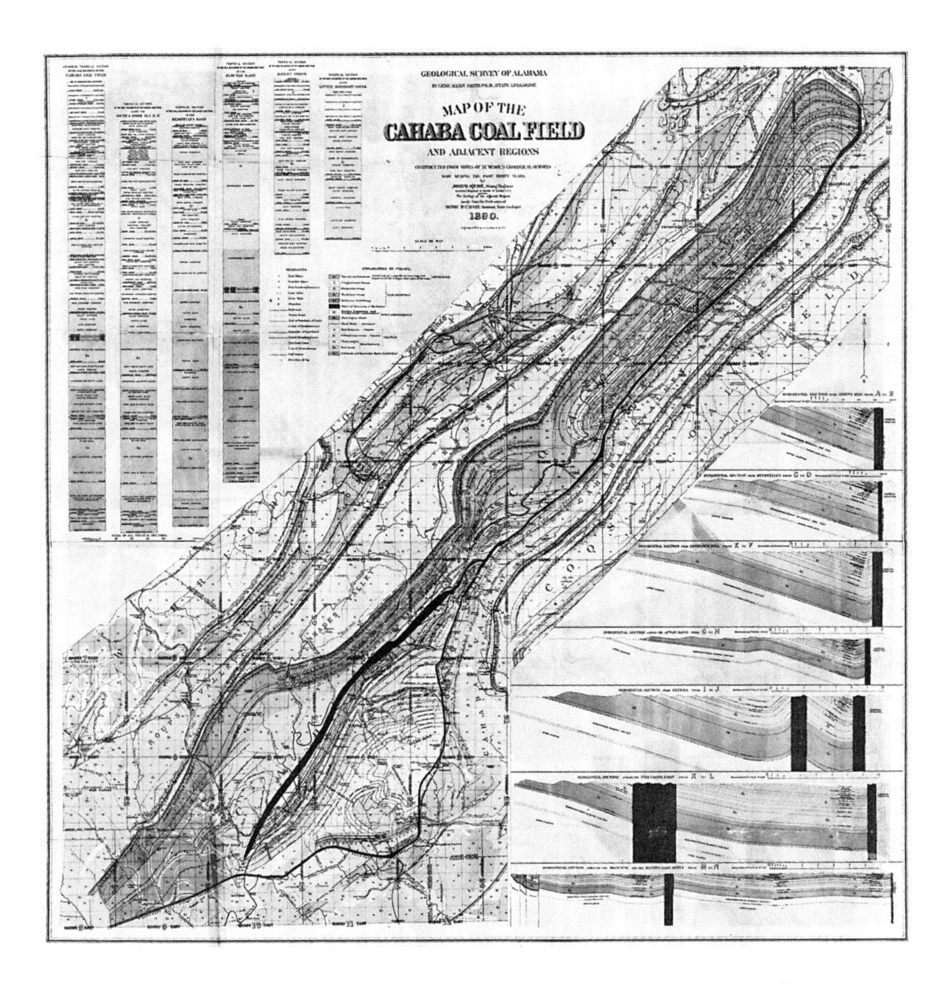

FIGURE 10.1 Map of the Cahaba coal field and adjacent regions, Bibb and Shelby Counties, Alabama, 1890. Lionel Pincus and Princess Firyal Map Division, New York Public Library Digital Collections. https://digitalcollections.nypl.org/items/6cd48a80-9722-0135-523e-436f181e71b7.

theorizing were practiced outside of and in contradiction to the normative modalities of materiality (where the "light" of Reason is a spatial and social distancing from the earth). Which is also to demand a redistribution of who gets to theorize and under what conditions in the "when and where" (M. Wright 2015) of life-forms, environments, and experience. The mockery of valuation that the prison mine enacts brings down the normative units of engagement, such as labor, work, and prisoner, even as it was narrativized through these terms.

As Clyde Woods suggests, "It has not occurred to the members of the neo-plantation school of social science that working-class Blacks have their own epistemology, their own theory of social change, and their own theories of class and ethnic depravity" (1998, 103). Incarcerated miners and their communities had to imagine differently, and with a largesse that defied those cramped spaces of internment in the earth, with both the racialized weight of deep time and a knowledge that near-surface environments are mutable in geologic time. It could be assumed that the earth, recognized as such (rather than as natural resource), provided other forms of connection and patience with the nonnormative—undergrounds that acted as a spatiotemporal disruption to rift the surface and the brutality of its conditions. Thinking against total enclosure is also a way to think with the rift as a term of the inhuman, as a way out of the dialectics of the race-matter category. A Relation with the inhuman can also be an escape hatch, a site of mineral drift or refusal. As with the escapees of January 1892 in Slope No. 2 who, utilizing their well-honed mining skills and refusing the one-way-in, one-way-out geography of the mine, dug their way through the old works of Slope No. 1 to make another exit strategy (Alabama 1888/90–1890/92, 46).

In the mine, light was found away from the free and living world, somewhere in the darkness of the dead matter of former geologic worlds. Those other suns must be something like Baby Suggs's idea of grace in Tony Morrison's *Beloved*: "She did not tell them to clean up their lives or to go and sin no more. She did not tell them they were the blessed of the earth, its inheriting meek or its glorybound pure. She told them that the only grace they could have was the grace they could imagine. That if they could not see it, they would not have it" (2007 [1987], 103–4).

To imagine the end of a prison mine became tied to the ability to imagine a sun beyond the cold, dark night and its deadly demotions into the subterranean earth. Lingering in that nonsolar world, in an oppositional configuration of Enlightenment values whose narrative arc of production, improvement, and progress configured its subjective domains through material designations, might be seen as a radical assault on substantiating freedom.

The prison was positioned at the entrance to the mines at Shaft and Slope No. 2.[4] Up before dawn, at 5:00 a.m., the incarcerated would eat and then follow the shaft down, often a mile underground, wearing the same greasy-coaled clothes every day(which led to both the increased risk of catching fire and health complications).[5] Incarcerated miners cut coal for twelve hours a day, in wet and poorly ventilated coal shafts, often barefoot

FIGURE 10.2 "The nation's first memorial dedicated to the legacy of enslaved Black people, people terrorized by lynching, African Americans humiliated by racial segregation and Jim Crow, and people of color burdened with contemporary presumptions of guilt and police violence": National Memorial for Peace and Justice, Montgomery, Alabama. Photo by author, 2018.

in ankle- to knee-deep water, and came up in the evening dark. In the underground, coal-encrusted, bruised, and sore bodies, breathing in the dank, explosive air, anticipating the sudden flammability and the quick lick of fire, in a geophysical state of permanent jeopardy: "killed by a rock"; "killed by falling coal"; "killed by falling timber"; "heart clot"; "killed by tram" at Coalburg, New Castle; "skull crushed by train"; "neck broken by falling slate"; "killed by tram car"; "not given"; "slate falling and breaking spine"; "injuries received in mines"; "haemorrhage from lungs"; "Blasts of rock producing internal injuries"; "Explosion of gunpowder in mine"; "killed while trying to escape"; "premature blast of rock"; "explosion of gunpowder in mine"; "chronic diarrhea"; "Blasts of rocks producing internal injuries"; "exhaustion"; "broken down constitution" (deaths in Pratt Mines in Alabama 1886/88, 49).

Returned to a prison house, of shared plank bunks that prisoners could not stand up in and where they often remained shackled in irons, miners were removed from the warmth of the sun and the surface of the earth. Placed underground for six out of seven days of their lives, in the near-complete darkness, they did not work on Sunday, the only day the mine was shut. If there was a song to be sung in this carceral spatiality, it might have been the Blind Willie Johnson gospel tune "Dark Was the Night, Cold Was the Ground" (recorded in 1927).[6] This popular Southern hymn, written by the English clergyman Thomas Haweis in 1792, was one that the incarcerated attending the newly formed Sunday service at Pratt Mines might have heard (although it is recorded in the Alabama reports that few people went to church, preferring instead to stay in the prison and resist the imposition of morality as part and parcel of the lie of a carceral civilizing project).[7] As if in recognition of this cosmic howl beyond the stretch of the dark night, Carl Sagan put a copy of Blind Willie Johnson's "Dark Was the Night" on the Voyager Golden Record *Sounds of Earth* that was on the Voyager spacecraft in 1977 and sent on a probe into the black sea of interplanetary space, as a calling card for future beings, with the inscription "To the makers of music—all worlds, all times." The gold-plated copper disk (mine and miners unknown) carried what Woods (1998) calls "blues epistemology" (made in the context of "plantation bloc epistemology") of geologic life into the far-flung solar night.

Physicians' reports on the convicts of Sloss Iron and Steel Company were often little more than apologetics for the brutality of the state apparatus. They suggested that the prevalence of "chronic diarrhae," dysentery, and fever, most commonly associated with unsanitary conditions of the prison and fetid water in the mines (access to fresh water was sparse, and often the only water available was collected in the mine where prisoners also had to relieve themselves), coupled with inadequate and uncooked food, could result from the "intemperance, fear and mental depression" of prisoners. Exhaustion was ascribed to a lack of work ethic, rather than to twelve-hour days mining ten tons of rock to extract four tons of coal. As F. P. Lewis, MD, suggests;

> The diseases which have been most prevalent here are simple continued fever, diarrhoea and anemic condition in which so many convicts are received. The cause of the first of these may be attributed, 1st, to the

> predisposing causes which are, intemperance, fear and mental depression the last two of which are put into effect when the convict is arrested and sentenced to the coal mines (which all new men appear to have a terror of), and 2d, to the exciting causes, which are over fatigue and errors in diet. As regards the first, they are entirely beyond our control, nor do I see that much can be done for the second, for convicts usually are not hard working individuals and consequently at first it takes but little to fatigue them. (Alabama, 1886/88, 239)

Lewis concludes that such high rates of sickness derive from "4th. The moral influence arising from their changed condition, from freedom to prison, with the deprivations of the privileges of the former and the environments of the latter. 5th. The change of climate, local and personal hygiene, etc." (Alabama, 1886/88, 190). Climate determinism, again, was the culprit for a state of being rather than being subjected to the geophysical states of Overseers' earth. The theft of a solar world compacted souls and terrorized the *terroir*.

The scarred lungs and weakened hearts from mining, inhumane conditions, and the choking dust of coal extractions scarred the miners with a duration that exceeded even the fabrication of sentencing, as the veins of coal and bodies became intimate. In a secret letter to the governor, working conditions at the Shaft were described as so bad that prisoners called it "murder in the first degree." As prisoners emphasized, the fear of deadly gas made everyday life terrifying: "We are in danger of loosing hour lives hear by gass. . . . Please don't let us be buried alive in Pratts mine . . . please have mercy on us as We are poor convicts" (quoted in Curtin 2000, 106). Miners threatened to strike, and attempted suicide and escape, but they also offered to work another year of their sentence if they were taken out of the mines.[8] The inspector noted a hurt deference in his response to the claim, dismissing the prisoners' concerns, while privately noting his own concern at the high death rate (twenty-one county and nineteen state convicts had died at the Shaft in the previous year). State and company physicians continued to insist that mining was healthy and thus could be classified as a progressive form of labor; they routinely dismissed any forms of disease associated with mining, particularly those of a respiratory and cardiac nature (which were well known from the mining experiment in the other Birmingham in England). For example, in the report of "R M Cunningham Phy. of P. C. & I. Co." and "Comer & McCurdy Min. Co." in the *First Biennial Report of the Inspectors of Convicts*, he suggests, "From the foregoing we can readily understand that working in the mines, generally has no effect upon the health of the convict, though there are some exceptions to this. Some suffer from the depressing effects of fear which they appear never to get over" (Alabama 1886, 259). Inhuman intimacy is scribed as depression rather than attributed to the proscribed geophysical world. A frequent refrain was the claim that illness was contracted because the incarcerated are "Negroes" and thus in poor physical and mental health. The argument was of "underlying conditions." Asthma ("pernicious asthma"), pneumonia ("Pueumonia"), and bronchitis and emphysema ("chronic pleurisy") were common. Coughing was the most frequent respiratory symptom. What was diagnosed as pneumonia, and the cause of death in 50 percent of cases in Pratt Mines, became known as the honeycombed pneumonia of black lung disease (a disease first described in the sixteenth century by a German mineralogist, Georgius Agricola, in coal miners). Coal workers' pneumoconiosis is caused by long-term exposure to coal dust. Shortness of breath, persistent cough, heart strain, chest pains, vascular damage, and vulnerability to tuberculosis can take up to ten years to develop. Besides the chronic injuries, there were specific forms of injurious impacts that affected miners daily, from the impacts of coal dust, the mining "itch," sores, muscle and abdominal distension, and the threat of poisonous gas. Carceral subjects were often returned to the outside world unable to physically support themselves and with long-term health complications from mining activities. There was a complex porosity of geologic life in the afterlife of the mine in Pratt City that continued long after release.

The internal indexing of the underground occupied breathing, circulation, and forms of cellular communication in the withdrawal of sunlight. Lack of sun exposure is widely accepted as the cause of low vitamin D, a nutrient that contributes to the maintenance of a calcified mammalian skeleton.[9] Vitamin D plays a dominant role in geologic living and in calcium and phosphorus metabolism, ensuring adequate levels of these minerals for metabolic functions and bone mineralization. We are solar creatures, too. Exposure of human skin to solar UVB radiation (wavelengths of 280–315 nanometers) leads to the conversion of 7-dehydrocholesterol to previtamin D3 in the skin. As a vitamin with hormone-like efficacy, it has numerous extraskeletal effects in a variety of nonmusculoskeletal diseases, including cancer genesis, hypertension, diabetes, and cardiovascular disease ("organic heart disease," "heart clott" [Alabama 1986/88, 97, 129–32, 247]); it is also involved in a host of other hormone regulations that affect depression, stroke, asthma, and regulation of autoimmunity. The effect of the lack of daylight on a body is total enclosure, both mental and physical. Thus, vitamin D is a crude cypher for what the sun animates and what incarcerated persons were made to endure. Since vitamin D is both a hormone and a regulator of inflammation and oxidative stress, the epigenetic legacy of deprivation directly impacted the aging process and DNA damage, which is to say that the prison mine continued to mine the afterlives of those that were stolen from the sun. Geotrauma both has lateral effects on the interdependence of lives and haunts the longitudinal flows of health (creeping along the edges of weighed-down senses of being). Its trajectory has a migrant temporality beyond the mine.

To live (and survive) without the sun performs the counterevolutionary act, organizing the material conditions of possibility for what the ideologies of racial hierarchies have previously described, of a form of life locked in the strata. To be removed from the sun is to have orientation, warmth, circadian rhythm, and sense of gravity (which is the blaze of perception that the sky enacts) taken away. If gravity is also something like the promise of another day and a grounding in a planetary orientation—Samuel Beckett's "the sun shone, having no alternative"—it is also the tether that holds the earth to the cosmos in exorbitant relation. That is, the sun gives the fullness of the earth as planetary, uncontained by the limits of a finite earth, or its modes of political oppression. Understanding incarceration as a nonsolar event draws attention to the ways in which physical deprivations are not just the erasure of bodily and kin relation but also the erasure of environmental, cosmic relations and how these conditions are lived. To remove the sun is to remove (inhuman) alterity, that which is outside of an oppressive relation and, as such, offers the imaginary and diversity of forces that give sense to freedom, tumbling outside of oppressive (human) strictures. I think here of Morrison's writing on survival in the afterlives of slavery. Paul D is talking to Sethe in *Beloved* on his experience in the chain-up in Georgia living in a trench in the ground: "Listening to the doves in Alfred, Georgia, and having neither the right nor the permission to enjoy it because in that place mist, doves, sunlight, copper dirt, moon—everything belonged to the men who had guns" (2007 [1987], 191). The development of a craft of survival was being able to love small but not thinly, as a passage of endurance to freedom:

> So you protected yourself and loved small. Picked the tiniest stars out of the sky to own; lay down with head twisted in order to see the loved one over the rim of the trench before you slept. Stole shy glances at her between the trees at chain-up. Grass blades, salamanders, spiders, woodpeckers, beetles, a kingdom of ants. Anything bigger wouldn't do. A woman, a child, a brother—a big love like that would split you open

> in Alfred, Georgia. He knew exactly what she meant: to get to a place where you could love anything you chose—not need permission for desire—well now, *that* was freedom. (191)

As the sun is withheld, so is the world that it brings to light, in all its wayward desires. As one goes deeper into the earth, into another stratum of time, the darkness of the Carboniferous and its baked tissues of plants that compressed in the earth are a geologic displacement in time that enacted and performed racialized temporalities.[10] The different geologic time zone of the mine, 360 million years before the present, set a different clock to post-emancipation Black time. The carceral mine can be seen as another kind of subterranean hell, kin of the slave ship, where the mine doubled as abyss, hold (Sharpe 2017), and a door of no return (see Alexander 2005; Brand 2001; Hartman 2016; Smallwood 2007). The mine reproduces in a different variant the physical geography of the transformation of person into the commodity form of the slave, now convict. The underground functioned in its slop of piss and shit and its depths of darkness as a spatiality that reenacted the conditions of fungibility of these geographies of enslavement. The imposition of the commodity form was realized via the spatial coordinates of the underground (in relation to the ground/horizon as the degree zero of value on the plateau). In the racial stratification of capital, to be underground was to be ungrounded as a legitimate subject in time and space.

Convict lease labor was both enclosure and destruction of a right to geography in its fullest sense as an ability to move freely in space, to be placed *as and with* freedom through experience, incarcerated *as and with* the minerality of the earth and its deep times. The systematization of soul compacting is alive and well in solitary confinement today, in the withdrawal of air-conditioning or heating in state penitentiaries, in the inundation of climate change effects, hurricanes, storms, and everywhere that the state refuses a livable outside to enact its violent forms of weathering (Sharpe 2017; see also Pellow 2021; Holt 2015). The pervasive use of darkness developed in the mine also extended in new forms of state punishment. Some undergrounds became fugitive spaces in the volume, others were utterly fungible undergrounds of no return. John Floyd, for example, received twenty-one lashes for failure to work in 1887. He hid in the mines for ten whole days, was recaptured and shackled and sent back to work, but he did not relent on his refusal. A former brickmaker, he was sentenced for life. This illegible mode of resistive action (to the inspectors) spoke to a capaciousness of escape, impossible without the solidarity of others, made as a form of holding with the underground, temporalizing it differently to open space to undo the script of the subterraneous. In response, the warden built a "dark cell" for him (Curtin 2000, 133) to increase the recursive modalities of subterranean pressure, to hammer the message of enclosure harder still—a dark box outside of time constructed to take the mine inward.

To be outside of the sun and its orientation is to be outside of sense and its configuration of a recognizable geography. Dark places induce sensory deprivation, collapsing the boundaries of the world into the soul. A mine in your head. Shafted all the way down. Flat-top prison, operated by Sloss Mines, had a long and low-topped shaft that went straight out from the prison walls into the mine. One cave to another. The penitentiary report puts it bluntly, "Since the convicts have been concentrated at Pratt Mines, Jan. 1888, there has been no escape from that place" (Alabama 1886/88, 8). The prison mine established one modality of carceral geophysics, the weight of sensory deprivation through solar death. The logic of conversion brought value and urbanism to the surface, levitating from the Carboniferous on an anti-Black gravity.

The Killing of Jere Ford

In another realm of the solar spectrum, a different kind of undergrounding, achieved through a kind of bureaucratic exorcism, exists and can be witnessed in the beating and killing of Jere Ford in 1886.[11] In this other solar apocalypse, his death was attributed to sunstroke. Mindful of Hartman's rebuke at the beginning of *Scenes of Subjection* and Black feminist work that questions the desire at stake in the reproduction of Black pain as a form of public consumption, I approach the violence of Jere's death mindful of the reproduction of the scene of subjection in the present. I narrate it here because of that present. I want to chart the geophysics of sense that makes such accounts possible within state reports (and continues to do so in state and environmental violence), which is also to attempt to address the grammars of the normative as they are articulated through understandings of labor, debt, regimes of suffering, and discourses of abjection in the practice of undergrounding carceral Black subjects. The account of Jere stands in for many minor accounts that never "make it" into the reports. But this is the point. Jere is here, in the archive, in this account, not because he is dead but because of the spectacular nature of his inhuman-inhumane death. The spectacular in the scene of subjection raises the question of the place of the quotidian to which it provides the limit-experience, while it is actualizing a distractive assembly of accountability when there is none. My focus here is to show the reciprocal agentic work of the inhuman and inhumane in the psychosis of materiality, which positions Black forms of life in a deadly collaboration with minerals and metals.

Jere was classified as a short-term prisoner, sentenced to thirty days for assault and a further eight months for costs accrued for the price of his conviction. He was sent to clay and fire the bricks that built the industrial towns of the New South. A too extensive report of his "accidental" death (murder) is made. In its account of violence, the report evidences the contortions of trickery that witnessing is used to legitimate, while it dissimulates the actuality of violence through recourse to transparency. All parties to the event are called to testify (except Jere, who is dead and gives testimony as an example only of his own undoing). The effect is a multiperspectival scene of subjection, yet (and this "yet" in the narrative is important to the outcome) violence is affectively regulated through this seeming exposure to the facts, its very daylighting as a shaft of Enlightenment reason. The account testifies to "the gratuity (as opposed to contingency) of violence that accrues to the blackened position" (Wilderson 2003, 20)—a gratuity that the exorbitant relation of the sun alone will be made responsible for, but which has a much more prosaic function in the histories of obscuring the extractive technologies of the earth.

In the torture transcript of Jere's murder, we witness the "serial bureaucratic grammar of the archive" and its functioning in the "state practices of social regulation" (Campt 2017, 90), exposure as erasure. How can we understand the excessive denial of a report that says one thing and concludes another, when it lays out the violence in such a graphic and unflinching manner, yet somehow evades its jurisdiction? Wherein the "transparency" of violence is spoken to nullify it? What causal geologic in the chains of signification allows a medical officer to conclude that the cause of death is sunstroke when a man has been repeatedly beaten? How do such reports perform "justice" as simply a reproduction of racist violence? A question that is still alive today is how the "acceptable" racial praxis of "environmental" violence is secured through practices of administration that manage incidents through "due process" without ever opening those events to any possibility of justice. The point I want to examine in this discussion is not simply that justice is complicit with a violent colonial matter economy in the materialization of race, but that between materialisms, the glare of violence is used as a mask, gaslighting the subjects and the sun.

The inspectors' report on Jere records, "He had been in jail three months before his conviction. Sent to brick yard to roll clay. Was sick. Was sent back to the pit and struck with a plank several licks across his body. Couldn't work" (Alabama 1886, 330). The facts are relayed as such:

> Mr. Robinson sent to a harness-maker's shop and procured a piece of leather twenty-six inches long, two inches wide and three-sixteenths of an inch thick, which was fastened upon a wooden handle about fifteen inches long and half an inch thick. He then tied Jere's arms around a tree and fastened his legs to the tree with a rope and struck him thirty or forty blows upon his naked persons of which drew blood. Jere was then released and went back into the pit where he work [*sic*] for a little while. About 4 o'clock he was stripped and bucked and about the same number of blows were inflicted by Mr Robinson. A short time afterward he was again whipped, and from the evidence before us, although there is some conflict upon this point, we think he received 100 lashes in all during the afternoon of that day. Between 5 and 6 o'clock Jere came out of the pit, and, as described by Mr. Robinson "walked like a horse with the blind staggers. . . ." He died between sundown and dark. The nature of the trouble I apprehended when I told Robinson to hold him up, was congestion of the brain. He was not rational. (331)

After the intense description of the Jere's torture, the report goes on to note,

> It will be recollected that the 30th of June was an exceedingly hot day, and Jere appears to have suffered intensely from heat from the early morning. . . . He also appears to have been dull and stupid during the afternoon and a part of the time certainly he was not rational, and it is certainly most singular that neither Mr. Robinson nor any of the other persons who saw him that day, and there were quite a number, should have detected that he was sick. . . . The physicians appear to think that Jere Ford died of heat, apoplexy or sunstroke. (332)

The inspectors state that Mr. Robinson had failed to read the laws that he had been furnished with by the state that it is unlawful to whip a convict more than fifteen lashes, or naked, or more than once in one day. The report states in Mr. Robinson's defense that apart from this anomaly, "it is only just to Mr. Robinson to state that, except in this instance, his treatment of convicts has been kind, and this strap used in this case, is the first thing of the kind." Mr. Robinson is backed up by the testimony of Joseph W. Winston, who stated, "I am Superintendent for Mr. Robinson of his convicts. . . . I thought this convict was a very stubborn, sulky kind of darkey; it was that or he did not have good sense" (Alabama 1886, 338). The testimony of

C. N. Robinson of his actions was as follows: "I struck Jere Ford, some time in the forenoon, with a short piece of sealing plank about two or three feet long and four inches wide. Struck him with flat side on his rump about two or three times. It hurt my hand and I dropped it, as it was hurting me more than the convict. Didn't hit. Him over the head with a plank" (345–46).

The evidential record of due process, in its minute details of brutality, is ameliorated by a range of social technologies, of physicians, witnesses, accounts, and thorough bookkeeping. The accumulation of stories and the visualization of a procedural account in the bureaucracy of legitimating violence are enacted to screen the evident violence on show. The multiple voices that speak to the violence of Jere's undoing are another ruse of white justice in the regulatory power of the carceral condition. Yet, the witnesses give credible accounts; they are not stooges in the traditional sense, ventriloquizing a script for the state, but they perform the work of an assembly that is gathered to disperse "legitimately" what an all too broken body testifies to. The action is to sublimate the subject into relations of force and to disperse accountability. The integral relations established between paternalism, punishment, and white patriarchy come together to dissimulate the crime in the "Death of Jere Ford" (Alabama 1886, 328).[12] However, there is something that is not transparent in how this total scene of violence is stitched together to disavow what the scene of reality is able to accommodate, through the conclusion of sunstroke (a conclusion that is at utter odds with the "witnessing" that is recounted even by the perpetrator of that violence).

Understanding the role of inhuman matter in its intimacy with the inhumane is important. The purposed wooden paddle, made to specification, so as not, we are told (by his own admission), to hurt the hand of the overseer. The instrument is one fashioned for the pleasure of the user, for the comfort of beating, for its jouissance. The one being beaten is registered as resistant matter only insomuch as to acknowledge how the imagination of the beating injured the inflictor, rather than the inflicted. Which is to say, without fuss, that the one being beaten is imagined as unable to feel or as brute matter. Only the pain of an improperly purposed instrument was worthy of comment and redress, which is another way of saying that the inhuman object (the paddle) was more recognized as having agency than the incarcerated subject. Jere Ford is simply the body on which the example of a good or bad paddle is played out (and a morally "good" or "bad" inspectors' report). The inhuman implement is the protagonist, as it were.

The inhuman properties of the paddle outdo the personhood of the prisoner rendered as inhuman. One materiality is available to being crafted, the other is rendered as matter who is coercively deformed through extreme violence. The relation to working materiality in this scene of subjection is the point of the extensive pages given to its unfolding (in the report and here). Thus, to fashion an elemental subjectivity of Blackness as another kind of natural resource, Jere becomes subject to the violence of the elemental, as contiguous and continuous. Origination is transferred so that only productive fashioning is ascribed to the narrative of material redemption. While the intimacy of the inhuman and inhumane is forced into the material substantiation of the production of the plateau, the origin story is what secures the tether of agency and its erasure. In the clay pit, Jere is *as* earth, subject to the firing of the sun.

While the beatings are savagely described over multiple pages of the report, they seemingly do not stick, are not made sticky as an event that binds the inflictor, but rather to the inflicted, and to the weather. Jere's condition is attributed to an unfortunate atmospheric event, a cosmic exuberance, the heat, lack of water, and everything but the beatings he received. Despite the procurement of an instrument specifically designed to beat someone to a pulp, the medical professionals defer to atmospheric forces. *Climate was the culprit*. The mode of the report itself, rather than what it reports, is prioritized to obscure the very thing it purports to show in a brutal exercise of the legitimizing power through new and old instruments of control. The inhuman is a constant presence, as witness, resolve, and purpose. It is the ruse of redress. Equally evidenced in the witness statements are the blunt refusals of Jere to show deference to that power and to be made submissive by it.[13] Becoming "useless" defied the narrative of utility and thus asserted agency of another form of living. Unable to represent or accept the challenge that the negation of another scripting of Black life beyond extraction poses, the scene is already obscene. The point of "regulating" spectacular violence is to make a structural adjustment to the vicissitudes of violence only to examine the most extreme, while legitimating the rest as unseen ("nothing to see here"—point and erase). It is evident that such negation, variously named laziness, sluggishness, stupidity, and insubordination, was the most extreme form of resistance to this state-sanctioned apparatus.[14] This transgression is named repeatedly through epithets of sullen refusal, yet the breaking of Jere Ford is "violence as *pleasure* without purpose" (Wilderson 2020, 91). The plank that is specially fashioned so as not to cause distress to the hand of the beater, the care in the organization of excess, suggests not only a practiced arrangement of contingent violence but also that excess does not exist in the libidinal organization of anti-Blackness. Obscenity is the point, to suggest the singular, extraordinary violence that is mobilized against the refusal to work and, in doing so, to suggest that such refusals were extraordinary, such that violent compliance in the category of labor signaled the slow and everyday "acceptable" landscape of violence. The excess of inhuman violence functions as in Kant's sublime to marshal white subject formation through the rush of the abyss, which in turn keeps the tensions between the category formation of the inhuman-inhumane in play as an active racialized engulfment.

The inspectors' reports were introduced to engage with and demonstrate paternalism and reform.[15] As Hartman comments, these attempts at "benevolent correctives" often "intensified the brutal exercise of power upon the captive body rather than ameliorating" it. She suggests that rather "than declare paternalism an ideology, understood in the orthodox sense as a false and distorted representation of social relations, I am concerned with the savage encroachments of power that take place through notions of reform, consent, and protection" (Hartman 1997, 5). The scene of subjection is distorted through kaleidoscopic mirrors; the bouncing of violence within this prism exposes everything in sickening detail yet conceals its proper attribution, so that the conclusion of death is an inhuman solar force. This elaborate maneuver is Enlightenment's shade, the quotidian and structural improvisations that constitute its ongoing materializing of Black subjects for targeted subjugation and undergrounding. The point is that narratives of materialism, and its structural forms of agency and modes of attribution, cover over the complex extraction of geopower and the forms of life it crafts in that pursuit.

Jere is an accomplice in his own brutalization—sticky in the sun, violence gravitates to his Black person. This deadly way of being transgresses his expected silent, laboring functioning as muted flesh, fashioned for the work of accumulation through racialized geophysics. Jere's transgression is to claim a space for consent and refusal within this enforced inhuman intimacy of the pit of clay. Transgression is not the point as such; violence marks the limit of compliance in the negotiation between the recognition of "will" or agency and the submission to a status of matter. As the slave block is refashioned into the podium of the legislature, within the slave mine/slave foundry, the ploy of reform and reparation in the guise of the inspectors delivers only a newly fashioned version of the master's prerogative. The return of undergrounds in the convict lease released another version of the Black ground that haunts the recesses of the normativity of materiality after

Reconstruction. The patriarchal desire of the inspectors to substantiate themselves as "just" gentlemen (akin to the "fathers" of geology), to reassure that violence is not without purpose (and thus give the libidinal an unguarded range), is part of a disciplining toward work, but beyond that, this is how colonial earth was made, through narrativization that draws matter toward the telos of white supremacy.

We know the name Jere Ford, and of his brutal and prolonged death, precisely so as not to know his name in the recognition of his subjecthood. Which is to say that he is recognized as an individual person to whom things were done precisely to dissuade us from the structural nature of the violence that he died under. The individual account and its minute details construct a particular case, an individuated act of transgression, when the violence imposed was ubiquitous and stratal. Violence in spaces of earth extraction was a racial formation that produced the social formations of whiteness up above. When Morrison says racism is the distraction, the question is how Blackness is used and made. Which is not to say that the anti-Black violence has a point or rationale rather than a psychosis per se, but that this is a *material psychosis* that organizes the conditions for the extraction and maintenance of geopower on the plateau. The obscenity of violence is engaged to distract from the normativity of control exerted. It prepares the ground and conditions of undergrounds. Like the blinding of the sun, the procedural accounts are a distraction from the establishment of organized, institutionalized forms of spatial enclosure and subjective erasure of the state made under the white fantasy of inhuman improvement. The slave mines function both as an affective infrastructure, racial stratification, and as a claimant for containment—mine, *Mine*—spatial possession and institutionalization of the inhuman propertied form. Mine/mine—the arc of possession—the claim and the conquest in natural resources. M-I-ne in the possessive, as a possession of subjects and subjectivity in the case of slavery. In the scenes of redress, the artist Tia-Simone Gardner gives the mine/mine dialectic another space in her work *Chronotopophobias*, where she names the "mine" in a mode of reclamation of her histories. She says: "'Belonging' is not some colonialist fantasy of possession by conqueror or by price, but possession as some sharper, unbreakable ferocity, with *punctum*, the *demonic ground* on which the colonial conqueror could not build his house. Fairfield and Birmingham, Minneapolis. They are mine, and so these landscapes, with the full weight of the verb, I mine them" (2020, n.p.). The mining relation is key. The inhuman as a subjective category was not a category mistake but a way to mIne—as plantations mined personhood and earth for the extraction. The figure of the mine and mining constitute a crucial geoformation of racial capitalism that imparts a normative mode of extraction, which has subjugating and planetary effects. The mine is a planetary archive of relations, but it is also a construction site of redress, as Gardner reminds us.

Underground Suns

Another sun is not just a question of survival; other suns overturned the whole structure of stratal relations in a implosion of value that sustains normative geographical imaginaries. Another sun is the underground reinventing surface conditions, orientations. Not just escape, refusal, or retreat but a tipping of gravity (listen to Leanne Betasamosake Simpson's lyric, the "sky is falling up," on her 2021 song "OK Indicts").[16] Just imagine, suns underground. Two kinds of buried sunshine were at work in the prison mine: the deep and buried seams of the Carboniferous and the geographic imagination that made other suns possible where none could be found "free in the daylight of the earth which is its own earth" (Césaire 1969 [1956], 12). Those other suns were testament to the work of discrete vernacular imaginaries of the possibility of geography in impossible conditions. To internalize the sun was to take the solar exuberance in as duration and warmth held against the clamor of the day. Like a song of the underground ("Dark Was the Night, Cold Was the Ground"), etched in gold, floating in the cosmic debris of outer space, sending its signal into the imagined geographies of extraterrestrial worlds. Making other suns is taking in the most asymmetrical of relations that dissolve the dialectic entirely and blast apart the strata. Overseer be dead! Plateau perspectivism no more! The sun's gifts are made present by other means. Buried sunshine, indeed! Colonialism's confinements are destroyed by exuberance, and blasts of the furnace set fires to the surrounds. This impromptu recognition of a solar utopianism is the burden of innovation and creativity; to be excessive is the deadly weight of nonrecognition in the normative and juridical register of being that denotes a subject of rights. The anticolonial will returns as a tacit geophysics that methodologically innovates within deadly materialism. Talking to Manthia Diawara on the *Queen Mary 2*, Glissant says that "Columbus had left for what he called the New World and I'm the one who returned from it . . . as something else: a free entity, not only free but a being who has gained something. . . . In relation to the unity of the enslaving will, we have the multiplicity of the anti-slavery will" (Diawara 2010). Glissant's geographic understanding of the anticolonial will schools us to be attentive to the collaborative work that inhuman epistemologies are mobilized toward, not just in coercive registers. "Stealing suns" points toward a need for a methodological scrutiny in how inhuman materials are weaponized but also instructs, in the context of environmental determinisms, to not overdetermine the efficacy of those violent scripts of matter. And to see the missing scripts of ecological epistemologies that theorize differently in the sun, and in the given presence of the day (see Kodwo Eshun's 1998 *More Brillant than the Sun*, for example).

The removal of liberty was a form of social death in Orlando Patterson's (1982) terms for those incarcerated and in collective terms of the extraction of predominantly young Black men from social space. The convict lease system targeted poor and displaced persons, relying on and intensifying the erosion of social ties, which often was not the case, as families and community support sustained incarcerated persons both within and outside the prison mine, even when they were not able to come up with the excessive fees to release people from the burden of additional hard labor for these administrational charges. Freedom was a matter of timing then; of the geologic time of the carceral mine that conditioned the contraction of the Black outdoors through a meshwork of macro and micro tools for the reduction of social space. Conditions of liberation had to engage not only with the assignment of stratal placement and its relation to the political present but also with what it meant to be placed in the deep time of the carceral mine; people had to carry the mine as a prison inside. Coal may be compressed sunlight of buried biomass and swampy forests, but underground within these carbon reservoirs it was dark. Freedom has a temporal spatiality, as thick time, and vertical time.[17] What matters is not just which time you find yourself in but into what racialized geology you are cast, and how that stratal placement assigns ideologies of movement and environment, or what McKittrick (2006) calls "cartographies of struggle."

Subterranean cartographies do not just organize space and time in a conventional sense but arrange depth, density, and deformation in the *stratal imagination* of overground and undergrounds, and in the pressure inflicted there; this is the access to the political present, to the surface, to life as a spatial expansion rather than an enclosure and incarceration. *Undergrounding* is a geographic process designed to remove subjects from the surface, which is to say, from the present, the social, and the sensible. I use the term *white surface* to designate the way in which the surface organizes relations to

FIGURE 10.3 Cans of Sunland motor oil, Kern County Museum, Bakersfield, California. Photo by author, 2015.

the underground in structural, symbolic, and material ways, and how that relation is racialized in the history of geology and the political presents of extraction. The underground is also a form of concealment from space, and as McKittrick shows in her work, "we make concealment happen; it is not natural but rather names and organizes where racial and sexual differentiation occurs" (2006, xi–xii). The space of intrasolar subjectivity—fungibility to the materialities of the underground—effected a slippage between material and metaphor in subterranean racial hierarchies. This is "stolen life," in Moten's terms, Black life stolen from social space and from the light. The special kind of anti-Black gravity that pushes Black bodies so deeply into the geophysical and psychic recesses of the earth. The renegade-ation of the placement of Blackness holds with the submerged location of its prior given through natural philosophy (from Kant onward) and the geologic elaborations of the races. Persons caught in the carceral institution of convict lease labor are captive *across* terms—terms produced by natural and scientific philosophies of race and by the gravity of white geophysics that would put Blackness underground.

The spatial organization of territory was linked with the geophysical configuration of race—undergrounds that must be carried internally and externally as a form of pressure that conditions the light of day. The collaborative script of the inhuman points to how the racial dimensionality of subjectivity may be thought as riven through the geophysics of power. It also points to sites of its undoing. The underground is a hot and churning maelstrom of gas and liquid and sometimes fire. It is the site of crashing plates, past cataclysms, pressure, stratification, upwelling force, and fury. These geophysics of compression make the world above and condition the politics of the surface. The Alabama penitentiary reports of the 1880s detail both a continuance of conditions of slavery and a new armature in establishing normative architectures of enclosure through specific forms of spatial theft and experimentation with environmental depravation—the solar guarantee, as the condition of life was transformed into a mode of execution.

In the case of Jere Ford, the issue of environmental determinism is raised to dissimulate any responsibility and to attach a solar causality to the violent events. It was an environmental determinism that literally conflated Black life and the sun as mutually exposed matter, thus predisposed in an imaginative frame of similitude. This total weather, as Sharpe names it, is a violent material form of governance that reinscribes the geologic determinism of the eighteenth century, while shifting the coordinates through the incarcerated body. This account of Jere in pain is not given here to reinvigorate these determinations or their legacy of effects, but to give space to a history in which proximity to the conditions of the inhuman is part of what Marriott calls the "black law of fungibility" or the "racist codes [that] can legally be named as *dying while black*," which "is as fatal now as it ever was" (2018, 327). The inhuman is another proxy for the flattening of Black life and the complexity of its emergence, yet it is not a closed space because of its cosmic dimensions, despite its deployment as a carceral property.

So why reinvest in these coordinates of materialisms or reinvigorate inhuman intimacies?[18] Embracing the inhuman as a parallel orphanage and the potential of cosmic materialism is perhaps what Arthur Jafa had in mind with his burning sun at the end of his video *Love Is the Message, the Message Is Death* (2016) or what Alexis Pauline Gumbs is thinking when she says, "Once upon a time the core of the earth made magma solid, built crust around itself where dreams could safely plant and grow. We are of that lineage" (2018, 78).[19] Moving further into the intimacies that history forged is a way of reckoning with the precise coordinates of the Overseers' affectual architecture, without trying to repossess the plateau, which belongs to the category of the human. If the inhuman is everything outside, then a solidar-

FIGURE 10.4 Coal storage bins stand like sentinels in the trees; the entrance to the mine is covered over and submerged. These antimonuments are all that is left on the mine's surface. *Brookside Coal Mine, Washed Coal Storage Bins Foundation, Mount Olive Road, North of Five Mile Creek Bridge, Brookside, Jefferson County, Alabama*, 1968. Historic American Engineering Record. Library of Congress, Washington, DC. https://www.loc.gov/item/al0956/.

ity with this "putting outside" of the human is a reaffirmation of a place of freedom that might unlock the incredulity and disaster of difference as it is parsed under colonialism and its afterlives. The inhuman might then be a place in which the beauty of difference and its movement are already germinated and disclosed otherwise, as a site to mine abolition across the stretch of the cosmic, in the inside-outsider time of the universe.

11 Inhuman Matters IV

Modernity, Urbanism, and the Spatial Fix of Whiteness

Carceral Mines, White Law, and the Reconstruction of the Abyss

> An Act to amend section 3794 of the Code, relating to vagrants.
>
> Section 1. *Be it enacted by the Senate and House of Representatives of the State of Alabama in General Assembly convened,* That section 3794 of the Code, which reads as follows: "The following persons are vagrants," be amended by adding the following: "Any runaway, stubborn servant or child; a common drunkard, and any person, who, depending on his labor, habitually neglects employment." Approved, December 15, 1865 . . . In addition to those already declared to be vagrants by law, or that may hereafter be so declared by law: a stubborn or refractory servant; a laborer or servant who loiters away his time . . . the common jail of the county may be used for that purpose. —Penal Code of Alabama, 1866

Any (Black) person found to be unproductive in social space, to be without purpose of arrival, destination, or direction was subject to be "hired out such as are vagrants to work in chain-gangs or otherwise, for the length of time for which they are sentenced" (15). Any person found unenthusiastically engaged in labor or taking space and time by loitering was subject to the same discipline. Common infractions were transmuted into felony convictions: "If, upon examination and hearing of testimony, it appears to the justice that such person is a vagrant, he shall assess a fine of fifty dollars and costs against such vagrant; and in default of payment, he must commit such vagrant to the house of correction, or, if no such house, to the common jail of the county. . . . The commissioners' court may cause him to be hired out in like manner as in section one of this act" (15).

Thus, people under the Black Codes were sentenced to *civiliter mortuus*, or civil death. Civil death was the loss of all or almost all civil rights by a person due to a conviction for a felony or due to an act by the government of a country that results in the loss of civil rights. While civiliter mortuus was not instigated in all Southern states, it was the de facto category through which Blackness was parsed. For example, the Georgia case *Dade Coal Co. v. Haslett* (1889) defines civiliter mortuus this way: "If the convict be sentenced for life, he becomes *civiliter mortuus*, or dead in law, in respect to his estate, as if he [were] dead in fact."[1]

The removal of post-emancipation subjects from social and juridical space became increasingly gendered and narrativized through the governance of sexuality (Carby 1985; Wells 1892). Governance of sexuality was concomitant with the changing forms of energy extraction, and both involved spatial epistemologies that placed race as a tactic of governance between the plateau and the rift. The theft of predominantly young Black men and children from the outdoors is paralleled with an enclosure of Black women in domestic servitude indoors.[2] The organization of social space after the plantation created differently gendered spatial forms and new forms of policing sexuality for the capitalization through racialized geopowers and the management of geography (see Weinbaum 2013). The sexual division of space was related to the newly emancipated position of Black men as voters (and thus a political social force) after the Civil War, but equally it related to the governance of "excess," whereby one excess (sex) was substituted as a distraction for the stealing of another (energy). Sentence to the mine made incarcerated subjects into felons, thereby erasing a substantial proportion of the Black vote. The political effects of disenfranchisement after the Reconstruction Act of 1867 (Fourteenth and Fifteenth Amendments [1886, 1870, respectively]) saw the Alabama voter registration in 1867 record 87,536 persons classified African American registered to vote and 69,065 persons classified White.[3] While some counties were missing in this count (registration books "destroyed by mould"), by 1883 there were only 3,742 registered "Negroes." Thus, there was a political efficacy in the racialized policing of the polis through questions of sexuality that served both the enrichment of corporations (geopower) and the assuaging of white hatred (psychic power), burying democratic questions in a libidinal and lithic racialized underground.

Black codes policed Black sociality by constraining Black mobility, while the lease system privatized that desired enclosure as a surplus value for extraction for the enrichment of the plasticity of public space ("Whites only") and the reproduction of the white settler state. During the instigation of these codes, the earnings of the state from October 1, 1885, through April 1, 1886, from the convict system in the state of Alabama were $21,088.16 (Alabama 1886). Thus, the Black Codes functioned like scatological interventions in freedom's possibility, wavelengths of indiscriminate force that ungrounded Blackness into the subterranean, while returning white revenue. Continuing the utility of Blackness in white settlement, the process of the erasure of personhood was a form of surfacing in which whiteness could remaster itself through new conditions of fungibility. This would be to understand the carceral mine as a shifting formation (a stratal racial order that produced and was produced by geophysical affects), in the enrichment of white settler accumulation above ground through a juridical subterranean Black stratum.

The carceral mine became a profitable model of relation on which to make racial undergrounds that could grow, move, and reproduce. In other words, placing carceral mines in their historic racialized geographies disrupts the narrative of national "progress" that is articulated through architectural infrastructures of iron and steel, industrialization, and commodification as something that was distinct from slavery and plantation epistemologies (and the connection of the plantation to European industrialization that was realized through the interlocking profits of the triangular trade and empire). Rather, the carceral mine alerts us to how the geologic grammars and spatial code of undergrounds transmute across different material praxes, while maintaining a common thread in the extractive principle and the enclosure of racialized human energy as carceral resource. As McKittrick addresses, the spatial formation of the plantation in its afterlives was a material location, a form of relation, and a form of mapping into economic systems of commodification: The "plantation spatializes early conceptions of urban life within the context of a racial economy: the plantation contained identifiable economic zones; it bolstered economic and social growth along transportation corridors; land use was for both agricultural and industrial growth; patterns of specialized activities—from domestic labor and field labor to blacksmithing, management, and church activities—were performed; racial groups were differentially inserted into the local economy, and so forth" (2013, 8).

Rather than the carceral mine being just another expression of the plantation, it functions as a concomitant site in the experimentation with enduring structures (and strata) of extraction through the specific geo-logic relations of the accumulation of *geopower through racial enclosure*. In the geographies of colonialism since 1492, the mine both preceded and exceeded the "life" of the plantation (the first colonial infrastructures were mines). Through extraction, the plantation and mine are linked as Black (and Indigenous and Brown) spatial forms in the context of the occupation and transformation of stolen earth and the arrangement of carceral persons-as-bodies for material transformation, which in turn systematically rearranged racialized relations of geopower.

In practice, the convict lease organized the sale of personhood to private companies. The "unrestricted use" given under the lease paralleled extraction without end. The sale of personhood for those imprisoned for life as convict labor made citizens symbolically "dead" in the eyes of the law, and without rights of recourse to the state.[4] In the removal of state-based rights, Alabama accorded the state of "civilly dead" to incarcerated persons: "Civilly dead is the state of a person who, although possessing natural life, has lost all his civil rights and as to them is considered dead."[5] As Alabama's incarcerated persons were separated according to strength and ability (according to the old slave categories), and the state was paid accordingly, the calculative apprehension of a Black bodily aesthetic of "young and strong" was reincentivized as profitable for sheriffs. The law was spring-loaded for entrapment of young, poor Black men. In a table on the deaths at Pratt Mines in the Alabama *Biennial Report of the Inspectors of Convicts to the Governor*, most of the men recorded had been sentenced in their early twenties with short sentences of a few (three to eight) months before fees. A child as young as ten (Jas Reynolds) was sentenced to three weeks. Many boys in their teens had sentences of less than three months that were transmuted into more than a year with additional fees (Alabama 1886, 214). Focusing on convict lease labor in the coal mines of Alabama in the late 1800s, this chapter looks at the racial empirics of mines in the surfacing of an urban form—the Magic City of Birmingham—to show how urbanism does not just require mines to make cities but gives rise to experimental forms of incarceration and racial modes of environmental experience.

If every building is also a mine in its relation to material extraction, it is also a racialized volume and geophysical event. This chapter advances in three directions: (1) demonstrating the affectual architectures of race in mining as paradigmatic to modernity and its gendered and sexualized relations; (2) situating urbanism in the Anthropocene as a vertical event; and (3) advancing a theory of the "geophysics of race" as a distinctly subterranean-surface axis of material development that conditions the material dimensions of racial experience. In conclusion, the chapter argues for addressing the vertiginous architectures of geology that are involved in deep-timing subjectivity and temporalizing racial dynamics.

Gendering Carceral Life

Women were in the minority of incarcerated persons and generally convicted for forms of sexual transgression, such as miscegenation, disrupting the peace, stealing, or defense against sexual violence.[6] Which is not to say that white law was not continually challenged under the rubric of justice by Black Alabamians, especially by women fighting against sexual violence.[7] Engaging the law, when the law was (and often is) a racist instrument, was a difficult yet necessary way to navigate the production of Black social space. The inadvertent middle-class investments in addressing the narrative of Black criminality via calls for morality (see Carby 1998), rather than via the geosocial conditions of its structural production, often tethered Black social reformers to discourses of improvement and investment in liberal individualism and its conduct-based policing rather than challenging the imposition of criminality and punishment per se. As much as emancipation was meant to inscribe into law the Black male citizen, this politics of recognition did not survive long enough to make it into practice in any meaningful way that addressed the inhumane grammars of Black subjugation. Law was, and continued to be, White Law, for the promotion and protection of white supremacy. The insertion into an emancipatory framework (built on slavery) and the organization of Blackness into the formation of juridical subjecthood were short-lived. The return to the earth in the carceral mine can be seen as a forced removal back into the classificatory logic of the inhuman.

Talitha LeFlouria, writing in the context of Georgia's convict camps, explores how Black women's lives were shaped by the convict system in ways that depart from and challenge white accounts of gender, skill, and labor, as well as the criminalization of African Americans. Taking up Mattie

Crawford's story—the "Only Woman Blacksmith in America Is a Convict"—LeFlouria shows how Crawford utilized her great strength to become a blacksmith, first at the Chattahoochee Brick Company and then as the sole artisan at a Georgia state prison farm. As LeFlouria comments, Crawford's "reputation as a blacksmith made her a luminary in Georgia. Yet, she and others like her remain elusive in the historical narrative" (2011, 48). Reports note, albeit through negative inscription, how Black women prisoners exercised their freedom of speech, movement, and sexuality in ways that alarmed the inspectors and challenged the inhumane effects of incarceration (see Curtin 2000, 129). The idea of a female blacksmith no doubt conformed to racist ideas of Black women's sexuality, playing on the curiosity and sexual fetishism of Black women as ungendered or outside of gendered expectations about white womanhood (see Carby 1985, 270).[8] Incarcerated women, although a much smaller percentage of the prison population, were still sentenced to hard labor (see Curtin 2000; LeFlouria 2011, 2015). While they were rarely down in the mines, their labor involved strenuous farm work, highway maintenance on chain gangs, and cooking and washing for miners. Many cases of incarceration result from either resisting or punishing the imposition of sexual violence. Crawford, for example, was sentenced to life for having killed her stepfather, who she claimed abused her. Once women were inside the carceral system, their experiences in prison mirrored the sexual stereotyping and violence of society at large. While many prison reformers seemed to have a heightened concern for the pliability of women prisoners, their obedience, and their effect on the male prisoners, the women's "moral" status is most often referred to as a concern in the inspectors' reports, read through the rebellious attitudes of Black women prisoners. The seeming lack of enclosure of Black women within white gendered expectations (through the disciplinary effects of heteropatriarchy) was of concern to both white and Black middle-class reformers (see Hunter 1997).

While sexual violence against Black female prisoners was legitimated and located in a justification of their already beleaguered position in society in the afterlives of slavery's sexual assault, it is evident from prison records that incarcerated women were seen as the most "difficult" category of prisoner due to their active resistances to impositions on their freedom. The inspectors lamented, "It is exceedingly difficult to determine what is best to do with the women. . . . These women are almost all colored and have the loose ideas of their race as to morals. . . . Decency and self respect require that the debauchery that had prevailed, and that will prevail where women are put out on contracts, should not be tolerated" (Alabama 1886, 21; see also LeFlouria 2011, 61, 62).

Inspector Henley noted that the women at Coalburg were "the most unruly and disorderly convicts that I have ever had to manage . . . & it will be necessary to enforce discipline at the cost of a good deal of punishment I am afraid" (quoted in Curtin 2000, 39). *A good deal of punishment.* The instability that women were seen to introduce within the predominantly male social space of the prison shows up in the imagination of the political economy of carceral conditions as a gendered geologic life, established primarily to police the polis from Black juridical power. In the prison governance of white gendered norms (see Carby 1998; R. Gilmore 1996), carceral life remained punishing to Black women, as their presence is coded as doubly excessive (sexually excessive and excessive to the function of the lease).

Labor and Lease

The relation between convict lease labor and slavery has methodologically been addressed from the perspective of carceral studies (Davis 1998; R. Gilmore 2006), labor and regional histories (Blackmon 2008; Curtin 2000; Lichtenstein 1996; Mancini 1996, 21; Oshinsky 1996), and gender, which either organizes convict lease labor in the comparative frame of the political economy of free labor, penal histories, or through ontological status of slavery.[9] Matthew Mancini assumes that "control of black labor was a leading motivation behind every significant effort to establish and maintain convict leasing for fifty years. Just as plain is the similarity between the brutal hardships of convict life and the oppression of slavery times" (Mancini 1996, 20), whereas Tiffany King warns of the dangers of the focus on labor. She suggests that the formulation of labor fails to repudiate the settler narrative while reinforcing the selective emergence of the poor white as extraction's postindustrial subject (and thus pre/foundational subject), whose masculinity and pride must be uplifted in the context of nostalgic deprivation of labor. Resisting the valorizations of labor and its settler grammars in relation to both Black fungibility and ongoing settler colonialism, King argues: "Within this Lockean formulation, Indigenous subjects who do not labor across the land fail to turn the land into property and thus fail to turn themselves into proper human subjects" (2019, 23). She suggests that it is Black fungibility rather than labor that "index[es] the imagined (surfaces) and actual sites of colonial spatial expansion and, in turn, the space of Indigenous genocide" (23). The efficacy of these discourses of improvement through geologic practices and Indigenous rights are recurrent in contemporary debates about fire-stick farming in Australia, the question of farmers or hunter-gatherers, forest farming in the Amazon, and all the ways in which Indigenous practices are legitimized through settler colonial frames of value. Alongside acknowledging how the settler frame did have liberatory possibilities for free Blacks (and is a multiracial project), another analytical route into the carceral mine is to understand extraction as a continuous site of investment that narrates and positions power relations and racial encounters, through the economy of geophysics. In a parallel spatial move, convict lease labor emerges as a concomitant strategy (to the gendered enclosures of domestic space) of containment and enclosure of Black masculinity in the afterlives of slavery.[10] The continuation of *geologic acts of spatial execution* enforced proximity to forms of extraction, which suggests the fungibility between the spatial construction of value in energy extraction and the lexicon of inhuman matter (which is not to reinvigorate this category but to understand its embedded effects and how it functioned as an affectual architecture). Black energy becomes a site of violence because of the recognition of its value and the historic conditions of fungibility of the Black subject (chattel slavery) that have set up in advance the conditions for the transferability of that value,[11] thereby devaluing Black life by the same calculus in which it becomes apprehended.[12] Conjoining material and energy economies through Black incarceration, "prison mining had become an enormous source of private and state profit, as well as the primary source of coke for steel" (Curtin 2000, 130). As such, carceral mines were a structuring materiality through which industrialization and social space emerged and took form. At the same time the problem of a historical "surplus" of white supremacy (as patriarchy, police, white women [see Bonds 2020], plateau) found a target.

From the initiation of slavery onward, the accumulation of energy for the settler state has relied on extraction from racialized persons coded as bodies to expand and consolidate its economic, philosophical, psychic, scientific, and cultural practices. The settler state is a *geophysical formation* as much as a territorial one; geophysics regulates the states of being, temporal and material states of whiteness, and forms of life through the production of race.

Carceral labor was used to depress wages of both white and free Black workers and to break strikes.[13] Use of such labor regulated interruptions to production and was strategically deployed as a divisive mechanism to break up any would-be bonds of solidarity that might have been established across the color line, in the context of unions and the social possibility of intimately shared crawl spaces underground. That is, racial differentiation was a mechanism for the continued control of sociability and spatiality, especially in the collective activity of mining (you cannot mine alone). Moreover, the normative lens of labor diverts attention from how subjects were enrolled within the mining system through the segregated status of labor and criminal. The question is not whether convict leasing was like slavery or not, as the two mobilize different categorizations of personhood, made at different historical geographic junctures; rather, it is important to recognize how each furnished a form of subjective enclosure in the calculus of racial capital that consolidated the ongoing colonial project of building the white plateau of settler space. There are clear differences in size of the enslaved population and convict labor, of geography and geotrauma, and in differences of "work," from agricultural to industrial, but a continuance in forms of racial capitalism through geologic transformation that embed racialization in industrial modernity. Wilderson calls out the inability of social movements grounded in Marxian or Gramscian discourses that coalesce around labor as the figurative movement, to think of white supremacy rather than capitalism as the base condition. He says: "The mark of its conceptual anxiety is in its desire to democratize work and thus help to keep in place and insure the coherence of Reformation and Enlightenment foundational values of productivity and progress. This scenario crowds out other post revolutionary possibilities, i.e., idleness" (2003, 21).

Wilderson's argument is that "the slave demands that production stop, without recourse to its ultimate democratization. Work is not an organic principle for the slave" (2003, 22). The worker "calls into question the legitimacy of productive processes, while the slave calls into question the legitimacy of productivity itself. . . . A slave does not enter into a transaction of value (however asymmetrical) but is subsumed by direct relations of force" (22–23). His analysis of the antagonistic subject forms that are possible within the structures of capitalism leaves little room, or desire, to approach labor and its narrative forms as anything other than a structuring principle of Life on the plateau. Yet, what is true ontologically does not always hold in the lived experiences of the day that made Black place through labored geographic claims in space. While the deficiency of Marxian categories is clear in their universal scripting of time and space as they relate to a concept of labor, the articulation of geography requires not just an ontological account of forces but a parallel material geophysics that adheres to specific place-based (or vernacular) relations and their temporal narratives as a *granular geologic life* made against more overt displays of geopower (and as a site of methodological repair structured against colonial earth).

There is a much more complex history to tell around the politics of protest, labor, and citizenship during the Reconstruction period in Birmingham, Alabama (see Curtin 2000; R. L. Lewis 1987; Wilson 2000). In the collapse of a liberatory geophysics, we can see a lineage of concrete and imagined relations that William Phineas Browne (the celebrated Southern "pioneer" in the coal business) named as the development of his "coal property" in 1857, referring to the imbrication of minerality and enslavement, whereby he collapses his "properties" of coal and slave into one another: "You know perhaps that for a time it was *sine qua non* with me to do my mining with slave labor to be owned by myself or by an association to be formed on the basis of my coal property" (Browne in R. L. Lewis 1987, 11). Enslavers had often been reluctant to "lease" the enslaved because of the danger of the mine work to their "property," whereas the lease engendered a policy of "unrestricted use" to the "property" of the state. Even before the lease labor system, slave-mine work was seen as too dangerous for the company's own enslaved persons, so enslavers hired and insured the enslaved from outside their "property." From the inception of the extraction of Southern coal seams, enslavement is present. Alabama's coal reserves remained comparatively undeveloped during the antebellum period and were not mobilized as natural resources at scale until Black labor was "released" from plantations and enclosed anew in the mines. As Mancini suggests, "Convict leasing, in fact, is best understood not as part of the history of prisons but as part of the elaborate social system of racial subordination which had previously been assured by the practice of slavery. That is, the lease system was a component of the larger web of law and custom which effectively insured the South's racial hierarchy" (1978, 339). Seen in this way, convict leasing can be understood as a crucial part of a more systematic approach to subjugation, rather than a new aberration of economic development (which partly explains why Birmingham was such an important site in the civil rights movement). The lease, an experimental technology at the time, presented a map to constraining the right to geography, through the criminalization of loitering by formerly enslaved persons while mobilizing the minerals "idling" underground to rebuild the Southern economy.

The terminology and imagination of slavery's afterlives were a contemporaneous form of speech in relation to mining in Alabama across a range of social orders, as the constant refrains of slavery's enclosures called attention to the organization of relations based on the previous racial order.[14] To make a case for the rationale of inspection (and the employment of inspectors), Russell Cunningham, who compiled the first statistics on prison mortality rates in Alabama and acted as physician for TCI, notes in the *Biennial Report of the Inspectors of Convicts to the Governor*:

> Until the act of 1883, requiring the state Inspectors to visit these people, there was little or no attention paid to them, and, if half that we have heard of their treatment is true, they suffered horrors shocking to every instinct of humanity. When the contractor, upon receiving them, paid up the costs, they were abandoned to their fate, with no law but his will. They were slaves without the restraints that were thrown around the slave-owner by association, by society and by interest, and the celebrated horrors of the Middle Passage are rivalled by the stories of the former treatment of county convicts. Few knew, and fewer cared, what became of them. . . . They were kept as long and discharged when the contractor pleased. In 1883 we found many held over their time, and some whom the officers of the counties they came from did not know had ever existed. (Alabama 1886, 23–24)

While convict lease labor was not in the same ontological order of slavery in its forms of natal possession, Du Bois described it as the "spawn" of slavery, insomuch as it replicated and extended new avenues for the geophysical, psychic, and spatial modes of extraction that were institutionalized during slavery (albeit in haphazard forms of subjugation, as early slavery had done in its time before forms of systemic "regulation" in the plantation economy). The very lack of regulatory norms under the "lease" shuttered its violence within the underground of the mine or furnaces, while the judiciary was arranged around the state motivation to curate a prison population of Black young men and boys for profit. As a spatial and social anomaly, convict lease labor created an arena of redescription, where specifically Black-identified misdemeanors of social conduct, often geographic, were reclassified as felonies punishable by fines that then translated emancipated subjectivity into privatized bodies of labor to which anything could and was done. "The white public . . . expected prisoners to make a profit for the state. . . . The state needed the profit from the mines" (Curtin 2000, 133). The collaborative work of the judiciary and the lease system created another taxonomy of the movement and fungibility of Blackness. Once they were sold to the lessee, the welfare, safety, food and nutrition, clothing, health, and discipline of the incarcerated were handed by the state to private contractors.[15] In another reformation of white space and Black enclosure, the lessee was master in all but name. In Alabama, a new principle of industrialization was demonstrated in which both the state and private corporations could prosper in the discover of new Black "grounds" for extraction. Blackness was made into a corporate commodity form of geologic life.

"The Spawn of Slavery"

Convict leasing did not just inhibit freedom of movement and attachments in life but also curtailed control over labor as a possible experience of freedom. In the racial calculus of space, the convict lease served as the regulatory function of that unfreedom, which severed Black life from the newly acquired rights that adhered to the category of the Human and were to be guaranteed by the state and its judiciary. Commenting on the way in which labor was associated with bodily and spatial possession, LeFlouria suggests the connection between labor and freedom: "The right to define the terms of one's labor, to move freely without the threat of the lash, and to build and expand black institutions, free from white authority, further defined the freedman's agenda to permanently disavow 'the badge of servitude' that marked the genuine and emblematic control whites exercised over each aspect of their existence" (2011, 48).

Freedom of space and the experience of free labor became reterrorialized under convict lease and its apparatus as one form of spatial contraction that reasserts the surfaces of white power and prioritized its material production. Those surfaces pertained to the very geography of existence, not just the economics of its perpetuation, standing by a tree, sitting on the grass, sleeping in the outdoors with the sun on your face, feeling the gravity of the earth free from the forces of subjugation. Living freely without the requirement to justify your existence. This is what Glissant (1999, 189) refers to as the right to opacity, an interregnum of space as a site outside of scrutiny and surveillance. Writing in the post-emancipation era, Mary Church Terrell, of the National Association of Colored Women and the National Association for the Advancement of Colored People, and an advocate for the Black women's club movement (which focused on the problems faced by poor Black women), named the convict lease system as "bondage . . . more cruel and more crushing than that from which their parents were emancipated forty years ago . . . men, women and children by the hundreds forced into involuntary servitude . . . accused of some petty offense such as walking on the grass, expectorating upon the side walk, going to sleep in a depot, loitering on the streets, or some similar misdemeanors which could not by any stretch of the imagination be called a crime" (Terrell quoted in Curtin 2000, 178).

Terrell identifies geographic acts of mobility as the challenge and experience of freedom, as racial segregation and disciplining of the body became operationalized through spatial and judicial grids of containment. The Black Codes punished vagrancy and so hit at the way in which migration and movement were forms of practicing and affirming freedom during Reconstruction in the afterlife of plantation containment (understood across a range of scales, from local "loitering" to migration across the Black Belt and interstate). That the world is a tacit negotiation with statecraft and its juridical-aesthetic acts is but one dimension of experience; it is also a negotiation with more intimate matter and its histories in the body that can only be named—dew-dressed grass, wide insect-circling flight, twitching-muscle sprung, bust-up broken-down thing, moonlight, and so on. How such things press hard on what propels thought and feeling in the lonely hours of the mine escapes the ledger. To be able to move outside of zones of detention, to possess and traverse geography, to "'*joy* my geography" (per T. Hunter), is to become acquainted with, but denied, the white spatial environmental frame (which is also the frame of settler colonialism and the permissible right to place). The Black Codes made ambulatory gestures a white activity.[16] The methodology of walking the city as an urban *dispositif* and narrating self and landscape was plateau activity. As Morrison states, "Anybody white could take your whole self for anything that came to mind" (2007 [1987], 295). A grid was put around sexual freedom ("Carnal Knowledge," "Felonious Adultery," "Sodomy," "Ass't to ravish"); mobility ("False Pretences [*sic*]," "Aiding an escape"); economic independence ("Br'g cotton in seed" or selling cotton seed after sundown, "Burglary to steal a cow," "Bribery," "Robbery," "larceny") to stabilize the white supremacy of space at a time of radical change (Alabama 1886, 186–222). Thus, the institutionalization of the carceral mine and its juridical net ushered in a new modality in the constraint of Black life that targeted the experience of geography as a freedom.[17]

In Du Bois's article "The Spawn of Slavery" in the *Missionary Review of the World*, he uses a genealogical association to name the convict lease labor and crop lien system "direct children of slavery," stating that they "to all intents and purposes are slavery itself" in the "arrangement of chattel mortgages" (1901, 737). Mortgaging lives was a way to develop property in its primary form as natural resource. Political discussions of convict lease labor occurred in a broader culture of penology, most notably by the National Prison Association (NPA), founded in 1870, which sought to reform perceived criminality through incarceration (see Curtin 2000, 170). As a lesson to be lauded, labor provided a moral narrative and practical resolution to narrativizing prison time. The virtue of work in a teleology of "improvement" was endorsed as an antidote to listlessness and crime. Both the NPA and middle-class reformers often tacitly accepted the dominant narratives about Black criminality and its ideologies, embedded in arguments of racist comparative biology that claimed Negro criminality as "regressivism" by nature, and thus placed Blackness in a state of eternal genealogical return (to barbarism). As Curtin comments: "White racists looked upon black criminality as a genetic trait" (2000, 177). This genetic imaginary was informed by the paleontology of Agassiz's racial geography and encouraged by white geographic gains. In such a placement, the Negro was rearticulated as a set of geographic coordinates (see King 2016a, 1029) (subterranean) and a set of geologic coordinates in hierarchies of deep time (as child to a paternal whiteness). While carceral conditions were often criticized, and the system made more "accountable" through newly established statistical and administrative institutionalizations, the organization of Black criminality (that was actively produced through these narratives and the Black Codes) was not always directly challenged. A number of Black political speakers, such as Ida B. Wells, Mary Church Terrell, Booker T. Washington, W. E. B. Du Bois, and Kelly Miller, campaigned and wrote about convict labor, locating it "within the context of slavery, racism and the southern need for cheap labor" (Curtin 2000, 174–75). But even Du Bois stopped short of dismissing Black criminality as an organizing structure of social space. The imagination of punishment had yet to give way to care in the context of political discourse of racialized geotrauma.

Convict lease labor, according to Du Bois, was how the "South sought to evade the consequences of emancipation" (1901, 738). He frames this evasion as a kind of resistance, but one that he, making the moral argument, suggests "in the long run weakened her moral fiber, destroyed respect for the law and order, and enabled gradually her worst elements to secure an unfortunate ascendency" (738). He continues to say that the South believed in slave labor and "was thoroughly convinced that free Negroes would not work steadily or effectively" (739). The problem in actuality seemed to be the opposite: that the formerly enslaved were deeply committed to grasping freedom and to securing work and land to enact ecological transformations of their own. What is telling is that Du Bois makes his arguments in the context of labor and moral fiber, suggesting an acceptable discourse of what it means to be free in relation to the questions of work, which can be read as a broader colonial discourse of improvement in relation to personhood, land, and the transformation of earth. The transmutation of matter was a condition of the recognition of subjecthood and political participation. He even goes so far in his commitment to this imaginative framing as to suggest "the inevitable tendency of many of the ex-slaves to loaf when the fear of the lash was taken away" (739).

Du Bois betrays his middle-class sensibilities and adherence to the project of Civilization, but he also anticipates his audience in the need to show his condemnation of the poorer classes.[18] *Loaf, loiter, lethargy, on the lam*—the vagrant is not productively engaged in a recognizable script of progress. While Du Bois is concerned to uphold Black respectability and with investments in normativity, he also demonstrates paternalism in the misdirection of violence that travels so intimately through the immediate histories of the formerly enslaved. He states: "The small peccadillos of a careless, untrained class were made the excuse for severe sentences. The courts and jails became filled with the careless and ignorant, with those who sought to emphasize their new-found freedom, and too often with innocent victims of oppression" (1901, 739).

Freedoming Geography

To be poor and Black without a support system was to be vulnerable to mobile categorizations of Black criminality and its geologic determinisms, where reform and extraction of personhood were indistinguishable practices of the state.[19] Such carelessness is understood by Du Bois as an absence of proper training, yet we might see that carelessness differently through Hartman's lens in *Wayward Lives* as an expression of freedom, a campaign made outside and against the expectations and scripts of the state. As Hartman describes it, "Waywardness articulates the paradox of cramped creation, the entanglement of escape and confinement, flight and captivity. Wayward: to wander, to be unmoored, adrift, rambling, roving, cruising, strolling, and seeking. To claim the right to opacity" (2019, 228). Echoing Glisssant's concept of the political right to opacity (understood as a possibility of freedom against scrutiny), Hartman places the wayward in its refusal, but also in the necessity of intimacy against the cold gravity of anti-Blackness: "It is the directionless search for free territory . . . it is a queer resource of black survival. It is a *beautiful experiment* in how-to-live" (228). For what Hartman articulates is loitering as a Black rhythm for being-in-the-social, freedom-ing attachments to space, a place from which to build a different ground. Without the white privilege of social space and a juridical or collective mode of address, the pressure of Blackness in the context of systematic but arbitrary violence rendered modes of survival and refusal that did not exist in the lexicon of whiteness. Loitering makes a different temporal script of environment in the noticing of time and space against the demand of business and purpose. As La Marr Jurelle Bruce captures: "I would rather loiter under moonlight than march into the sunset" (Bruce 2019, 352). Middle-class reformers were particularly alarmed at how Black workers would spend a large proportion of their meager wages on "unnecessary" travel and travel deemed to be "without purpose," taking buses to other counties and coming home again, suggesting desire for the pleasure of movement, without destination or the intent of arrival.

One thing Du Bois is clear on is that convict leasing continued the rule of white law. He acknowledges the stakes of impossibility in that the "testimony of a Negro counted for little or nothing in court, while the accusation of white witnesses was usually decisive" (Du Bois 1901, 739). The result, Du Bois observes, is a sudden and large criminal population in the Southern states, and the problems associated with this surplus to the state, that did not have the means nor the inclination to "house or watch it even if the state had wished to. And the state did not wish to." He sagely concludes, "Thus a new slavery and slave-trade was established" (739). After slavery, the state assumes a regulatory function in governing the geophysical "state" of the Negro population under the guise of morality. It assumes and consolidates the role of the master, and the former policing units of slavery transmute into new spatial watchdogs that continue practices of control and surveillance. As Du Bois comments, when "the masters' power was broken the patrol was easily transmuted into a lawless and illegal mob known to history as the Kuklux Klan" (738). Lease ~~labor~~ was extraction without the binds of property that would see an investment and return on the maintenance of the "property" of the enslaved as a form of capital accumulation. Du Bois suggests: "It [convict lease] had the worst aspects of slavery without any of its redeeming features" (1901, 739). Here he presumably is referring to a financial incentive to maintain the life of the enslaved, which no longer existed with the convict lease.

Du Bois articulates a metaphor of disease, between innocence and depravity, whether for a Christian doctrine or as a way to persuade the morality of his white audience. The infection that he narrates is concerned with the corruption of "herding together" the innocent and guilty, a crime of not differentiating properly. While Du Bois does not argue against the normative narrative of improvement through work (no doubt the basis of appeal to white audiences), he seeks to show the creation of new forms of wretchedness in the context of that work that ameliorate its intended effects (see Du Bois 1903, 1910, 2013 [1935]). The effect of the systemic subsuming of Black citizens into the police state, Du Bois suggests, is the utter breakdown of the judiciary and the establishment of criminality as purely a condition of color, with whiteness being a license for being beyond the law (except in the most extreme cases).[20] That he makes his arguments through labor (and its degradation) and through morality (shame) is evident in the prevalent discourses of reason that governed debates.[21] He is clear, however, to recognize the role of the state in producing the criminality it supposedly sought to govern.

Surprisingly, perhaps, his caution to the Southern states is around the emergence of the "real Negro criminal" that has "stirred the South deeply"—a fearful criminality produced by the system that "broke all bounds and reached strange depths of barbaric vengeance and torture" (Du Bois 1901, 742). Du Bois's imaginary of the "strange depths" of Black criminality, which was promoted by a system that eschewed justice, reinvests the identity of a specifically Black criminal identity and its abyssal fearfulness in the white geographic imaginary, while pointing to where responsibility for this sociological invention lies. Even in Du Bois's imaginary, racial hierarchies produced different stratal relations ("depths" of the real Negro criminal) and sociological explanations ("produced by the system"). He suggests, "The state became a dealer in crime, profited by it" (741). Speaking of the effect on consciousness, he argues: "The effect of the convict-lease system on the Negroes was deplorable. First, it linked crime and slavery indissolubly in their minds as simply forms of the white man's oppression. Punishment, consequently, lost the most effective of its deterrent effects, and the criminal gained pity instead of distain. . . . Worse than all, the chain-gangs became schools of crime which hastened the appearance of the confirmed Negro criminal upon the scene" (742).

Du Bois concludes that in the South, there exists "a false theory of work" that was assumed as a result of the continuing racial order of the slave system that the slave existed for the master's benefit.[22] According to Du Bois, "The black workman existed for the comfort and profit of white people, and the interests of white people were the only ones to be seriously considered" (1901, 742). Such was the normalization of this racial order that Du Bois observes, "For a lessee to work convicts for his profit was a most natural thing" (743). While challenging how the racial order was being remade in the lease, Du Bois remained committed to a trajectory of improvement (see Moten 2007). Stopping short of condemning labor as a moral good, he qualifies: "Without doubt, work, and work worth doing—i.e., profitable work—is best for prisoners." The nature of the work, he suggests, should be not in contest with free labor and not purely for financial gain, and for the benefit of the criminal, "for his correction, if possible" (744).

Thus, Du Bois is not quite able to give up on his own scripts of a progress narrative through labor even as he deftly navigates expectations by seeming to endorse the white logics of work while rerouting that narrative through the historic nature of work under slavery. Hear Hartman's rebuke here about the chastising of those who refused the progressive scripts: *"They have ceased to be slaves, but not in order to become wage labourers . . . content with producing only what is strictly necessary for their own consumption* and embraced wholeheartedly *indulgence and idleness as the real luxury good"* (2019, 230). Idleness is a frequency of spatial praxis under anti-Black geophysics. If the term *work* eschews the racialized orders of labor that constitutes and refashions racial capitalist endeavors in the mine and thereby skews the focus to resist an antiracist analysis, the genealogy of slavery to lease labor illuminates its juridical and category limitations in the ongoing production of calculative inhuman life. In the progress narrative of labor, the question of subjecthood—criminal or corrected—is both reinstated and offered up for contemplation. If Blackness is the geophysical formation that secures the state, it simultaneously offers up an antithesis to this labor in its antagonisms.

Inspector Dawson complained of the "inordinate self-esteem" of convicts, pronouncing, "In fact, he is enclosed in a shell of self importance through which it is hard to penetrate and which almost absolutely forbids improvement" (Dawson quoted in Curtin 2000, 28). Curtin suggests that "more likely, black prisoners' 'shell of self importance,' which led them to reject white efforts at personal control and reform, had its roots in African-American community life after emancipation" (2000, 28). The continued rejection of white control of space and sociality, even during incarceration and always on the verge of violent retribution, is a refusal that permeates official reports, labeled variously in the ledger as stubbornness, laziness, retardation, sullenness, or pride. When freedom of mobility was curtailed, resistance became organized not just through escape and the geotrauma of outright refusal to work but in more subtle negotiations with stolen life and nonparticipation in white narratives of "improvement," coupled with the achievement of artistry at becoming skilled miners and blacksmiths and learning and advising on mine conditions and practices, often under conditions of extreme duress.

Likewise, Henry Colvin Mohler, writing in the *Journal of the American Institute of Criminal Law and Criminology* in 1925, observed: “The numerous negro convicts . . . seemed like heroes to other members of their own race” (583; quoted in Mancini 1996, 22). That such an observation was made after the appointment of inspectors and the notation of state-held records for the status of the incarcerated suggests that the endurance of such labor was a herculean survival against an organization of enclosure that had little concern for survival.[23] Mancini reads this quote as evidence of the difference between convict lease labor and slavery, referencing Orlando Patterson’s definition of “dishonored persons” (1982) as a mark of slavery, suggesting that convicts were not dishonored in the same way. This is plausible, but it perhaps misses the cadence of what it meant to survive a system that was at best indifferent to survival. What becomes visible within those mines is mutual aid, not primarily as a political organization (as it also became) but as the grounded fact of the emergence of forms of collective life that were crafted in the context of extremely dangerous material mining practices.

There can have been no return to the surface without such mutuality because the mine would have continued to propagate its deadly undergrounding, but this did not entirely happen (although miners continued to be medically impacted long after they left the mine and through epigenetic transfer). Black miners returned to the mines (whether in Birmingham or farther north in West Virginia) as free men, not necessarily on their own terms but intent on the political organizing of their conditions.[24] The proffering of sociality that was deemed to be associated with being incarcerated that is described in Mohler’s observation says much about the resistance and recovery of Black life as its own anti-gravity-defying event, evidencing a communal solidarity against the stratal pressure of existence in the shadow of the plateau.

At the entrance and exit to the mine stood the armed wardens; inside the mine the miners were “free” in their sociality, their bodily conduct, and their communication with one another. The mine itself was a space that had the possibility to hold onto other forms of collective living, an underground that had within it an internal space to counter the site of the carceral cage above ground. Such spaces might be thought of as something like what Hartman (2019, 308) calls (talking about the lady lovers who assembled in Harlem flats) “freedom in the sequestered zones,” “a clandestine space, a loophole of retreat” in the policed and surveilled zones. Curtin argues that forms of collective action pervaded Black life, for those incarcerated and formerly incarcerated persons: "In order to promote solidarity, the freed people practiced a form of social control not usually recognized by historians and social scientists" (2000, 35). Survival was the brokerage of a will that endurance in the carceral mine required; being put underground without the certainty of the sun names an illicit site of emergence that refuses the obliteration of relation that extractivism tries (and often fails) to enact.

The racialized geophysics of the mine had a gravity that encodes matter and the organization of space—practices that orient spatial possession toward the consolidation of white property that conditions the pressures of space in a gravity of anti-Blackness, a Black abyss that feeds the reflective steel surfaces of white modernity. In a shift from the focus on categories of bodies under slavery, labor inadvertently redesignates Black experience in a nonracialized field that assumes a history of volition that was incongruous with slavery or the convict lease. The analytic of labor as economic rather than corporeal relocates violence into a value relation that obfuscates the location of its origins in slavery rather than in white industrial modes of production. And it obscures the passage and racial patternings of colonial earth from the earliest mine to its current iterations. In this granular version of geologic life in Alabama, Overseers in white law join with private business interest, labor foremen, white supremacist groups, and geologists to regulate and enact geographies of containment and reinforce the whiteness of the plateau. The geographic imagination of the plateau is of a magic city that is birthed in elemental origination. The mines—as a form of riftwork—tell us something beyond negation, of the human event of living inside an inhuman category and overcoming the inhumane to make something beautiful and tender in the survivance of a day. Looking at Arthur Rothstein's photographs of coal miners' homes in Birmingham tells us a different story about the work of living.[25]

Fungibility Not Labor (after King)

The terminology of *labor* has a selective normalization that is inappropriate (in the context of Black incarceration in mines), even as modifiers are attached to it, such as *carceral*, *unfree*, *Black*, *enslaved*, *coerced*, and so forth. The pole that these modifiers circle, in the New South, is the imaginary and ideal of white labor, which is in turn underpinned by the desire for "cheap" Black labor to do the actual work. While different from slavery in its forms (legally and ontologically), convict labor actualized some of the codes and actions of coercion, by the Overseer (now turned industrialist), through a range of physical and psychic economies: random and targeted violence; petty reward systems and inducements; the promise of salvation; physical and mental degradation.[26] Normative assumptions that run along the liberal embrace of labor as evidence of subjective achievement and appropriate striving have a white masculine body as center and type. Any small-town coal museum will tell this story of the commemorated white miner. This representative norm is racially coded into the present.

Assumptions of labor as improvement are tied to the ideals of freedom in the liberalist mode that Shona Jackson (2012) calls the "metaphysics of labor," as individualized and humanist, which consecrated in advance the Black miner as outside that order, while offering the lure of a contradictory path to recognition within that mode. Stephen Haymes argues the designation of labor as a category for Black work is ontologically nonsensical: "In the antiblack world of (post)settler societies, *black labor* is an oxymoron. . . . It is simply a tool in the perpetual service and extension of white desire and its will to power" (2018, 44). Black labor, then, is a rifted category, where Blackness performs dialectically in negative relation to the claim of labor rights, functioning as the included exclusion that substantiates the improvement of matter but from outside the full terms of the category. In this equation, mortgaged chattel (in Du Bois's terms) cannot improve matter because only labor that can function in the category of the human can undertake improvement. Both Tennessee Coal and Iron lease in 1888 and Sloss Iron and Steel in 1893 stipulated to the state the requirement for "productive" labor in the lessee, that is, labor that can produce beyond the cost of "its" [human] maintenance. In this formation Black labor is "surplus" to produce value *from* the earth, without the primary capacity for the creation and accumulation of value.

Fungibility becomes exercised in the total extractive logic in the carceral mines, and raw material becomes "natural" and national resource through the application of labor as raced. Thus, the transformation of matter through labor is only symbolically available to whiteness as an "originary" generator of value through geopower—and author of the origin story of natural resources—within the geosocial context of the lease. The prioritizing of economics as a frame of analysis obscures the very social and spatial function of how the convict lease system worked collaboratively as a quiet violence in the systematic and pathological disciplining of spatial expression through extreme forms of Black exposure/enclosure. The ploy of labor shortages after emancipation, often argued as the economic motivation that provided a point for analysis by historians, denies the regulatory function in social space of the lease system in the South, while also not adequately accounting for the availability of free Black labor (Mancini 1996, 53–54). Rather, economic motivation and the creation of surplus value (beyond that of the plantation) without capital expenditure (save for the lease) still does not attend enough to the reassertion of geoforces of segregation.[27] As described by King, "Under slavery and conquest, the Black body becomes the ultimate symbol of accumulation, malleability, and flux existing outside of human coordinates of space and time. Rather, Blackness is the raw dimensionality (symbol, matter, kinetic energy) used to make space. As space, Black bodies cannot also occupy space on human terms" (2016a, 1028–29).

Blackness, after emancipation, was denied the spatial and temporal claims of whiteness and the contours and identity of humanist placemaking.[28] Thus it was rendered ontologically vagrant in a model where place is established through hierarchies of the human and its racialized arrivals within time ("You Are Here," arrival, property, and territory vis-à-vis "You Are Not-Here," nonarrival, placeness, fungible without property or territory). King's (2016a, 1028–29) analysis of Blackness as flux and "raw dimensionality (symbol, matter, kinetic energy) used to make space," gives a mode of apprehension with which to transfigure the fixity of the stratal hold on race, particularly if we understand the attachment to the placing of Blackness as a compensatory or aspirational claim made to arrest the flux of time as a medium of movement that claims freedom. It might seem perverse to dismiss the category of labor, a double erasure even, when so much backbreaking labor was done for others' enrichment, but the rejection of the redemptive value of labor is necessary in overcoming the material valuation systems of racial capitalism and its subterranean racial debt (made in the context of the predatory debt of fines and the rupture of relationships which sought to further isolate the racialized poor from liberatory geographies).

The argument of wage value under slavery and the convict lease system shows differences in the status of the enslaved and the incarcerated.[29] But this form of accounting conceals more that it reveals insomuch as neither form of enclosure can be considered as labor or even symbolically waged, and it is erroneous to make analogies that replant the theft of personhood in categories in which this is not recognized as primarily an inhumane category of extraction. John DuBose's (1886) "false theory of labor" establishes a politics of recognition that makes a tool of utility without a corresponding understanding of the broader implications and functioning of the system of regulating subjectivity and space, specifically in the reactivation of a space of captivity in the spatial margins of the underground. Inspectors too asked after the question of carceral subjectivity, attempting to make connections between physiology and capacities of criminality in the body of the "convict." But these questions are simply posed in the context of the utility of driving the incarcerated harder, and recognition is a tool of furthering oppression and the exercise of power. The inspectors ask:

> What influence, if any, does prison life have upon the health of convicts? This is a big question. . . . What is a convict? . . . He is a citizen deprived of his personal liberty, and denied his social, civil, political, industrial, religious and last but not least his sexual privileges. All these must of necessity and actually do work to the detriment of bodily vigor; physical capacity and health of convicts and intrinsically tend to shorten life. . . . The difference between death rate of free miners and convicts is truly very great. Why this difference? Whoever answers this question, solves the problem of working convicts in coal mines. (Alabama 1886, 246–47)

The role of labor in shaping questions of political affairs and governance can be further disrupted, if we take on King's (2016a, 1022) suggestion to see "Black fungibility as a spatial analytic." Disposability was enacted as a bodily condition of the fungibility of energy and its depletion, and the question the inspectors pose is merely one of extending the longevity of extraction. Thus, labor itself is a deeply inadequate term for this geophysical exchange and its ongoing praxis as a pedagogy of geologic life in colonial earth. The "agency" that labor denotes is erased in any meaningful sense in the context of convict leasing because the depletion of energy is directly related to the increased dispersion of energy on the surface of the earth. ~~Labor~~ undergrounds are a geophysical analytic of race.

While social historians have worked to show the often-vexed paths into free labor, and the establishment of Black working-class miners and artisan metallurgists that instantiated the postbellum actualization of the economic subject within racial capitalism, the transition is misleading in its assumptions around a desire for and toward an individuated liberal subject as the ideal. The rubric of labor that imaginaries of Blackness are mobilized within has often served as a ground to resource extraction—extraction that is tied to the elision of the social and ecological worlds of Black place-making. The agility of analysis that King brings to bear on the question of labor expands the concept of fungibility beyond its subjective form into a spatial theory that explicates Spillers's designation of the body as a "territory," in which the

"Black fungible body as 'territory' becomes linked to the space-making project of the 'New World socio-political order of conquest'" (quoted in King 2016a, 1026). As settler territory is expanded, the mode of racial conversion is across the stratal relation of overgrounds and subjugating undergrounds to produce spaces for whiteness. As chain gangs became a spectacle of social space, simultaneously disciplining Black conduct and establishing racial criminalization of the poor, the mines were at a spatial remove and below ground, involved in the sedimentation of value. Spatial isolation fed racial capitalism.

In a report from Dr. R. Greene, New Castle coal mine, to Colonel R. H. Dawson, president of the Board of Inspectors, Greene raises the specter of the inhumane only to suggest that it is a necessary evil that lightens the load of the white society (white man's overburden): "The questions of inhumanity and large death-rates cannot, I think, under the present management, be invoked to overthrow a system, which, while it affords humane treatment and protection to the life of the convict, at least equal to any other, at the same time produced a revenue that lightens the burdens of the good, law-abiding citizens, imposed by the conduct of this lawless class of society" (Greene in Alabama 1886/88, 273).

These spatial executions facilitated brutal forms of inhuman occupation. Although Curtin comments that the extraordinarily high death rate in mines was common knowledge, inspector W. D. Lee suggests that this was "in deference to a vitiated public sentiment which demands all the revenue out of the convicts which can be had. That is the whole truth & you know it" (Lee quoted in Curtin 2000, 10). If chattel named the objectification of personhood as property under slavery (Spillers's flesh unto which anything can be done), in the carceral mine, personhood was redefined as a body of energy—the *flesh of geology*—in the conversion of geopower, demanding "all the revenue out of the convicts which can be had" to lighten the (over)"burden" of whiteness. The twinning of natural resources and the Black body as human resource in the carceral mine exercises control *across* bodies of earth and personhood that are imaged, imagined, and made to be (under duress) "productive" for the maintenance and enrichment of the settler colonial present and plateau. Coal and young Black men become the means for spatially expanding the geographies of settlement, as a second wave of white geopower was established through fossil fuel extraction and the consolidation of new urban material forms of steel and iron.

While social historians have emphasized the transition from incarceration to free labor, documenting that 50 percent of the free workers in

Birmingham mines were formally incarcerated (Curtin 2000, 2), this does not mitigate the forms of enclosure and larger spatial dynamics at play. That mining became a way to live after incarceration does not explain anything but the limitations and severe restrictions on labor for free Blacks and the precarity in which social space was traversed (there was a parallel geography to this in South Africa) . Returning as "free" labor to Pratt Mines was held in an entangled set of economic, social, and psychic geographic attachments. "You can't eat freedom," as the saying went. To return freely and with control, to make something better out of the uncontrollable, must have had no small kernel of cease-fire nestled within it. Being broken down and being lifted up by the mine was one and the same; it was familial. When one returned to the underground, whether staying in the same place or heading up north to West Virginia for the promise of "equal pay for equal work," the subterranean had become a collaborator, of sorts, to an "alternative present" (Hartman 2007, 100). Maybe, after Hartman, the "hope is that *return* could resolve the old dilemmas, make a victory out of defeat, and engender a new order" (100). And maybe it was possible to make a new way into the mine, along seams that were "bound to other promises," "revisiting the routes that might have led to alternative presents . . . crossing over to parallel lives" (100). In another engagement with the Carboniferous, the time and space of Reconstruction held the possibility of difference in abeyance for the future perfect.

Coketown

By the late 1880s, many free miners who had been formerly incarcerated made up a substantive part of the new residents of Pratt City, which was known as "Coketown," and set up between Slope No. 2 and the Shaft mines near the new architecture of the Pratt Company Mining Prison Camp. As Pratt City grew, the geographic proximity and shifting status of incarcerated and free miners influenced political atmospheres in the mine. As Curtin comments: "The antagonistic relationships between free miners and coal companies in Birmingham also influenced prisoners' behavior" (2000, 146). Shared concerns between free and incarcerated miners created points of contact and scripts of dissent. While formerly incarcerated persons were denied voting rights, they nevertheless constituted a strong political organizing force. Since they had often been schooled in mining through the most extreme form of apprenticeship (as most "miners" were former farmers), it was a testament to the collective organization within prison mines that incarcerated young men both survived the dangerous work and became skilled practitioners. This mutual aid and craft mentorship (as there was no training provided for "shooting" the coal or any other aspect of mine work) was a crucial part of collective internment in a category.[30] Where the inspectors recorded accidents,[31] they tended to blame the prisoners for their lack of expertise or carelessness rather than seek to examine the inherently unsafe conditions and practices.[32] For example, as chief mine inspector of the state J. Hooper reported in the *Pratt City Herald* on September 9, 1899, "For the month of August I have to report an accident at Belle Sumpter, causing the death of two men. Dan Tolliver and John H. Jones (colored), miners. These two deaths seemed to be caused by the reckless and careless use of explosives, which is hard to guard against." While young men and boys had to learn how to set dynamite, pick coal, detect poisonous atmospheres, and haul coal, often neck deep in water through underground tunnels, these skills at manipulating and responding to materiality—inhuman tactics—were collaboratively shared across the zones of life and death. What might be said of Black miners is that they collectively developed both skilled mining and liberatory practices; as Curtin comments: "The political activities of black miners, fresh from the prison mines, showed that former prisoners were not discouraged but organized. . . . Ex-prisoners seemed to be especially active in party politics in precincts near the Pratt mines" (2000, 198).

Politicians appealed to white miners to vote according to the color line, as racial fervor was used both to campaign against the solidarity of free labor and to keep Black politicians out of government. In this familiar slippage between recognition and devaluation, Black miners suffered the lowest wages and the most dangerous jobs without financial reward for hard-honed rock skills. Reading the lease through the rubric of labor, rather than the afterlives of slavery, ignores how the lease often intensified and fed off the precarity of that passage during and after Reconstruction. The young, Black, able-bodied carceral subject became recognized in the coal and iron industries as the "good" laboring subject—the subject who could not organize to strike and could reliably produce (under the whip) coal without interruption, which in turn fueled an industrialization that was not under the siege of labor negotiations or power.

As an experimental iteration of capture for racial capital, the lease stemmed from the same structural stratigraphic impulse that the relation between genealogy and territory established under slavery, in the continuing relation between Blackness and subterranean subjectivity. It was an arbitrary form of policing social space for the plateau and achieved psychic and material consolidations of geopower for whiteness. The naturalization of subjugation and privatization of personhood in the judiciary under the lease system created, in conjunction with lynching, a totalizing attempt at the exercise of white power as a spatial execution of Black agency in the postbellum South. While emancipation recognized the existence of rights, in principle, the lease system nullified them in practice. The always-possible foreclosure of the social created the prison mines as a physical and psychic pressure. In Glissant's words, it was to "escape the abyss and carry the abyss's dimension" (quoted in Diawara 2010). The disruption of the spatial tense of undergrounds-overground (the prison mine) unsettles any easy settler colonial recuperation of labor as a structural collective, as well as resisting an easy incorporation into class-based social struggles of capitalism. The suspension of labor as an analytical frame animates another set of understandings in the kinds of relations that are being enacted between Black people and their environments (without the lie of autonomy and within inhuman and nonhuman lifeworlds), tacit knowledge practices that do not cohere to the script of the nation-state produced by geology. Being off script in nature-nation discourses bends toward more open-ended geographies of intimacy with the earth that are not prefigured around deadly encounters and the reproduction of colonial earth.

Modernity, Urbanism, and the "Spatial Fix" of Whiteness

> The people generally have come to realize that Birmingham is the biggest thing south of the Ohio river, and that whatever new light is shining for them has its origins in the mineral development. —*Weekly Age-Herald*, May 1, 1889

Racial strata made urban grids. A magic city called Birmingham levitated from minerality and black coal. Vulcan, the god of metallurgy (and the largest cast-iron statue in the world), stood on the hill overlooking Birmingham, as pyro-overseer of the plateau of iron and its white settler futurity. Produced from the iron ore in Pratt Mines (which became U.S. Steel), he was made to script Birmingham's origin story, its "roots" in mineral ores, and to perform the biggest booster statement a town could muster "south of the Ohio river." Displayed at the Louisiana Purchase Exposition in St. Louis, Missouri, in 1904, this Roman god of iron, complete with hammer, anvil, and spear, represented Alabama in the Palace of Mines and Metallurgy as a symbol of Birmingham's spectacular engagement with geology. The 120,000-pound (54,413 kg) Vulcan came "home" with the grand prize. The Magic City was rocking. His return was a little less glamorous. A spear was lost en route, and Vulcan was abandoned in pieces alongside the railway tracks due to unpaid freight bills; the 11,000-pound (4,990 kg) head loitered at the edge of town for a while until resurrection at the fairground brought him back to life and into circulation, and eventually to his own visitor park to commemorate an origin of Southern industrialization. The clean historiography of the mineral city was mapped into a Roman god in this forging of a nation, *Birth of a Nation* style.

There are no monuments of "nonnatural" history to those convicted to the lease, in which Blackness was made anew as a libidinal space of submergence and a space of amalgamation between seams of core, iron ores, phosphate, and steel production; the fertilization of earth; the construction of railroads; and the smelting of iron. In multiple ways the Black subject is spatially subtended to the underground as miner, resource, and dark matter. In Rivers Solomon's science fiction novel *An Unkindness of Ghosts* (2017), even in outer space, world-building must be performed by subterranean dark-skinned labor in the lower slum decks of HSS *Matilda*. A dark hand always seems to deliver capitalism's objects of desire and its escapes, as the background color of the category of value, as mineralogical origination seemingly brings it forth. As Wilderson comments, "First, capital was kick-started by approaching a particular body (a black body) with direct relations of force, not by approaching a white body with variable capital. Thus, one could say that slavery is closer to capital's primal desire than is exploitation. It is a relation of terror as opposed to a relation of hegemony" (2003, 22).

In my work in the archive as I searched for some visual trace of the prison mines, and around the town of Pratt, the only spatial markers I could find were entries in the fire insurance map of Birmingham, a stark reminder of the insured pyric life of the overground that accompanied the deadly life of the explosions underground. As the most profitable prison state, Alabama set itself up as a financial model for the nation and as a blueprint for white settler colonialism (that could migrate internationally, to South Africa, for example). It was a fitful experimentation by its own admission: "In our experience we have passed from very crude experiments to a condition which enables Alabama the most humane administration of the lease system known to any of the States" (State of Alabama 1891, 27). The emphat-

FIGURE 11.1 At the feet of old Vulcan, Palace of Mines and Metallurgy, Louisiana Purchase Exposition, St. Louis, Missouri, 1904. Keystone View Company, Meadville, Pennsylvania. Library of Congress, Washington, DC. https://www.loc.gov/item/95508082/.

icness of the tone and the bureaucratic exuberance of reports are markers of the overcompensation for an idea not yet normalized in the transition from human-as-property to human-as-disposable-energy. All the learned markers of the plantation system are in the report, the governance of sexuality via pronouncements on morality and health of the population that pivoted around the eugenics of the white settler family.[33] Yet, the social order of the carceral state as a state of entombment is not yet instituted; it was in the process of being established, pressure tested as economically, pragmatically, socially, and morally (and thus politically) viable. The coupling of a business model of carceral labor and a "just" tale of Black enclosure in the mine articulates an incorporation of subjecthood into an architecture of affective enclosure that is still actively working today in the prison-industrial complex.[34] In the fruition of the magic experiment, Alabama enacted the conjoined geographic relation between architectures of lease labor and the production of the materiality of urbanism, through the development of Black criminality as an urban affect. Vulcan stands in for the clean white dream of achievement through geology that doubled down on the subterranean status of the convict-as-slave and prison-as-mine in the carcerality of geophysics. The trajectory is such that after slavery built the foundations of racial capitalism, sustaining its growth and expansion into modernity, Alabama demonstrated how the prison could labor in a field of fossilization and force Blackness into new inhuman intimacies with fossil-being. It is

FIGURE 11.2 “Heaviest Corner in the World,” the intersection of First Avenue North and Twentieth Street. Birmingham, Alabama, 1910.

here that geology prepares the quiet context for racial capitalism and its material grounds.

In 1901 as U.S. Steel was floated on the stock exchange as the largest business enterprise ever launched in America, it made 65.7 percent of all steel produced in the United States and 30 percent of global production, thus solidifying US claims to achieving the biggest industrial revolution in the world.[35] Capitalized at $1.466 billion (213 plants; 41 mines, including Pratt and Fairfield, Alabama), U.S. Steel was listed as the first billion-dollar corporation in the world. Its founders were Charles M. Schwab, Gilbert Gray, J. P. Morgan, and Andrew Carnegie—familiar names in capital’s terrain. As *Steel: Man’s Servant* (Reed 1938), a United States Steel Corporation film, put it: “When labor and intelligence are applied . . . U.S. Steel builds the nation, because what would we become without it?” Steel, automobiles, skyscrapers, defense structures, flat-rolled sheets, tin plates, tubular steel, iron ore, earthworks of the settler nation and infrastructure of Anthropocene earth. The steel is still there, the railway tracks too, the ingots, the sheet, and the strong beams that hold everything up, like Vulcan. As racial capital floated, the convict lease subtended. The material structures that allowed buildings to rise into the air, railways to crisscross the nation and crossroad in Birmingham, Alabama, all the upward push and outward expansion of infrastructure as the

FIGURE 11.3 U.S. Steel advertisement, *Saturday Evening Post*, September 4, 1935.

sedimentation and consolidation of white settler space created urban forms and sleek materialisms that forgot their actual origin story. They erased the carceral mine in the earth and the hardscrabble incarcerated miners who did the work that made the nation materially manifest its date with destiny.

This is not to advocate for recognition of the part played by laboring toward the national project but rather to see the racial geo-logics of that gravity that was built on the backs of a Black stratum that got pushed deep

FIGURE 11.4 *Tennessee Coal, Iron and Railroad Co., Company Furnaces, Ensley, Alabama*, ca. 1910–20. Detroit Publishing Co. Library of Congress, Washington, DC. https://www.loc.gov/item/2016815606/.

within the earth to ground these edifices of white supremacy and their floating billion-dollar shares. What held up also pushed down. Dependable as U.S. Steel, so the script runs of material transcendence in the social text of matter, dependable as US racism in the color line of geology, in the operative geophysics of race. In the materiality of settlement, coal became part of the migratory possibility of white supremacy (as it did in England), making undergrounds that could move and grow, the projectile force of the extractive principle *enacting settlerism by depletion* through racial stratigraphy.[36]

The quieted social histories of convict lease labor in Alabama showed the emergence of a spatial and material fix for white Southern achievement, consolidated through "natural" resources. What starts as an experiment emerges as a proof of concept, placing Alabama as the richest Southern state. The intrusion into Reconstruction that convict leasing enacted quickly turned into something with the flexibility and muscle of a structure that becomes applicable in other places and spaces, that had little to do with the "fortunate geology" of Birmingham. The lease, while signified through questions of labor and economic geography, actualized a crisis and curtailment of the polis (as political space). Methodically, this shows how the maintenance of white surfaces in a terrain of racially segregated freedom had an equally (although obscured) verticality. The Southern descriptor of the problem denotes a particular politics of maneuver and prejudice, but it also distracts from how the convict lease system—the privatization of personhood for corporations—was about the "look north" toward capitalist investment.

Alabama's experiment became a key moment in the development of racial capital. It also promoted a handy resolve to the "population problem" of the South that geologists Lyell and Agassiz promoted. This late nineteenth-century precept of today's prison-industrial complex blueprints a carceral geography that takes its roots in an anti-Black ground and quietly expands beyond the mine (and takes the stratigraphic work of the "fathers" into twentieth-century capitalism). As fossil fuels represented the expansion of value and accumulation on the surface, the relation between the subterranean geology and the surface became one of enrichment from a racialized debt paid from the underground (and the deep time of the Carboniferous) to a white supremacist nation-state. While paleontologists such as Agassiz campaigned for a mythic origin to race, located in spatialized temporalities of deep time, these origin stories erased the quotidian race work that specific practices and regional politics enacted. Bobby Wilson (2000, 33) suggests that the transformation of race must be understood geographically as well as historically and that race was remade through those spatial relations of "local regimes, changing economic situations." As Birmingham, England, was an economic model of development through geology and the relocation of a rural workforce to urban centers, so Birmingham, Alabama, became a model for accumulation through racialized extraction of rural spaces. While Birmingham followed its British namesake in the industrialized depletion of the earth, the particularities of the region produced a very specific form of experimentation with industrialization and race that cannot easily be accommodated to the Marxian model of relations. It is not that the prison mine was a junior partner to the plantation but that it literally made the edifices of modernity, materializing the nation through a set of relations between the surface and underground that was stratified by race. The Southern issue of newly errant Blackness that challenged the slave-owning and planter economy—the stratigraphic Overseers—produced a regional urban theory that experimented with a novel solution to the regulation of resource capitalism and social space, which was the racialized underground. Birmingham demonstrated to the rest of the United States how to maintain the molten core of the slave model of production through new stratal relations, imagined through vertical rather than horizontal models of racialized capitalist debt relations in the institution of the prison mine.

Against the grain of the one-dimensionality of the categories that mark the convict ledger—race, county, and crime—to decimate the multidimensionality of Black life, inhuman witnessing, and alluvial materiality, another form of testimony might be called to think with this astringency without renaturalizing elementality. At Sloss (the heritage site) you can get a poster of a strapping Black miner, ladle over his shoulders, ready to pour an ingot in front of the ironworks. Proud, John Henry style.[37] Ready to take on the machinic age, to pay the ambiguous price of strength, the mythical hammerman of the Chesapeake and Ohio Railway's Big Bend Tunnel in the 1870s. The great antihero of African American masculinity who beats the drill, only to have it beat him. In the geo-logics of Reconstruction's incarcerations that structured Southern masculinity, a different kind of dismemberment faced the carceral mine and railroad worker, one that "latches on to the steeldriver as an ideal of black masculinity in a castrating country" (Whitehead 2002, 189). While John Henry, as the song goes, might be strong enough to beat the drill, the beating beat him into the ground.[38] Claudia Rankine refers to the medical term *John Henryism* to explain the strategy for coping with prolonged exposure to the stresses of racism while expending high levels of effort that result in accumulating physiological and psychic costs: "They achieve themselves to death trying to undo the erasure" (2014, 13).[39] In diagnostic terms, ratings of John Henryism are a measure of high effort coping among African Americans in the context of racial discrimination.[40] In Lauren Berlant's (2011) parlance, John Henry is an icon for "cruel optimism," of trying to outrun the thing that is trying to kill you, only to kill yourself in the process.

As the song proliferated on chain gangs and in mines, workers breaking rocks to the temporal refrains, it was a way to hold one another's brokerage down the line. "John Henry" was a song for geologic life, a stratal refrain, collectively carrying the vulnerability of rocky engagements within the earth.[41] As the carceral system had its expansion, so too did the song as a form that repeated across distances and places to be owned and recast by its singers and hammerers. "John Henry" is a sad and defiant rock ballad, of the deferment of freedom that pitched up against the rock and then the new machines of modernity. As one skilled route into labor opened and seemed to offer some path out of the convict lease system for Black workers, however tortuous, it was closed down by new technologies that sought to render the scope of passages to power obsolete. In the book *Steel Drivin' Man* (Nelson 2006), the mythic John Henry is claimed for the convict lease. An incarcerated person of the same name was entered in the ledger and buried among the bodies found in grounds of the Virginia State Penitentiary, one of those who had worked the Chesapeake and Ohio line, laid the iron track, and drilled the Lewis Tunnel in West Virginia. I have a poster of the Sloss furnace on the wall above my desk, with its John Henry–esque figure. It is in the everyday of this writing. There is no mention in the self-guided tour of Sloss of the lease and its legacies. Pratt too is closed, the entrances to the Shaft and Slope No. 2 mines grown over; a small bust of Daniel Pratt overlooks the river. The graves are in unmarked earth, and nobody has gone digging.

In many of the John Henry songs, Polly Ann, his wife, picks up the hammer and finishes the work. As Scott Nelson suggests on the post-Reconstruction sexual divisions of labor: "The story of a hammering woman seemed to speak a fundamental truth about black life in the South. As fewer and fewer black men could vote in the years after Reconstruction, black women carried on 'race work' in female institutions and clubs. They hammered away when it was dangerous for black men to do so" (2006, 107). In a Black feminist retelling, the musical group Our Native Daughters do a version of the song called "Polly Ann's Hammer," in which Polly Ann picks up the hammer, wields "it harder than any man can," leaving with a message of liberation in the myth of labor: "Throw down the hammer, and you'll be free."

The geomyth of Henry speaks to the axis of racialization in the convict lease that produced forms of mineral death and *its* afterlives in the inundation of geologic exposure. There are many John Henrys in the prison records. If we shift the perspectivism from fossils to processes of fossilization, then other sediments can be drawn out—reanimated, in Glissant's terms—as alluvium in the ghosts of geology. The muscular minerality made in the metallic bonds of incarceration, in which lives were forged in the erasures of coloniality and the profitable brutalities of extraction, draws out the inhuman as witness; we can name what is in plain sight, out on the surface, in the light of day—what stands as value masked through the disappearance of its devaluations. The "mineral isolate" points to the broken bones buried away, the foul "methano" gas, hydropericarditis, the beatings, the "blasted rock producing injuries" (Alabama 1886/88, 49), the infections that crept up on lungs and the dust that coated them, the short breath of silicosis, all undergrounded—broken events of nonnatural history that no one goes looking for, and few want to find. Bringing these histories to the surface is part of the redress of the colonial earth and the configuration of material redistribution. I think here of RaMell Ross's elegiac film *Hale County This Morning, This Evening* (2018) and his recurring question, "What is the orbit of our dreaming?," as the light shifts and the camera holds with the time of the sun, as questions of escape and duration linger through the changing day.

The earth has reclaimed much more than its inhuman materials. The inhumanities are in the shaft too. The erased is never erased; there is always resistance in relations of geoforce. The inhuman testifies to the inhumane. It is a site of traffic. A metal bender called Jim Crow. A song called "John Henry." Broken ground. Being condemned to the Carboniferous is a form of time travel, backward into the swampy lignin forests of compressed sunlight. Black in time is back in time in the lithic script but not in the makings of being Black in the day (see Summers 2019). The present always presses on the past, which exerts its own accumulated weight on what lies beneath. This is how the bedrock hardens. Each stratum has been imprinted from events that mark its geologic passage. In geology, surface is present time, from where things are read. From a "here" to a "there," the degree zero of the now is the index that marks the depth of relation. The "here" of the settler colonial present is also a ground of exclusion and denial of other modes of Relation, calling the shots on time's pasts. The mineral city can be seen as a mode of governance and filiation of white surface and time. This political present sits on racialized grounds and occupied lands, and natural resources are a political script that uses the semantic field of geology to naturalize its settler colonial violences. *Undergrounding* is an operative ontology of exclusion, a mode of subjects under erasure. *Un*grounding is a methodological amelioration. The earth's gravity is not just a metaphor, an allegory for the pressure that exerts itself on Black and Native life, but a material condition—a geophysics of sense—that impacts the possibilities of racialized persons to remain upright and in the sun, grounded in the freedom of the day.[42]

The mine and the furnace literally tempered Black exuberance and hammered a new gendered form of inhuman existence into being. Inhuman intimacies were once again the answer to surfacing a white earth. Geology underpins the imaginary of national projects of settler colonialism, as well as imaginaries of modernity and futurism. Political relations are enacted as spatial and subsurface enclosure produced by the desire for psychic and material control of space *for* whiteness. At the same time, geology establishes the grammar for a series of repetitions and material transformations in the constitution of concepts of value, materiality, ground, and the extensive transformation of earth into resources through extraction. It is a first-order language of materiality and natural (and political) philosophy. These repetitions are inscribed on persons-as-bodies and on and with the earth. Both geophysical states (race and earth) are transformed through forms of deformation (to consolidate other forms of enrichment). The proof of concept achieved in Alabama prompts a broader question, in the context of these racial capitalist modes: How do these subterranean spaces define the racialization of *all* space? That is, how are the white surface and the racialized underground, as a violent discourse and unequal material exchange, interpolated through all subjective (colonial) relations to the earth under racial capitalism where Black bodies are "bodies occupied, emptied and occupied" (Brand 2001, 38)?[43] That is, histories of fungibility, marked by the *extractive principle*.

If, as Fanon observed, the Black body was a vehicle toward the materialization of Man, then the shaft was the mode of settlement to make the plateau. Thinking urbanism through its shafts and extractions—every building a mine—is to understand urban architectures through what is taken rather than what is given, the residual and affective architectures that barely remain rather than the built environment, the stratal relations rather than surface exchanges. If we looked at the extraction sites rather than the buildings, not just commodity chains and ethical sourcing but the very conditions in which extraction is made across multiple disciplines, bodies, and their exhaustions, and through inherited conditions of exposure that pass in the mine from inorganic matter to the organic, there is a way to think colonization in reverse, to decolonized futurity and understand the ruination as a path to somewhere else.[44]

12 Inhuman Matters V

Trees of Life (and Death), "Strange Fruit," and Geologies of Race

> An aesthetics of the earth? In the half-starved dust of Africas? In the mud of flooded Asias? In epidemics, masked forms of exploitation, flies buzz-bombing the skeleton skins of children? In the frozen silence of the Andes? In the rains uprooting *favelas* and shantytowns? . . . In mud huts crowning goldmines? Yes . . . Imagining the idea of love of the earth. . . . Aesthetics of rupture and connection . . . under no circumstances could it ever be a question of transforming land into territory again. Territory is the basis for conquest. Territory requires that filiation be planted and legitimated
>
> —ÉDOUARD GLISSANT, *Poetics of Relation*

In 2018, the National Memorial for Peace and Justice opened in Montgomery, Alabama, as a monument to the victims of American white supremacy. Eight hundred rusting Corten steel columns hang from the roof, each listing the county and number of lynching victims, many categorized Unknown or Unnamed Man.[1] The names and places are drawn from over four thousand documented killings by the Equal Justice Initiative, from the end of Reconstruction in 1877 onward. The suspended ingots materially mark the steel and iron industries that were fueled almost exclusively by convict lease labor in the coal mines of Alabama, around Selma and Shelby, at the end of the nineteenth century, and the extraction that was first mined by enslaved persons.[2] As the walkway descends under the suspended steel, the viewpoint becomes subterranean, taking the viewer below the hanging bodies of metal to embody the view of a spectator. An architectural perspectivism hangs the optic below the imaginary of a strange tree of life, the symbolic bodies of metal elevated only as a marker of death. Inverted skyscrapers of the Magic City. As Billie Holliday sings: "Southern trees bear strange fruit / Blood on the leaves and blood at the root." As at the National Museum of African American History and Culture—that takes the visitor down in a large lift underground from the present to the past, through time, to the year 1400, where the doors open into the conjoined question of "freedom and slavery"—the perspective is a *lithic-eye-view*, a subterranean take on the present (see Gómez-Barris 2017).

As the elevator door opens on the beginning of the slave trade, a mineralogical exchange abounds that underpins the colonial impetus that made Africa a continent of extraction through global geographies of accumulation, displacement, and enslavement: on display are sugar ornaments in silver; gold, iron, and copper coffles; and an ornamental pair of little silver slave fetter earrings that signify property. Pressing the audience together in physical proximity, the exhibition carries the visitor through to the "door of no return" (see Brand 2001) and inside the hold of a ship, and onto a walk-through of Black historical time to the ground level of the present and up to the culture and community galleries on higher floors. The racial relation between the underground and the overground is posed by the spatiality of both these exhibitions, history as strata, geotrauma hanging as a question in the present tense.

Inside the Legacy Museum: From Enslavement to Mass Incarceration in Montgomery, down the street from the steel ingots of the memorial, stand jars of earth collected from the ground beneath the lynching trees. Different-colored earths, standing in tall jars, collected by relatives of those who were lynched or by community members, record a geology of racial violence. The memorials of earths and the hanging steel monuments connect the quotidian enactment of hierarchies of species life as they were set out in the nineteenth-century racial paleontology of *the* "Tree of Life" (as origin, genealogy, filiation). Thinking the praxis of both these acts of *Southern Horrors* (Wells 1892), of trees and their prehistoric forebears sedimented in the mine, highlights Black subjugation in proximity to an entanglement of geologic acts that cultivated the renewal of white social space in slavery's

FIGURE 12.1 Detail of the enslaver Captain John W. Anderson's Mason County, Kentucky, "slave holding pen" (early 1800s). National Underground Railroad Freedom Center, Cincinnati, Ohio. Photo by author, 2016.

afterlives.[3] And how Black vulnerability was kept "present" in space as a violent antagonism that questioned spatial autonomy and self-determination after emancipation.[4] In response, recent institutional Black archives have collected erasures of earth and challenged the violent placings of white supremacy in its pursuit of claims to the surface and present. As the white supremacy of matter and geologic philosophies buttressed ideas and actuations of racial ordering, an "aesthetics of rupture" is charged with making a different surround and earth. The record of those who were not permitted into the genealogical account of the master were subject to temporal arrest or geographic abandonment. The poet Dionne Brand marks this in her novel *At the Full and Change of the Moon* (1999) by a family tree that contains people labeled "the one unrecalled," "the ones left in the sea," "the one taken by hurricane" (n.p.). Their marked absence produces a map of noted silences and another geography that has no archival account. Her geography challenges the patriarchal lineage of white geology (and its selection of successors of those who could claim paternity and those who could not) and scribes the geotrauma of the diaspora.

Visually mirroring the geologic survey cores and mineralogical samples that preceded Birmingham's rise as the "Magic City," the named jars insist on understanding the earth as inhabited. The colored earth pays attention to the illegitimated geologies of race in the formation of whiteness and the nation-state. As both a statement and a practice of digging down to materialize the disappeared Black archive, the aesthetics of the earth are a counterpoetics that demand a different sensory geo-logic of the earth. Counterpoetics not only are counternarratives of history that do not presume, as in the case of forensics, the possibility of a transparent and complete truth, but the poetics reside in the tension of incomplete and incompletable histories of geotrauma. Taking ownership of the soil as both an individuated (named) and a community project reinscribes the ground with its subjective and communal life and caring for the sites of experiences outside of the eugenically inflected geologies of whiteness. In the gathering there is the refusal of a fiction that there is nothing to see in this ground other than resource or white futurity. Recalling the dead is a deliberate cultural occupation of the violent and sensual nature of earth. It claims another relation of material witnessing.[5]

The memorial soils address the chasm of correspondence between that which is there and that which is gone, and the questions that this provokes for receiving the disappeared and hearing their testimony as an ongoing shock that travels with the present. The soil in its forensic potentiality as

FIGURE 12.2 Exhibit of underground "holds" of fifteenth-century Elmina Castle slave trading post on the Gold Coast, Ghana. National Museum of African American History and Culture, Washington, DC. Photo by author, 2019.

material witness mobilizes a psychic attachment to the possibility of worldbuilding in the absence of a world. The community action holds the fact of the gathering that the shock-forward of erasure initiates. While the forensic promise is just that, the gathered earth marks a possible way to open materialisms beyond the geo-logics of territory.

As Glissant makes clear in the epigraph to this chapter, territory is defined by its limits, and those limits must be expanded: "This is the reason it is worth defending against every form of alienation" (1997, 151). What Glissant suggests in the return of force is to craft an aesthetics that is a "variable continuum, of an invariant discontinuum"; such a counterpoetics risks itself against the idea that an aesthetics of the earth will seem "anachronistic or naïve: reactionary or sterile" (151). Like Wilderson's (2010) concept of "negative heritage" as the interruption of historicity rather than history, Glissant advocates that "we would inhabit Museums of Natural Non-History" that "re-activate an aesthetics of the earth" (151). In the same geography of "strange fruit" hung to reify those symbolic trees of racialized life, innumerable shallow graves of predominantly Black men and boys who died in the mines of Alabama are located but unlocatable.[6] As with the recently discovered

FIGURE 12.3 John Babayak installing models in the museum's Oil Geology Hall (1955). American Museum of Natural History Library, New York. Asset ID#323426

graves of incarcerated persons of the convict lease at the Sugar Land Plantation in Texas, men were buried with the refuse of the mine, incinerated in the blast ovens of the iron and steel furnaces. Abandoned mines full of abandoned histories, museums of natural nonhistory of carceral undergrounds. Attempts to concretize mineralogical traces of the strata of the unliving tie together the genealogical and the stratigraphic in the Black archive of geologic life.

As mass incarceration served the dual purpose of attending to the fear of the reproduction of Blackness in space (or the "Black outdoors" as a political antagonism) while furthering the eugenics of the settler nation, the work of community collection effaces the indifferent histories of geology in that region. Earth is made in an altered aesthetic register to the samples and cores that were, alongside the lease, part of the technology of the material expansion of white surfaces. The charge of these jars of earth aesthetically parlays with technologies of the earth as mineralogical sample, proof of concept, biopsy, or soil library for differentiation and identification. The indivisibility of this soil as history exposes the fiction of its scripting. There is the resonance with handfuls of earth gathered up in clenched fists of remembrance. The soil is not meant to lie. It is part of an objective science. It is both place, forensic archive, and in its negative utility (as in blood and soil), a political resource.[7] As Zakiyyah Iman Jackson argues, "The scene of the empirical can also be a site of the performative reiteration of sociosystemic aporia and opacity. The 'metaphysical' and the 'empirical' are not distal from one another but entangled" (2018, 627). The sincerity of these geologic acts for community groups is part of "placing" lynching in an erased landscape of memory, naming a violent geography, transforming the perverse spectacle into an empirical mark of geotrauma. In a call-and-response aesthetic mode, these acts organize and communicate with other geologic acts in mimetic relation to produce a different kind of visibility of earth as a property of subjective formations. As the "bodies" of Corten steel that hang in the Legacy Museum in Montgomery visualize and inscribe a landscape of lynching, they also make connections to the violent material histories of the convict lease and its production of iron and steel that shaped the urban modernist landscape. The aesthetics feels wrong. There is something beautiful in earth cores, in Corten steel, in all the geologic materialities, but this is perhaps the point: geology hides the hurt. The urban fabric carries the flesh of geology as much as these monuments.

The retrieval of the memory of inhumane geographies will always be incomplete, yet it represents an *intra*materialism, shuttled across thresholds of spatial and temporary categories that dissemble the enclosure produced by the praxis of white geology. While too many names and experiences remain unknown, this caring for names, locations, and earths does provide another geophysics that challenges subtending narratives in their unearthing. The soil is made sentient through these memorials, disrupting the historic role of geologic grammars in dispossession and slavery, and in the ongoing extractivist projects of settler colonialism that continue to impinge on Indigenous and racialized communities. The jars in their uniformity connote archival forms; the earth probes at the material proximity of lives in a cartography of inhuman enclosures, to and *for* the political present, in grammars of care that develop an aesthetics outside of that violence. While mirroring an uncanny aesthetics of attachment (to geologic cores), the jars of earth offer methodological repair work on these epistemologies, even as they retain the aesthetic allure of their geological counterparts in making territory. Marking the possibility of material witnessing as a possibility raises the libidinal relation of materiality and geotrauma. As a principal site of its historical and cultural production, the legacy of geotrauma demands the imperative to understand decolonizing in and through a geologic register (this is also the basis of earth reparations that address the environmental legacy of these racial geologies). Which is also to say that if there can be no understanding of geology without race, reparations, repair, and remembering must be made within the geosciences and the broader field of geophysical relations.

Like a hundred other museum exhibitions of geology, geologic survey stores, or practical museums of geology for field sampling, the jars are a locative device of subsurface aggregate, a mineralogical core into the temporal sedimentation of materiality. What is read off these temporal lines is perhaps less important than the acts of unearthing, as a practice of disturbing the normalcy of the surface, strata, and its "innocent" succession or past geologic lives. The persistence of mimicry in relation to the colonial archive challenges the reiteration of earth as deracialized geology. The colonial archival aesthetic works alongside the recent forensic unearthing of Black persons on plantations and of Indigenous children in residential schools, to populate the rift of coloniality, but it also refuses the easily ascribable storying that valorizes the hidden labor that brought the nation into being, as further consolidation of the settler state. Simultaneously, in the case of the lease, it problematizes the white masculinity of the miner and coal as an ongoing site of white supremacist rhetoric. While the reconstitution of knowledge is an ongoing struggle in how the present is narrativized, there is no easy settlement in these new archives. In thinking with the monuments of steel, mine, and earth, I want to notice alongside the visibility of these memorializations the attention they cast on the materialities of race (Saldanha 2006). As sites of recovery, and erased geographies and geologies of nation-building, these Black archives speak to what Césaire understood in the context of Caribbean experience to be the impossibility of knowing the earth in its present without Black and Brown life—a racial geography that "plunges into the red flesh of the soil" and "without whom the earth would not be the earth" (Césaire 1969 [1956], 59, 57). Which is to say that there can be no understanding of the earth without an understanding of the pasts of this racial geography.

As the jars of earth set up a dissonance in the mimesis of mining samples, they set up a conversation to another earth that constituted the soils of the present, and thus mobilize a temporal rift in histories of remembrance. Temporal rifts possess liberatory qualities through the redefinition of the violences of colonial histories and its account of the earth, offering inhuman tenders for building another earth. This archive of unthought strata—*not-fossil*—is what ties the soil to the hurricane, the levees built by convict lease labor to the Louisiana Superdome, the wildfires being fought by the incarcerated in California to the burning buildings of the Watts rebellion, to all the strata of Black life that insulate white bodies from the earth's geoforces. This is also to think about the materialization of ontological displacement in its forms of quotidian subjugation, as enforced mining that erases the possession of mine.

In a counteraesthetics, the "aspiration (the pretention) to the universal must be interned in the dark secret soil where each lives his relation to the other. The poetics of this shared quotidian is fastened in the succulence" (Glissant 2010 [1969], 16). In the dark and secret soil, attendant histories of subjugation are selectively erased to promote a clean version of extraction that naturalizes and buries its racializing violence. Visit at any small-town mining museum, in Watford, North Dakota, or Madison or Beckley, West Virginia, in Bakersfield, California, or Dora, Alabama, and you will see the narrative (and naturalization) of white heteropatriarchy and the fossil fuel industry writ large: coal and oil extraction serving in the heroics of white nationalism ("Trump Digs Coal," as the election campaign ran to speak selectively to the debt relation of deindustrialization and its dejected conditions, through the identity politics of white as the authentic color of mining [M. Thomas and Yusoff 2018]). It is a narrative that denies the complex geographies of interrelation between natural resources and race, while simultaneously producing those fossil fuels as sites (particularly in the context of postindustrial poverty) of investment for white supremacy in the nostalgia for "productive" masculinities.

As the reification of rocks was systematized in the same orders of elemental essentialism (or the search for mineral purity) as the geo-logics of race, Black and Indigene were rendered fungible categories of economic unitization within the project of territory, who did not arrive on the scene on their own terms as (universal) subjects able to "work" metals and minerals, but as categories of value within the projection and realization of natural resources. *Pochahontas Mine No. 1, West Virginia*. Once the inhuman is understood as subjective space, the fiction of Life as autonomous, agentic, and independent falls away, alongside the bifurcation of the *bios* and *geos*. The Blackened archive of earth challenges a subjectivity raised outside the epistemological remit of the inhuman, wherein the purified ontology of Life is pulled back to the racialized propositions that constitute its historic ground and grounding, or another earth. Thus, the Blackened archive functions as a circuit breaker in the recursive dialectical grounding of humanism in the inhuman (scripted through the differentiation of genealogical lineage and isolation). It breaks the script and unmasks the fictions of narratives of materialisms' relation to subjective forms and their violence of emergence. Taking seriously the ways in which geology encloses, as well as differentially subtends, forms of subjectivity and modes of living through the *bios-geos* split, I turn now to the Tree of Life as a diagrammatic tool for bifurcating the movement and directionality of Life and its grounds.

The Purposeful Tree of Life

The categorical division that began in the project of geology in the paleontological imaginary continues its structuring work as a material and theoretical praxis of white supremacy of matter through the nineteenth and twentieth centuries, primarily in biologism-as-possessive-individuality that severs its "root" from the environmental grounds in which it emerges and is indebted to. Whiteness in the North American settler context organizes the unearthing of value for accumulation, which simultaneously destroys the ground of Indigenous relation through that unearthing. Reversing the direction of the inhuman-as-inhumane requires thinking about the question of how solidarity might be possible through the inhuman ground, coming through the earth to build affiliation in an expansive zone rather than configured around a projectile imagination (what King calls "conquistador humanism" [2019, 44]). I take up the materialization of the Black archive in relation to the "Tree of Life" both to account for the embodiment of genealogical thought (within paleontology and later within subsequent biologisms) and to show how within the division of *bios* and *geos* in the late nineteenth and early twentieth centuries, biocentrism is plotted along a geographic color line in the act of parsing Life to sediment racialized grounds. The paleontological racial ground that geology established became "replanted" in biocentrism, shifting the foci to biology (and the body) and the question of Life itself—a focus that becomes familiar through its biopolitical formations in life sciences and, in the humanities, coalesces around vitalism and humanism. As the stratigraphic imagination becomes sedimented into philosophical and social thought, it is backgrounded, understood as that which propels Life and its eugenic imaginations, but is separate from the organism. The Tree of Life becomes distinct as a narrative of purposeful vitality that continues to ground genealogy and filiation in inhuman modalities. So, while the individual emerges as the subject of Life, the geologic imagination holds that hallucination in a normative racialized category of matter. Thus, the biologism that populates the nineteenth and twentieth centuries remains broadly secured by seventeenth- and eighteenth-century racist paleontology, a paleontology that was used to justify quotidian and spectacular incidences of terror.

The eugenic instantiation of the Tree of Life transmutes a symbolic demarcation of anti-Blackness as an incarcerated binary to facilitate a progressive account of the biologism of Life that is coded through white geology. The continuation of the inhuman as a racialized ground that feeds the tree's trajectory suggests the tight intimacies of categories of exclusion in accounts of matter and subjectivity that govern the politics of nonlife. Thus, historically in the early 1900s the inhuman can be seen to be raised both as an ontological challenge to the separation of life and nonlife and as an occupied placeholder of that threshold that secures the ground of progressive (white) Life. Today, those two genres of the institutional and the spectacular (originating in the lease and the lynch) have collapsed into one another in police killings, where the immunity of state-sanctioned violence for white perpetrators is coupled with the regulatory carceral force of the prison-industrial complex. Resisting claims of either subjectless matter or the performativity of the inhuman in a vital or animist mode in materialist philosophies, I argue that the inhuman needs to be understood in its historic geologic grammars—metaphysical, geophysical, and paleontological—to look at how these different modes of engaging the inhuman collaborate as a specific praxis of world-making along a racialized axis. Matter, even in its abstraction as mineralogical object (without a subject), is a repository for a material-discursive formation of race work that continues to structure the nonhuman as ecology and eugenic tree (see I. Brown 2020).

The Black earth archive puts pressure on the concept and categorization of the inhuman as an antagonistic force and relation, that is conceived either in a nihilistic ontology or as undirected forces of the cosmos (i.e., nonlocal, undesirous of or indifferent to life, materiality without direction or purpose). Thus, to speak of the inhuman without the Black earth archive is to skip over its most intimate historical and ontological relation, while also extending the violence of the afterlives of those histories into the present through the affectual architectures of their reproduction. The negation of the racializing praxis of the inhuman is foundational to the establishment of the ethical humanist subject. Thus, summoning a poetry of the rocks without attending to the foundational internment of race is a deferral of its material and metaphysical instantiation.

The erasure of inhuman intimacies also has consequences for how environmental justice is understood in the present and in which registers and spaces of redress that justice is understood to be located. While race continues to materially buffer the valuation of whiteness (and its multiethnic material subject position) in environmental events and extractive practices, there is a need to understand the relation between the individuated transcendental subject of Western metaphysics that contemplates inhuman nature from the ascended viewpoint (as in Thomas Moran's, Albert Bierstadt's, and Thomas Cole's landscape paintings, or Lewis and Clark on the plateau again) and the position of the ground (as subject, resource, and geophysics). A poetry of the rocks, without the map of relations that its geologic grammars make possible and erode, gestures toward a nonattendance to colonialism and its ongoing extractive formations that constitute the praxes of colonial-Anthropocene earth. By the same accord, if geologic grammars make colonial extractivism and subjugating relations possible, they are a crucial site in unmaking the reproduction of those forms and relations into the future. Thus, there are liberatory possibilities in the ontological proximities of the inhuman to differentially inscribe the strata and switch up its stratifications (Fanon's the first shall be last and the last, first), for the necessity of earth reparations that pay attention to its racial inscription.

Returning to the jars of soil: they render the abstraction of the ground into a different sense act to take place and make an inhabitance of the subsoil. The different-colored earth activates and echoes the different strata of geologic time that were keyed on the geological survey maps of Alabama (the first geologic map of Alabama was produced in 1835, and extensive mapping was undertaken from then until the late 1890s).[8] While geologically specific to the violence in Alabama, the soil campaigns on the sense for a broader theorization of the nonrecognition of forms of life lived below the surface—in both metaphoric and material terms—as an erased history of place on a symbolic level and as specific racialized and gendered material history of extraction and experience. These named jars will have a very different resonance and pain in their proximate historic and community relations, but these are not unrelated to the broader geologic histories of racialized geologies (their parallel cores in the state geologist's office). As particular moments of material inscription in the categories of matter and narratives of materialism, the soil is part of the lexicon that makes Blackness adjunct to mining, terror, and extraction. The material transformation of Blackness and the transformation of geologic materials are an amalgamation.[9] The jars of earth offer a counterstatement to the disappearance of the Black archive, a resistance raised in racial geologies' grammar but refusing its inhumane instantiations. Earths burn in the fissure of the proximity in geologic life, intimacies that run along geologic epochs and racial lines.

Trees of Life and Death (or, On What Grounds Necropolitics Grows)

If eighteenth- and nineteenth-century geology was marked by an increasing desire to chart beings through time and to stabilize the "events" of geologic time into profitable seams, the ascent of evolutionary theories allowed previously static Beings to move more readily across temporal thresholds (and thus imputed a more mobile geographic imagination of strata). In short, Life became migratory following the geographic liberation of whiteness as empire. In the context of colonialism, Human origins theory began charting a geophilosophical way outta Africa, as fast and furiously as it could, while moving farther into the "dark continent" to ransack its resources and codify its people (according to the premise of the evolved apex of the arrival of *Homo sapiens* in Europe). The first hominids, as they are routinely exhibited in the museum context, are African. While the Great Rift Valley had yet to be "discovered" as the geographic origin of life, Africa was never imagined as the hominid end point. The Indigenous African stands as sentinel of the species trajectory, but the geographic trajectory is outward, a projection into a New World. Blackness stands at the gateway to the cosmos and becoming human, but it is a threshold presented for improvement, a portal for a white futurity. That is, the African must go through a transformation—genealogically and geographically—to become the civilized subject of European contemplation and enterprise. Africa in the colonial imagination is past source, not present relation. Root, but not tree. An engraving from Sierra Leone in the Museum of London Docklands makes

FIGURE 12.4 Artist unknown, *Cutting Down the Fetish Tree*, 1850. Museum of London Docklands, West India Quay, London.

this point directly. In *Cutting Down the Fetish Tree* (1850), we see a suited colonial overseer, hand outstretched, showing his work to the crowd and spectator, as a muscular Black woman cuts into a tree that is laden with animal skulls, knives, various implements, handkerchiefs, and arrows. In line with the monocultures of plantations, the fetish tree and its magic possibilities (and their cultural and ecological affiliations to category crossing) were cut down so that the plantation could take its place.

In the Tree of Life, the presupposition of an aboriginal state of matter and its projection into Indigenous and Black personhood holds subjective life, literally, in the ground, while liberal individuality is being forged as transcendental, via the Tree of Life, as epic nature mastery. The modern hominid left the past behind and arrived anew in another geography. A static account of Being in time broke with absolute stratification, and ideas of "progressive" evolution began to be articulated against a backdrop of extinction from dynamic earth events. First used by paleontologists, the species tree became a way to both articulate and visually map the movement and connection of beings across time and to constrain them in a new hierarchical structure or network that controlled movement and maintained a progressive account of Life's arrival at the present (to reinforce its politics around inclusion and exclusion in polity).[10] The tree was arboreal but could

FIGURE 12.5 Species tree of human evolution diagram: "You are here." David H. Koch Hall of Human Origins, Smithsonian National Museum of Natural History, Washington, DC. Photo by author, 2017.

still have an apex to hide the telos. Nobody was stratal bound, but neither were they at the top of the tree, so to speak.

While the Tree of Life was a model, metaphor, and diagnostic tool popularized by the former geologist Charles Darwin's *On the Origin of Species* (1859a, first published 1837), which articulated a common line of descent with modification and adaptation through evolutionary change, it had been in use since the early nineteenth century. Phylogenetic tree diagrams were employed to represent genealogical relationships through time (for Darwin, it was a species shift from ape to modern human). As a continuation of the concept of the last common ancestor theory in paleontology, which placed the "root" of life in the earth's geologic epochs, the tree symbolized the hierarchical and relational structure of life and nonlife as it was understood to exist. In keeping with the monogenesis claim, life was mobile but was still held together by its origination and arboreal organization. Thus, the work of time remained important for the understanding of mobility and spatial politics, and where that was understood to be located and originated. Species origin stories supplemented genealogy for actual duration in place, conveniently erasing Indigenous claims of being in place. Pocahontas is

FIGURE 12.6 Detail, National Memorial for Peace and Justice, Montgomery, Alabama. Photo by author, 2018.

entombed in 1882 in endless representation of racial capitalism in a coalfield and mine in West Virginia.

One of the first genealogical trees was a "Paleontological Chart" in the publication *Elementary Geology* (1840) by Edward Hitchcock, although the most famous "first" tree was Jean-Baptiste Lamarck's in *Philosophie Zoologique* (1809), which was an upside-down tree that starts with worms and ends with mammals. Lamarck was not a believer in a single common ancestor and instead believed in parallel but separate lines of progressive descent, from simple to more complex forms of life. While the Tree of Life underwent much scrutiny as a bad metaphor and material imaginary for the relation between different forms of life, it persisted as a way of understanding the movement of beings through geologic time, within the trajectory of arrival, the "You Are Here" in the polis of the present. Take, for example, the exhibit *From Fish to Man* (1949) in the Biology Hall of the American Museum of Natural History. The plaster casts move from fish head through reptiles, lemurs, monkeys, and gorilla heads to the Indigenous man (racially coded African/Aboriginal) to the end of the species line, a classical Greco-Roman bust of civilized man (racially coded European). The pairing of geologic and human timelines in linear development conflates the multiplicity of events and durations in time with an account of monolinearity in space, which reinforces the idea that "dead ends" are unimportant and that continuance (or growth, as it is rendered in liberal fashioning) becomes the marker of species accomplishment. It is a form of life that is cut off from its emergence in place and the systemic organization of nonemergence. The diagrammatic implies a sequential development, and as geology proceeds by inference, this visual grammar does work in consolidating a progressive life story imagined in line with the planet's own. Before the tree model became a map for genetic divergence and relatedness, it was a diagram of paleo difference and similarity read through colonial genealogy. *Not-fossils* stay in the ground.

While geologists and the genetic sciences have all but abandoned the tree, the accumulative movement through time has stubbornly persisted as an imaginary, hanging about in public culture and science communication as a go-to narrative frame.[11] Before the tree morphed into its antiteleological branched relation (mapping onto imaginations of cybernetics and kin, in its later historical networked forms), it was a hierarchical division of life that paralleled the racial ideals of the mine, plantation, and settler colonialism—that of white growth through conversion of the "ground." The security of "You are here" is clearly more compelling than the acceptance of a less confident exceptionalism in the universe. And the confidence of being in the right place and time was never that secure for some, depending on their relation to the symbolic tree of life, as spectator or spectacle. Genealogical trees perform what Michelle Wright (2015) calls the linear, unbroken course of genealogical relation, that was/is a racialized space-time of colonialism.[12] Wright demonstrates how the focus on Middle Passage histories, European colonization, or African civilization creates "either a linear progress narrative or, when reversed (as in Afropessimism), a reverse linear narrative indicating that no *Black* progress has been made because of the continual oppression by white Western hegemonies that began with slavery [and] moved through colonialism" (2015, 8).

This linearity of time narrates Life and its imagined forms and possibilities (as well as its antagonisms) along the color line and its switchbacks. That is, Life (as it was becoming in its particular biocentric and biopolitical forms) cannot be taken as something immanently self-explanatory, and if it is, this obscures its temporal and geographic historicity as a certain material or geologic *form* of life: across registers that span from the material conditions of labor and privileged social positions, to long-term population longevity through to the use of fossil fuels (beginning in nineteenth-century Britain), to the development of medical procedures (with their attendant histories of testing on Indigenous, African American, and Pacific and Marshall Island peoples without consent). As the tree guarantees an account of a striving, progressive, purposeful directionality in the iconography of life, nonlife gets left in the ground as nonagentic, static, and inert matter that relies on the operation of Life to author, accumulate, and ontologize it. Life in the rift remains obscure to the plateau and deep in the ground.

There are three broad questions in the episteme of the Tree of Life that I engage to explicate the politics of the inhuman (as *geos*, non-life, not-life, inhumane life) in what becomes a century of biologisms.[13] First, what analytics allow Life, as a mineralogically situated artifact and ontological historical object, to be cut off from its inhuman ground? And how does this establish a concept of an autonomous time for biology free from the gravity of earth time and environmental conditions of possibility (i.e., in ways that make Life transcendental)? Second, how does the positioning of the border and boundary of Life/*geos* come to adjudicate on social and political forms of racialized life through the figuration of the inhuman (i.e., in ways that position the earth as the structural guarantor for Life yet a partitioned participant in its agency)? Third, in this relation of figure (Tree of Life) and *geos* (earth), how does a hierarchical structure of verticality coupled with a concentration on the horizontality of social and reproductive flows get replayed as a racializing axis that continues a deep time trajectory of the geophysics of race? My concern is with the deadly antagonisms between geology and racialized life and its geo-logic matrix, rather than to take Anthropocenic geologic subjecthood as a self-evident category of explanation. The image of the origins of life in the La Brea Tar Pits Museum in Los Angeles makes the point: The painting is of a continuum of life from the bacterial origins to the end with man in a spacesuit. Man has shaken off the shackles of life as a trajectory of mammalian accomplishment and sits in between on a precarious threshold of cosmic beyond and floating in space (entirely reliant on the infrastructures of technology, which is the mark of Man's accomplishment and escape from nature) (see Yusoff 2019).

In the Tree of Life, the thread of Life gets traced through geologic time as an accumulative event. As Life moves through time, from its humble bacterial beginnings, it is growing, branching out, reaching, becoming the "referent-we." Life and the micro-operations of Life Inc. become an operative category of accumulation and improvement in a different kind of nationalist body, what Wynter calls *Homo oeconomicus* or *Man 2*. *Man 2*, as the profitable accumulator, continues in the wake of *Man 1* (*Homo politicus*), as an analytic that divides the "naturally selected" (Europeans) and the naturally "dysselected" (the symbolic Other as Negro, native, poor) through the practices of material, psychic, and scientific (genealogical) divisions (see Wynter 1989, 640). Instead of this racialized division, Wynter suggests the need to move toward a different mode of subjective relation and autonomous social formation, which she calls "transphysics." By this, she argues a counterconcept of subjectivity as collective and specific to a place in the world (or what might be called a locative autonomy rather than autonomy that is pernicious in its colonization). She says: "The call for a 'transphysics' is linked in Glissant's poetry and fiction to a poetics whose primary referent/topic is, rather than the subjective and intimate life of an individual, that of the blocked individuality and fulfillment of a people, the Antilleans, of their realization as a new collectivity" (1989, 639).

Glissant's account of alluvial citizenship as a collective that builds up, again and again, through Relation (and thus must counter against the erasure of Relation) counters the Tree's monofiliation or single root metaphor. Decolonization requires belonging in the material landscape in ways that do not repeat colonial claims of territory in their geographic imaginary or in the syntax of their materialisms: those imaginaries of geologic life are genealogy, filiation, and origination, and they are organized through natural resources (as elementary, value, and white geology). Both Glissant and Wynter are concerned with how places and practices that have been covered over by colonial practices (such as forgotten rivers and mountains of marronage) recur in Caribbean poetry as imaginaries that replant culture in ways that do not have a colonial perspectivism as their point of departure and scope of reference. Wynter's "replanting" of cultures of the enslaved in a new land is different from colonial supplanting of roots (that literally replace the fetish tree with a plantation). If one thinks a rooted methodology that is alluvial rather than genealogical, the counterpoetics of the suppressed emerges in a way that uncovers that which is secreted in the underground by passages of people that have been bound to the earth ("the earthbound") and must live in the intimacy and alienation of that inhuman Relation, and in the gravity of colonial duress. Explicitly arguing against the division of inhuman and human, Glissant moves against the colonial conflation of cultivation and civilization as the technology of terraforming earth.

"You are here" commands the assertion of origins through genealogic relation (the enslaved bound to colonialism's temporal and spatial frame), while erasing the violence of that geologic grammar on other registers of sense. This kind of "analytical framework is unsettling because it simultaneously archives the violated black body as the *origin* of New World black lives just as it places this history in an almost airtight time-space continuum that traces a linear progress away from racial violence" (McKittrick 2013, 9). McKittrick argues that when this violence is named, it is read from a "here" in which the racial hierarchies of humanness are deemed to be socially and economically open (and, to some extent, already incorporated in a "we" that improvised spaces—slums, the Global South, waste sites, and marginalized people, i.e., new plantation sites—in a striving *toward* some kind of imagined arrival). Thus, the reading of "then" is made from a shared locational "now" that reinforces the sociobiological eugenic narrative of progress. These conversions to the normative geographies of whiteness (as they are realized through multiracial participation) simultaneously deny Black geographies both in their historic forms and contemporaneously in the Black outdoors (see Solomon 2019) whereby "Life, then, is extracted from particular regions, transforming some places into inhuman rather than human geographies" (McKittrick 2013, 7). The migrational legacy of the inhuman-inhumane couplings of material and epistemic orders that plant Black life in the mine (and the strip mine of the plantation) intern a recursive inhuman spatial production, a geo-logic loop of being earth. Planting, then, becomes a way to live and grow a passage through the earth via consent, not violence.

Archive: *Two African American Men Standing next to a Tree in Georgia*

In one photograph by W. E. B. Du Bois, two African American men stand next to a tree in Georgia. In another, a young boy is in the flux of movement. His shadow self is captured in the slower cadence of film as he turns toward the tree. In these two prints, among many assembled for the Paris Exposition of 1900, a black-and-white photograph by W. E. B. Du Bois hits against the hard Corten steel monuments in a gentle liberatory push.

It is perhaps the simplicity of these photographs that first takes you apart. In the middle ground, a grainy image of two men standing by a tree, and then the affect, the double vision of histories of violence that would have

FIGURE 12.7 W. E. B. Du Bois, *Two African American Men Standing next to a Tree in Georgia.* In album (disbound): Negro Life in Georgia, U.S.A., compiled and prepared by Du Bois and exhibited at the Paris Exposition Universelle, 1900. Library of Congress, Washington, DC. https://www.loc.gov/pictures/item/99472429/.

them caught in a particular kind of anti-Black gravity that is recorded in photographs that sit historically alongside Image 328 in the Library of Congress archive. There is something in how the man on the right has his hands behind his back, the vertical lines behind him, and the grasp of the tree that directs vision upward from the foreground. Or how the eye travels up the slope into the too familiar representational amalgamation of eugenic tree life and Black death. Maybe it is the centering of the tree, the lower perspective of the viewer, the smokestack of ironworks in the background that barely captures your eye. But maybe that is not enough to find the pretense of innocence in a reading that continues to privilege the ocular genre of an overdetermined nature aesthetic: the all-too-ready suture that ties the Black outdoors to a remembrance of disfigurement and prefiguring of subjective deformation. How has the white gaze, that my brown eyes slip through and double back, defined a normative historic grasp that has already made the outdoors a scene of foreshadowed violence? What is the rubric of representation in the structures and practices of remembering that Du Bois's photographs quietly petition against? And what methodological

FIGURE 12.8 W. E. B. Du Bois, *African American Boy Leaning against a Large Tree in Georgia*. From Du Bois's photograph albums of African Americans exhibited at the Paris Exposition Universelle, 1900. Library of Congress, Washington, DC. https://www.loc.gov/pictures/item/99472429/.

repair is needed in these inhuman worlds to de-police these geographic imaginations and spaces of nature?

As race enacts geophysics that pressure geography's inhuman archive, this tension holds together with the liberatory intent of the photographer (Du Bois), in his picturing of Black social life and its recasting of the visual sociologies of pathology. Geology is a visual process, a way of looking that had a collaborative counterpoint in more overtly aesthetic practices such as landscape painting and later in photography. In the archive of vision, the retention of a white optic across disciplinary genres of the inhuman claims forms of Black and Indigenous life as ghosting the scene of nature. Unlike the geographies of the convict lease—buried, unmarked, mines without trace—these photographs seem to offer a different testimony of space, one that speaks across the registers of relation between spatial expression, nature, race, and the eugenic impulses of whiteness. It is tender in all the wrong places and quietly explosive in how it insists on the right to leisured freedom in the outdoors. The appeal to the normative has a liminal switchback in the histories of racial violence involving nature (understood as the scene for scripts of white belonging and geologic ownership). The recoil to a violence that is not even there tells us about the overburden of white optics of geography that use spectacular violence to police more mundane rural exclusions. The photograph tells a story of the afterlives of anti-Blackness while firmly locating responsibility for the social investment of the reproduction of dehumanization in the view from the plateau and its epistemologies of matter.

Black photography in North America was one of the most prolific sites of the early adoption of the technology. African Americans were quick to set up studios to document Black life (especially the Great Migration and "arrival" in the northern states), realizing its power to produce Black subjects and the very real stakes of holding the means of production and the modes of representation. Du Bois's photographs are explicit in their address of the sociology of criminology and the mapping of Blackness as a visual and spatial "problem" within photographic images (from Agassiz's daguerreotype to the creation of criminality of young Black men in the convict lease). Amid Du Bois's collection of predominantly middle-class Black life (although the suit could denote Sunday best), in the flip and dissimilation of associative trees, the photograph seems at first to be claiming a different ground. Yet, it is so deeply about the racial geo-logics of place and challenges the ways of looking that socialize space through undoing spatial precepts that hold the tree otherwise and rescripting the vicissitudes of racialized nature. The photographs bring Blackness into the European aesthetics of the propertied relation through the genre of landscape depiction. If the genre tells us what to expect, sets the geocodes of reading, the photograph nonetheless leaps into a parallel genre. It does not function in the propertied imaginary, where masculinity and possession of the landscape command the gendered tree of Judeo-Christian symbolism, Life, as its accomplishment. In the formula where Man functions as the portal to property relation and the tree, like the surrounding nature, is subsumed into patriarchal figuration. Whereby the tree itself is also gendered as a symbol of heterosexual reproduction as the Tree of Life (in its Judeo-Christian connotations of the origin story of Adam and Eve) replacing the tree's own reproductive sexual-asexual intersex bio-logics.[14] Yet, the photograph presses on another coordinate, in the push of the plateau for spatial hegemony and the pull of colonial cultural memory. How does the historical genre of the outdoors erase Blackness so readily? Like countless images of landscape paintings that sink Indigenous life into the invented wilderness or Lewis and Clark–type vistas from the plateau or as sublime titans of nature waiting to be conquered by muscular geographers. All of these nature aesthetics matter in the compiling of an environmental imagination that perpetuates an unequal earth in climate futures.

The branches of the large tree ride down, seemingly joining the young men in an arboreal embrace; the vertical tendrils caress their necks, but their feet are firmly on the ground, holding to gravity, against the anti-Black gravity of the plateau. They stand, upright, one looking at the camera, the other off to one side. They have the weight of a free life, yet the optical scene into which they step is complicated by the force of eugenic trees crafted elsewhere in the white geophysics of the earth. How does this specific materiality of the outdoors come to signify race and spatial processes of racialization, so that there is no nature that is also not an image of erasure and overdetermination? This doubleness of the experience of looking oscillates from a shape-shifting beauty to vulnerable repose. In a visual switchback, Du Bois reverses the flow of responsibility and puts it firmly on the viewer.[15] Thus, he relocates the libidinal investment firmly in a white optic and its policing of geography, unlike the visual claims made in Cuvier's and Agassiz's images that place violence everywhere else but with the geoscientist. The ambiguity of the photograph catches on the stark relation between the geologic theories of Life (enacted through racial trees and blackened grounds) and about how unfreedom still haunts the Black outdoors and libidinal investments in inhuman nature as a collaborator in various forms of *residual spatial violence*.

The double take in many of Du Bois's photographs re-visions the non-spectacle, whereby the normative is unsettled by Blackness: a tree with men standing underneath it, touched by the tree's branches, suited, leisured, in the outdoors. To reassign Hazel Carby's point, the "genealogy of race, nation, and manhood to be found in *The Souls of Black Folk* imagines its community by reversing the direction of the archetypal journey. . . . The journey determined the imaginative and symbolic landscape in which the conscious desires and ambitions of black humanity could be created and asserted" (1998, 16). The photograph does not redeem the anticipated violence against the male body, but it makes the viewer alone responsible for their own expectations around symbolic violence.[16] Thereby it signals the white socialized optic of geography in the archetypal national script where white patriarchy "settles" the nation and tames feminine nature through the policing of sexuality. The ambiguity and perhaps antagonism of Du Bois's photographs are their relation to the normative, whether the invention is understood as a gesture toward reconsecrating normativity (and its heteronormative relations) or a more radical cut of the nonnormative, unsettling the conscription of nature in the racialization of physical geography.

While trees have a weaponized history in the structuring of violence that operates across symbolic, paleontological, libidinal, and gendered ground, the photographs flip the script. There is opacity too, insomuch as the young men and tree can be what they are. Turning away in time. Unmolested. There is something tender in the tree's grasp, its nonhuman hold given by the tonal bleed. Removed from the extraction principle, the subjects are present in their time and freedom, having their photograph taken, standing by a tree. It is a tentative moment, unpracticed perhaps. The image of the young boy in flux gives this photograph its fugitive moment. A shift in the durational capacities of space unseats extractive time and allows for new occupations. Its beauty is what is barely depicted, a disruption to the representational cartographies, a blur in the normative frame, where only the tree stands as sentinel witness in the endurance of time. Beauty for Hartman (2018) becomes the loophole with which to "escape" the enforced determinism of social conditions, to live wildly and beyond one's means, to move for a while with a different gravity of grace within conditions of force.

In the background there are smokestacks, perhaps of a small factory run through the convict lease. Coal is burning. Two trees, one above ground, the other Carboniferous and below, two different geologic epochs 300 million years apart delivering the same spatial script. In the symbolic work of trees, the filial relation in the scene of lynching is imagined through an inhuman root, but one where race is not "innocent" in the same lexicon as the tree, but as a Lovecraftian tentacular that must be conquered and quelled.[17] The narrative and spatial anxieties may have shifted since the time of Du Bois's photograph, but the same story of policing the racial polity of nature remains. Put simply, as Willie Wright (2018) argues, environmental racism should be seen as anti-Black violence (as a counterpoint, see Barra 2020). The ambiguous coordinates in the photograph subvert and oblige an engagement with the Black outdoors through expanded registers of what might be understood as the "environment" and its racialized epistemologies.[18]

Sarah Jane Cervenak and J. Kameron Carter's concept of the Black outdoors is salient here in how it unmoors Blackness from property into properties of space and the elasticity of Blackness in relation to these coercive forms. These authors articulate the Black outdoors as "unhomed experimentalism" (2017, 48), a "reimagining of Black life unmoored from the axiomatics of (self) possession" as a potentiating free state, un/housed from the propertied state, in which Blackness moves "in the face of and against settlement's erotics and ecstatics" (47). Disaggregating the domestic from privacy in Black life, as a site of surveilled exposure, they understand Blackness as a propertiedness that unfences spatiality into a "free state" (48). Reimagining the Black outdoors as a geographic project of representation (alongside other knowledge-making practices) is a way toward other geosocial futures that are organized through the testimony of different inhuman pasts. Such "free states" require a shift in the structural organization of materialisms and the bonds of attachment to certain partial readings of the earth. In a broad sense, the overdetermination of white inhuman optics is a form of environmental determinism that has an active praxis. Loosening the ties of this overdetermination is a work of reclamation, in which aesthetics is a site of reparative geographies. Reassigning value in the flux as a mode of experience held against extraction is a way of resisting the residual spatial violences of racial geophysics.

Visual Taxonomies

If visual taxonomies are centers of regulation, claims of inhuman solidarity between lithic and liminal orders are a double movement against the flattening of Black life and the dimensionality of inhuman nature that is tied in the inhuman root. Engaging with the animation of placed geographies as an expansive realm of experience and nature relations reclaims Black ecologies' insurgency against enclosure through state geologies of the lease and industrialization. Small collaborations with trees are part of building Black worlds (see Roane 2018, 2022). The methodological resistance in the aesthetics of colonial geo-logics entails the nonmonumentalizing of Black life and trees, engaging histories that have been expunged from the archive that work against the abolition of enclosure. This abolition memory operates in the rift insomuch as it is a way to think against the foreclosure of certain nature relations, in aesthetics that stage and move outside of the spatial disaggregation of racial enclosures. These stories are not mine to tell, but the structural work of making space for their telling is the inheritance work necessary in the afterlives of colonial geographies.

Arboreal anomalous zones in the theater of white Southern geopower created and perpetuated racialized spaces in liberal democracy, reinvesting in the racial violence of the liberal state.[19] What is most disturbing and incisive about this image is that the *tense* of the Black outdoors is held in the photograph. I use this term in the context of Campt's (2017, 17) insightful work on the visual grammar of Black futurity, in which she defines "futurity as a tense of anteriority and a tense relationship to an idea of possibility. It is the tense of possibility that grammarians refer to as the future real conditional or that which will have had to happen." What Du Bois pictures is both the future and the *actual* real conditional, as an antithesis to anti-Blackness, the staged ordinary on the brink of the redefinition in Black spatial practices (see also Raine 2013). The photograph gives rise to the optimism of "new circuits of movement" (Campt 2017, 26) after emancipation, of the stretch of leisured temporality, and its potentialities. In conjunction with the convict lease the control of space equaled the control of freedom, where movement is dimensional as the possibility of learning and exercising that freedom as a corporeal and social condition distinct from slavery's containments. Curtin (2000) describes how social reformers worried about residents of Black Belt communities visiting other towns and spending their money on "unnecessary" travel.[20] Travel was a classed practice, and travel of the racialized poor was of concern because it did not heed those geographic category distinctions of the leisured classes. Unscripted Black movement was of social concern. Middle-class groups sought to discipline and regulate the seemingly excessive expenditure of money and energy on moving without purpose and utility, and not in the service of labor, accumulation, or heteropatriarchy. Movement for movement's sake, without destination, in financial defiance of accumulation, was an affront. Standing by a tree, lounging by a tree even, without the imposition of fear was an image of Black freedom.

If the archive of movement is read against a backgrounded grid of oppression that evidences freedom's expression, then this is a tentative photograph of practicing possession (Kelley 2002). It recognizes and reinvests in the "loiter" as a freedom-making praxis, an intimate and racialized geography of movement that undoes the racial scripts of purposeful existence (Hartman 2019). Inversely, the photograph crisscrosses "the routes that might have led to alternative presents . . . crossing over to parallel lives" (Hartman 2007, 100). Such environmental practices were also the actual routes to the present and fuller geographies. Du Bois appropriates a white geography of apprehension. This parallel visuality tells us something about both the historical claims on space and its inscription, and the modalities of memory. It is the clash of a casual loitering pose, the smart yet leisured clothes, the presupposition of *freedoming* geography, that crashes into the line of sight, which hits the mound and rises to greet the tree and its subjects, a perspectivism that refuses other remembered images of other things that can be done with beautiful trees and beautiful young Black men.

If the two men give pause to question the sociology of the scopic regimes, it is the next photographic frame that gives the release, the blurred image, as flux and the frequency of life that is not held within the authorizing (cauterizing) framing of Black life. This still-moving flux is akin to that which Campt (2017), McKittrick (2013), and King (2016b, 2019) alert us to as a modality of Blackness that provides a way of stepping out of the enclosures of colonial linear time and space, in a mode of geography otherwise. This is what McKittrick calls the Black ungeographic. It might also be recognized as a claim on another geophysics. In so many photographs I have seen in the archive of this period, in industrial sites in the United States and South Africa, Black men and women are blurred next to their white counterparts. There is a chosen turning away from a record of white property and prosperity, especially around sites of labor. A refusal to enter the future on these terms that blurs the trajectory of that futurity. Walking out of the picture, absconding from the archive, leaving no trace in the juridical record is a riff on what Hartman (1996) calls the "double bind of agency"—the freedom not to have an agency that is already withdrawn or violently compromised.

The agency of disappearance speaks to the cover and shelter of opacity, while the audacity of escape gives rise to a positive fungibility, a creative maneuver through space and time. The blurred moves strike the violence of the archival-like moments of unconformity in geology, in which strata break with the continuous rock record, shattering the uniformity of time and the readable materialism of space. Unconformities are a type of geologic contact—a boundary between rocks—caused by a period of erosion or a *pause in sediment accumulation*, followed by the deposition of sediments anew. This interruption to the plasticity of whiteness and its homogenizing reach disrupts the process of accumulation and signals a form of life under a different gravity.

A Conversation on Beauty

Among the most startling aspects of Southern geography are the trees. The life span of a plantation oak is between three and six hundred years, much longer than the current nation-state (and akin to the saguaro cactus in the Sonoran Desert). The beautiful oaks laden with moss, the sycamores that catch breeze in the hot and heavy air—people gathered under these trees to enjoy lynching and get married. This horrific ambiguity is not lost on Billie Holiday when she sings the pastoral and the monstrous.[21] Giving back to the Tree of Life the history of its strange fruit is to articulate the inhumanisms that ground the construction of Life. In Morrison's *Beloved*, the character Baby Suggs speaks to the resistances of a different kind of beauty: "Boys hanging from the most beautiful sycamores in the world. It shamed her—remembering the wonderful soughing trees rather than the boys. Try as she might to make it otherwise, the sycamores beat out the children every time and she could not forgive her memory for that" (2007 [1987], 7). I think Morrison is saying that memory is also about what can be borne, to make something beautiful out of the terror was a way to go on, to take something from the unbearable that made it bearable, to intervene in a narrative that would leave Baby Suggs with nothing. Beauty was a route to futurity; to sacrifice the boys' memory for that which would sustain the living, "the beautiful soughing trees" were what could be rescued from the terrible scene. As Hartman explains, "It is a way of creating possibility in the space of enclosure, a radical act of subsistence" (2019, 33). Aesthetics might hide the hurt, but it is also a site in its transformation. In Sharpe's (2019) work, beauty is a method, a way to make it through the afterlives of slavery, where care is the possibility of continuance.

Yet, a more cynical and obscene question dances around Morrison's painful arboreal observation, a question that I think haunts the environmental humanities and Western environmental movements: "What if" the beauty of the trees has already won out for the white imagination, and will always win out? In considering the beauty of the savannah oaks, the picturesque backdrop to the dramas of the Big House and plantation life, its "romances" as told by the plantation tour, there is a commonality that is presupposed that places the trees above and beyond Blackness. The "what if" future tense of the beauty of trees feeds the aesthetics of the plantation as set and scene of grim longing and nostalgia for dreams of white supremacy. The tree did not stand as an interdict to the monstrous imagination and its enactment. And, if the white imagination had paused there, what would it say about such a conceit that refused those young boys any possibility of consideration in the naming of beauty, like Kant's (2009 [1755]) dictum of the universal experience of beauty "except for the Negro"? The "what if" campaigns on an extension of trees to young Black boys, the extension of beauty from recognition of nonhuman life to that of a young Black boy, rather than the other way around. The explicit and implicit conceit is that white people have more in common with a tree, with nonhuman nature, than with those "too black for care" (Wilderson 2020, 17). This affiliation of whiteness and nonhuman nature slumbers within the concerns of the environmental humanities and its "difficult" relation with conservation.[22] With few exceptions (Green 2020), it seems easier to find a place of care for the origins of the universe, a starfish, or fungal species than it is to deal with the whiteness of the discipline and its racial histories of violence.

In conversation imagined between Hartman and Kant, the two are arguing about beauty. Hartman is redressing a notion of beauty from Kant's perception of beauty as available to every subject, universal, *except for* . . . with the universal imagined in a conditional tense. As Morrison succinctly puts it, "some fraudulent 'universalism'—a code word that has come to mean 'nonblack'" (2019, 199). Hartman has no use for Kant's sublime. You don't seek out limit experiences when you are placed beyond the edge of reason. In the background there is Arendt scolding James Baldwin for his attachment to beauty. She writes to him after the publication of "A Letter to My Nephew" in the *New Yorker* (November 17, 1962), the text of which became part of *The Fire Next Time*. Arendt compliments Baldwin for his address to the oppressed peoples (with whom Arendt confesses her solidarity) and rebukes him for his use of love in an account of politics. Baldwin had asked: "When I was very young, and was dealing with my buddies in those wine- and urine-stained hallways, something in me wondered, *What will happen to all that beauty*?" Arendt replied that Baldwin's article "certainly is an event in my understanding of what is involved in the Negro question. And since this is a question which concerns us all, I feel I am entitled to raise objections" (Arendt 2006 [1962]). Like her account of the Boers' murderous reason brought on by their encounter with the alien, Arendt starts from her fear to justify her violence to Baldwin's ethics of care: to love that which is not loved. She says:

> What frightened me in your essay was the gospel of love which you begin to preach at the end. In politics, love is a stranger, and when it intrudes upon it nothing is being achieved except hypocrisy. All the characteristics you stress in the Negro people: their beauty, their capacity for joy, their warmth, and their humanity, are well-known characteristics of all oppressed people. They grow out of suffering and they are the proudest possession of all pariahs. Unfortunately, they have never survived the hour of liberation by even five minutes. Hatred and love belong together, and they are both destructive; you can afford them only in the private and, as a people, only so long as you are not free. (Arendt 2006 [1962])

Arendt's rebuff is both to the radical politics of love as an epistemic revival and to the African American experience of the lack of division of private and public space regarding violence. She cannot see Baldwin's tender invocation of the Fanonian "invention" (Marriott 2018, 312) of being lit as a counter to the violence of a pyro-subjectivity. As Hartman's project *Wayward Lives, Beautiful Experiments* makes clear, to "love what is not loved" is the most radical politics of all. Fugitivity, queerness, and errancy in forms of Black sociality that are captured in the state ledger as waywardness, insolence, and much worse interrupt the state's apparatus of capture and carceral forms of internment in the social forms of the state—calculative racial life. Hartman writes of the politics of Black intimate and social life, the lives of young Black women after emancipation who shaped cultural movements and politics that transformed the urban landscape yet would never appear in a political tract or philosophical treatise: "Esther Brown's minor history of insurrection went unnoted until she was apprehended by the police" (Hartman 2019, 232). This granular inhabitation of the transformation of subjective life through intimacy, sexuality, and a political sensibility of defiance suggests other locations for the geographies of freedoming that hold space differently, against the grain of purpose, utility, patriarchy, and destination.

The liberatory acts that Hartman, like Baldwin, writes *toward* and *with* are grounded in a fierce love of Blackness, made in the brutality of broken earths that shift the epistemic and methodological registers, seen by Hartman as a general strike against the social categories that would otherwise contain Black life in a prison state of anti-Black reason.[23] Hartman refocuses the *when* and *where* of Black urban life, its different temporalities across the tight and punishing grids of white space, refuting the political enclosure of sociality in the apprehension of spatiality. She says it was "well understood that the desire to move as [Esther Brown] wanted was nothing short of treason. She knew first hand that the offense most punished by the state was trying to live free. To wander through the streets of Harlem, to want better than what she had, and to be propelled by her whims and desires was to be ungovernable. Her way of living was nothing short of anarchy" (Hartman 2019, 230).

In the constricted social realm of the outdoors, Hartman's political tract is of the queer movement of young Black women who practice freedom through loitering, dance, and chorus. Moving in and out of historical accounts, the strictures of sociology, and the punishing vocabulary of judicial capture, freedom is beyond the residues of the archival ledger and political theory: "Beauty is not a luxury; rather it is a way of creating possibility in the space of enclosure, a radical art of subsistence, an embrace of our terribleness, a transfiguration of the given" (2019, 17). Attentive to the spaces and places of Black women's beauty and the improvisation needed to make joy in a hostile environment, *Wayward Lives* argues for the recognition of a Black general strike made after the plantation, from the 1890s to the 1930s, a general strike that is inconceivable in Arendt's political theory and its spatial divisions. Geographically, Hartman poses questions about the "when" of Black urban life through the movement in the ambulatory aspiration for a loving Black sociality. Highlighting the coercive pathology of the anti-Black state in the arrest of movement, Hartman's wayward women trespass on Arendt's politics; she unsettles Arendt's designation of the private/public transactional spaces of politics that rely on a propertied and gendered form of access to the political and physical space. Arendt's designation of the properly political as public rather than private segregates space in the United States in the terms of white supremacy and its racialized control of physical geography.[24] The distinction between public and private is already collapsed in white policing and modes of governance that practice an ever-intimate reach. The prison mine gets at the verticality and vertigo of that relation. The Black queer wayward impedes the circulation of the divisions that Arendt makes between public and private, between love and politics, between the imaginary of ethics and its preferred bodies of appearance. As Joy James comments, speaking on Arendt, "State violence (through the laws, the police, the military) practiced in the 'private' realm shapes the practice and sites of power in the public realm but also incites communicative power and democratic action among those resisting oppression and exclusion from governing. The practice of voting disenfranchisement of African Americans through bureaucracy (poll tax), violence (imprisonment), and terror (lynching and/or police brutality) also suggests that the private realm was never truly understood in the United States as a site void of the practice of politics" (2013, 309).

James argues that "ignoring the historical and contemporary specificities of this democracy and its racialized state violence and dominance allows Arendt to construct a theory of power that floats freely above a foundation mired in racially fashioned domination" (2013, 310). Plateau politics. Baldwin alludes to a similar imaginary of the weight of racialization in his letter to his nephew, which Arendt finds egregious to political theory. Baldwin states: "I wondered, when that vengeance was achieved, *What will happen to all that beauty then*? I could also see that the intransigence and ignorance of the white world might make that vengeance inevitable—a vengeance that does not really depend on, and cannot really be executed by, any person or organization, and that cannot be prevented by any police force or army: historical vengeance, a cosmic vengeance, based on the law that we recognize when we say, 'Whatever goes up must come down'" (1962, n.p.).

Speaking to the effects of racial hierarchy on the political realm and its sensibility (understood as a political and punitive affect of histories of racialized violence), both Baldwin and James focus our attention on an anti-Black gravity and how it is made to appear as the gravity of the earth itself—a geophysics rather than a geophysics of race. If we relocate that argument in the realm of the geologic and what it means to be grounded, geophysics becomes a way to answer the surround of racialized geopolitics and its materializing geo-logics (which is the effect of how normativity becomes naturalized). In a mode of subjective politics this would be to theorize how certain conditions *ground* the enforced "inevitability" of political realities, but also become sites of their remaking, in ways that reshape the institution of *being* and *force* within the historical gravity of colonialism and its afterlives (which might also be called the Anthropocene).

In the tract of unreasoned ground, where inhuman nature is made to violently collide, the terms *being*, *subject*, and *nature* are exclusionary, produced historic forms; they come to stand as forms of ideation of a world without materiality, making normative claims about the earth on behalf of the partial "we" from a floating plateau maintained through an asymmetrical field of environmental justice. If the plateau becomes the planetary, then its unequal earths propagate the future through past modes of separating progeny. To joy geography is to be in search of a wayward belonging, looking for new geophysics of sense that would hold forms of life differently. The spatial coordinates are not just geographic in the conventional sense of being organized through territory and the desire to evict Black ownership from Southern geographies but traverse along lines of bodily relations to geophysical forces. The role of African American and Black and Brown international organizing in forging the environmental justice movement suggests an early practical appreciation for the connection between civil rights and environmental exposures, where environmental concerns are coconstituted by racial apartheids.[25] Black geographies are marked by the movement with and through conditions of the inhuman—as a subjective and material state—in metaphysical and material orders. Yet, a subjugating relation with the inhuman is still a relation with the world. It still produces something other than that subjugation because those inhuman worlds are materially open to the cosmos. That relation summons a far wider geography than the imposition of oppression would have us believe. The collapsing of two asymmetrical forms of geopower into one another (Blackness and earth forces) also made the earth home in a way that did not reinstate or have an immediate or stabilizing recourse to the earth as property, substantiating a form of address to racial capitalism that is thought within the earth, rather than over it, as the capture of surface or value—a rift in Overseers' earth.

Coda: Gravity and Grace in *The Tree of Life*

In Terrence Malick's film *The Tree of Life* (2011), the origins of the universe are visualized to tell a story of earth as the 4.54-billion-year-old platform for mammalian supremacy that ends with the white settler family on the beach of time. Mammalian ascendancy culminates in white hegemony. Life starts at roughly 3.5 to 3.8 billion years ago, as the film goes with the single-celled organism mating its way to microbial usefulness (viruses also codeveloped at this time, although that of course is another story), and as in every museum exhibition, the years of inorganic life seem a bit dull, so we skip straight on through to the profound moment of cell division (for an alternative telling, see Hird and Yusoff 2019). And the rest is history, or so the telescopic narrative goes. Bacteria innovate at learning to eat the inorganic stuff, converting geology into life, for a few billion years or so, and then *bam*, blink of a geologic eye, and the arrival of the dinosaurs 240 million years ago, a species that has some cultural mileage in the warm glow of mammalian blood politics and species war. A short CGI moment later we are poised on the brink of dinosaur ethics. Dinosaur learns it has the choice to not be brutally indifferent nature, to change its nature, and so it decides on mercy toward the about-to-be-ingested foe. Meat-eater dinosaur gives the herbivore a break, in the creation scene, another moment of grace punctuates deep time. Step gently on the neck of your foe, the film seems to suggest; look toward the telos of a higher nature. The dinosaur nationalism of the American Museum of Natural History is so old school, new dinosaurs realize that nature can be changed, not just brutally conquered. This is a film about grace, we are told.

Grace is white settler gravity. Grace is the role of the mother in subduing the dark, brooding masculinity of Brad Pitt (coded nature). The dinosaur's genealogical prodigy of consciousness populates; out of the cosmic abyss comes a human womb (watch out for that lizard brain . . . just saying). From the screenplay Malick narrates, "Reptiles emerge from the amphibians, and dinosaurs in turn from the reptiles. Among the dinosaurs we discover the first signs of maternal love, as the creatures learn to care for each other. Is not love, too, a work of the creation? What should we have been without it? How had things been then? Silent as a shadow, consciousness has slipped into the world." Metaphysics meets the geophysics of the cosmos, told through a troubled middle-class Texan family in the 1950s.

This ultimate natural history creation narrative starts with the Big Bang and ends up with the birth of a child in a Waco hospital, as white American suburbia appears as the genealogical heir to the questions of the universe. This is *the* Tree of Life, Malick's aesthetics insists. Right here in the garden, in suburban Eden. The universe is everywhere, of course, but the universal gets made only here. The cosmic narrative is our own, if we invest in the white family hard enough and support angry heteropatriarchy long enough. The saintly mother figure, Grace, is light and dignity, the eternal maternal, inheritor of the good dinosaur ethic, conduit for nurturing masculinity. She sets the scene: "We must choose between the way of grace and the way of nature. . . . You have to choose which one you'll follow." Grace is forgiveness, holding the patriarchal white line, supporting the man, through devastating loss. Nature is brutal, muscular, Werner Herzog–style. Nature is the void of meaning, unthinking and ruthless; it must be conquered by grace. She must keep going into this battlefield, like the painting *American Progress*, her beatific vision overlaying native land, flowing toward the horizon of Manifest Destiny. In one scene, the wife floats beneath the tree, transcendental. She is the levitating goodness of white womanhood, lifted up rather than pulled down, recalling Darwin's famous proclamation: "As buds give rise by growth to fresh buds, and these, if vigorous, branch out and overtop on all sides many a feebler branch, so by generation I believe it has been with the great Tree of Life, which fills with its dead and broken branches the crust of the earth, and covers the surface with its ever-branching and beautiful ramifications" (1885, 105).

No feeble branch is she. "Grace doesn't try to please itself. Accepts being slighted, forgotten, disliked. Accepts insults and injuries." Grace is a good wife and a good mother. She wants nothing for herself. The battle is manhood. It is the battle of Nature. It "only wants to please itself. Get others to please it too. Likes to lord it over them. To have its own way. It finds reasons to be unhappy when all the world is shining around it. And love is smiling through all things." Gases blown out of the universal vents turn into a DDT cloud sprayed for mosquitoes. Protozoa progeny morph into suburbia sublime. Fade out cosmic orb, fade in pregnant belly. She is of the tree, the leaf, and butterfly. "Help each other. Love everyone. Every leaf. Every ray of light. Forgive," she says. On the beach of geologic time, it all comes together; the kaleidoscope gives one final vision in the end days, of reconciliation, of the reunion of the family. Estrangement is not the way of grace; grace is in nature, after all. We saw it in the creek at dinosaur school. The film asks, Who are the referent-we? This is coupled with the question to God, "Who are the (referent-)we to You?" (i.e., how do we establish meaning in a cold cosmos, and what is "our" culture to your paternal nature?). Between the blink of a geologic eye and the fullness of time, the film seems to suggest that all this arrival of the universe is for the contemplation of the white liberal family on the shores of time. Origin is ownership. The world is there to receive us, it seems to say, as long as (referent-)we have enough grace to receive this geohistory (and a strong lineage on the tree, woe to those with a feeble branch . . .). Grace must be struggled for; this is the domestic scene, the mothering in the battle with violent nature. This is the white privilege of the earth, being able to imagine that the cosmos is whole and that wholeness—from the roots to the tree—is yours for the taking, as the very purpose of the universe.

In a different kind of ontogenesis, I hear Morrison talking about the grace you got to make, if you are the imagination of what grounds the Tree of Life, with no billion years of evolution to consecrate your apex or womanhood to guard the filial line. Morrison's grace is a salve to the geotrauma of slavery's abyssal birth, a different kind of continuance in a world that is not trajectory but broken earth. In the

> clearing—a wide-open place cut deep in the woods nobody knew for what at the end of a path known only to deer and whoever cleared the land in the first place. In the heat of every Saturday afternoon, she sat in the clearing while the people waited among the trees. . . ."Here," [Baby Suggs] said, "in this here place, we flesh; flesh that weeps, laughs; flesh that dances on bare feet in grass. Love it. Love it hard. Yonder they do not love your flesh. (Morrison 2007 [1987], 102–3)

Morrison's grace is an act of love and an act of imagination pitched against anti-Black gravity. As *Geologic Life* seeks to understand geology (in its broadest sense as a field of material relations and narrations) as a weaponized tool of raciality that has historically shaped the grounds of struggle, reconstituting the imagination of the earth is necessary to refiguring the racial arrangements of matter and their ongoing effects.

Imagination will shape the museums of natural nonhistory you inherit.

IV Paradigms of Geologic Life

13 Ghost Geologies

DIG. Geology mobilizes a *normative mode* of colonial materiality through its kin network of extraction, natural resources, architectures of geomorphic transformation (dams, river straightening, climate change), redistribution of geologic materials (kitchen sinks, nuclear bunkers, asphalt, etc.), and geochemical energy (carbon, nitrogen, phosphorous, etc.).[1]

ORIGIN (AND PRESENT TENSE). *Colonial earth.* In the historic conversion of earth through the grammars of geology, the materializing of epistemic violence enacts a world-building and world-shattering that are characterized by unequal access to geopower (accumulation from geologic value) and its inverse, exhaustion (exposure to extractive predation).[2] Here, two subjective-earth states can be distinguished: (1) those *given by* extraction—white man's overburden (now a multiethnic and racial affair)—and (2) those continuously *exposed by* extraction processes, such that they are subject to the weight of the overburden, as racialized and racist gravity, which makes digging yourself out of a hole and hitting rock bottom a constant threat, not to mention the collapse of the earth itself. Extractive predation involves the transfer of energy, and it is how racist patriarchal violence is spatialized, maintained, governed, and regulated. Geology does not merely map the earth; it transforms its planetary and subjective states. Colonial earth is the Anthropocene now.

INSURGENT EARTHS. Geology is also a medium of solidarity with the world. Geologic chronicles of the rift offer the possibility of time travel to examine the infrastructural inheritances of geology (the inhumanities) for the basis of earth reparations.

Mine as Paradigm and Racial Undergrounds

I have argued that colonial geology made an apartheid of materiality and materialisms that had distinct spatial forms (plateau and rift) and conditioned racialized geophysics. The mine is paradigmatic of the spatial exchange of this racial geophysics and the temporal conditions of the gravitational geo-pulls that structurally and stratigraphically support the continuance of the plateau and its futurity. Mines are subterranean spaces that are always under pressure and compression from what is above. In the division of surface and underground states, exercised through geo-logics, material analytics, temporal citation, modes of description, and geographic imaginations, geology is a tool of raciality. Yet, the rift (as underground and epistemic break) possesses the possibility and the spatial autonomy of a state of radical geologic epistemology and pedagogy. It is somewhere to *dig in* and plot. And, as geology is a mode of time travel, it is somewhere to re-script the expansiveness of temporal duration and language of time to overcome the strategic alienation of earth via property.

$$\frac{\text{mind over matter} \neq}{\text{undermine stratal overburden} \infty}$$

The mine is paradigmatic in terms of what it institutes as spatial and subtending relations, exhibiting what I have described as a geophysical approach to race that relies on depth relations to maintain surface flows of power. Mining makes temporalities that geophysically transform planetary processes, and these are analogous and materially manifested through subjective forms of embodied geology: the flesh of geology. Mind over matter is the grammar of Enlightenment geology and the practical geology of colonialism. Lithic and liminal materialisms became the overburden of the mine—that is, colonial materialism organized by the optic and imperative to extract and accumulate through subtending stratal relations. This is mine in the possessive sense, mine as the mastering of matter through the organization of the grammars of geology around the affective and geochemical work of purity. This move to purity is part of a more general division in colonial materialisms between *bios* and *geos*, and toward a narrative of "elemental isolates" and the erasure of inhuman memory. Matter is dividuated based on its ability to function as an autonomous mineralogical and metallurgical unit. This dividuation of geologic elements eradicates the supplementary actions of geology, as geochemically communicative and temporally ac-

FIGURE 13.1 Plantation Place, named for the building's former renown as the world's center for the tea trade, relabeled in October 2020 as 30 Fenchurch Street, City of London.

tive states of matter (or its queer geochemistry) that are already situated in the earth and inorganic relation (gold ♥ arsenic; mercury + silver, forever). Geochemical matter has different kinds of porosity and promiscuity. Under certain conditions, it mixes and makes new amalgams. It aggregates, often in toxic and atmospheric ways, changing state when it is unearthed or segregated from prior relations. This queer geochemistry moves through bodies: human bodies, bodies of earth, bodies of water, space bodies, bodies that begin and end in the earth (Clark and Yusoff 2018). Terracide was always terrorcide because it took away the ground and invented new modalities of who and what would ground its geographic imaginations.

The Enlightenment understanding of inhuman nature and its "properties" has continued to structure material legacies and futurity through the imperative of natural resources. And now it has structured the geologics of a planet called Anthropocene. The inhuman epistemologies of colonial earth are the product of colonial relations of conquest, of territorial geographies and their imaginaries of world-making, in the practical purification and policing of categories, in the geostories of scripted succession. Such carceral forms are a means both to property and to colonial earth's disavowal of what materially subtends and supports its emergence and imagined autonomy from the terra/terror. Attentive to the collaborative work that inhuman epistemologies are mobilized toward, the inhumanities aim at a different act of attention—a cartography of colonial earth that actions methodological paths to its disassembly.

Geologic grammars of extraction stabilize the material object for the extraction of value and the creation of the commodity form. The desire that launches the affective work of purity into the world is the same colonial desire that orders the earth from its perceived dishevelment as "undeveloped" matter (as it orders its subjects) through what Wynter (1996) called the "secular telos of materiality." Through a mastery of materiality, colonial earth promotes an ethic of accumulative value rather than attending to what conditions extraction's possibility (in terms of subjects and earth). And so, colonial earth practices are rapidly extinguishing those possibilities across a myriad of forms of life. It is a recursive tale. Disregarding durational geochemical attractions and disaggregation in this process of valuation upends and cancels worlds of being in all their complex organic and inorganic interrelation (climate change is but one iteration of the planetary breakage).

The psychic splitting of the world and its historical realization in the material segregations of inhuman and human materialisms launched the concept of "race" as a means of controlling and governing the *geos-bios* fracture. The mine was the archetype or industrial blueprint of those extractive relations and combined racial subjection with planetary changes of state. Colonialism began with dreams of the shiny, and the mine was the site of its practical geologic realization and realism. Extraction produced arrangements of matter that disfigured organic and inorganic relations alike, such as bodies and land, allowing some subjects to seemingly "float" above the earth, subtended by the others that were made to feel the actual weight of the earth (down the mine shaft). Unequal access to geopower was/is the struggle over *material sovereignty*.

A counterreading of geologic life requires a supplementary account, where the elemental is neither purity nor immutable segregated ground. Rather than the elemental, from which we derive a table of pure form that is singular only through extraction and erasure, a parallel geology engages the consequences for materiality across and through time—as a durational occupation of attachments and inhabitations—that is posed as an open question of language and Relation in inhuman epistemologies and material methodologies. If the geophysics of race was the outcome and process of racialized forces of geology (in the twinned arrival of the inhuman-inhumane), the spatial imaginaries of inhuman time can be materially forestalled and radically refused in an engagement with temporal and stratal rifts. The colonial control and containerization of the inhuman amount to the control of time—when the geologic is understood as a pathway to temporal agency—and thus an engagement with the narrativization of temporality is crucial to building different material worlds. Inhuman geochemical attractors not only are molecules of matter and amalgams of mineralization that pass through the earth but also arrange bodies and are a difference engine of temporal inhabitation and intimacy. This is another "time" of race. Change the narratives of time and decolonize space. I have argued that engaging the *not-fossil* is a passage into a different materialism, enacting *inhuman memory as a future praxis*.

Rocks are kindred, in the untended wounds of colonial geology, where geotrauma is a language that speaks across the rifts of worlds that racialized geologies tore asunder. In the weaponization of geology, rocks are also sentinels to the temporal possibilities of an otherwise, collaborators in the rift's brokerage with the stabilization of extractive forms and value. Amnesia about geotrauma is the refusal to acknowledge what is lost and what is created around that loss: those that fought back, remained attached and committed, refused extraction's gains, are lost to this erasure. Bones and rocks hold struggle and memory against this forgetting. Inhumanisms are a pedagogical practice that mitigates the narrowness of humanism and offers a charge and a cosmic materiality that breaches the breath of the world. While the mine names a colonial geography of extraction and its burdens, it also places the rift as a forgotten sight/site in the perspectivism of the plateau.

If mining unearths the exoskeleton of geologic life to produce the "nowness" (or the *now, now,* temporality of the *mine*) of contemporary life (its energy and communication networks, its highways, and climate pathways), it also creates openings and passageways of unintended fractures—fissures—that lead into other unearthings and ungroundings that belong to other worlds worth making. In the mine the Overseer remains at the surface, changing the forms of mutuality that the underground as a politically autonomous zone historically gave rise to (where the underground can be understood as a network of relations rather than the site of property). While accumulation is concerned with getting the stuff up and out the shaft, undergrounding is a condition of often unnoticed, traversal, underground railways, caves, hideaways, and secret shafts, undermining praxis. Undergrounding names a racialized spatiality that organizes the governance of the surface of the plateau and its forces of deformation as racial and geophysical deficit. In other words, you can time travel with rocks toward anticolonial earth futures through spatiotemporal inhuman tactics. The shock forward of this method of ungrounding geotrauma points to the future ancestral practices that might remake the earth in a different geologic mode.

In the consideration of architectures of attribution and accumulation, the mine must not be isolated from what it builds on the surface. Spatialities are not always contiguous in a geographic sense, but they do contain one another in relation. This is Césaire's call to see his "thumb-print and my heel-mark on the backs of skyscrapers and my dirt in the glitter of jewels!," his name in every city built by the trade in enslaved persons: "My name is Bordeaux and Nantes and Liverpool and New York and San Francisco . . ." and every place enriched by his exploitation, "Virginia. Tennessee. Georgia. Alabama" (1969 [1956], 29). Alerting us to the erased subjective marks of racialized geologies that have shaped the earth, Césaire concludes his poem with the shout that no geology is neutral; it is "Red earth, blood earth, blood brother earth" (30). The earth is kin and a nonnatural history museum of brutal silenced histories. Besides earth as imagined as resource, it is also a recourse: to memory, place, and the imagination of new futures. Invisible traces of race glitter in jewels, and ghost geologies rattle in the shiny worlds of metallurgy and mineralogy. Race and racism are the ghosts of *chasing the shiny*. Every morning the fingerprints have been polished away by an invisible army of nighttime workers who clean the marble and buff the steel, so no testimony can be made of the origination or the maintenance of materiality. The grammar of geology is not only the site of blackened, brown, and red earth monuments, but it is these accounts of earth that sustain the missing earths and grammar its geotrauma: the way in which race is made to function as something akin to an element, as immutable, as ethereal, as interchangeable, and from the earliest murders, stealings, and enslavements as a site of speculative valuation (as the role of insurance companies and debt financing made the trade in enslaved persons). Against the petrification of value—and the fossil subjects that it made—the inhuman is an interlocutor.

The relation between mines and centers of manufacturing in colonial and postcolonial worlds was organized to follow a Western-centric model, whereby the colonies provided the raw materials, through the mine and plantation, to the imperial centers that stamped their mark of origination on goods—silver from Sheffield, cotton from Lancaster, and so forth—and

FIGURE 13.2 Directional sign reading "To Shaft," Mahogany Drift Mine, Beamish, The Living Museum of the North, County Durham, England. Photo by author, 2016.

then often sold those goods back to the colonies at inflated prices. The colonies were the mine for imperial sites of manufacture, organizing the general movement of energy and value into European (and then American, and now Chinese, centers). These geographies of displacement of energy and of accumulation organized the basis of racial capitalism. Colonial centers avoided the "development" of the colonies as industrial centers, relegating them to functioning as the mines for manufacturing processes elsewhere, while selling the ideation of the telos of materialism. This mine-to-manufacture relation enabled the control of the accumulation and price regulation of capital goods. As recent research on the ecological economics of unequal exchange empirically elaborates:

> Value added per ton of raw material embodied in exports is 11 times higher in high-income countries than in those with the lowest income, and 28 times higher per unit of embodied labor. . . . On aggregate, ecologically unequal exchange allows high-income countries to simultaneously appropriate resources and to generate a monetary surplus through international trade. . . . Moreover, high-income nations obtain significantly

> higher revenues for the resources they export than poorer nations, which is mostly due to the positions occupied in global supply chains and their respective roles in the world economy. (Dorninger et al. 2021, n.p.)

The racial deficit model in the production of colonial spaces, as well as the trade and traffic in persons as fungible matter for extraction, doubly instantiated the conflation between place and matter as deadened until it is worked by Euro-Western centers, which in turn accorded geologic agency—or geopower—to those centers. Thus, the manufacture-mine relation spatially mirrored the master-slave aggregation in its temporal-spatial compressions. The master's house or prison mine enacted this microspatiality as an intimate geography. Neo-extractivist modes are now the means by which the primacy of geologic agency remains regulated within the confines of racial capitalism, as both a history and a present of colonial geophysics.

From Mine to Plantation

Racial capitalism is not just about the economies of keeping a stratum of racialized poor in their stratal "place" in relation to the plateau—that is, subtending surfaces through white geology—but about the whole affective infrastructure or geophysics that maintains this subtending relation (in ways that punctuate the plane of the plateau and often land with sudden force). The racial undergrounds that are created above and below ground are conditioned by carrying the weight of the plateau as a burden, and through the more insidious obscuring of these paradigmatic relations that keeps the surface racialized. The mine historically proceeds and manifests as the plantation, or rather, the plantation crisscrosses with the spatial relations and forms of geopower inherent in the factory mine, as it specifically institutes spatial-scopic forms that differ from the mine, such as the Overseer as spatially omnipresent (whereas in the mine the Overseer remains at the surface). Mine technologies such as the steam engine, first developed to remove water from the mine and to raise coal up the shaft, were introduced in Jamacia by enslavers who used them to intensify the production of sugar under the burden of continuous production of crushing cane. Sugar economies and industrialization are tied by this transfer of energy in the form of sugar, which synthesized field and factory along a geologic continuum.[3] The gendered forms of increased sugar production in the colonies were used to fill the "calorie gap" (Mintz 1985, 149) in women and children in the capital.

This domination by white capital at the expense of the recognition of mutuality was argued by C. L. R. James (1938b) more than a century later in his article "British Barbarism in Jamaica: Support the Negro Worker's Struggle" in *Fight*. He says: "Tate & Lyle, as everyone who buys sugar should know, make a fortune every year by selling to the British workers sugar grown by Jamaica workers. They must keep these two divided at all costs. Hence with that solemn shamelessness so characteristic of British capitalism, Mr. Lyle discovers that the West Indies laborer does not remotely resemble the English laborer. The real trouble is, of course, that he resembles the English laborer too much for Mr. Capitalist Lyle."

The imbrication of the flesh of geology, as energy (libidinal and lithic), animates possible forms of solidarity in ways that are governed by the heightening of racial disparities by white capital. The gendered imbrications of sugar and steam highlight the cyclical time to the relations of colonialism, yet the spatialities of colony-as-mine remain robust. The plantation is a strip mine, mining subjectivity and soil through the categories of the inhuman and the practices of oppressive monocultures (as geologists organized racial types, hydrology, irrigation, and the composition of soil). The plantation is a reduction in the interdependent forms of life, its complexity and diversity; in short, it's a flattening and a stratal move that compresses and compacts forms of life. Its predecessor and traveling companion is the mine.

In response to the flattened material histories of the shaft, the Caribbean and Americas appear as a geography in *Geologic Life* through the authors of Martinique, namely, Césaire, Glissant, and Fanon, and their theorizations of new modes of being and through contemporary Black feminist theory and poets of the diaspora. The scope of Caribbean theorizing as a decolonial praxis at a moment of rupture is audacious in its summoning of the abyss to build and to remember buried forms of being. They are miners unearthing geologic ghosts and submarine graveyards that rattle across the Atlantic graveyard, "fifty million men, women, and children ripped out of the Ledger to sink to the bottom of the ocean or wash up like foam along the shores of America" (Glissant 2011, 8). In this going down, not to reach up, but to pass through the molten heart of the volcano, they individually and collectively restore the conscience of sense that knows itself in the cosmic side of the inhuman, rather than its most pressed upon category of inhumanization. Or, rather, their materialist poetics plots what I identify as the three paths of the inhuman—inorganic, subjective, and cosmic—in dialogue, and claims that breadth as an expanded inheritance that denies the narrow margin of genealogical and geologic grammars. Glissant thinks with the volcano,

Fanon with the abyss and its buried sunshine, and Césaire summons from the depths of the ocean, as well as recording slavery's thumbprint on the urban fabric of polished cities enriched by the misery of enslavement and its afterlives. Their crucial awareness of what was required of them in terms of reinvention is as remarkable as it is beautiful, as it plots a path alongside difference that does not seek to destroy Relation, as colonialism does, but to bring its painful history into conversation with the grace and rebellion that have endured that passage. I see their thought as a reconstitution of cosmo-being beyond any cramped dialectic.

Engaging the inhuman was not a reinvestment in dehumanization—a negative ontology—but its difference was the reconstitution of the generosity of the earth and cosmos.[4] That was optimism in the breadth of life's inhuman cradle, at the scale of the universe, in the dark depths of the oceans, and in the very magma of being (in all the ways that negated the nation-state and its propertied material forms). Crucial to this cosmontology is Glissant's wrestling with the question of genealogy and its legacy, in the question of how identity is "planted." Crucially, he recognizes that thinking identity must be done with the earth, as a historical and geographic entity that was made by and through colonial geologies. Which is not to see everything as colonialism, as other earths were made in deference to and defiance of that coloniality. Addressing the legacy of geology and its grammars of race and difference, and how this parsed subjectivity as genealogically entitled or not, is a project that haunts the present and its calcification of racial terms.

Deborah Thomas articulates plantation affects in the present as a simultaneity of time between past and present and the way in which violence continues to shape the forms of political subjectivity and the hierarchies of what it means to be human in Jamaica.[5] What she calls "priorness" organizes a temporal orientation in the political field of modernity, which might also be understood as a temporal ontology of being. Bound to the priorness of those hierarchies that produce a conception of Being within the strictures of Western time, the suture of time to the inequalities of being a "property" of the plantation get reactivated through the temporal deficit of racial capitalism (i.e., the promissory freedom is always just that, a temporal subjunctive in white capital). Along with other writers in Black and Caribbean studies, Thomas shows how plantation time structures the Black subject as temporary and economically deficient. Through practice, she curates affective archives in which people shape their agency and innovation in the present as a reparative act. The shape of these archives is

> nonlinear and thus unobligated to the teleologies of liberalism, [which] can also shift the politics of reparations away from discretely local and legally verifiable events and toward the long and slow processes undermining our ability to forge social and political community together. They can urge us to be more skeptical about nationalist narratives of perfectibility whereby we triumph over past prejudices and injustices through a force of will and commitment to moral right, instead encouraging us to train our vision more pointedly to transnational geopolitical and sociocultural spheres and to the messiness of sovereignty at different moments. (D. Thomas 2019, 7)

Joining the temporal arrangements and material situatedness of Black (and Indigenous and Brown) subjectivity within the plateau and mine, sovereignty can be said to emerge under different geophysical regimes and cultural modes of production that form other archives of affect. And, if we follow Thomas in a different register, the slow processes of compounded histories of geophysical conditions can be understood alongside the actions of geopolitical regimes. Changes of state thus require attention both to the material and political conditions of subjectivity (in relation to sovereignties that exceed the nation-state) and to how these hierarchies are mobilized within and outside their sites of origination (i.e., the mine and plantation), with attention to the historic duration of those temporal states. Racial hierarchies involve geophysical states, depths of material relation and vertiginous effects, differentiating pressures, and accumulative burdens of gravity that create distinct historic affectual architectures of states of being that differ from, but are informed by, the political state. The task of unburdening obligations to material teleologies is one route into liberation from those oppressive geologic grammars. Reparations require the repudiation of the promise and priority of material development, alongside ecological reparations that reconstitute the livability of spatial dynamics and geographies of inhuman pleasure.

Mine as Property

Tia-Simone Gardner's (2020) prerogative over the possessive liberatory etymology of the word *mine* in her work reclaims the industrial mines of Birmingham in the intimacies of that lived possession of her history. Fred Wilson enacts the artistic and curatorial practice of "Making It Mine" in his exhibition *Mining the Museum* (1992) at the Maryland Historical Society.

Jennifer L. Morgan notes how one enslaved woman, Arrabell, on a plantation in South Carolina, called her child "Mines" in order to stake her claim, although, "of course, even as she staked this claim, she and her child were sold. The next sale may have been the one that irrevocably reminded her that Mines could not actually be hers" (2004, 132). The child called "Mines" enacts the future anterior (in Campt's [2017] terms); its refusal and claim are made all the more impossible by the trade in flesh-as-property that puts a Black mother's claim to mothering and to hold her child as her own in doubt. As Morgan writes, "Through the birth of their children, enslaved women may have seen a means to reappropriate what should have been theirs all along. Arrabell's child's name appears to indicate poignantly the struggle inherent in reproduction in this most unstable moment in the development of slave society" (132). Mine in the possessive sense, then, as relation, must be understood as racialized too in relation to the fathering prerogative of geology. In Arrabell's naming, a noun becomes a verb. Its assertion is a rebellion against the state of property relations. The claim to call one's own one's own was historically not a birthright under the genealogy of racial difference.

As a site of reproduction, the mine can be understood as part of the extraction and expansion dynamic of colonialism, whereby persons-as-bodies geocoded as Black, Brown, and Indigenous came to represent the frontier space of the geographic movement across and below yet are excluded from the geographies of its accumulation. The dynamic is one of subtending economic, social, physical, gendered, and sexed colonial architectures while remaining absent in how they substantiate the appearance of the colonial, settler, and neocolonial surface (as an interlocking temporal and material structure of the plateau). The geophysics of race refers to the gravity of anti-Indigenousness, anti-Blackness, anti-Brownness as a form of undergrounding racialized life, in mines, locked in material metaphors of somebody else's possession. The mine can be a slaughterhouse near the border, a beach in Morecambe, the late-night cleaning crew in the city of London, a prawn fishing boat, a Malaysian palm plantation, a suffocating lorry in Essex that holds the graves of young people fleeing the social consequences of leftover environmental devastation from multinational companies in Vietnam. Extraction is all around. The mine generates a set of hierarchies between what is undergrounded and what is accumulated on the overground, and how accumulation is managed across these geophysical states as racial gravity.

This project began in the impetus to map geophysical states in terms of extraction and its effects on planetary processes, alongside an account of the flesh of geology (to understand how the histories of disembodiment in geology—in the splitting of the *geos* and *bios*—have structured the possibilities to think or theorize across these forms of embodiment and earth), and then, at a certain point, I began to understand how race was a mode of governance and form of stabilization across this breach of *bio* and *geos*. And that the dynamics of the mine functioned well beyond the mine in the racial hierarchies that constitute and built racial capitalism. Alongside this mining into the earth, geology became a site in the production of narratives about subjects and time and subjects in time, which created a specific political valency of time, and directionality as achievement (a narrative that was paralleled in approaches to matter, that articulated production and the containment of categories as accumulation).

Colonial geophysics creates anti-Black, anti-Indigenous, and anti-Brown gravities, ones that press on and deform the possibilities of subjective life in structural and intimate ways. This process of unearthing and subtending, across the registers of subjectivity and earth, is a conjoined process of becoming that made colonial earth and missing earths. This racial gravity of the afterlives of slavery is there in the condemnation of Black and Brown life across carceral spaces, environmental justice, and toxic atmospherics, as well as the burden bodies must carry in racialized spaces. Which is not to say that Blackness or Indigeneity is reducible to the thingification that has been promulgated in that category. As Morrison and others suggest, the grace and poetry that follow contest this. It is perhaps not surprising, then, that one of the most targeted international political categories of subjects is Indigenous environmental defenders, who for the most part are fighting the development of extraction from corporations originating in settler colonial countries, such as Canada, Australia, and the United States, or companies that are running their accumulation through the stock exchange of London, New York, or Toronto. White geology was an idea that had a projectile reach and exportation geo-logic and could transmute along with it a set of subjective relations beyond the territorial designation of colonial-postcolonial states, operating under the grammar of natural resources, which, to paraphrase Edward Said (1994, 69–70), masked the "imperial taproot." Geologification accumulates through spatial and temporal mobility (as we have seen in Alabama). As geology went to America, a specific material form of colonial earth praxis was instigated, which was then exported all over the planet in the genre of natural resources.

The mine is paradigmatic because of what it institutes in terms of spatial and subtending geophysical relations, and what it methodologically requires in terms of a geophysical approach to race. Race is the glue that was invented to suture the splitting of *bios* and *geos*; a split that requires reconstitutional geologic acts of being. Understanding race as a geologic proposition is a way to open up the imbrication of inhuman materiality, materialisms and subject positions of extraction that exceed the configuration of environmental racism as a spatial organization of exposure to environmental harm to consider the temporal ramifications of deep histories of the inhumanities. Going beyond spatial surface referents as the frame of analysis is a means to decodify the geophysics of space, that is, to attend to the ways in which space is dividuated and enacted within a set of racialized dynamics that make the precarity of those spaces and the possibilities of movement, in terms of both expression and physicality, an often segregated experience of affective and material attunement. The geophysical axiom of the inhuman as a historically constructed material mode of being conditions what space *is* and *charges* it with differing pressures. A new kind of geopolitics—that understands states of geophysical being—might replace a political account of the subject as biopolitically bordered and understand the role of the earth in making forms of life: forms of life that are indivisible from a racialized account of the inhuman. Getting caught on the wrong side of the matter divide, coded-as-resource rather than Life, or getting subtended in the strata to be narrativized as a fossilized subject whose geologic occasion is prefigured as an imagined extinction, was a mode of subjective stratification. This is why I think there is no liberatory politics to be had through an account of biopolitics, because biopolitics is grounded in and through a colonial earth; the *bios* of its politics is already dividuated from the racialized and racist account of earth that prefigures its subject division (as subject not matter). To put it another way, the biopolitical figure as an account of political subjectivity relies on a grounding in an inhuman designation that needs ungrounded racialized subjects to build its emergence.

Understanding race as geophysics and geology as racialized paves the way to never consider subjectivity without the earth *and* to see earth processes as a site in a racializing stratum of subjugation that made colonial, now Anthropocene, earth. Alongside this acknowledgment is the most pressing issue of the consequences of subjective transformation of geology, and its impacts in the intimate and affective architectures of material relation (geology is part of racism's everyday garments that regulate its structure and possibility: young miners in opencast pits hauling coal and those in fuel poverty seeing breath). Natural resources are the normative or vernacular Western relation to materiality that encompasses land and country in the form of property (and access to subjective properties of Life).

The grammar of geology is this *monolingualism of materiality*, as the plantation is a monoculture of the soil and the mine a monoextraction of the deep and subsurface. These material projects of domination are formations that show the *sedimentary context of race*, as a structural mode of accumulation and a geophysics of being. This sedimentation of race—a shadow geology—is realized through historical geographies of colonial materiality and the imposition of geologic forms of capture that partitioned the world across earthly (*geos*) and selective subjective bodies (*bios*). Recognizing the grounding operations that make the rift and attuning to poetic dispositions of the languages of broken earth, the underground can be mobilized as an affective resistance of properties' predation. Unearthing and upending colonial earth's structural hierarchies promises a confrontation with the domination of the white geology and its oppressive gravities.[6] Before the move to the normative question of what is to be done, there needs to be a reckoning with what *has* been done, and for whom and through what languages of the imagination and materialism it was done: this is a history of geologic life.

Dispossession and the Mine, Mine, Mine

stratal compression through racialization
creation of value through desedimentation of geology + devaluation of Indigenous, Black, Brown life
(inhuman + inhumane)

whiteness
*
*
gravity
(*antiblackandbrown*)

$$\left(\mathit{dispossession}\,(\mathit{race}) + \frac{\mathit{possession}\,\frac{\mathit{pressure}}{\mathit{surface}}\,\mathit{property}}{\mathit{possessive}\,(\mathit{value} + \mathit{geologies\ of\ flesh})}\right)$$

mine
surface→∞

£
$$$

Race solidifies the geophysics of whiteness, pushing down to rise up—the surface-underground relation—to extend the plasticity of surfaces in the mutual consolidation of geology and whiteness as value. This plasticity is the *intra*material sphere of economic, social, cultural, philosophical, and psychic forces and their spatialization. Race is what holds the axis of racial gravities in subservience to the surface. Underground processes and the action of the rift are what enables processes of unearthing the possessive condition of the mine. Pushing down (Indigenous, Black, and Brown life) to lift (whiteness) has a counterforce that goes underneath these relations as a counterarchive of geoforces. As Leanne Betasamosake Simpson sings in "Caribou Ghosts and Untold Stories," "Meet me at the underpass, rebellion is on her way" (2017).

In the simplest equation, what matters about materialism in a colonial schema and syntax is its containment and reorganization of earth into another set of valuations (this is commodity fetishism 1.0). What matters more is not the commodity itself. The commodity is not the telos, nor are its circulation and commodity chains the geography, although it may be the focus; rather, the telos is severance of the Relation so that value can be controlled through dispossession. Dispossession provides a fertile broken ground for further accumulation, and its trajectory is severance across *all* fields. Broken earth is the precondition of accumulation. In the case of settler states, severance is then reconstituted in the affectual nucleus of the white settler family and nation-state, where the means of production is the earth (*geos*). It is severance from the idea and material Relation with the earth, as the unity of all possibilities of emergence, where being is situated in the continuance of material energetic geoforces that make a world, organized around its tender rather than its extraction. *Geologic Life* has addressed how the bones of geology and the stratigraphic imagination structured Enlightenment thought and its dreams of a floating world. Alongside the emergence of the humanist subject, geology ushered in a subject birthed within its elemental grammar: the Inhuman. Interned in the geographic imperatives and desires of colonial conquest, the dual natality of the inhuman as both a category of matter and a race set in place a material praxis of anti-Black, anti-Brown, and anti-Indigenous geophysics. The grit and grind of being intimate with the earth are not all about imposition. It is a space of knowledge practices of the earthbound. The inhuman is a praxis. The perspectivism enabled by the rift is different and durational. It is a site where the collisions of history are always on display, stratal, fossilized, and present. The historical modes of surface production are always in tension with what they resist, subsume, and attempt to erase, in the racial hierarchies of strata that are designated below.

In the abyss of erasure, Glissant puts remembrance in the "rocks dug out of time" (2011, 145) in his novel *The Overseer's Cabin.* He weaves together one woman's knotted stories of Martinique's histories, incepted by the arrival of colonial violence and her fighting against the black holes of time and forgetting. The earth archive that Glissant imagines is not metaphoric; it is just that, an archive of earthed stories, grounded in inhuman intimacies, pushed into the hard minerality of the world, as a site of witnessing and repair. As he states, "It's the word put into the earth you're turning over" (205). Glissant calls the last chapter in the book "Rock of Opacity." The dead inhabit the rock; neither is a dead matter, nor gone. Extraction is not the end or beginning of the story. "Patrice Celat inside the rock, and Odono Celat. . . . They are going back along the Trace of Time Before. It's to make it clear for us, to light it" (205). The "Trace of Time Before" refers to the all too painful routes of history that have made the present possible and everything that was lost to the vicissitudes of the inhumanities. Glissant argues that Martinique's history begins with its trauma. There is no other past than this trauma that is geographic and geologic. The Indigenous people were exterminated by the enslavers. Everything begins in and with this trauma and the coming of these slave ships, he says. Remembering requires a medium that is strong and hard and holds fast, like the rock. The rock hides its duration in plain sight. It is the opacity of deep time and its redress, a materiality that will stand up to erasure and forgetting and not be reduced to the imperative to make transparent (and thus extractable for its elemental core). It may be necessary for that which we do not remember to have a rock in the family.

If the grammars of geology stabilized the object for extraction through the racialized languages of natural resources, Glissant reclaims the rocky world from this erasure and reification, as a site of memory and relation held against the apartheid of that meaning. Parallel to his cry for the right to opacity for the colonized, as a demand of the oppressed to not be made transparent to colonial languages of description and dehumanization, the *rock of opacity* locates this concomitant cry in the earth as a holder of histories and durational difference to the colonial clock (and its erasure of before and after). The way in which Glissant understood the black volcanic sand of Martinique and the magma of the islands was by no means without regard to the actual geopowers that shaped its geography and most recent history. In 1902, the Mount Pelée volcano wiped the colonial town of Saint-Pierre and its inhabitants off the face of the earth in a short breath. The tragedy was fresh in the memory of the island and its writers. The temporal frequency of the volcano swept away Saint-Pierre *as if* it never existed and delivered a terrible idea of freedom that outlasted the colonial monuments (with Ludger Sylbaris in a cave prison as earth witness).

The volcanic archipelago that joined the islands and continued its communicative ruptures under the sea and between magma-fired mountains was a geography that spoke endlessly in Glissant's texts ("Our landscape is its own monument: its meaning can only be traced on the underside. It is all history" [1997, 33]). When Fanon (1986 [1952], 94) returns to Martinique and complains that all this whiteness is burning him up—"the earth crunches under my feet and sings white, white. All this whiteness burns me to a cinder"—he is clearly not speaking metaphorically. While he is referring to the violence of the imposition of Blackness through a white gaze, and what this being Black for others does to a subject's sense of themselves, the white heat that burns its geophysical self into being is actual. It is a gaze that renders the subjugated other, materially and psychically, in the ongoing presence of the planters and their plantation-raised white geopower. The "clearing" of the broken earths that enacted the extermination of Indigenous peoples, established the mine, the plantation, the Middle Passage, the replantation, and the forced migrations that made diasporas, all of these are geotraumas: specific material-ecological events that enact trauma through a geophysicality of being that is the erasure of Relation, constituted in the interregnum of actual places and material geographies. These are geographies constituted by fracture and geophysics enacted by the pressure of white man's overburden. In the hold of the ship, which is the bottom of the ship, the geophysics are different—the air heavy with bodies and their effluents reeks the buoyancy; closer to the friction between water and hull, below the waterline the pull of the ocean floor creates a different force of gravity, another kind of residuary resistance.

The recognition of the violence of matter in the racial project of colonialism, what I call white geology, opens a space to challenge the racializing and racist gravities of material geographies. As with the splitting of the atom, the implosion of space and time has distinct coordinates in the severance of *bios* and *geos*. The cut of *bios* removed Life from the binds of earth, while making other subjects earth-bound as *geos*. A particular historic junction is enacted with the export of European geology—as a nascent science—to colonial states, which set the terms for settler colonial transformations of subjects and earth beyond colonial territories. The grammars of geology that underpinned the change of planetary state that has its inception in colonial earth became multiethnic and multiracial but are predicated on the spatial and temporal dimensions of racial capitalism (as plateau and rift, as racial undergrounds and mines). The geophysics of race continued and continues to underpin the nomenclature of natural resources as a change of world-state called the Anthropocene. To understand anything about environmental change necessitates a reckoning with this deep time history and a reparation toward its nonrenewal.

Much is at stake in the refusal of a deadened inhuman world, especially in Indigenous, Brown, and Black thought. It is the assertion of the living against the compacting categories of colonial experience of environments and their afterlives. The inhuman is an axiology of resistance and a counterforce of defiance. Resisting a narrative of overcoming or redemption, Glissant's call is one of openness to these proximities and an assertion of the political need to build up a *poetics of reconstitution* that materially builds through those erasures—*ghost geologies*—putting words on long-abandoned bones and taking up stories from the silt of the earth where they lay. This alluvial subjectivity of dark oceanic abyss replaces the divisions and segregations of the world and specifies them in their geographies. Contra the imperial archive, *inhuman solidarity* is the basis of a geohistory and present of Relation in geologic lives. Recognition of the porosity of being with the inhuman highlights the imbricated worlds of extraction across subjective forms of life and earth frequencies and the injunction to their renewal.[7]

In the action of joining a critical reading of the history of geology, its scripts of temporality, and the spatialities of race, I have argued that structural-stratigraphic geophysics of race organizes a racialized mode of extraction of the earth that creates the experience and actuality of racial gravities. As geologic grammars petition for a singular geography indexed on value—natural resources—tied to the white supremacy of matter, the rift draws attention to the density and difference that a fuller geographic apprehension of the geophysics of being brings. Connecting this colonial environmental history with the poetry of Caribbean writers and Black feminist theorists, the languages and modes of imagining counterforces are elaborated on in the context of the material histories that shaped the syntax of dispossession. Attentive to the geophysics of power, I have used a *sedimented discursive method* to pay attention to all that is rendered *not-fossil* in the context of stratal violences and to offer an affective architecture of the accrual of geopower that provides a cartographic unraveling. Sedimentation is the residue of many ended worlds. As the fossil secures the empirical knot and claims the colonial archive, the *not-fossil*, its alluvium, is always in dialogue with its own erasure, building nonhistories without end that travel back and forth in time, an aesthetics claimed as durational rift, insurgent in the geophysics of the earth.

> *"Beauty discloses to us the movement and meeting of difference."* (Glissant in Diawara 2010)

Acknowledgments

For the Orphans (Not Prisoners) of Geography

This book took a long time to write, which may be fitting for a book about the racial affects of deep-time narrativizing. But the span of ten years has meant that I have not quite found an adequate way to write the acknowledgments, given the volume of talks, engagements, chats, and questions that made other paths grow. I was finishing the final edits on this book when my dad died in 2022. Among the minimal paperwork that was left after his exuberant passage across the world were his papers relating to the British Nationality Act of 1948. Under "Race," he was coded Arab. He had traveled across the world to London as a young boy with the British Royal Navy, whose white officers had a penchant for watching Malay Thai boxing. My parents never mentioned race, and yet a "mixed" marriage in the 1970s deeply shaped the lives of us all. I was nine before we met my dad's family, and on that trip, I saw my first mine in Sri Lanka. In the holes in the ground, the informal gem mines, children the same age as my brother and I worked covered in wet clay earth. Experiences have strange undergrounds that surface unexpectedly in the work we do. There are many people who help guide that process with generosity and companionship in the ideas we eventually settle in. It is not straightforward to acknowledge these surfacings.

This project has a broad geography which eventually landed in a transatlantic conversation. My first official academic trip to the United States, in 2013, was at the invitation of Janet Walker to visit the University of California, Santa Barbara and a conference on rising sea levels, alongside many Indigenous scholars. This was the beginning of an intellectual journey that frames this book. Janet has remained a wonderful interlocutor ever since, and my formulation of geotrauma owes a debt of gratitude to her understandings of trauma and the work of repair that she does in environment and media spaces. I am also grateful for early influences on my inhuman thinking from the extraordinary independent thinkers Elizabeth Grosz and Elizabeth Povinelli, as well as Jennifer Gabrys and Myra Hird, all of whom are also amazing people. Since then, I have had two wonderful visits to Arizona State University and am grateful to Ron Broglio for those generous invitations and where I had the good fortune to meet J. T. Roane, among others. I also benefited hugely from residential fellowships at Durham University in 2017 (with special thanks to Andrew Baldwin and Divya Tolia-Kelly) and a Senior Fellowship at The Internationales Kolleg für Kulturtechnikforschung und Medienphilosophie, Bauhaus Universitat Weimar, in 2019 (which was an occasion to learn in depth about Enlightenment inhuman epistemologies and see Goethe's rock collection), a Summer Intensive at ICA Miami in 2019, and virtually at Digital Earth in 2020. During that time I have been to many US universities and institutions and have been engaged with many different disciplinary and interdisciplinary scholars, from whom I have learned much and whom I would like to thank profusely. More recently, the Summer Institute at Colby College was a joyous emergence out of the more solitary days of COVID-19, and I am grateful to everyone there for such intense and creative conversations, especially my seminar group—Zeynep Oguz, Waqia Abdul-Kareem, Christina Zenner, Eyad Houssami, Jordan B. Kinder, Moritz Ingwersen, Janet Walker, Chris Walker—and colleagues, Liz Deloughrey, Mel Chen, and Sunil Amrith. In 2022 I was a Drake Scholar in English at Vanderbilt University and had occasion to engage with a fantastic PhD cohort; I thank them too for their extensive culinary tour of Nashville. Additional thanks, along the way, in no particular order, to Angela Last, Nigel Clark, Arun Saldanha, Rory Rowan, Jamie Lorimer, Stephanie Wakefield, Karen McKinnon, Makalé Faber Cullen, Lesley Lokko, Eray Cayli, Jill Didur, Justine Bakker, Thandi Loewenson, Charlotte Spear, Maddie Sinclair at Warwick University, M.E.D. students at Yale Architecture School, and, for brief but transformative conversations, Deborah Thomas, Saidiya Hartman, Tina Campt, Cajetan Iheka, Jonathan Howard, Francoise Vergès, and Hazel Carby. Artists that I have been lucky enough to work with include Myung Mi Kim, Esi Eshun, Otolith Group, Caecilia Tripp, Ivy Monteiro, Sophie J Williamson, Kiyan Williams, Camille Norment, Sarah Roselana, Cave_Bureau, Phoebe Collings-James, Sandra Mujinga, and Sondra Perry.

Academic life in the UK is not quite what people might imagine it to be. However, at Queen Mary University of London, I have been lucky to be surrounded by a collegial crew, smart and questioning students, and exceptionally bright and creative PhD students: Helen Prichard, Kate Lewis Hood, Charlotte Wrigley, Mariana Reyes-Carranza, Giovanna Gini, Alexandra Boyle, Fran Gallardo, Matthew Beech, Therese Keogh, and Imani Jacqueline Brown. My research time at QMUL has been transformed by the

"Planetary Portals" group—Kerry Holden, Casper Laing Ebbensgaard, and Michael Salu. The ongoing portalista extravaganza has been a joy and a privilege that is underpinned by trust and friendship in the work. Finally, Mary Thomas both read and shared a considerable part of the journey of this work, in West Virginia, North Dakota, and Alabama. I am indebted to her insights. During the passage of this work, Helen, Andrew, Lisa, Torin and Lorcan, Amanda, Robyn, Tim, Bubbles, and Lupa have all been stalwart friends. And Nick, Tess, and young Disco. Sarah Truman read this manuscript with great attention, sage advice, and encouragement. She helped make the writing clearer and supported the formation of the tome from across the continents. Finally, to my friend who has taught me much about the orphans of geography, Jennifer Gabrys.

The entire process has been overseen by exceptional editorial guidance and unwavering support from Elizabeth Ault. Her responsive and clear communication, and that of the expert team of Lisa Lawley and Susan Ecklund, at Duke has guided this manuscript into being.

Notes

Epigraph. a geologic dirge

1 See Dionne Brand's "besides the earth's own / coiled velocities, its meteoric elegance" (2006, 100).

2 Sylvia Wynter creatively misreads Marx, taking his call in the *Manuscripts of 1844* for the senses to become theoreticians in their own right in "Black Metamorphosis" (n.d.), by which he suggests that senses are not receptors but, like labor, work to transform matter and create objects in their own right. Thanks to Jennifer Gabrys for this connection.

Introduction

1 In this book, I use the term *geology* in its broadest historical sense as a formation of a set of ideas about the Earth (object), Being (ontology), and thought (stratigraphic and geophilosophic); understood in its formative disciplinary constitution as a science of the earth, extraction, and fossil and anatomical epistemologies rather than a discrete earth science of inhuman matter and material processes. In the rogue geology of the Anthropocene, this interlocking geosocial formation of geologic subjectivity comes once again to the fore.

2 Sylvia Wynter (1996, 301), in her account of the secular telos of material redemption as the narrative lure of the "underdeveloped," describes how the deceit of a racialized body, produced in a state of debt, is positioned to be "redeemed" through material development. She uses the term *archipelagoes* to connect the Black jobless in the United States and the correlated prison system to shantytowns and favelas in the Third World. She thereby gives us a geographic imaginary and social reality of transnational solidarity.

3 Arjuna Neuman and Denise Ferreira da Silva use the term *implicancy* in their film *4 Waters-Deep Implicancy* (2018).

4 The *shock forward* can also be thought as the living affect of erasure, in the relations built around *who* and *what* is taken away. An environment grows in response to absences as much as "cultivation"/growth. This is also to see

erasure as a white supremacy fantasy—a desire for a "clean" extraction, a development without anticolonial insurgency and environmental entropy.

5 If the colonial and settler "family" is a unit that organizes extraction and genealogy of affects and the inheritance of property, the question of how to reconstitute the "family" arises as a site to undo white heteropatriarchy and to queer the reproduction of the world by having a rock in the family, which is also to see these infrastructures of kin-making as a reproductive structure of division and segregation of matter. Having a rock in the family disrupts and differentiates the genealogy that secures the white supremacy of matter. Having a rock in the family disrupts the *geo-bios* separation and the reproductive futurism of the heteropatriarchal family that is built through extraction. If we think about having a rock in the family as a normative proposition, how would language need to change?

6 The temporal tactics of fungibility are performed today in the work of Black feminist scholars, such as Black Quantum Futurism, N. K. Jemisin, Fahima Ife, Kara Keeling (2009a), and Tavia Nyong'o, among others. See James Scott (1992) and Jovan Scott Lewis (2020) for discussions of retooling the "transcript."

7 Traditional forms of Western ethics are secured on a biopolitical-juridical subject, raised on what Sylvia Wynter calls a partial humanism (a genre of the human[-ism] that is limited), or what Frantz Fanon might call a parochial ethics. This biopolitical focus (on a biological rather than geologic subject) is a product of Enlightenment-colonial thinking that participates in the division between the *bios* and *geos*, which is at stake in planetary precarity. Thus, there is also the need to reconstitute the role of the body beyond the individuation of a rights-based subject to see earth reparations as integral to sovereignty.

8 Race is bound up in specific geographic relations that map into existing social structures and complexities to frustrate genealogies of accounting, even as they are mapped into the larger arcs of the "imperial intimacies" (Carby 2019) of colonial lives and afterlives. In the nineteenth century, race became articulated as geography (environmentally determined) but not tied to a specific relation of place, thus leaving land and "untimely persons" available for theft.

9 In referring to "affective infrastructures," it is useful to note (per Lauren Berlant) how affect reproduces infrastructures that present themselves within the coherence of the commonsensical or normative. Kara Keeling understands affect as a form of labor that "underscores the extent to which our efforts to assimilate that which moves us are bound to the ethico-political context of our times and available to capital and its normative structures of command, as well as to the related yet distinct operations we know as racism, homophobia, misogyny, and transphobia, among others" (2009b, 565).

In the relation between infrastructure and futurity, thinking *sense-as-infrastructural* is a way to come at the reproductive potential of how sense arranges a political and epistemic scene of geopower and its futurity and to find an analytic that helps break with forms of reproduction that reconstitute normative (oppressive) social conditions, in order to revolutionize new modes of production that redirect and reconstitute geologic forces.

10 For example, the Carboniferous became conceptualized by President Donald Trump and the Department of Energy in a 2019 speech as "molecules of U.S. freedom" and "freedom gas" in the geologic now of fracking petropolitics (liquefied natural gas), populist movements, and white settler masculinities.

11 I use the term *Global-World-Space* to capture how in 1492 the aim to create a global space of extraction/transaction/value was made through the objectification of the world and the conquest of geographic space and the indices of geologic value. While Global-World-Space is an ongoing imposition of an idea rather than an achievement, it sought to make its totality through material and spatial languages of captivity in geologic grammars, as well as in its more familiar geographic projections. Imperialist geologic practices were potentiated alongside a foundational inscription of a race (see Ferreira da Silva 2007). What the production of the universality of this historicity achieves, Ferreira da Silva argues, is "to construct the racial as an improper aid to otherwise appropriate strategies of power" (2007, xxiv), rather than recognize how these analytics of globality and their spatial agency are sustained by the foundational inscription of race. Blackness is thus catastrophic at a metalevel of globality rather than a crisis in a particular narrative.

12 At the insistence of locating this scholarship and in resistance to the crude and calculative resedimentation of racial identity forms, I should state that I was born in the geographic rifts of colonialism and its afterlives. Both my parents were born in the throes of World War II; my father, born in the Japanese-British colony of Malaya-Singapore, was classified in his papers as Arab and my mother as British-Irish. My parents were "mixed-race" with Brown babies and caught between many worlds. The year I arrived, 1974, was a high point of National Front activity, and the party ran its MPs on an agenda to "return" nonwhites to their imagined origins. I grew up in that white landscape, with every bus stop and underpass tagged *NF*, bone deep in British racism and classism.

13 I understand "thick time" as the multiple temporalities of Black, Brown, and Indigenous struggles for racial justice and peace in negotiating and securing practices outside of the imposition of colonial time.

14 The eighteenth through nineteenth centuries saw the transition from an (in)human(e) enslaved economy to an inhuman fossil economy to substantiate the plateau. Rather than shift the racial dynamics of whiteness, racial undergrounds underpin this transition, from the Slave Compensation Act of 1837 that funded the Industrial Revolution in the United Kingdom to the

convict lease labor mines in the United States that built the modern urban centers and the political dynasties of industrialists.

15 Ferreira da Silva conceptualizes the "analytics of raciality" as a particular strategy of power that "has produced race difference as a category connecting place (continent) of 'origin,' bodies, and forms of consciousness. The primary effect of this mechanism of power/knowledge has been to produce race difference as a strategy of engulfment—a modern scientific construct whose role is to reveal how the 'empirical' is but a moment of the 'transcendental'—used in the mapping of modern global and social spaces" (2001, 423).

16 Identity classification coordinates which bodies become visible and how. It can also be seen as the masking of the territorial imperative and its relocation into a subjective order that submerges the theft of land and places relation into another body that is not earth. This politics of substitution is paralleled in another double move, whereby the hierarchical organization of identity as race was used to sanction the theft of personhood—a theft that is designed to fulfill the demands of extraction that is its territorial imperative. The relocation of the territorial referent—geology—organizes bodies as well as land in the ongoing realities and forced diasporas of settler colonialism and racialized geographies.

17 Seventy percent of revenues generated on the London Stock Exchange by FTSE 100 companies (such as the largely Australian mining conglomerate Broken Hill Proprietary) come from overseas, thereby detaching the exchange from the UK economy and implicating it in many other material geographies and relations.

18 I write from geography with an attention to its spatial modes toward Black feminism as possibly the most radical theory in the cosmos. It builds with love and changes the conditions of the desert. In this sense, I think of Black feminist theory as a geopower and a force field. Toni Morrison and Audre Lorde were building shelters; Christina Sharpe, Dionne Brand, Tiffany King, Saidiya Hartman, Michelle Wright, Katherine McKittrick, Hazel Carby, and Tina Campt, to name a few, build geographies of passage and possibility, handholds to different surfacings. I also write during an exciting time of the emergence of young scholars in political geology who are writing with empirical depth and epistemological questioning about the racial formations of geology, and in the flourishing of the field of Black environmentalism and Black ecologies.

Geologic Life Analytic

1 "Relatants" could be used to describe a possibility/property of relation in Édouard Glissant's sense as a durational sensitivity cultivated in antagonism to the political order ("a world in which one is, quite simply, one

agrees to be, with and among others"; Glissant 2010, 128). See also Saidiya Hartman and Fred Moten's reworking of Glissant's concept of the "Consent not to be a single being." And the consideration of *being as breach* across matter relations and their divisions into Life/Nonlife, Human/Inhuman, etc.

2 Biocentric modes "grounded" in geologic forms and forces as the "unthought" foundation that give biocentric modes stability, resilience, and duration.

3 Geologic cookery: magma as subterranean aesthetic force

1. *adj*: cooking life on the surface; Rift Theory
2. *noun*: "geosocial formations" (Clark and Yusoff 2017) or stratigraphic

thinking as social theory, example: Kant, Althusser, etc. > Marx, working class as spring-loaded magma camber; Foucault, archaeological method and formations, so on . . .

4 Genealogy as grounded in geology; Marx, Freudian unconscious; social stratification, etc. Filiation as imperial lineage; sexuality and the Other.

5 Kant's "Essence of Man"—Universal, but . . . "absolute racial difference" positions Blackness falling beneath the Human in his theories of moral subjecthood. Addressing humanism's Universal as it is passaged through anti-Blackness.

Geologic Life Lexicon

1 I began thinking about geotrauma over ten years ago in the context of climate change, and in conversation with a trauma scholar, Janet Walker. I had been circling around the idea of geotrauma in the context of the immiscible and indeterminate harms of climate change, often to Indigenous, marginalized, and/or racialized people, that failed to register or be redressed through global models and affectual politics.

Chapter 1. Insurgent Geology and Fugitive Life

1 Where I capitalize the word *Geology* in the text, this is to signal a structural use of geology in terms of the discipline rather than in terms of the broader field of geologic relations. Similarly, the capitalized word *Being* denotes the Enlightenment formation of subjectivity rather than a shared ontological and indeterminate state of existence.

2 Geologists lick rocks to identify mineral species and confirm identity. Also, licking gets the rock wet so the surface can be seen.

3 Jared Sexton argues that "Black existence does not represent the total reality of racial formation—it is not the beginning and the end of the story—

but it does relate to the totality; it indicates the (repressed) truth of the political and economic system" (2010, 48). He understands Blackness as a pivotal axis that puts pressure on all disciplines (see also P. Douglass and Wilderson 2013; Weheliye 2019).

4 What extraction and its quaterization of matter mobilize is forces of exuberance and recombination. Pollution is better understood not as a defilement of categories of purity but as a problem of aggregation in the resettlement of geochemical bonds. Flocculation of microplastics, for example, can be understood as a mobilization of attachment to sediments or microorganisms that cause biofouling, whether by drift or dispersal. What is at stake are the modes of attachment that are instituted by industrial extractive processes and the new communities that are formed, which are often hostile to previous life-forms. Geochemical effects are held in check from these new formations by stratal positioning. Exploding the strata does not just "liberate" tight oil but liberates many other buried possibilities to the surface, often possibilities with promiscuous geochemical intent.

5 Hortense Spillers identifies flesh as the production of Blackness through slavery and its afterlives as "that zero degree of social conceptualization that does not escape concealment under the brushes of discourse" (2003, 206).

6 It is worth noting that when developing his theory of gravity, Isaac Newton consulted tide readings from all over the globe, and one of the most crucial sets of readings came from French slave ports in Martinique. Access to the numerical observations of the Atlantic slave trade informed the development and provided the empirical basis of his natural law.

7 The subject is clearly a problematic unit and form of address and has been a figure that has substantiated the terms of the *geos-bios* divide. Thinking with subjective forms and pluralized forms of subjectivity releases some of the neoliberal bonds of individuation, but the history of the subject is tied to the production of Life in ways that corroborate the validity of certain subjects and not others.

8 Nahum Chandler's (2013) critical practice of "paraontology" argues that Blackness needs to be thought of in terms of nonbeing: a disruption to ontology that Moten spatializes as the undercommons (Moten 2016, 28).

9 Clyde Woods (1998, 103) argues that blues epistemology was a form of social action and explanation that countered what he calls plantation bloc epistemology—the historic legacy of structures of racial inequality and its cultural cover of paternalism.

10 The date 1492 is usually used for the invasion of the Caribbean and the Americas as a marker for the inception of a "New World" and its broader reconceptualization of the geographic imagination of Global-World-Space. The year 1452 is also referenced as the beginning of slave sugar production by enslaved Berbers in Madeira. King (2019, 1) suggests that the inaugural

time-space to decenter the Western-Americas axis be considered 1441, when gold and the first enslaved Negroes were brought to Portugal from what was then Guinea (now Senegal). The date may be less important than the temporal architectures that the origin stories create and the efficacy of their telos in the present (especially for challenging white supremacy and the homogenizing impulses of its narratives of progress). Above all, purity claims and monogeographies should be resisted.

Gold is formed in supernova explosions or by the collision of neutron stars. It is found in small quantities in the earth's core and on the surface through the flight of asteroids. It is one of the heaviest metals and least corrosive but most malleable. Historically and contemporaneously, gold is the measure or degree zero of value, or the axis of all valuation. Thus, gold is a site of origination in economies of value. Gold is "cleaned" with mercury and arsenic. It remains one of the deadliest forms of pollution to Indigenous occupants and land around gold mines.

11 Kantian aesthetic thought seeks a stabilization of forms and categories of experience through hierarchical languages and universal subject-object forms to establish modes of valuation and judgment.

12 Aesthetics offers a way of speaking across forces, structures, and provocations rather than framing insides and outsides to conceptual arguments, thus challenging the dominant historicity of a single-world history—the imagined plateau of Global-World-Space. Aesthetic modes are crucial in the creation of new spatial imaginaries. I stick with aesthetics to reformulate it, to reexamine its structuring work in the indices of the political, and its structures of sense and relays. This writing is also inspired by poetry and its open-ended form. Dionne Brand (2018) says that poetry actively seeks to overwrite the capacity for nonbeing in the diaspora within the narrative incapability of bringing the tomorrow of language. Poetry offers a way of unearthing the exclusion of terms by making language strange, thus disrupting commonsensical cadence and retooling sensibility to other geographies of meaning—poetry for the revolution (both quiet and world shattering), aesthetics for the time being.

13 Geologic grammars arrange understandings of agency, politics, mastery, and harm, which in turn constitute and animate futurity through the reproduction of selective life-forms.

14 Often Black and Brown subjects enter the Anthropocene ledger in their propertied or numerical form as a question about reproduction, population, or consumption, or through redemptive strategies of salvation in environmental policy statements that exhibit amnesia about the structural relations of vulnerability. Such paternalistic and statistical framing of Black and Brown life replicates the colonial afterlives of enslavement, invasion, and indenture, what Hartman called the "racial calculus and a political arithmetic" of slavery (2007, 6).

15 This follows a genre of naming in ages (rather than epochs) of dominant material modes, such as the Stone Age, Bronze Age, Iron Age, and Carbon Age, and political geologies such as gunpowder, coal, cotton, and steel empires that homogenize material culture over large areas to suggest a trajectory of the colonial empiricism of matter.

16 The analytic of "tense" is taken up in a feminist mode by Berlant (2011), Braidotti (2011), Povinelli (2011), and Campt (2017), and while it is differently populated, all authors use tense as a way into the affectual architecture of durational infrastructures of sense, feeling, and relation.

17 Within the racial cartography of colonialism, the human (white), as Wynter has forcefully shown, is bifurcated and occupied by overrepresented European man (Wynter 2003).

18 In *Demonic Grounds*, McKittrick names an overarching architecture of enclosure across different geographies that visualizes the spatial project. She says: "If we imagine that traditional geographies are upheld by their three-dimensionality, as well as corresponding languages of insides and outsides, borders and belongings, and inclusions and exclusions, we can expose domination as a visible spatial project that organizes, names and sees social differences (such as black femininity) and determines *where* social order happens" (McKittrick 2006, xiv).

19 Richard Owen, the first director of the Natural History Museum in London, made a paleontological map in 1861 of geologic time and put man in the strata, signified through the identity fossil of "weapons," thereby creating an early iteration of political geology as weaponized in the economy of time.

20 The modes of endurance offered in responses to a crisis of geologic life often involve vicious circles of more extraction and desperate stopgap measures of environmental expenditure, locked into normative modes of living that are exhausted, yet prolonged as a feat of endurance; an ecocide of squandering and saving that never changes the tense of its engagement with the world.

21 The earthquake disrupted the primacy of the civic and maritime functioning of Lisbon, and consequently Britain increased its trade in human persons. Portugal and Britain were the two largest enslaving nations, accounting for about 70 percent of all enslaved Africans transported to the Americas.

22 The disaster became mapped into the governance of infrastructures and imaginations of urban planning as a sociality of risk and preparedness.

23 Diamonds were first "discovered" by Europeans in Brazil in the 1720s, and by 1743 a diamond region, Tijuco, was created.

24 While various theories have proceeded with the dismantling of the nature-culture binary, they have often left the *geos-bios* binary intact because the cost for that disaggregation involves dismantling whiteness.

25 Critiques of environmental determinism and geography in colonial discourse are well established, but less so in the context of deep time geographies and hominid research (see Coombes and Barber 2005). Methodologically, the idea is not to produce a new environmental determinism via geology, especially in the context of an appropriative turn of epistemologies/ontologies that are positioned "outside" Western thought and then resutured onto the future without due diligence in what it means to "borrow" without consent and without a commitment to reparation or to pluralize without the question of politics. At the same time, there is a need to break down the outside/inside fiction of Western thought, given the histories of incorporation and erasure in knowledge production, and the production of thought through antagonisms.

26 Such was the racist cultural sedimentation of the blackface aesthetics of this material analogy of black and coal that a cartoon film, *Coal Black and de Sebben Dwarfs*, was released by Warner Bros. in 1943 (directed by Bob Clampett) to compete with Walt Disney's *Snow White* (1937). The film trades in racist and highly sexualized gendered caricature figures. For a redress, see Audre Lorde's poem "Coal" (1997, 6), where she says, "I/Is the total black, being spoken / From the earth's inside."

27 King demonstrates how Blackness as an geographical imagination of a different matter-space-time was used by British plantation owners "as a form of raw material" for "spatial expansion" that enabled colonial geographic projects, whereby Blackness could be understood as a spatial analytic (King 2016a, 1025; also see M. Wright 2015).

28 Consent has a double valency in the context of colonialism, often used to deflect subjection onto the subjected. For a discussion of the deployment of notions of consent in the context of Sara Baartman, see Sharpe 2010, 83. Simpson's formation, as I understand it, seeks to embed reciprocity across relations for care and responsibility against the segregation of worlds.

29 Lowe: "I use the concept of intimacy as a heuristic, and a means to observe the historical division of world processes into those that develop modern liberal subjects and modern spheres of social life, and those processes that are forgotten, cast as failed or irrelevant because they do not produce 'value' legible within modern classifications. Just as many observe colonial divisions of humanity, I suggest there is also a colonial division of intimacy, which charts the historically differentiated access to the domains of liberal personhood, from interiority and individual will, to the possession of property and domesticity. In this sense, I employ the concept of intimacy as a way to develop a 'political economy' of intimacies, by which I mean a particular calculus governing the production, distribution, and possession of intimacy" (2015, 18–19).

30 Foucault addresses race through the terms of antistate struggle, calling fascism and Stalinism the "two black heritages." Nowhere in his discussion in

Society Must Be Defended (2003) does he discuss the biopolitics of race that is his milieu.

31 In this tack, I also draw toward Ann Stoler's understanding of Derek Walcott's poetry as an attempt to disrupt the "facile distinction between political history and poetic form, urging us to think differently about both the language we use to capture the tenacious hold of imperial effects and their tangible if elusive forms . . . to track the uneven temporal sedimentations in which imperial formations leave their marks" (2013, 2).

Chapter 2. Rift Theory

1 My thinking about the rift started in conversations with Nigel Clark about the limitations of a special issue we had put together on "geosocial formations" in *Theory, Culture and Society* (2017). Its theory was white, and in its rush to join the social and the geological, it failed to conceptualize the distinct racial rifts of those formations. We also had a conversation with N. K. Jemisin about her "Broken Earth" trilogy at the New School in New York, organized by Stephanie Wakefield. We put together a session at the annual conference of the Association of American Geographers in 2018 in New Orleans titled "Geosocial Rifts." Katherine McKittrick presented a paper in one of the AAG sessions, which was later published as "Rift" (2019). She reads the rift against the grain as an interruption, positioning Black studies as the rift or opening into theorizing normative narratives, as an interdisciplinary conversation with race, geography, and liberation. I am indebted to the participants of the "Geosocial Rifts" sessions for their multiple theorizings of the rift.

2 In Marxist theory of metabolic rift, often deployed in environmental thinking to capture the disassociation of social processes from nature, the rift is an irreparable alienation from nature resulting from human labor and agriculture. Marx was writing during a time of the intensification of plantation agriculture and exhaustion of soils, which resulted in the augmentation of the chemical composition of earth, predominantly through the imperial annexing of countries and islands rich in guano (see the Guano Islands Act of 1856, which allowed the US government to take possession of islands rich in guano and to defend this theft with the military). The metabolic rift is a way to approach the political ecology of separability, but it already presupposes the universality of separability and its accomplishment.

3 The Great Rift Valley in East Africa is framed as a site of origination for hominids, the "Out of Africa" that predates the "Out of Earth" white patriarchal desire for other planets in Anthropocenic end-time fantasies. As "birth" and "cradle" of the plateau, as well as pit and mine of human origins

and humanity, the rift is also where the "new" is imagined to be born and migrational invention is substantiated. The point to note is that there is no single origin of hominids and that Africa was a "hotbed" of evolution precisely because it was a site of dynamic geographic, climate, and tectonic change. The geologic "discovery" of the Great Rift Valley was bound up in the colonialism of East Africa. The geologist John Gregory, who coined the term *Rift Valley*, worked with the rock and fossil collections at the British Museum for Natural History, and he joined the expedition to British East Africa in 1892–93. An Austrian geologist, Eduard Suess, had described part of its tectonic structure in 1891 from the German "side" of East Africa, but Gregory is credited with understanding the plate-splitting and tectonic movements of the earth, with his work published in the journal of the Royal Geographical Society in 1920. He described the rift not as two sides being pulled apart but as a breach due to a subsidence between two series of rents. Gregory was also a proponent of Galtonism and an antimiscegenationist (he asserted that intermarriage between members of the three identified racial groups—Caucasian, mongoloid, and negroid—produced inferior progeny and recommended racial segregation). He was also president of the Geologic Society of London from 1928 to 1930 and delivered "Race as a Political Factor," the annual Conway Memorial Lecture for 1931 (also published as a book). With Henry Hutchinson and Richard Lydekker, Gregory produced the two-volume popular book *The Living Races of Mankind* and also published *Human Migration and the Future: A Study of Causes, Effects and Control of Emigration*. Gregory's career involved positions of mining (such as the directorship of the Geological Survey in the Mines Department, Victoria, Australia) and professorships at Melbourne University (where he became a supporter of the "White Australia" policy) and the University of Glasgow, Scotland. He led mineral-prospecting expeditions across the British Empire, including to Southern Angola in 1912, and another group to Palestine to examine its suitability for establishing a Zionist state.

4 See Yusoff 2013 for the genealogical implications of new human origins theories about hominid evolution and mixing, and the queering of the human story. Human origins are also defined by the "fossil-not-fossil" relation in fashioning the trajectories of species.

5 See Beth Greenhough, Jamie Lorimer, and Kathryn Yusoff, introduction, "Future Fossils Exhibition," *Society and Space Magazine*, August 28, 2015, https://www.societyandspace.org/forums/future-fossils-exhibition.

6 The geologies of race that I outlined in *A Billion Black Anthropocenes or None* (2018a) that simultaneously established the co-natal value of Black and gold compel us to think geology in its captive mode, as a language of valuation and delimitation that binds subjects in a liminal space through lithic apertures.

7 As Blackness is named the ontological limit and therefore inadvertent guardian of the spatiality of the rift zone, Indigeneity is named as the rift

in "settler time" (see Byrd 2011; Rifkin 2017; A. Simpson 2014; L. B. Simpson 2017). In both placements, Indigenous and Black subjects are stretched between temporal accounts of colonial identity toward the accomplishment of a material telos of extraction that underpins the state and its reliance on political geology for state power.

8 Wilderson (2020, 304) suggests: "For the Slave, the implement [from Marx's reference to slaves as 'speaking implements'], exploitation and alienation are trumped by accumulation and fungibility. Slaves *themselves* are what is consumed, not their labor power."

9 As Hartman argues, "The slave is the object or the ground that makes possible the existence of the bourgeois subject and, by negation or contradistinction, defines liberty, citizenship, and the enclosures of the social body" (1997, 62).

10 In parallel, Greece has just appointed its first chief of heat in Athens after the devastating climate-induced fires in the summer of 2021. Amid the arrival of abrupt climate change, we can imagine that politics will take on a more elementary representation with fire, fluvial, and solar ministries, for example.

11 McKittrick names the rift as Black studies (2019, 243), in the interdisciplinary terrain of geography, which raises several sharp questions about methodology given the rapaciousness of academia; her *Dear Science* (2021) offers some clear directives here.

12 Those experiences in the flesh of geology that have not been validated by the plateau as a knot in the event of human history.

13 The question which Black feminist work on archives has asked, in a sustained and creative mode over the last forty years: Why reside in a language that produced and sustained nonpersonhood, erased the inability to recognize the complexity of other relations and forms of value, and beauty in the world? (See Carby 2019; Campt 2017; Fuentes 2016; Hartman 2008; and M. Wright 2015, to name a few.) The Black feminist revolution in the archives forces a reckoning with redress, refusal, and care as pedagogical historical practices.

14 For a geophysical application of the mobilizing of a "hydrosocial" approach using sediments within the context of "colonial hydrology," see Mukherjee and Ghosh 2020.

15 The shift to geographic explanation away from ontology has unspoken implications for the recognition of Indigenous claims to origination.

16 For Glissant, passage to the New World is three times an abyss: the abyss of a boat that "dissolves you, precipitates you into a nonworld"; the abyss of the depths of the sea, "a beginning whose time is marked by these balls and chains gone green" where human "cargo" was thrown overboard to lighten the boat and where "underwater signposts mark the course between the gold Coast and the Leeward Islands"; and the memory of all that had been

left behind in "alliance with the imposed land" (1997, 6–7). The oceanic rift sears its beginnings into the transplanted, as the forced routes leave their sediment of the many dead on the ocean floor.

17 When I refer to "earth" in counterdistinction to "Global-World-Space," I am not invoking a Heideggerian or Arendtian Earth/World split, where Earth is essence and Worlding is the elaboration and possession of human technology. Instead, I am invoking earth as Relation (in the spirit of Glissant), where all matter is considered involved in the production of geologic lives and thought cannot reside in an immaterial body. In a parallel mode, Derrida (1999) called the desire to order and contain through the genesis of beginnings and endings the "prosthesis of Origin."

Chapter 3. Underground Aesthetics

1 Agassiz's response to Black servants in a Philadelphia hotel is that he "could not take his eyes off their face in order to tell them to stay far away," as he wrote to his mother in Switzerland in December 1846 (in Gould 1981, 77).

2 It was not simply that certain geologists had racist attitudes and bent their science to that end, but rather that the whole enterprise of geology had its roots in the colonial expansion of whiteness and deformation of Indigenous, Black, and Brown life. Geology was the material science of empire. The integration of geology and race as the precursor to extraction can be seen most directly in the publications of geologists themselves, such the British-American geologist and geographer George William Featherstonhaugh's *Excursion through the Slave States: From Washington on the Potomac, to the Frontier of Mexico* (1844). With his integrated survey of manners and geology, Featherstonhaugh pronounced on the social geology of Southern geographies, where slavery and geology went hand in hand in the category of the inhuman to underpin a political geology of energy extraction. Featherstonhaugh became the first geologist to the US government, surveying portions of the Louisiana Purchase. See also the discussion of Charles Lyell's *Travels in North America* in Yusoff 2018a.

3 The "American school" referred to Louis Agassiz, George Gliddon, Samuel Morton, Ephraim Squier, and Josiah Nott and their collective racial views centered around polygenism. Agassiz's reference to the "imperial body collecting" project (Fabian 2010, 171) is most likely to Josiah Nott or Samuel Morton (1849).

4 Agassiz to his mother, December 1846, translated by S. J. Gould (1981) from the original letter in Houghton Library, Harvard University, Cambridge, MA.

5 The archives I have assembled are positioned to interrupt the neat histories that organize materiality discretely in human and inhuman categories.

Where such overlaps are discussed, such as between Agassiz and Ralph Waldo Emerson and the Harvard group, they are reduced to biographical affiliations rather than disciplinary and systemic structures of production. A discontinuous archive that "roams" across disciplines offers the possibility of witnessing new relations and affectual architectures.

6 As an antidote to this white settler vision of the Jurassic, read Danez Smith, "Dinosaurs in the Hood": "little black boy on the bus with a toy dinosaur, / his eyes wide & endless / his dreams possible, pulsing, & right there" (2015, 39–40).

7 Note how Toni Morrison in *Sula* reroutes alabaster: "Oh yes, skin black. Very black. So black that only a steady careful rubbing with steel wool would remove it, and as it was removed there was the glint of gold leaf and under the gold leaf the cold alabaster and deep, deep down under the cold alabaster more black only this time the black of warm loam" (1973, 135). Or, how Dionne Brand writes in the left-hand margin in *The Blue Clerk* (2018).

8 Samir Amin's work *Unequal Development: An Essay on the Social Formations of Peripheral Capitalism* (1976) addresses the formation of edges to capital accumulation through the designation of the "underdeveloped" as a differential, which set up sacrifice zones for capital extraction and environmental damage as peripheral to accumulation zones. Amin observes how "geographers are content to juxtapose facts, while the basic question of their discipline—how natural conditions act upon social formations—remains almost unanswered" (10).

9 James Scott's work on the political states sees space outside political society as a "shatter zone," a geographic zone where hill people were mutually constitutive of the states they sought to avoid and escape (2009, 326–29).

10 The mural's legend refers to "Forty Thousand Years," a song by the Australian reggae rock band No Fixed Address. The mural's construction in 1983 was facilitated by community artist Carol Ruff with participation by Aboriginal artists Tracey Moffatt and Avril Quaill.

11 The concept of marron is derived from the Spanish word *cimarrones*, which translates into "mountaineers" (who lived in the hills or interior of islands, in marginal and broken grounds). It denotes a political act of subjectivity through an associative geophysical entity and a subjective name of a peoples that is in active identification with geologic features. Glissant characterizes marronage as "the political act of these slaves who escape to the forested hills of Martinique," which "now designates a form of cultural opposition to European-American culture. This resistance takes its strength from a combination of geographical connectedness (essential to survival in the jungle and absent in the descendants of slaves—alienated from the land that could never be theirs), memory (retained in oral forms and vodou ritual), and all the canny detours, diversions, and ruses required to deflect the repeated attempts to recuperate this cultural subversion" ("Glossary" in Glissant 1997, xxiii).

12 Bataille's project was to address "the movement of energy on the earth—from geophysics to political economy" (1991 [1949], 10; see also Yusoff 2018b, 257). This book goes in the opposite direction, from political economy to geophysics.

13 An example of this would be to seek to step into the normative forms of subjectivity that are a result of the plateau-rift relation, such as patriarchy (which is a product of colonialism and, more recently, fossil fuels extraction, which sustains and is sustained by violent masculinities).

14 Hegelian aesthetics is concerned with aesthetics as a mode of synthesis that is conditioned toward an end, rather than aesthetics being the end in itself.

15 See Tina Campt (2017, 17) for a discussion of the future anterior in Black feminist refusal.

16 In other words, underground aesthetics must retain its subterranean force as it discloses itself in order that it continues to worry at the very structures that it is trying to break apart (which might mean hiding its beauty). As soon as the aesthetic mark is settled, it no longer refutes the languages that repress its being in the world.

17 Aesthetics offers a way to come at the temporal dimensions at play in affectual architectures, the tense, eventfulness, and poise of the materialities of experience and the matter-time of emotion.

18 Where aesthetics is not directly involved as part of the theoretical equation, it tends to be invoked through the scene of address—by the formation of problems—and, specifically, how this prepares the ontological space for the subject/thing to emerge in a particular form of life or death.

19 Blanchot, writing between the wars, demanded refusal as a constitutive politics of the traverse of language and its renewal.

Chapter 4. "Fathering" Geology

1 The Jardin des Plantes was where Cuvier encountered Baartman and where the Galerie de Minéralogie et de Géologie, the Muséum National d'Histoire Naturelle (1889), and the Galerie de Paléontologie et d'Anatomie Comparée (1898) were located.

2 The National Park Service website text charts an unbroken line of "American History, Alive in Stone . . ." from unnamed "first inhabitants" to the giant carved political figures that "tell the story of the birth, growth, development and preservation of this country" (https://www.nps.gov/moru/index.htm).

3 Baartman is not mentioned in Rudwick's (2008) epic account of geohistory, *Worlds before Adam*, or his (1998) biography of Georges Cuvier. We read in detail about Cuvier's life and of his personal tragedy at the death of his daughter, and of a "celebrated human skeleton from Guadeloupe" and "another skeleton from the Caribbean Island, which had later been acquired by

the Museum" (Rudwick 2008, 226), but not about his work on race and the atomization of subjects, except as celebrated cadavers.

4 Descent is a key question in the genealogy of species, whether they were understood to be regularly branched or "irregularly branched," as Charles Darwin suggested. The contestation between these trees of life represented much more than a geologic question, but map onto the ontology of the human and its biopolitical orderings. Hereditary traits, even in Darwin's writings, quickly turn into markers of social power in a notion of evolutionary progression in races and families (e.g., he was enthusiastic about the view of his cousin Francis Galton, the founder of the eugenics movement, regarding hereditary patriarchal genius).

5 James Hutton initially introduced a notion of deep time in the late 1780s, describing it through cyclical geologic events, rather than as a linear and progressive set of events that Lyell described later.

6 For example, the first geologic survey map of the United Kingdom, "Tabular View of the British Strata," was completed in 1790 (published in 1815) by William "Strata" Smith. It was published as an identifying device for the purposes of coal extraction and mineral surveying.

7 Geognosy was a branch of geology that examined the material constitution of the earth and the relation between surface materials and the earth's interior.

8 Rachel Lauden describes the dual trajectories in geology as follows: "One is historical: geology should describe the development of the earth from its earliest beginnings to its present form. The other is causal: geology should lay out the causes operating to shape the earth and to produce its distinctive objects. This distinction corresponds closely to the distinction between 'historical geology' and 'physical geology'" (1987, 2). Broadly, historical geology is concerned with processes, such as crystallography, mineralogy, petrology, sedimentology, structural geology, paleontology, geomorphology, geochemistry, and geophysics, as well as with generalizations about the historic geologic record, constructed through the eventful earth and its revolutions.

9 Darwin, for example, engages in this question at length and in his correspondences to the geologist Charles Lyell. He says: "The theory of natural selection . . . implies no *necessary* tendency to progression. A monad, if no deviation in its structure profitable to it under its *excessively simple* conditions of life occurred, might remain unaltered from long before the Silurian age to present day. I grant there will generally be a tendency to advance in complexity of organisation; though in beings fitted for very simple conditions it would be slight & slow. How could a complex organisation profit a monad? if it did not profit it, there would be no advance" (Darwin 1859). And, Lyell wrote to Darwin: "I feel that Progressive Development or

Evolution cannot be entirely explained by Natural Selection[;] I rather hail Wallace's suggestion that there may be a Supreme Will & Power which may not abdicate its functions of interference but may guide the forces & laws of Nature" (Darwin 1869).

10 Darwin: "It is very true what you say about the higher races of men, when high enough, replacing & clearing off the lower races. In 500 years how the Anglo-saxon race will have spread & exterminated whole nations; & in consequence how much the Human race, viewed as a unit, will have risen in rank" (Darwin 1862).

11 Time travel in Afrofuturism is employed as the means to uncover and repossess origins (and thus futurity) before slavery, to imagine beyond and before the natal foreclosure of the Middle Passage and its epistemes.

12 Étienne Geoffroy Saint-Hilaire (1772–1844) was a naturalist who established the theory of the "unity of composition" in biological life, suggesting that all life was formed of the same elementary composition, and he was seen as a supporter of Jean-Baptiste Lamarck's evolutionary theory (even though his ideas around materialism differed markedly). He was chair of zoology at the Muséum d'Histoire Naturelle. A two-month Georges Cuvier–Geoffroy debate was held in 1830 at the French Academy of Sciences to debate ideas of change in life-forms. Frédéric Cuvier (1773–1838) was chair of comparative physiology and head of the Muséum d'Histoire Naturelle in Paris (1837) and a foreign member of the Royal Society of London.

13 Difference in this context centers on the visibility of certain bodily traits, so Venus was not a name given to a subject but a racial identity category and atomized body part. This is why there were many Venuses, and they were interchangeable; they were preformalized as exchangeable objects that represented a "type."

14 As Hartman argues, "Seduction makes recourse to the idea of reciprocal and collusive relations and engenders a precipitating construction of black female sexuality in which rape is unimaginable. As the enslaved is legally unable to give consent or offer resistance, she is presumed to be always willing" (1997, 81). The key issue, Hartman argues, is the production of "nonautonomy of the field of action" (1997, 61) in the object constitution and thus subject formation of the enslaved.

15 Both Sharpe and Hartman retain a relation to speculative narratives as a way of working through the politics of knowledge production and refusing the prescription of Black social death in the telling of archival traces. Both writers ask difficult questions about care and responsibility in relation to the archive, in ways that acknowledge how the archive will always write blind in its notation of history, erasing and substituting some subjects for ciphers, writing a script of (white) supremacy in which subjectivity cannot simply be rewritten in the spaces left by those erasures.

16 I am grateful to Brenda Thomson at Arizona State University for raising the question, during my 2020 visit, of why these scientists were interested in recording her "resistance" when violation was already a given.

17 As McKittrick says, "Baartman's life and afterlife provide an almost perfect collaboration of science and spectacle. Sketches, newspaper advertisements, diaries, and medical documents delineate narratives so racist that Baartman became, and in some cases still is, the iconic representation of the abject, less than human, perpetually condemned and fetishized racial body" (2010, 117).

18 Qureshi (2004) recounts several equally violent accounts of the obsession with Khoikhoi women's inner labia, suggesting that it was an area of interest, if not a genre in comparative anatomy, rather than solely Cuvier's pathology and those of the audiences viewing Baartman at the Museum of Man for 150 years.

19 From Glissant's conversation with Diawara (2010) on the *Queen Mary 2*.

20 There is no answer to how Baartman lives this rift in the geologic, geographic, and psychic text of her story, through all its iterations as Khoisan type in nineteenth-century science to repatriated remains, and cramped life into a specific sign.

21 In a lewd reference in Darwin's *The Descent of Man* (1871) he returns to the fetishized Hottentot Venus: "It is well known that with many Hottentot women the posterior part of the body projects in a wonderful manner; they are steatopygous; and . . . this peculiarity is greatly admired by the men. [A] woman who was considered a beauty . . . was so immensely developed behind, that when seated on level ground she could not rise, and had to push herself along ground until she came to a slope—[T]he Somal men are said to choose their wives ranging them in a line, and by picking her out who projects farthest [to the rear]" (371).

22 White womanhood was not immune from being violently served in this patriarchal frame, but the ability to step into the normativity afforded a different kind of protection. Wynter suggests that the "Hottentot Woman" becomes the signifying category of the atavistic mode of human sexuality. She continues: "The European prostitute, like all lower-class women to varying degrees, was therefore assimilated to the abject category of the Hottentot woman in medical and criminological discourses" (1989, 644).

23 In his letter to Lyell, Darwin gives a de facto justification of colonialism through theories of life: "I suppose that you do not doubt that the intellectual powers are as important for the welfare of each being, as corporeal structure: if so, I can see no difficulty in the most intellectual individuals of a species being continually selected; & the intellect of the new species thus improved, aided probably by effects of inherited mental exercise. I look at this process as now going on with the races of man; the less intellectual races being exterminated" (Darwin 1859b). The approach is reiterated

in Darwin's letter to William Graham: "Looking to the world at no very distant date, what an endless number of the lower races will have been eliminated by the higher civilised races throughout the world" (Darwin 1881).

24 French anatomist Henri Marie Ducrotay de Blainville also published notes on Baartman's dissection in 1816. These were republished in 1817 by Cuvier in the *Memoires du Museum d'Histoire Naturelle*.

25 Sharpe critiques emancipatory narratives that continue to subjugate Black personhood. She argues that in Baartman's "rescue" by abolitionists and repatriation to a postapartheid South Africa, she becomes a subject and name that operate as pure signification, continuing to claim Baartman in a symbolic mode that is "no more *for* Baartman than are the other, more explicitly objectifying practices" (Sharpe 2010, 73). Dialectic reversal confirms the trajectory, using Black women's bodies to theorize new futurities, while flattening them through metaphoric usage. In the affective dynamics of cultural memory, her social reproduction has depended on the desire to subsume her into a different social order while maintaining this right to look, which is in effect to continue to objectify her even as she flies through time (see McKittrick [2006, 17] on the "spatial grammar" of Black time travel).

26 Letter 4510: Darwin to A. R. Wallace, May 28, 1864.

27 Cuvier observes "that her movements were marked by a quickness and capriciousness which reminded one of those of the monkey tribe. She had moreover a habit of pushing out her lips in the manner of the orang-outang. Her disposition was gay, her memory good, and she remembered after several weeks, a person whom she had seen but once: she spoke Dutch tolerably well, which she had acquired at the Cape, knew a little English, and began to speak a few words in French; she danced after the manner of her own country; and shewed some ear for music. Personal ornaments pleased her, but that which flattered her taste more than anything was brandy, drinking which during her last illness, probably hastened its fatal conclusion" (1827, 199).

28 The prehistory of genealogy is produced through stratigraphic thought (popularized by Hutton and Lyell) that secures a notion of species-beings in time (Cuvier). This prehistory matters because it produced durational objects secured in natural history that were seemingly neutralized by the deep time of their production. In the context of colonialism and chattel slavery, genealogy was used to mobilize relationality through the settler state and the figure of the (hetero)normative white family as the archetype of its transmission and regulation (a type in biology is the organism against which all other variations and iterations of the species are measured and to which the species name is attached as the anchor to a taxon). This foundational human "type" has its origination in the geologic ground of

species-subjective life and constitutes its biopolitical arrangements through the organization of temporal lines.

29 Interest in the reproductive organs of Black women was taken in the contemporaneous context of the capacity of enslaved women to "regenerate" labor through forced reproduction under slavery, thus reproducing (racial) capitalism on the plantation.

30 Cuvier asserted that Mongolians had established empires, but their "civilization always remained stationary"; that the Negro race "have always continued barbarous"; and that the Caucasian have "extended by radiating all around." He elaborated how it was by the European stock that "philosophy, the arts and sciences, have been carried to their present state of advancement" (Cuvier 1827, 50).

31 See Sharpe's discussion of "stop-and-frisk" for the political present of this formation of spatial arrest and anti-Blackness: "Stop-and-frisk is one rite of passage that marks the space/race/place of no rights and no citizenship (à la Dred Scott v. Sanford [1857]) in one direction and, in the other, the space through which the rights to free passage are secured for non-blacks" (2012, 829).

32 Phylogeny and geology mirrored each other in the construction of top-down sequencing of higher- and lower-order relatives. The new stratigraphy imagined "from the Superior (Tertiary) formations . . . as far down as the base of the 'Carboniferous' formations," which "involved above all the determination of the order" of formations (Rudwick 2008, 36–39). Both stressed the "progressive" yet stratified character of the history of life. The precise nature of this genealogy was a question that ran across and within species, organized through the imagination of stratigraphic-like distinctions.

33 The question of genesis and geology has been presented as the advancement of natural causation over divine origination, largely understood to have been articulated through the theory of evolution. The question of the age of the earth "before Adam" and the "interpretation" of natural events (such as the Lisbon earthquake of 1755) as no longer the articulation of divine will, but of a dynamic planet, predate the scandal over evolutionary claims and, as such, do much of the groundwork for an ontological shift in the understanding of the genesis of the world.

34 Lyell opened the second season of Lowell Lectures in Boston in 1841 and participated in the third annual meeting of the American Association of Geologists and Naturalists in 1842. Édouard de Verneuil was a French paleontologist and president of the Geological Society of France in 1840, 1853, and 1867.

35 As with Cuvier, the concept of "missing links" established an internal logic of lineage (mostly vertical) coupled with hierarchy.

36 Lyell's nickname was "Pump" because of his verbal grilling of opponents in their use of nomenclature and interpretation. He also was very good at

copying other geologists' work. James Hall of New York was North America's preeminent paleontologist and geologist of the nineteenth century, for example, and he published an anonymous letter (signed "Hamlet") in a Boston newspaper charging Lyell with geologic piracy.

37 To set this fathering in its social reproduction in imaginaries of the present, see Louis Menand in his Pulitzer Prize–winning nonfiction book on pragmatism, *The Metaphysical Club* (2001). After his "soaring" biography of Agassiz's patronage and increasing privilege through Cuvier, Lyell and his discussions with Emerson, William James, Thoreau, and so on, Menand quotes Agassiz's frequently referenced "Letter to His Mother" (Rose Agassiz, 1846, Harvard University Archive) in all its full racial abjection. Menand suggests as an interpretive frame for this violent white supremacy, a naturalization of a reflex to difference: "And it is surely almost instinctive, in most people, to find beings of a kind one has never encountered before unpleasantly alien" (Menand 2001, 105). In a single sentence, he excuses and generalizes Agassiz's abjection.

38 Southern polygenist Josiah Nott (who referred to his specialism as "n[*****]ology") was concerned with interbreeding. Agassiz went to Mobile, Alabama, to Nott's home to lecture on the topic. He used his idea of a fixed race to argue that the white race would potentially extinguish itself through miscegenation and commit to its own extinction. His ideas continue in the "Great Replacement" theories of white supremacists today.

39 The concept of reorigination in Africa set the scene for the later political claims that Agassiz (and Lyell) would make about paternalist geographic "returns" of formerly enslaved subjects by the state to an origin point outside of America.

40 Samuel George Morton was a prominent craniologist and founder of American invertebrate paleontology and author of numerous popular texts on geology and anatomy. His work was crucial in substantiating ethnology as a science. He also analyzed fossils brought back from the surveyors Lewis and Clark's expedition across the American West.

41 See also the work of African American author J. F. Dyson (1886), who put forward an explanation for the unity of the human race in which he suggested that Adam was a man of color (red) and that Eve was white. All their progeny were therefore of differing colors and were the originators of the various racial groups.

42 The "practical amalgamation fostered by slavery" was the coerced sexual reproduction and assault on men and women on the plantations as a means of increasing what was referred to as "stock" and that of the fathers' prerogative. This sexual economy intensified in the Southern United States after 1801 (the time of cessation of the Atlantic trade in human flesh), which in effect internalized enforced "reproduction" and rape.

43 Questions of escape are equally fraught in this context, falling off the edges of the strata of society or changing geographies to become racialized differently (e.g., Stuart Hall [2017] comments that in Jamaica he was considered anything but Black, referring to his privileged upbringing and its mutual reading through his "light" complexion; arriving in England was, of course, a different story).

44 To counter Agassiz's claim of the reality of the house servant, see Harriet Jacobs, *Incidents in the Life of a Slave Girl* (1861); for discussions of the role of this narrative in Black feminist thought, see McKittrick 2006, 39; Hartman 1997, 105.

45 Attempts to "humanize" these photographs suggest a problematic sense of repatriation that forgets its subjects, reiterating the original status of their dehumanization, seeking to persuade us of a subjectivity that was never lacking, and should never be presumed to be so. As Wynter reminds us, the category of the human originated in the bifurcation of the Human and His others (as witnessed in Agassiz's division). Thinking this division as a spatial and geophysical relation between figure and ground, as I have argued, means that there is no restoration, only the possibility of an earth-based reparation, repair, and refusal of reproduction.

46 See Carrie Mae Weems gallery label from 2010, https://www.moma.org/learn/moma_learning/carrie-mae-weems-from-here-i-saw-what-happened-and-i-cried-1995; http://carriemaeweems.net/galleries/from-here.html.

47 Susan Schneider is critical of the acceptance of "Agassiz's empirical alibi" (2012, 233) as a foil of the disinterested scientific gaze, understood as a turn away from questions of Black sexuality and its coerced collusion in the reproduction of white male patriarchal power systems. The point may be that empiricism was never without a racializing and sexualizing frame because that was the material condition of its birth and development. Something about the continued right to look in this way, like all the visitors who viewed (and consumed) Baartman in the Museum of Man, highlights the repetition of forms of racialization that scaled down Blackness to a series of parts as part of the commodifying process and foreclosure of the subject position. There is something else about how these photographs might be cared for in a way that doesn't seek restoration through fixing Black sexuality, again, but rather repairs a responsibility implicit in disassembling the structure of their viewing (see Hartman 2008), one that repudiates the continued positioning of subjectivity without consent.

48 Charles Hartt's more prosaic account of the geology of Brazil made during the Thayer Expedition maps a regional cartography of power and "underdevelopment" through the lens of geology, of rivers, mountains, sections, and gold mines. There is one appendix, "On the Botocudos," which describes, alongside the chapters on coal basins and extraction techniques, the hair-

styles, rituals, grammatical structures of language, and customs (including procuring fire and practicing cannibalism) of the Indigenous group (Hartt 1870, 577).

49 See the Thayer Expedition papers at the Museum of Comparative Zoology, https://library.mcz.harvard.edu/thayer.

50 The idea of migration of free Black people to Central America rather than Liberia was discussed by Abraham Lincoln in an article in the *New York Tribute* titled "The Colonization of People of African Descent." He also asked Congress for $600,000 for an emigration fund for the express purpose of deportation outside the United States. The president drew particular attention to the availability of coal and mining in settlement: "I attach so much importance to coal. . . . It will afford an opportunity to the inhabitants for immediate employment till they get ready to settle permanently in their homes" (see "From the Anglo-African: The President's Interview with a Committee of the Colored People of Washington," *Pacific Appeal, A Weekly Journal, Devoted to the Interests of the People of Color*, September 20, 1862).

51 Mrs. Agassiz is largely oblivious to the power dynamics of their patronage by the emperor. She sees the accommodations afforded her as curiosities: "Though it has no longer the charm of novelty for me, I am always glad to visit an Indian cottage. You find a cordial welcome; the best hammock, the coolest corner, and a cuia of fresh water are ready for you. As a general thing, the houses of the Indians are also more tidy than those of the whites; and there is a certain charm of picturesqueness about them which never wears off" (Agassiz and Agassiz 1868, 364).

52 Dom Pedro II visited the United States in 1876 for the world's fair in Philadelphia and spent time with Elizabeth Agassiz.

53 W. James 1865. See also II. Brazil, William James drawings, MS Am 1092.2, Houghton Library, Harvard College Library, accessed December 23, 2020, https://id.lib.harvard.edu/ead/c/hou02066c00005/catalog.

54 The presentness of this inability to perceive a full affective life for Indigenous, Black, and Brown life is an ongoing moment.

55 Photographs available at the Peabody Museum, Walter Hunnewell (photographer), Prof. Louis Agassiz (collector), Object number: 2004.24.7378/7668/7579, https://collections.peabody.harvard.edu/objects/details/545677?ctx=404aeae3259980235d7e170ff75ee64ee9b3b123&idx=1.

56 The racial effects of hemispheric Southern slavery in the transatlantic axis of race were markedly different in Brazil than in the United States (with Brazil having a much longer and larger trade in enslaved persons and complex well-established systems of racialized lease labor that predated the industrialization of colonial labor, as well having more mine, mineral, and race-based extraction alongside the geologic praxis of plantations, all of which effected the racial dynamics on the continent).

57 Outside whiteness, Agassiz establishes racial difference as a cascading hierarchy, from Indian to Negro: "In the example of negro and Indian half-breeds we have seen, the negro type seems the first to yield, as if the more facile disposition of the negro, as compared with the enduring tenacity of the Indian, showed itself in their physical as well as their mental characteristics" (Agassiz and Agassiz 1868, 139).

58 Full quotation: "The *status* of slavery embraces every condition, from that in which the slave is known to the law simply as a chattel, with no civil rights, to that in which he is recognised as a person for all purposes, save the compulsory power of directing and receiving the fruits of his labor. Which of these conditions shall attend the *status* of slavery, must depend on the municipal law which creates and upholds it" (Taney and Supreme Court of the United States 1857, 922).

59 Spillers argues that if slavery complicated motherhood by producing it as historically voided, then it asymmetrically rewrote notions of fatherhood as enslaved persons were "twice-fathered, but could not be claimed by one and would not be claimed by the other" (2003, 232). She shows how the "name" (of a slave) is the rupture point between the break from African identity and a placement within an inchoate American one, which produces a "social subject in abeyance, in an absolute deferral that becomes itself a new synthesis" (232). This deferral of the possibility of genealogy established through both blood and the relation of names scars the operation of genealogy as a possible mode of accounting for the very alienated rift it was made to analytically suture together (on behalf of whiteness).

60 In the schema between the negative horizontal axis of forced miscegenation by plantation owners and the positive recognition of an oligarchy of lineage in deep time along the vertical axis, time's arrow was rendered as neutral ground. Through the semiotic reductionism practiced in geology, race became implanted in natural history as a force of nature that obfuscated its construction to justify and resolve the "troubles" of colonialism and its violent natalities.

61 For Du Bois's diagrams of racial color lines and his photograph collections, see "Du Bois Albums of Photographs of African Americans in Georgia Exhibited at the Paris Exposition Universelle in 1900" (https://www.loc.gov/item/99471631/ and https://lccn.loc.gov/99471631) and "Charts and Graphs Showing the Condition of African Americans at the Turn of the Century Exhibited at the Paris Exposition Universelle in 1900" (https://www.loc.gov/pictures/item/2005679642/), Daniel Murray Collection, Library of Congress, Washington, DC.

62 From Paris, the exhibition went to Buffalo, New York, to be included in the Pan-American Exposition (1901) and then to Charleston for the South Carolina Inter-State and West Indian Exposition (1901–22), after which it was deposited at the Library of Congress.

63 Shawn Michelle Smith's (2004, 22) research seeks to restore the visual significance of the color line in the context of a visual culture of scientific racism (in part, initiated by Agassiz's racial inscriptions as an evidential and affectual base for theories of mankind made in the context of geologic reasoning) and the importance of the visual lexicon in race talk.

64 Hartman argues that Black feminist archival methodologies involve "thinking about subsistence as a radical process of collective survival and thinking about the wayward and queer resources of black survival . . . thinking about young black women as radical thinkers, which no one ever does, because they imagine that thought is only the capacity of the educated, or the endowment of elites. What is it like to imagine a radically different world, or to try to make a beautiful life in a situation of brutal constraint?" (2018, n.p.).

65 As Shawn Michelle Smith convincingly argues, "Du Bois not only utilized visual images to describe racial constructs but also conceptualized the racial dynamics of the Jim Crow color line *as* visual culture" (2004, 25).

66 See Dionne Brand's *At the Full and Change of the Moon* (1999) for the possibilities of a different kind of fragmented genealogy, a queered ancestral line that ricochets through colonial legacies to form a countervision of the passage of relation. Her genealogy, down the maternal line, maps the diasporic "non-arrivals" of those who wander in slavery's afterlives.

67 Hartman asks: "Is it possible to exceed or negotiate the constitutive limits of the archive? By advancing a series of speculative arguments and exploiting the capacities of the subjunctive (a grammatical mood that expresses doubts, wishes, and possibilities), in fashioning a narrative, which is based upon archival research, and by that, I mean a critical reading of the archive that mimes the figurative dimensions of history, I intended both to tell an impossible story and to amplify the impossibility of its telling" (2008, 11).

Chapter 5. Geologic Grammars

1 While Emerson came out as a supporter of abolition in 1844, he moves in and out of racial essentialisms in his private and public comments on race and retains a clear racial hierarchy and depiction of Blackness in nonhuman and inhuman registers. He suggests that Negroes might be able to form a coming civilization, arriving in the form of men like Toussaint Louverture or Frederick Douglass, but only "if he is pure blood" (Emerson quoted in Patell 2001, 56). Troubled by miscegenation, he accords racial progression and hierarchies through the faculty of reason, stating, "Ideas only save races" (55). When Emerson does begin to consider the injustice

of slavery, he does so like many others, from the point of view of morality, which seeks as its primary concern to confirm the superiority of the white subject as an ethical being. Emerson's racial ideas follow Agassiz's in terms of their adherence to national traits (and nationalism) as a science of inherited nature, or "Permanents Traits of the English National Genius," as his 1835 lecture had it.

2 Like Agassiz, Emerson imagines the Negro as an ancient life-form in the geologic order, pre-Adamite, next to the trilobite. He says, "In Africa the negro of to-day is the negro of Herodotus. In other races the growth is not arrested" (1862, n.p). The duration of Black and Native life is acknowledged and then dispersed in geologic time through a reference to a fossil.

3 Mark Maslin and Simon Lewis, "Why the Anthropocene Began with European Colonisation, Mass Slavery and the 'Great Dying' of the 16th Century," *The Conversation*, June 25, 2020, https://theconversation.com/why-the-anthropocene-began-with-european-colonisation-mass-slavery-and-the-great-dying-of-the-16th-century-140661. See also Koch et al. 2019.

4 Sharpe defines monstrous intimacies as "a set of known and unknown performances and inhabited horrors, desires and positions produced, reproduced, circulated, and transmitted, that are breathed like air and often unacknowledged to be monstrous" (2010, 3).

5 National newspapers reported on the wilderness trip, noting in racist tropes that the men were living like "Sacs and Sioux." The trip became immortalized as a founding moment in the intellectual movement of linking nature, art, science, and literature in American natural philosophy. Emerson wrote a poem, "Adirondac" (1904a [1858], 182–94) about the trip, and it formed part of his reflections in his essay "Nature" (1909 [1836]). Emerson's transcendentalism was said to be taking away the "Father of the Universe" and replacing it with a direct intuitive experience of nature that has parallels with later vital materialisms.

6 J. Hamilton Couper Plantation Records #185-z, Southern Historical Collection, Wilson Library, University of North Carolina at Chapel Hill, https://finding-aids.lib.unc.edu/00185/#folder_2#1.

7 Lyell visited North America four times in the twelve years from 1841 to 1853. He gave the Lowell Lectures in Boston in 1841 (named after another industrialist, John Lowell, whose textile mills were fed by the enslaved) and lectured in New York and Philadelphia. He ascribes the evils of slavery to the quickening value of capital: "The evils of the system of slavery are said to be exhibited in their worst light when new settlers come from the free states: Northern men, who are full of activity, and who strike to make a rapid fortune, willing to risk their own lives in an unhealthy climate, and who cannot make an allowance for the repugnance to continuous labor of the negro race, or diminished motive for exertion of the slave." Lyell also

participated in the third annual meeting of the American Association of Geologists and Naturalists in 1842 (see Dott 1996, 101–40).

8 For a discussion of Lyell's geologic surveying of America and his racial views on slavery and the "Southern problem," see Yusoff 2018a.

9 Macarena Gómez-Barris argues against epistemic difference and embodied knowledge being merely resistive, as a "more layered terrain of potential, moving within and beyond the extractive zone," which requires the formation of new methodological approaches to track the less apprehensible worlds; to "explicitly challenge the frames of disciplinary knowledge that would bury the subtlety and complexity of the life force of the world that lie within the extractive zone" (2017, xv).

10 His conclusion is that instead of "befriend[ing] the cause of humanity," any legislation should resist this "desire to assist" "men fully grown it may be, in years, but mere children in experience" (Ansted 1854, 298).

11 "Aunty, the cook, wants to go out on Sunday to a dipping service or preaching some ten miles off—no uncommon event—the family must either breakfast at six or not at all. If Pompey has been out on Saturday night rather later than usual and has had an extra half-dollar for whiskey, the boots and shoes, and often horses, must remain unattended to; and as there is no punishment short of flogging, and few masters resort to that for any but serious cases, the offence is overlooked and soon becomes a habit" (Ansted 1854, 295).

12 Attentive to the fears of miscegenation and obsessions with purity, Ansted assures his British audience that they should not be concerned about mixing: "Nor must it be imagined that any amalgamation that may be going on alters the conclusions. It is notorious, that any stain whatever of Negro blood at once places the victim into the coloured class" (1854, 297–98).

13 Conforming to the hierarchical placements of race via geography, he says: "In point of fact, the physical peculiarities and natural *habitat* of the negro, and those peculiar qualities which characterize him, are such, as not to form a useful amalgamation with the Anglo-Saxon race in North America, even when allowed such fair play as seems possible. It is to all appearance physically impossible that the two should be on a true level; and, as is universal in such cases, the stronger must drive out the weaker" (Ansted 1854, 298).

14 Sharpe discusses Brand's rendering of the "Door of No Return": "'For Brand that un/known door is the frame that produces black bodies as signifiers and bearers of enslavement and its (unseeable) excesses. It is the ground that positions them to bear the burden of that signification and that positions some black people to know it" (2014, 190).

15 The American Colonization Society was a group established in 1816 to "support" the migration of free African Americans to the continent of Africa. The society founded a colony on the Pepper Coast of West Africa (Liberia)

in 1821–22, called Mississippi-in-Africa. The society met with objections from African Americans and those living in the areas being colonized. It was supported by a broad range of interests, particularly from the Southern states, which feared rebellions. Of the 4,571 emigrants who arrived, only 1,819 survived until 1843 due to disease outbreaks. Aware of the high mortality rates, the American Colonization Society, which published the *African Repository and Colonial Journal*, continued to send people and by 1867 had "assisted" in the immigration of more than thirteen thousand African Americans to Liberia. A parallel organization in Britain, the Committee for the Relief of the Black Poor, attempted to establish a province, Freetown, in Sierra Leone for London's Black poor.

16 Text of a talk by Dr. Parr at the opening of the Hall of Oil Geology, the American Museum of Natural History, March 22, 1955 (my italics), press release, Archives of American Museum of Natural History, New York.

17 While geologic grammars have historically marked the acquisition of territory and ongoing settler colonialism (Bebbington and Bury 2013), they also underpin the languages of possible repudiation within a political sphere. A methodological tension arises here in relation to the earth sciences as a praxis that is both a potential space of instigating social justice in questions of environmental harm and a site of perpetrating and defending it.

18 The first geologic society in the world, the Geological Society, was founded on November 13, 1807, in London. As a society, it distanced itself from the mining and mineralogical societies, as well as the more speculative "Theory of the Earth" groups. Instead, it tasked itself with "the purpose of making geologists acquainted with each other, of stimulating their zeal, of inducing them to adopt one nomenclature." There were thirteen founding male members, several of whom owned plantations.

James Laird, the founder of the Geological Society, was born in the parish of St Elizabeth, Jamaica, where his father, Henry Laird, owned a sugar plantation of 102 acres—Prospect Estate, listed with 102 enslaved persons and 51 "stock" (*Jamaica Almanac* 1811). The Slave Registers of Former British Colonial Dependencies lists over 100 enslaved people registered to Laird's elder brother Henry, who then owned Prospect Estate. James later became a joint owner and collected a private income from the estate, which he used in part to fund the publication of the Comte de Bournon's treatise on mineralogy. The official register of compensation claims following the British abolition of slavery indicates that James Laird submitted two claims for compensation as an enslaver: Jamaica St Elizabeth 559 (6 enslaved), 560 (1 enslaved); the value of the two claims is given as £136 15s. 1d., £26 12s. 2d. ("Jamaica St Elizabeth 559," accessed December 30, 2019, http://wwwdepts-live.ucl.ac.uk/lbs/claim/view/23948).

James Franck (1768–1843)—one of the thirteen inaugural members of the Geological Society, a military physician, and a mineralogist—is named (without further biographical details) in the same register as (potential) slave owner of St Kitts 634 £395 0s. 1d. (24 enslaved) "St Kitts 634" (http://wwwdepts-live.ucl.ac.uk/lbs/claim/view/23766). Richard Philips (1778–1851) was a Quaker and part of the Society for the Abolition of the Slave Trade; he published *Case of the Oppressed Africans* and *For the Abolition of the Slave Trade*, both in 1787. William Allen (1770–1843) and William Philips of the Geological Society were also active (Lewis and Knell 2009, 129). A William Phillips and a William Allen are listed on the slave register as receiving payments for enslaved persons, but it not clear whether these are the Geological Society members, although being Quakers did not preclude them from accruing wealth from the trade in human flesh. George Wear Braikenridge, a fellow of the Geological Society, owned 256 enslaved and received £4,673 9s. 7d, from the Bagdale Estate and Fullerswood Pen (https://www.ucl.ac.uk/lbs/claim/view/14586).

The thousands of specimens collected through the slave trade still reside in geologic institutions and are used in mineralogical research. The minutes of the Geological Society of London, August 19, 2020, indicate that a review is underway to address "slave trade and racial superiority and a report on links to eugenics and colonisation may follow."

The Geological Society, London, invited "those to join them as the true sons of science, who have a desire, and a determination, not so much to adhere to things already discovered, and to use them, as to push forward to farther discoveries; and to conquer nature, not by disputing an adversary, but by labor; and who, finally, do not indulge in beautiful and probable speculation, but endeavor to attain certainty in their knowledge" (Hitchcock 1840, 307), thus consecrating the father metaphor and patriarchal lineage. The Council of the Geological Society selected a quote by Francis Bacon to appear as its motto on the title pages of its inaugural series of *Transactions* in 1811: "We invite him as a true son of Science to join our ranks, if he will, that, without lingering in the forecourts of Nature's temple, trodden, already by the crowd, we may open at last for all the approach to the inner shrine" (H. Woodward 1907, 44).

19 See Caroline Lam, "Decolonizing Geoscience," *Geoscientist*, March 1, 2021, https://eoscientist.online/sections/features/decolonising-geoscience/.

20 For a biography of Sir Henry Thomas De la Beche in relation to the legacies of British Slavery, see Center for the Study of the Legacies of British Slavery, accessed September 20, 2023, https://www.ucl.ac.uk/lbs/person/view/2146633718.

21 The organic-inorganic story becomes further broken apart in the event of Darwinian evolution, and the congruity between these spheres is disrupted as Life becomes the protagonist and the earth is backgrounded as infrastructure. However, despite this backgrounding, stratigraphic structuralism retains its grip on the sociological imagination, functioning as a transtemporality that continues to organize concepts of the political present and societal change—or the social mantle of perception.

22 The racial troubling of white supremacy remains through the interior inconsistencies of that thought, while pointing to a broader ontological terror about inhuman nature itself.

23 Hartman uses the term *ungovernability* to articulate a speculative narrative of feminist fugitivity that plays outside the modes of recognition of resistance. She asks: "Could they ever understand the dreams of another world which didn't trouble the distinction between man, settler, and master? Or recounted the struggle against servitude, captivity, property, and enclosure that began in the barracoon and continued on the ship, where some fought, some jumped, some refused to eat. . . . [Where] the aesthetic wasn't a realm separate and distinct from the daily challenges of survival, rather the aim was to make an art of subsistence, a lyric of being young, poor, gifted, and black" (2019, 230, 235).

24 The field of paleontology was arguably launched by two papers delivered by Cuvier in 1796, one on the comparison of elephant and mammoth bones, which he later names the mastodon, and the other on the identification of the giant sloth, using contemporaneous sloths. Both papers used fossil finds and comparative anatomy to build a theory of extinction (or catastrophism), which became widely accepted. The mastodon and giant sloth have become emblematic guardians of geology and often adorn geology and natural history museums as archetype specimens (in the Orton Geological Museum, Ohio State University, Columbus, for example).

25 The racialized axis of extraction that underpins the ascendancy of geopolitical power through geographic conquest gave rise to the ability to accumulate identity from, and direct forms of, geopower. We could understand the historic organization of the inhuman under colonialism as the *occupied elemental subject position*. Such a subjective-subjugated occupation forever changes the lexicon of the elemental as a mark and medium of that history. The earth remembers this passage too. Its environmental histories and historical geographies are also changed by and through this passage, as are those who move through it.

Chapter 6. Stratigraphic Thought and the Metaphysics of the Strata

1 Minerals both return to an elemental condition and never return. There are different geochemical attractors. Endless recombinations build new versions of earth. In scenes of mining that produce strata's orphaned displacement of materiality, there is no return of mineralogical matter to its beds and fused relation. Once on the surface, relieved from the geophysical pressure of subsequent earths that came after it in epochal time, pressing down on its agency, minerals and matter have new lives, transforming

atmospheres (in the case of fossil fuels) and rearranging processes of life (in the case of nitrogen). In the "life" of strata, the process of orphanage at the surface makes something geochemically different from its life in the strata. There is no going back before destratification. As human-made mass on the planet surface now weighs more than biomass, new geophysical architectures emerge (Elhacham et al. 2020). Conceptualizing this within geologic life means that human architecture—or geopower—is both a deep time and a deep earth subsidy to the surface and the regulatory arrangements of racialized relations.

2 These projects in turn were the legacies of British slavery and in dialogue with the plantation and its geologic practices.

3 In an opening speech for the new Ilmenau mine on February 24, 1784, Goethe the paleontologist frames the unearthing strata through the shaft as the means of deep time promotion to the surface: "This shaft, which we are opening today, is to become the door through which one descends to the hidden treasures of the earth, through which those deep-seated gifts of nature are to be promoted to daylight" (quoted in materials from the exhibition *Adventures in Reason: Goethe and the Natural Sciences around 1800* at the Schiller Museum, Weimar, 2019).

4 "Us" refers to those schooled in French theory, including Glissant, whose PhD supervisor was Gaston Bachelard. Bachelard's book *Poetics of Space* (1994[1958]) transformed spatial imaginaries, and *Psychoanalysis of Fire* (1938) introduced the elemental as a structure of thought and material hermeneutics—which is also to say there is no neat separation between the "West and the rest." Glissant's *Poetics of Relation* (1990) is in direct correspondence with Bachelard's *Poetics of Space* and *Poetics of Reverie* (1960).

5 Briefly, the concept of the geosocial in the Anthropocene brings the following into view: (1) *Socializing geology; geologizing the social* (no more "standing stock" subtending society). The idea that humans and their activities are fully engaged in the Earth System, albeit in a haphazard and unthought way, complicates the already compromised notions of a human-nature duality (of a world "out there," resistant to human extraction). (2) *Rematerializing social practices as planetary and lithic* (e.g., communication, consumption, fossil fuels, capitalism, political and economic system, etc.). The earth as a space of action is changing shape. (3) Obliterating a *notion of history* that posits these two concepts—Nature and the Human (natural history and human history)—in opposition as temporally and materially distinct. It challenges the idea of how we think about history, deep histories and future histories, and their material basis. The Anthropocene also releases a nonhistoric moment. For what is an epochal claim if not an opening up of history to the claims of something beyond history, beyond life itself, and beyond a sensible temporal? Yet it is also important to remember that colonized subjects had their claim on historic time erased through racial

theories of space-time and were relegated to the historicity of natural history, opened to selective narrativizing within colonial development narratives. So, the collapse of human and natural history holds only for those who get to function in the category of the human and write and function within the category of history, not those denied historical capacities. (4) *Conjoining political states and planetary states* (political thought must now engage with a full range of geologic forces rather than background these concerns to the technologies of geopolitics, natural resources, and natural disaster). (5) Establishing *energy as a geosocial system* that changes planetary states. The Anthropocene also taps into the more fluid understanding of the material stratifications of life as the expression of geologic forces; as plants are an expression of the sun, it petitions the imagination to personify an elemental subjectivity of coal fired people, fracked people, nuclear people, and so on. Energy accumulation clearly enacts unequal geophysical effects, for both individuals and populations, and geopolitical agency, for corporations, nation-states, and colonial states. (6) Embodying *inhuman planetary agency* through the reanimation of the Carboniferous and Devonian epochs (which raises complicated questions of control and agency and thought about the interagency of life-nonlife in the move away from biocentric models). (7) Establishing *political geology*, a new regime of power as a thresholding of geophysical forces; it gives a body and a life to geology as the substratum to human life that pertains to the conditions of survival. Political geology indicates an active form of stratigraphy that inscribes life and geosocial formations, rather than humans just inscribing the stratigraphy of the earth with their waste signatures. (8) Fomenting a *"revolution of thought"* in human-environment relations. The Anthropocene brings into view the structures of exchange between geologic strata and social worlds. In all, the Anthropocene produces a much more complex understanding of geology if the interconnection between furious surface political, material, and economic actions and deep lithic and racial processes is understood as productive of a contingent earth that is marked by a change of state through the withdrawal of habitable space: a "change of state" and a change of the contours of subjectivity are predicated on change of geophysical states.

Chapter 7. Geopower: Materialisms before Biopolitics

Portions of chapter 7 appeared in an earlier form as "The Anthropocene and Geographies of Geopower" in Mat Coleman and John Agnew, *Handbook on the Geographies of Power* (Gloucester: Edward Elgar, 2018): 203–16.

1 These questions currently present themselves in the context of petro-populist authoritarianism and climate change.

2 The most engaged writing on the limit and excess in Western philosophy (by writers such as Bataille, Blanchot, Walter Benjamin, and Pierre Klossowski) emerges out of Europe's own abyss of World War II. Theodor Adorno introduces the idea of the inhuman society in *Minima Moralia: Reflections from Damaged Life* (1951), asserting that life does not live in the society of industrial war, whereby intimate acts of life should be thought in relation to the catastrophes of the twentieth century. While Adorno, along with many contemporary critics, sees the descent into the inhumanity as located in his immediate historical context in World War II, the fall already anticipates its redemption in the negative project of utopia.

3 McKittrick argues that **** the formulation of Man produces a biocentric coordinate, across the Cartesian axis of "man-native-n*****-n***** woman-jobless." See "Man = 0,0 = freedom): origin/pivot/definer, Man-as-human (the closer to the origin, the more human)" (2015, 153).

4 Regarding Wynter, McKittrick argues: "In her work, the racial-Other does not *haunt*, but rather is a fully present figure who *enables* emancipatory cognitive leaps. . . . What emerges is a co-relational figure that Wynter . . . describes as articulating both *bios* and mythoi: a figure who is a physiologically organic *and* cognitive *and* creative being that *authors* the aesthetic script of humanness" (2015, 144).

5 Wynter argues that the "perspective of 'otherness' from which to explore the issue to which we give the name of race, as the issue that the Black American intellectual W. E. B. Du Bois identified in 1903 in its more totalizing, more absolute form as the 'Color Line' (Du Bois, 1903) . . . as Aimé Césaire of the Francophone Caribbean pointed out in his letter of resignation from the French Communist Party in 1956, the 'Color Line'—i.e., of race as the Western-bourgeois analogue of Latin-Christian Medieval-Europe's feudal principle of caste—was the issue whose historically-instituted singularity could not be made into a subset of any other issue. Instead, it had to be theoretically identified and fought on its own terms (Césaire 2010)" (Wynter 2015, 186).

6 Clearly, an engagement with geopower requires unsettling the "we" that is positioned by capital as an expansive category, and a confrontation with the history of immigration to America as "getting white" as quick as you can, to paraphrase James Baldwin. Sexton suggests that the native-born Black population of the United States "suffers the status of being neither the native nor the foreigner, neither the colonizer nor the colonized. The nativity of the slave is not inscribed elsewhere in some other (even subordinated) jurisdiction, but rather nowhere at all" (2010, 41). Yet, "the racial circumscription of political life (*bios*) under slavery predates and prepares the rise of the modern democratic state, providing the central counterpoint and condition of possibility for the symbolic and material articulation of its form and function" (40–41).

Chapter 8. Inhuman Matters I: Black Earth and Abyssal Futurity

Epigraph: Nina Simone, "Sinnerman" lyrics © Sony/atv Music Publishing LLC, Warner Chappell Music, Inc, Kobalt Music Publishing Ltd., Warner Chappell Music Inc., 1956.

1 See artist Kiyan Williams's work at their website, https://www.kiyanwilliams.com/.

2 "A person beside a person inside a life. That's called parataxis. That's called the future. We're almost there. I'm not telling you a story so much as a shipwreck—the pieces floating, finally legible" (Vuong 2019, 190).

3 Warren says: "My use of *ontological terror* is designed to foreground not only the terror the human feels with a lack of security, but also that this fear is predicated on a projection of ontological terror onto Black bodies and on the disavowal of this projection. Thus, humanism does not exhaust ontological terror, and an antimetaphysical understanding of it is necessary to analyze antiblackness" (2018, 173). The migration of this terror is beyond the US nation-state because the US is a superpower and its script is geographically expansive.

4 Hear Warren alongside Arendt: "What is hated about blacks is this nothing, the ontological terror, they must embody for the metaphysical world. Every lynching, castration, rape, shooting, and murder of blacks is an engagement with this nothing and the fantasy that nothing can be dominated once and for all" (Warren 2018, 9). This nothing is the source of terror.

5 In Europe, race becomes deployed as a specific technology to undercut economic bonds across difference that might unite factory workers with indentured plantation labor. As whiteness is transformed into a transcendental category (a connection across economic classes), economic subjugation is negated for the white worker through the ability to impose racial subjection.

6 I gave a version of this chapter at the conference Arendt on Earth: From the Archimedean Point to the Anthropocene at Northwestern University in Chicago in 2019. There the suggestion that Arendt's ethics were concerned only with white subjects was vigorously challenged.

7 Grosz categorizes the "nick" as a seismic rupture through which the future comes into being: "It is not the predictable, foreseeable continuation of the past. . . . Only if the present presents itself as fractured, cracked by the interventions of the past and the promise of the future, can the new be invented, welcomed and affirmed" (2004, 257).

8 In the sublime, the inhuman is what Kant calls "rude nature" that first impresses itself painfully, and then through representation the Kantian subject is saved; thus representation is the imagined site of resolution. Alongside the sublime contours of rude nature, Kant also imagined a mathematical sublime that was based on a calculative dimensionality, of

physics and its statistical body. In this sense, Kant inadvertently describes the violent exposition of the void and the calculus of slavery through the inhuman.

9 Quoted in Dom Phillips, "Brazil Museum Fire: 'Incalculable' Loss as 200-Year-Old Rio Institution Gutted," *Guardian,* September 3, 2018, https://www.theguardian.com/world/2018/sep/03/fire-engulfs-brazil-national-museum-rio.

10 Listen: Bessie Jones and the Georgia Sea Island Singers, "Sink 'Em Low"(1959), from the album *Get in Union,* YouTube, accessed September 22, 2023, https://www.youtube.com/watch?v=AiRetIBwKu4.

11 For Arendt's response to Fanon, see Marriott 2018, 153.

12 Michelle Wright (2015) argues a way out of this impasse, suggesting a fuller geographic engagement with Blackness as a methodological necessity to counter the strictures of normative accounts of Blackness. When Blackness is opened to the difference of geography, it slips its confinement on an ontological plane and returns with the specificity of the experience of Black geographies to insist on a location that is not conceived in the syntax of the universal.

13 Marriott: "The abyssal is what gathers the universal and particular precisely by pulling them apart, by assigning each the limited transcendental coordinate of the other, coordinates that can only be misrecognized from outside the void by which each remains unseen by, or at the furthest reach from, the other" (2018, 315).

14 Voyage iron, as was the name given to northern European iron that was a significant part of the exchange economy of the slave trade, arguably underpinned a transformation of metallurgical practices in coastal Africa—where iron was not found in great quantities and with different properties (Evans and Rydén 2018). West Africa was a recognized place of venerable metallurgical traditions and blacksmithing, and iron was embedded in spiritual and cosmological cultures that understood forge work as a form of inhuman communication.

15 Fanon poses this problem of identity in his book *Black Skin, White Masks* (*Peau noire, masques blancs* [1952]). Blackness is created as a mask for whiteness, oriented by the demands of whiteness and its hierarchies of power. It is the gaze that grazes. The violence of the colonizer is the annihilation of body, psyche, and culture and the demarcation of space (Black and Brown bodies become rendered out of place in colonial space and time). Fanon concludes that the Black man does not exist, nor does the white man. They exist only to the extent that they create one another in the course of the sterile dialectic that pits a superiority complex against an inferiority complex.

16 Whiteness of course manages liminality and predates it, often immune (seemingly) to its affectual power (liminal affects).

Inhuman Matters II: Deep Timing and Undergrounding in the Carceral Mine

1 With thanks to Gean Moreno for the referral to Cameron Rowland's work, including *Norfolk Southern (Tennessee)*, 2017, steel relay rail, 18 × 489 × 73 cm (7 × 192.5 × 28.7 in). Relay rail is rail that has been removed from its original line and resold, often by railroad companies to mining companies for pit railways. The steel rail is made using coal and iron ore. In the late 1860s, the Alabama and Chattanooga Railroad; the Georgia and Alabama Railroad; the Selma, Rome, and Dalton Railroad; and the Macon and Brunswick Railroad were constructed using convict lease labor. The original line was constructed from Birmingham to Pratt City, Alabama, in 1876 to haul coal. By 1895 all these lines had been consolidated into the Southern Railway, which built hubs in Birmingham, Chattanooga, and Atlanta, allowing it to transport coal and iron ore throughout the Southeast. It was later sold to Tennessee Coal, Iron and Railway Company. The subterranean contract of the convict lease carried overground to further participate in the social affects of the apartheid of geology.

2 For more information, see Convict Leasing and Labor Project, "Black Lives Matter. Black Graves Matter, Too," accessed September 20, 2023, https://www.cllptx.org/campaign. The Convict Leasing and Labor Project aims to lead a national conversation on the history and impact of forced labor, including chattel slavery, convict leasing, and the modern crisis of mass incarceration.

3 Rowland's work addresses the continuities between slavery and incarceration, and the commodification of prisoners from convict lease labor to present. Their work *1st Defense* NFPA *1977* (MoMA 2011 and 2016) displays two Nomex firefighting suits that are used to clothe California's 4,300 prisoner firefighters. The accompanying text reads: "'The Department of Corrections shall require of every able-bodied prisoner imprisoned in any state prison as many hours of faithful labor in each day and every day during his or her term of imprisonment as shall be prescribed by the rules and regulations of the Director of Corrections' (California Penal Code § 2700)" (https://www.moma.org/collection/works/216232).

4 Orlando Patterson's (1982) concept of a social isolate in his formation of the condition of natal alienation in the social death of the Black subject under slavery can be extended into the conditions of the material execution—as mineral isolate—that were operated in the inhumane worlds of the prison mine.

5 In a contemporary continuance of the unequal geo-logic of race and earth, the for-profit company geo Group Inc. is one of the largest privately owned prison companies. Based in Florida, geo Group is a "diversified service platform" with subsidiaries in the United Kingdom, South Africa, and Australia.

6 Sloss-Sheffield Steel and Iron Co. had a one-in-ten death rate of convicts.

7 The monument, which was sponsored by the Southern Poverty Law Center, is made from black granite, possibly mined from the Piedmont Upland, in the same region as Alabama Marble. Alabama Marble, designated the state rock, was used to build criminal justice and insurance buildings, statues, cotton exchanges, national banks, the Lincoln Memorial, and the US Supreme Court Building, making the material a fabric of political life and its exclusions (see https://quarriesandbeyond.org/states/al/al-structures.html). Geologically, the region is also "home" to the oldest gneiss, dating from 1.05 billion years ago, and a gold-mining extraction zone of the Brevard Fault Zone.

8 Sloss ceased production in 1970 and was designated as a National Historic Landmark in 1981. Its decline in the 1920s was complex, but historian Gary Kulik (1981, 28–31) suggests it was due to the lack of modernization in Southern mining, which was bolstered by the cheap labor of the convict lease system utilized by the Sloss Company, combined with precarious free Black labor. Lichtenstein (1993, 120) notes how the furnace was stoked by hand from the top until 1927, some thirty years after machine modernization was introduced in the northern plants. When modernization came to Southern mines and railways, free Black labor was most affected because it was the most precarious and lowest paid, and thus subject to be laid off most readily.

9 The Alabama Mining Museum in Dora, Alabama (just north of Birmingham), contains no mention of the lease. The museum is stuffed full of the dusty memorabilia of white miners told as a quaint, half-forgotten hard-luck story of the past. On the museum's Facebook page, people post pictures of Confederate flags that are on shown en route to the museum, and people "heart" the posts accordingly.

The Equal Justice Initiative in September 2019 erected a sign at Sloss as part of the Community Historic Marker Project. The marker recognizes two Black coal miners lynched in Jefferson County (while defending other Black men at the convict leasing site of Brookside Mines, Tom Redmond and Jake McKenzie were killed by white mobs in 1890 and 1897, respectively). Equal Justice Initiative, "Jefferson County, Alabama, Memorializes Lynchings with Marker Dedication," September 14, 2019, https://eji.org/news/jefferson-county-alabama-memorializes-lynchings/.

10 Lichtenstein argues there is no indication from the historical boards at Sloss of how unionization or workplace segregation "was translated into changes in Birmingham's racial climate, or alternatively, how labor movements helped perpetuate segregation both within and beyond the plant gates, as some historians have suggested" (1993, 122).

11 Alabama operated a two-tier prison system; prisoners convicted of misdemeanors were sentenced and leased by county courts, and felons were sent

to the state penitentiary. County convicts were made responsible for all court fees, including the fees incurred by the sheriffs and jurors and their own, and these fees (often more than fifty dollars) were added to their sentences and paid for by serving extra time at hard labor. Given the substantial fees that sheriffs received for arresting able-bodied Black men, this system was lucrative for many parts of the state. "In addition to serving as a means of racial repression, the lease undoubtedly underpinned racial segregation and capitalism in the New South" (Curtin 2000, 3). It specifically targeted the poor, who could not afford to pay the excessive fines and were not able or willing participants in middle-class projects of moral and economic reform.

12 Prisoners were not mute in challenging operators and made complaints about the scales that were used for weighing the quantity of coal, health and safety concerns, and the increase in "tasks" (or targets) for each miner (Alabama 1886; 1886/88).

13 Coal mining remained a steady source of employment for free Blacks during the first thirty years of the twentieth century. In 1910, over 40,500 performed work associated with coal mines, of whom approximately 29,000 were miners. As the 1930s arrived, however, increased mechanization hastened the of end of Southern Black mining. Incarcerated persons were not removed from Alabama mines until 1928, occasioned in part by the murder of James Knox. Knox was unusual insomuch as he was a white man convicted of passing a thirty-dollar bad check. He was tortured to death by guards at Alabama's Flat Top Mine on August 14, 1924, because he was unable to meet the daily ten-ton quota. Protests over his murder, alongside a waning coal industry, eventually forced Alabama to shut down its slave mines. It is significant that it was a white convict who died of "suicide" shortly after being leased to the Sloss Flat Top Mine in 1924. His death, recorded as heart failure, turned out to be from torture, after he was whipped for "malingering" and held upside down in a barrel of scalding hot water. In a cover-up of his murder, it was written that he died of ingesting mercury, but the pills were found by a lawyer to be still lodged in his throat three years later when his body was exhumed. Public outcry drew attention to practices in the prison mines and led to closure of the system in 1928, but it was only the death of a white man and the prosecution of a white lawyer that effected this. By 1930, there were over 55,000 free Black coal miners in the vicinity of the Magic City. That year, African Americans accounted for 53 percent of Alabama's coal diggers. In contrast, 22,000 African Americans were employed in West Virginia's mines in 1930. Unlike in Alabama, Black and white miners fought side by side in the Paint Creek–Cabin Creek strike of 1913–14 in West Virginia. The Black union man known as "Few Clothes" Dan Chain—portrayed by James Earl Jones in John Sayles's film *Matewan* (1987)—became a focal point for his courage and organizing. The scene from the film in which the white miners discuss the admission and support

of Black miners is telling in how it scripts the divisions of labor; the white foreman says (as the James Earl Jones character stands silently by): "You are not men to that coal company, you're like equipment, like a shovel, a car, a hunk of wood brace. They'll use you until you wear out, until you break down, get buried under a slate fall and then there'll get a new one. . . . They don't care what color you is, or how much coal you can load . . . when there's only two sides, they that work and them that don't." The appeal that is made to the white miners is to recognize themselves as conceived as inhuman materials too.

14 Historical records suggest that coal mining in the United States, from its earliest incarnation, involved the enslaved working in coal pits. As early as the 1700s, enslaved persons worked in Richmond, Virginia. While the Civil War abated the development of mines in the South, several sprung up to supply the Confederacy, and a bill was passed to exempt mines with more than twenty enslaved persons from active service (R. L. Lewis 1987, 10).

15 According to Letwin, "During the 1850s, industrial boosters redoubled their efforts. Growing awareness of the mineral potential of central and northern Alabama, spurred by a visit from the English geologist Sir Charles Lyell and a series of reports from state geologist Michael Tuomey, generated optimism among financiers" (1998, 13)

16 Visits by speculative financiers made it into the local papers. The *Birmingham Iron Age* (February 23, 1881), for example, announced: "Several prominent Northern capitalists are in the city." Alongside the announcements of the visits of capitalists, mines announced their hires of incarcerated persons ("The Newcastle Coal and Iron Co. have hired the convict labor of Pickens County at $9 per head per month") and the gossip section traded in the racialized slurs of Cuvier's and Agassiz's paleontology ("The boys over the way are disgusted with the 'monkey show.' Well, we thought we had the 'missing link,' but we give it up.")

17 Subterranean fiction was a popular genre in the late nineteenth century. Common themes were the depiction of a primitive underworld of archaic humans and beasts in counterdistinction to the technological development on the surface. In the twentieth century, "hollow earth" and underground fiction transformed the subterranean into a white supremacist hideout for lost white races and Nazi commands accessible through the polar regions.

18 Full quotation: "Upon the whole their treatment by the company is wise and humane and no doubt many are better cared for than when at liberty. With these facts your readers can decide as to the justness of the charges that have been made against the Southern convict system by a certain renegade sentimental novelist" (*Birmingham Age-Herald* 1888).

19 The resilience of pioneer discourses in contemporary extraction economies, such as in North Dakota, is a reminder of the enduring white settler

colonialist formations of geology, with the fantasy of futurity and family at the core.

20 There were a number of differences in the social geographies of coal mining in the United States. In Appalachia, for example, especially West Virginia, there were more free Black miners than in any other coal region because of the availability of employment and "equal pay for equal work" in these fields. Black miners in West Virginia were not segregated in the mine, but Black miners' families attended segregated schools and company activities such as picnics, and racial segregation was evident in the hierarchization of labor.

21 For significant gains in landholdings after emancipation, see Muller 2018 (for analysis of the relation between Georgia's Black-owned landholdings and the broader relation to convict leasing) and Higgs 1982.

22 My discussion of white settler surface politics and fossil fuels—"eugenics of the rocks"—was developed through conversations with Mary Thomas during fieldwork in North Dakota and West Virginia.

23 See John Witherspoon DuBose, 1836–1918, Papers, Alabama Department of Archives and History and W. S. Hoole Special Collections Library, University of Alabama, Tuscaloosa.

24 The *Weekly Age-Herald* goes as far as to suggest that mineral development in Birmingham has brought a new glimpse of modernity, a modernity that is secured only because of white rule: "This Birmingham is doing a wonderful thing for Alabama. It has brought us our first glimpse of the nineteenth century, and is the first proof we have been able to get that we are not floundering about in the era of reconstruction when all that men lived for was political redemption, while they let their fields and fortunes go to waste. . . . The only political redemption we fought for was white rule, and we have had its substance for fifteen years. Now Birmingham accepted white rule as a fact and reasoned out that the political advantage from it was to make investments secure and give the people good conditions under which to go to work" (May 1, 1889).

25 The "offshore-ness" of race is an imaginative resolution of the denial of a claim to be placed. It can be seen in the representations of the "Windrush Generation" in the United Kingdom, whose image is always associated with a ship that never gets to "land." Similarly, in the engineered "crisis" in the Mediterranean, the boats that carry migrants and refugees are never meant to beach in fortress Europe. Or refugees end up being housed in inhumane conditions on a floating barge by the British government after traversing the life-threatening English Channel. For an alternative reading to the foreclosing of landing via the shoal, see King 2019.

26 Birmingham newspapers often reported on Agassiz and human questions of evolution. For example, see a report on Prof. Agassiz's findings of "The Gulf Stream. Some New Points in Regard to the Formation of Its Plateaus," *Birmingham Iron Age*, February 2, 1881.

27 See Kim TallBear (2013) for a discussion of whiteness and science in the context of Indigenous racial ordering.

28 The racial prefiguration of labor as white and not subject to conditions of enslavement meant that solidarity among miners was open to the exploitability of these rifts. At Pratt Mines in 1889, for example, "several hundred white miners lay down their tools and leave the mines, with the go-ahead of the mine boss, to pursue a black miner accused of raping a white woman. Upon capturing the suspect, they lynch him" (see Letwin 1998, 31). There was some solidarity among miners across color lines, but often this had more to do with the way in which convict labor was seen to drive down wages for free labor, and how it was used to break strikes. The "Coal Creek War" saw three hundred miners with guns free prisoners at TCI's Briceville, Tennessee, facility on July 15, 1891. The following week fifteen hundred miners returned to free more prisoners. H. H. Schwartz of the Chattanooga Federation of Trades reported that "whites and Negroes are standing shoulder to shoulder," armed with 840 rifles (Lewis 1987, 24).

29 The first unit of the museum was that of comparative zoology, under the direction of Agassiz, established in 1859. The museum was extended under the direction of Alexander Agassiz, Louis's "footstepping" son, and developed into the Geologic Museum. The Geologic Museum and the Mineralogical Museum were eventually joined, whereby Harvard was established as one of the leading institutions in mineralogical research.

30 Also see Waterhouse, Benjamin, 1754–1846, "Lecture: Prefatory to the Lecture on Mineralogy," *OnView: Digital Collections & Exhibits*, accessed August 27, 2023, https://collections.countway.harvard.edu/onview/index.php/items/show/17142.

31 As a caveat, in his lecture Waterhouse adds: "The Peruvians and Mexicans were found by the Spaniards half civilized, they manufactured gold, silver and copper, but not iron."

32 Social reformer Clarissa Olds Keeler reverses this axiom of white ascendency in her pamphlet, *The Crime of Crimes; or, The Convict System Unmasked*. "Torturing them in the prison or in the mine recesses is a sin against high heaven," she states (Keeler 1907, 8).

33 *Birmingham Iron Age*, November 26, 1879, reports: "Birmingham is now in the midst of a season of unexampled prosperity, we presume no one will deny. . . . Real estate agents are over run with applications for houses which they cannot supply. . . . Confidence in the grand future of the magic city is thoroughly established, everybody is busy, and everything is booming. The princely profit now being realized from the iron furnaced in this region, is creating the wildest excitement among the iron men of the country, and bids fair to result in a development of the resources of this district which will eclipse, in its extent and magnificence the wildest dreams of Birmingham's most sanguine friends. . . . This is a land of wealth beyond all that is

elsewhere known. Here the world's demand for cheap iron can be supplied, and in the near future . . . the smoke of a thousand industries will arise." Not unlike the boomtown rhetoric found in North Dakota with the advent of fracking, the notion of the boom functions on the order of a biblical rebirth for settler colonialism via the rocks. On the same page as the discussion of the Magic City's ability to conjure "cheap iron" at a princely profit, there is an advertisement for "Hard Labor Convicts" that states, "Sealed bids will be received by the Court of County Commissioners of Sumter County, for the hire of hard labor convicts of said county, for the year 1880. J. A. Abrahams, Judge of Probate."

34 Maurice Evans, a South African, traveled to the US South, in an anthropological capacity to observe and meet with representatives of the South, to understand the future course of racial struggle. The South reminded him of "a block of ice, inert, immovable, without thought or life" except on the question of race, which "saturated the minds of whites with fear and hatred" (quoted in Worger 2004, 65).

35 See, for example, the website of the New Orleans Deep South Center for Environmental Justice, accessed September 20, 2023, https://www.dscej.org/.

36 King argues that we must see Black bodies as part of the geographic imaginary of whiteness. She says: "Blackness, as expansion and spatial possibility, becomes a constituting feature of the spatial imagination of the conquistador/settler rather than just another human laborer exploited as a mere technology to produce space" (2016a, 1023).

Chapter 10. Inhuman Matters III: Stealing Suns

1 The sun is an inhuman relation that gives without receiving (in Bataille's formulation). Thus, solar excess is not predicated on reciprocity and is not subject to predation or humanistic incorporation into origins and filiation, despite the scripting of elemental origination. The solar bind of asymmetry gives alterity to the possibility of more generous worlds.

2 The full quote is: "I was trying to nail solar keys to the measured door of the New Domain. And I gave over to what was elemental . . . / Elements (1949) / Oh extinguished suns!" (Glissant 1997, 39).

3 Curtin's book *Black Prisoners and Their World: Alabama, 1865–1900* (2000) is one of the few texts that addresses the history of convict lease from prisoners' letters and petitions to the inspectors and families, rather than reading for the force of repression in official documents (as much labor history does). Through the figure of Reginald H. Dawson, the chief inspector of the Alabama Department of Corrections from 1882 to 1896, Curtin reads the archive of letters to rethink the assumptions and practices of libera-

tion that prisoners spoke of, tried negotiating with, and actively protested. Dawson distributed stationery to prisoners for the express purpose of their writing every two weeks (Curtin 2000, 31), which they did, to family members, the state, the judiciary, inspectors, and the outside (although this applied only to state and not county prisoners). Alongside short-lived schools and various educational activities, from 1882 to 1891 a literary record exists for the communications and complaints of the incarcerated. The institutionalization of the carceral mine into one company, TCI, and its ten mines brought in a new architecture of the lease system, with improvements in prison buildings, sanitation, and night schools for a while, run by the prison reformer Julia Tutwiler at Coalburg, the Slope, and the Shaft. A note on the present: in 2012 at the Tutwiler Prison for Women in Alabama an EJI investigation exposed the widespread officer-on-inmate sexual violence against women (see https://eji.org/cases/tutwiler/). Not dissimilar to the way in which TCI pushed back on the incarcerated, scandals regarding horrendous conditions of abuse often resulted in increased violence and retribution rather than redress.

4 Conditions at Shaft and Slope No. 2 Pratt Mines are listed as below:

Injury cases treated from April 1, 1887, to October 1, 1888: admitted to hospital, Shaft, 235, Slope No. 2, 221; treated in cells, Shaft, 880, Slope No. 2, 995; total, Shaft, 1, 115, Slope No. 2, 1,176 (Alabama 1886/88, 188).

A Shaft prison labor chart identifies the dangers as follows: loading coal, 52.85 percent admitted to hospital; shoveling slack, 22.22 percent; gobbling rock, 28.28 percent; furnace, 69.23 percent; railroad, 120 percent; building coal ovens, 38.86 percent; dumping coal, 55.55 percent (Alabama 1886/88, 193).

Total deaths at Pratt Mines: 80 for two years (average number of convicts at Pratt Mines about 600 [300 Shaft, 300 Slope No. 2]). Place of death: Shaft, 45; Slope No. 2, 33; Rock Quarry, 2 (Alabama 1886/88, 201).

5 For example, the Banner Mine explosion in 1911 that killed 122 convicts and 6 free men was due to an explosion caused by gas buildups, and the fire was facilitated by greasy coal-impregnated flammable clothing.

6 "Dark Was the Night, Cold Was the Ground" was one of twenty-seven songs, collected by Carl Sagan to represent diversity of life on earth, included on the Voyager Golden Record. The record, *Murmurs of the Earth: The Voyager Interstellar Record* (published in outer space in 1977 and here on Earth in 1978), included a wide range of earth recordings, including the sounds of insects, animals, volcanoes, and laughter, as well as Blind Willie's song. The accompanying message stated: "We cast this message into the cosmos. . . . It is likely to survive a billion years into our future, when our civilization is profoundly altered and the surface of the Earth may be vastly changed. . . . Here is our message: This is a present from a small distant world, a token of our sounds, our science, our images, our music, our thoughts, and our feelings. We are attempting to survive our time so we may live into yours."

7 The Sunday day of rest was introduced by the inspectors to influence the "mental condition" of convicts, thereby, it was claimed, rendering them more physically able (Alabama 1886/88, 238).

8 Glenn Feldman shows how complicated, but also unexpected, race and labor relations were in Alabama during the 1908 coal strike with regard to free Black labor. Ultimately, Alabama's governor, Braxton Bragg Comer, violently broke up the two-month strike by ordering the Alabama National Guard to tear down the sixteen hundred tents that housed an estimated thirty to forty thousand people. He justified his actions through the racial political imaginary: "You know what it means to have eight or nine thousand n****** idle in the state of Alabama" (quoted in G. Feldman 2011, 175). As Feldman argues, "Although the power of the state aligned with that of corporate interests ultimately overwhelmed the miners' ability to resist, race was a huge factor in the strike and in the state's willingness and propensity to intervene on the side of capital" (176). "Thus Comer's decisive, and—it may be argued—ruthless intervention on behalf of capital points to the uncomfortable fact that in Alabama (and much of the South) progressivism often stopped at the color line. Far too often, and in far too much of the South, it was for whites only" (176). While Alabama biracial unions made some meaningful gains through labor activism, appeals to white supremacy remained dominant on both sides of the dispute. Feldman suggests that "because the strike contained elemental threats to Jim Crow and white supremacy, it faced long odds" (187). Yet, the framing here suggests that race and capital were somehow separate conceptual entities, when the history of enslavement tells us that the racial origins of capital emerge through the invasion of the New World and the establishment of the regulatory and racialized action of colonialism through white geology.

9 One study assessed current deficiency of vitamin D among African Americans as the highest in the US population, at 82.1 percent (Forrest and Stuhldreher 2011, 48–54).

10 Coal is not a pure mineral. It is a spectrum of carbonaceous rocks derived from the accumulation of vegetation sedimented under swampy conditions.

11 The *Biennial Report of the Inspectors of Convicts to the Governor* by the Alabama Board of Inspectors of Convicts records names, race, incidents, illnesses, and deaths. Unlike the elaborate investigation into the death of Jere Ford, most entries are minimal in the record: "Year ending October 1st 1885—Bob Nelson was killed at Blount Springs November 7th, 1884 by a rock failing upon him; Tillman Miles was killed at Pratt Mines January 3rd 1885, by an explosion . . . etc." (Alabama 1886, 8).

12 The crushing weight of the mine reports from the Alabama penitentiary is more brutal still because the reports came after a time of intense vio-

lence in the early "experimental" days of the lease and its pyric erasure of records where people were simply "lost" in the system and in the mine (and might come to light only in an excavation of the Pratt Mines and the surrounding landscape). The language in these reports is horrifyingly detailed as it attempts to establish a systematic and paternalistic system of extraction through reporting, whereby the reporting of violence becomes its mode of repression and control, which is another form of being lost.

13 Resistance to conducting the work of forced labor gets displaced into a justification for deathly punishment, thereby confirming both the resistance and the modes of its shattering. This nonrecognition is established through the rule of normativity with historically situated white paradigms of value. To echo N. K. Jemisin, "One person's normal is another's shattering." For a discussion on negation, see Hartman and Wilderson 2003, 185.

14 Hear Morrison's account of Paul D's experience in *Beloved* hammering on the chain gang: "Singing love songs to Mr. Death, they smashed his head. More than the rest, they killed the flirt whom folks called Life for leading them on. Making them think the next sunrise would be worth it; that another stroke of time would do it at last" (2007 [1987], 128).

15 For example, Keeler laments, in her antilease pamphlet, "Gross ignorance concerning our penal institutions, and the inhuman treatment of many of the inmates, especially of those whose labor is hired out to heartless contractors, is not wholly confined to any particular part of our country. However, I shall at this time endeavor to show something of the evils resulting from convict leasing in our Southern States" (1907, 7).

16 Leanne Betasamosake Simpson, "OK Indicts," YouTube, accessed September 21, 2023, https://www.youtube.com/watch?v=hoLpvvW33BA.

17 Captive Black persons are subject to the "times" of the white settler judiciary that differentially locates Black bodies in the system of spatial occupation, sentencing, and duration of sentence.

18 Katherine McKittrick discussed environmental determinism in the conflation of the Black body with environmental conditions, as a making of the toxic body, in her talk at the Holberg Symposium. Her point, to put it briefly, is that focusing on the Black toxic body reinvigorates colonial environmental determinisms and their sedimentations of Black death rather than livability (Katherine McKittrick, "From Double Consciousness to Planetary Humanism," Holberg Symposium, University of Bergen, June 4, 2019, https://www.youtube.com/watch?v=HQF2SsOhbCg).

19 Arthur Jafa, *Love Is the Message, the Message Is Death*, YouTube video, 2016, accessed September 21, 2023, https://www.youtube.com/watch?v=lKWmxoJNmqY.

Chapter 11. Inhuman Matters IV: Modernity, Urbanism, and the Spatial Fix of Whiteness

1 The term *civiliter mortuus* had a range of interpretations, most notably in *Dade Coal Co. v. Hazelett* (1889) and *Henderson v. Dade Coal Co.*, 100 Ga. 568 (28 S.E. 251) (1897).

2 In Atlanta, for example, more than 90 percent of Black female wage earners were employed in domestic work, as servants or laundresses or in childcare, and thus generating the conditions of social reproduction for whiteness (see LeFlouria 2011, 47). LeFlouria argues that incarcerated Black women worked like men in the convict lease in Georgia but were often sexually assaulted, linking enslaved women to those who were assaulted by state-sanctioned violence in the lease.

3 Voter registration books were created in accordance with the Second Reconstruction Act of March 23, 1867. https://digital.archives.alabama.gov/digital/collection/voter1867.

4 See Wynter, "No Humans Involved" (1994), a letter to her colleagues on the beating of Rodney King.

5 See § 7637, Code of 1907, ref. in *Quick v. Western Ry. of Alabama,* Supreme Court of Alabama.

6 Entries for women in the Alabama ledger include, for example: "Avery Alice from Jefferson Ass't to murder May 2 '84 Life Cook Age 21"; "Martha Johnson from Wilcox May 26th, 1886 28 year old labourer Attempt to Murder 5 years"; "Annie Tucker, from Perry Murder second degree April 6 '83 10 years Age 20 (one bad conduct roll notation)"; "Fanny Hagin from Montgomery Grand Larceny April 12 '86 2 years Washwoman Age 22"; "Selina Thomas form Montgomery Grand Larceny April 12 '86 2 years Laborer Age 22"; "Molly Guy, White, Miscegenation, 2 years, 108 days for costs"; "Lizzie Keller, Negro, Vagrancy July 24 '85, 12 months 30 days plus 3 months 14 days for costs" (Alabama 1886/88, 21, 44, 52, 59, 87).

7 Du Bois notes that the judiciary system went from being only for the prosecution of whites to an instrument of white power: "The slaves could become criminals in the eyes of the law only in exceptional cases. The punishment and trial of nearly all ordinary misdemeanors and crimes lay in the hands of the masters. Consequently, so far as the state is concerned, there was no crime of any consequence among Negroes. The system of jurisprudence had to do, therefore, with whites almost exclusively" (1901, 738).

8 The issue of white women in proximity to convicts was of particular concern to the inspectors: "Women should not be confined in the same prison as men nor even admitted in the same enclosure. They are the cause of most of the quarrelling between men. . . . White women should not be confined in the same enclosure as negro men. No one who is not brought in constant observation of this abomination can realize the degrading

effect it has even upon the low class of white women that usually compose convicts. I am satisfied there is not a right feeling white man in the whole State who after seeing the working of this custom would not denounce it" (F. P. Lewis, MD Physician, to convicts of Coalburg Coal and Coke Co. in Alabama 1886, 261).

9 There are other social histories of Black labor in the South and in the coal mines of the North. For example, West Virginia, having had the highest proportion of free Black miners, is often a touch point for examining the integration of free Black miners and whites with "equal pay for equal work" (although, often in unequal labor relations). Yet, visit the small mining museums in West Virginia and ride through the towns, and these "Friends of Coal" present a heroic and distinctly white mining history. Against the highly polluting backdrop of mountaintop removal, declining jobs, and opioid epidemics, the heroic mining years of coal are made anew through the local history objects and stories that adorn the volunteer-run museums, in the call for the white supremacy of matter, for a former president who supports coal, and for the white worker who has built the nation (see Yusoff and Thomas 2018).

10 The libidinal economy of power and its violent optics made requirements on the aesthetics of racialized masculinity via the fullness of muscles, the symbolic containment of latent power, and its connections to regimes of property under slavery and the lease. Hartman talks about the muscularity of Abolition Man as he is depicted on his knees, begging for recognition in the category of the human. She says: "Of course, once you have assumed the position of supplicant and find yourself genuflecting before the court or the bar of public opinion, then, like the strapping man on the medallion, you have conceded the battle. It is hard to demand anything when you are on bended knee or even to keep your head raised. And you can forget trying to counter the violence that had landed you on your knee in the first place. Being so low to the ground, it is difficult not to grovel or to think of freedom as a gift dispensed by a kind benefactor or to imagine that your fate rests in the hands of a higher authority, a great emancipator, the state, or to implore that you are human too" (Hartman 2007, 124).

11 N. K. Jemisin (2019) unpacks this relation between the recognition of value and the source of devaluation and degradation in her *Broken Earth* trilogy, through the account of orogeny children who can sense and buffer seismic events, acting on the energy of rocky matter. These children are kept immobile in wire chairs plugged into "nodes" that regulate seismic tremors, as enforced labor that stabilizes the political economy and the earth.

12 According to Inspector Greene: "The want of information in mortuary statistics, both in prison life and out of it, has given rise to much prejudice against the present system of convict labor. The fact that the death-rate of the negro in prison is everywhere, under the same circumstances, twice or

thrice times as great as that of the white man, is not generally recognized or known. In the Newcastle prison the death rate of the negro has been more than ten times greater than that of the white convict." Report from Dr. R. Greene, New Castle coal mine, to Col. R. H. Dawson, president of the Board of Inspectors (Greene in Alabama 1886/88, 272).

13 Confirming the importance of the lease to devalued capitalist labor regimes, TCI President George Crawford declared in 1911, "The chief inducement for the hiring of convicts was the certainty of supply of coal for our manufacturing operations in the contingency of labor troubles" (quoted in Mancini 1996, 108).

14 For example, in protestation against a proposed mine safety law (that in fact did nothing to protect miners and did everything to compel them to the will of the company, from restricting the use of private doctors to the enforcement of purchases in the company store), the United Mine Workers argued, "It is a clumsy attempt to deceive the legislature and fasten a condition of slavery on the Alabama miners by false pretenses." Officials charged, "It is not for the protection of the miners, but solely for the protection of the operators against any liability for the maiming and killing of those whom they send down . . . into the earth" (quoted in Ward and Rogers 1987, 70–71).

15 For the developments of frameworks of mine safety and its regulation, see chapter 4, "'At the Mercy of the Earth': The Mine Safety Law," in Ward and Rogers 1987, 65–76. As Ward and Rogers comment, "Alabama was particularly vulnerable to charges of legislative neglect: between 1900 and 1910, mine explosions claimed 1,180 lives in the state" (65). The largest of these was the Banner Mine incident in which 128 miners were killed. Of 22,003 coal miners in the state (free and convict), 209 lost their lives in 1911 alone, and there were nearly three thousand reported accidents. The expansion of accidents was due in part to the expansion of mining operations but also to the depth and increasing complexity of mine shafts: "Deeper penetration into the earth meant exposure to concentrations of explosive methane gas. Increasingly sophisticated equipment expanded the amount of coal dust, which was both a health and safety hazard. Conversion to electricity created added danger, and emphasis on the amount of coal mined (rather than on the size of the pieces) led to the misuse of explosives" (66). Due to the national expansion of coal mining, in 1910 a coalition of miners, inspectors, and coal and metal mine operators created the United States Bureau of Mines, and a mine safety law was passed in 1911.

16 Alabama was one of the first states to establish Black Codes (1865–66), which covered all manner of social and sexual codes designed to restrain African Americans and severely limit their physical, economic, and social geographies. The codes enshrined anti-Blackness in the judiciary as a geographically and socially active form of governance over Black conduct. The 1865 code governing vagrants and vagrancy underpinned the operational

foundations of the lease. Curtin gives the example of the mayor of Mobile, who sentenced a group of Black women to ten days in the workhouse for "disturbing the peace" and engaging in a "war of words" (Curtin 2000, 6). Embracing the newly designated "freedom" of speech became a path into criminality.

17 Although the Slope and Shaft mines provided some opportunities for privacy, with surveillance being enacted at the mine head in the form of the count of the "take," the incarcerated had little privacy and remained a collective in both the prison and the mine.

18 Curtin comments: "The black intellectual community of the late nineteenth century, varied as it was, agreed on certain basic middle-class assumptions that took a dim view of black working-class culture and pleasures" (2000, 182).

19 This is where there is a divergence: the myth of racial progress is mapped into the materiality of middle-class acquisition, namely, as property, whereas being poor and Black is mapped into a kind of inverse material trajectory, into the underground.

20 The carceral system went from being 99 percent white during slavery to an almost entirely Black system of enclosure during Reconstruction.

21 As an antidote, Hartman's brilliant account, *Wayward Lives* (2019), is a tract written in defiance of the gravity of shame, its projection onto and into Black forms of women's lives.

22 Incarcerated subjects also recognized the continuance of slavery. Martha Aarons, for example, was sentenced for seventy-five years for the death of her child. She says: "And when I think that I have for sixteen years been a faithful Slave for the State of Alabama, the time has come for when she might be generous and forgiving to me" (quoted in Curtin 2000, 128).

23 In the United Kingdom, the Inspectorate of Mines was established by an act of Parliament in 1842, as part of the Mines and Collieries Act. Initially reluctant to intervene in mining health and safety because of the influence of mining interests, the inspectorate was introduced to bring incremental regulation and placate moralistic Victorian reformers. It prohibited all children under ten years old from working underground in coal mines. It was not until 1872 that the Coal Mines Act assigned clear responsibility for safety and health to mine management and stipulated specific mine safety rules (see Mills 2010). The US General Mining Act of 1872 had little focus on health and safety and refers predominantly to claims in the governing of the properties and the process of staking and acquiring mines.

24 For further discussion of the development of political organizing and mining, see Robin Kelley (1990, 1–10), who suggests that in Birmingham, Black miners created a "tradition of militant, interracial unionism" (4) that left a "remarkable record of labor activity: of 603 strikes initiated by Alabama's workers between 1881 and 1936, 303 took place during 1881–1905" (5). Black

convict labor was used to progressively undermine unionism in Alabama (Brown and Davis 1999).

25 See Arthur Rothstein, *Coal Miners' Homes. Birmingham, Alabama*, photograph, February 1937, Birmingham, Jefferson County, Alabama, United States, https://www.loc.gov/item/2017775915/.

26 A Black Code that existed in Alabama, established in 1866, was known as "the Act to Define the Relative Duties of Master and Apprentice," which said that "masters" could take responsibility for Black minors, "apprentices" under the age of sixteen who either were orphans or had no means of sustaining themselves. If the minors ran away or refused to work, they could be tried in a court of law and sent into the convict lease system as vagrants.

27 For an overview of the "for whites only" thesis, see C. Vann Woodward 1981. For the broader perspective of coal and race relations in the South, see Foner and Lewis 1980; B. Kelly 2001; Letwin 1998; and Worthman 1969. And, for the epic coal strike of 1908, see G. Feldman 2011.

28 Convict lease labor happens within the scene of a material crisis of humanism, which accompanied Reconstruction in the antebellum South, whereby those designated and treated as less than human (and outside social and political life) are recategorized within the human, a category that is built on their exclusion. The collapse of the Hegelian dialectic of Human-Other had more immediate practical and intimate forms than the distance of theorical objecthood can account for. In Alabama, convict lease labor is part of the emergence and negotiation of the "social" in its spatial and subjective modes; a social space that now includes Black life in its polis where hence it was excluded but which white law seeks to render immiscible though can no longer do so through its established categories of the human. The convict lease is an attempt to designate a new category for Blackness that performs de facto the prior exclusion but within a *new* lexicon of geologic life—specifically, a form of subjugation that performed to categorize on behalf of the law, but for the profit of the corporation.

29 Mancini argues that slavery and the lease cannot be seen as equivalents, even though they function within the larger category of labor (1996, 24).

30 "Shooting" was drilling holes in the coal seam and then using explosive powder to break the coal up into pieces for loading.

31 For example, the recognition of and response to accidents did not necessarily make conditions better for prisoners but often led to more scrutiny and policing: "There were formerly frequent accidents in this mine, many of them serious, and some fatal in their results. The President of the Board, believing that these were the result of ignorance and carelessness on the part of convicts working in the mine, as soon as he took charge, demanded of the contractors that there should be an increase in the number of pit-bosses in the mine, so that convicts could be better instructed and more closely watched at their work" (Alabama 1886, 33).

32 In "The Pratt Mines Disaster, and the Causes Leading to It. Inefficient Ventilation and Dangerous Gases. The Full Official Report," *Birmingham Age-Herald*, June 10, 1891, Inspector Hooper concludes: "From my investigation of this occurrence I wish to say that if this explosion occurred as claimed by officials of the Tennessee Coal and Iron company, as one of those unforeseen accidents that could not be guarded against, then this is a dangerous and unfit place to work convicts, who are ignorant of the dangers of mining and not competent to protect themselves and who have no option as to where or when they shall go to any place to which they are ordered; that if this explosion could have been guarded against then there is very great neglect somewhere in the management of the parties in charge of shaft No. 1. . . . I also wish to state that in view of the dangers incident to the working of gaseous mines and of the ignorance of the convicts (who are generally recruited from the cotton fields of South Alabama) that no convict should be allowed to work in a gaseous mine; that this is especially dangerous when there is only one avenue of escape from said mine. . . . I think in view of these facts that No. 1 shaft is a dangerous and unfit place to work men who have but little knowledge of the dangers to be guarded against in mining." Hooper bemoans the lack of an escape shaft from the mine, but preventing escape attempts was exactly the point of there being only one way in and out of the mine. The scene of the explosion was about one mile underground from the hoisting shaft. The men killed were recorded as Philip Page, larceny and burglary, Jefferson County; Tom Hamilton, grand larceny, Montgomery County; Tom Hare, burglary and larceny, Jefferson County; Charles H. Robinson, forgery, Montgomery County; W. D. Mayfield, murder, Lawrence County; A. M. Hays, murder, Jefferson County; J. G. Davis, gaming and concealed weapons, Lamar County—all convicts; Thomas Moore, freeman, working as carpenter in the mines. These men died from explosion, burns, and gases.

33 The sexual economy of abolitionist movements was marked by the "concern" of British women for the moral standing of their absentee husbands, who spent half the year in the Caribbean and who supposedly returned emboldened with sadistic forms of sexual pleasure and disease. As young men were sent to the plantation from England, to be sexually "licensed," abolitionist rhetoric balked at the imagination of the sexual excesses of Black women. In parallel, women filed between 40 and 45 percent of the claims for "property" to the Slavery Compensation Commission (established in 1833 to facilitate the process of awarding compensation to enslavers).

34 The spatial enclosures of the lease condition are not just incarcerated spaces but condition the Black outdoors. To be put outside, in Toni Morrison's terms, placed the Black outdoors as irrevocable, precisely because it was an abandonment to white spatial codes. She says: "There is a difference

between being put out and being put outdoors. If you are put out, you go somewhere else; if you are outdoors, there is no place to go. The distinction was subtle but final. Outdoors was the end of something, an irrevocable, physical fact, defining and complementing our metaphysical condition. . . . Dead doesn't change, and outdoors is here to stay" (Morrison 1999 [1970], 17). As in a call-and-response to Morrison, Moten writes: "Fugitivity, then, is a desire for and a spirit of escape and transgression of the proper and the proposed. It's a desire for the outside, for a playing or being outside, an outlaw edge proper to the now always already improper voice or instrument" (2018, 131).

35 By the late 1920s, the company had only a third of its production in steel; the rest was in oil and gas mining, chemicals, construction, transportation, and real estate.

36 I echo Katherine McKittrick's analysis here of the idea of the plantation as something that is migratory: "With this, differential modes of survival emerge—creolization, the blues, marronage, revolution, and more—revealing that the plantation, in both slave and postslave contexts, must be understood alongside complex negotiations of time, space, and terror" (2013, 3).

37 For a discussion of the myth of John Henry, see Whitehead 2002; Tettenborn 2013, 271–84; Bradford 2008). In historic terms, John Henry's race against the pneumatic drill is not just a race against the technology of modernity, but against the closing off of a "way out" that many former convict leasers found in labor. While John Henry has been claimed for the Big Bend Tunnel of the Chesapeake and Ohio Railway in West Virginia, Dr. John Garst, a chemist from the University of Georgia, argues via eyewitness accounts that the famous steel-driving contest took place on September 20, 1887, in the Coosa and Oak Mountain Tunnel, Alabama (Garst 2002).

38 Bessie Smith also recorded a song with hammer tones beating out the cadence of "this old hammer killed John Henry." Leadbelly, or Huddie William Ledbetter, incarcerated at the infamous Sugar Land Plantation in Texas, recorded "Take This Hammer," whose lyrics go: "Take this hammer, carry it to the captain / Tell him I'm gone / If he asks you was I runnin' / Tell him I was flyin.'"

39 An excerpt from Rankine: "When you arrive in your driveway and turn off the car, you remain behind the wheel another ten minutes. You fear the night is being locked in and coded on a cellular level and want time to function as a power wash. Sitting there staring at the closed garage door you are reminded that a friend once told you there exists the medical term—John Henryism—for people exposed to stresses stemming from racism. They achieve themselves to death trying to dodge the buildup of erasure. Sherman James, the researcher who came up with the term, claimed the physiological costs were high. You hope by sitting in silence you are bucking the trend" (2014, 11).

40 See Bonham, Sellers, and Neighbors 2004; Hudson et al. 2016; S. A. James, Hartnett, and Kalsbeek 1983.

41 Listen: Moor Mother, "Chain Gang Quantum Blues."

42 Using language reminiscent of Michelle Wright's (2015) idea of the Middle Passage to the New World as a "big bang of blackness," Ishmael Reed calls it "an Atlantic of blood. Repressed energy of anger that would form enough sun to light a solar system. A burnt-out black hole. A cosmic slave hole" (quoted in Weheliye 2002, 21).

43 Sharpe understands all modern subjects as constituted by slavery, but not all bear the weight of this identification. She says: "While all modern subjects are post-slavery subjects fully constituted by the discursive codes of slavery and post slavery, post-slavery subjectivity is largely borne by and readable on the (New World) *black* subject" (Sharpe 2010, 3).

44 In a parallel vein, speaking to Derek Walcott's poem "Ruins of a Great House," Ann Stoler characterizes the project of her book *Imperial Debris* as attuned to how "the process of decay is ongoing, acts of the past blacken the senses, their effects without clear termination . . . the eating away of less visible elements of soil and soul. . . . Walcott's caustic metaphors slip and mix, juxtaposing the corrosive degrading of matter and mind" (2013, 1–2). In the rot that remains, Stoler reads "a provocative challenge to name the toxic corrosions and violent accruals of colonial aftermaths, the durable forms in which they bear on material environments and on people's minds. . . . Walcott refuses a timeframe bounded by the formal legalities of imperial sovereignty over persons, places, and things" (2).

Chapter 12. Inhuman Matters V: Trees of Life (and Death), "Strange Fruit," and Geologies of Race

1 James Baldwin names the dynamics of this environmental sensibility: "For what I did not know so intensely was the hatred of white America for the black, hatred so deep that I wonder if every white man in this country, when he plants a tree, doesn't see Negroes hanging from its branches" (Baldwin quoted in Wynter n.d., xx).

2 Small mining concerns began in the 1850s with nascent industrialists like William Phineas Browne, who started extraction with two enslaved men, hauling by hand and then wagon to the Cahaba River. By 1856, Browne had leased enough enslaved persons to remove five tons of coal per day from his mines. In 1862, Browne was contracted with the Confederate government to deliver four thousand tons of coal to feed the iron and steel furnaces that made the cannonballs and steel armaments for the Confederate army. Browne's biography lauds him as a great Southern industrialist, a corporate father of geology. The discovery of coal along Alabama's river in 1815

identified the southern end of the Appalachian coal field, which spanned an estimated seventy thousand square miles from Pennsylvania and Ohio to central Alabama. Coal mining became active again in the 1870s, when the Louisville and Nashville Railroad constructed a line that connected the Alabama River at Montgomery with the Tennessee River at Decatur. Railroad officials hoped to link Red Mountain iron ore with the coal of the Cahaba field to promote iron production. The Pratt Coal and Coke Company began in 1878 with several slopes and a railroad link to supply coking coal to the furnaces of the Birmingham District. These mines would feed the Birmingham iron industry and eventually spawn the development of the mining community of Pratt City.

3 While the psychic control of social space might seem more readily tied up with the performance of human hierarchies in trees in the "non-productive" expenditure of lynching, it equally performs its quotidian work in the "productive" expenditure of convict lease labor.

4 During the 1890s, white Alabamians lynched more Black people than any other state in the nation (between 1889 and 1899 lynch mobs murdered 177 Black citizens in Alabama). Alabama was also the site of the last recorded lynching in the United States; in 1981, Michael Donald was murdered by the KKK and found hanging from a tree in Mobile. In another grim Alabama "last," the last known (illegal) transportation of enslaved Africans to the United States on board the *Clotilda* in 1859–60 came to Mobile Bay and headed up the Mobile River with 110 to 160 enslaved persons from Benin, West Africa. Some of the descendants from the *Clotilda* formed the free Black town called Africatown. For an alternative narrative from the plateau, read Zora Neale Hurston's *Barracoon* (2018) and Sylviane A. Diouf's *Dreams of Africa in Alabama* (2009). In 1959, to commemorate one hundred years since the voyage, a steel shaft was sunk one hundred feet into the earth in front of the Union Missionary Baptist Church, and a bronze bust of Cudjo Lewis, one of the last known survivors, was placed onto a pyramid of bricks that had been made by the *Clotilda* captives. The monument was forcibly pulled out of its placement in 2002.

5 Mass incarceration and environmental degradation are brought together by the Prison Ecology Project, https://nationinside.org/campaign/prison-ecology/.

6 In July 2018, the shallow graves of ninety-five convict lease laborers were discovered at Sugar Land Plantation in Texas.

7 Hear Will Harris's question: "I know that *blood* stands for race and *soil* for nation but *blood and soil* makes me think of bloodied soil. Do some people imagine themselves in the same relation to their place of birth as a scab to a wound?" (Harris 2020, 20).

8 According to Alabama state geologist Michael Tuomey, Little Cahaba Iron Works used enslaved labor as early as 1849 for mining.

9 See Nyong'o (2009) for a discussion of the performativity of amalgamation and Blackness.

10 The new "mycelium turn" and its rhizomatic thought might be seen in contradistinction to the above-ground/underground distinction of the eugenic tree.

11 One of the biggest revolutions in evolutionary theory has been the "Third Kingdom of Life" proposed by Carl Woese in the late 1970s. The "Third Kingdom" did not dispense with the tree structure but made it more of a bush with three distinct areas: bacteria; eucarya (everything else that is not bacteria); and archaea (newly identified ancient single-celled life-forms that shift genetic material across organisms, thereby complicating and participating in any verticality of lineage; archaea also process minerals, so are part of the life-nonlife exchange.) Because of the identification of horizontal transfer of genes, archaea break with the "parental" form of vertical transfer that was established in geologic accounts of lines of descent (that formed the basis of Darwin's and subsequent theories of evolution). This is not to promote another "better" biological metaphor but to acknowledge how horizontal gene transfer made the "tree" metaphor redundant. The ability to reproduce sideways, as it were, organized a new imaginary of relatedness and exchange without sexual reproduction, referred to as transduction or shuffle DNA. This "infective heredity," with DNA coming from outside the vertical form of ancestry, campaigns for the sense of a very different kind of worlding to reconstruct paleontological viewpoints (see Quammen 2018).

12 As a counternarrative, Michelle Wright (2015, 74) suggests "epiphenomenal time," a time that recognizes the manifesting of the past in the present that runs along an axis of asymmetry rather than a linear back and forth.

13 This is a biologism that swerves away from the conceptualization of the roots of "Civilization" being born out of materiality (Stone Age, Bronze Age, "fossils as natural resources," Industrial Age, etc.). In the narrative arc of twentieth-century biologism, the distinct temporal lines of life and nonlife play an increasing role in backgrounding geological and environmental relations as ground, backdrop, and mineral resource.

14 I think here of Thomas Gainsborough's *Mr and Mrs Andrews* (1750), in which the wooden woman doubles, alongside the tree in her immobility, as a perspective point for the masculine possession of property, as land, agency, and patriarchy.

15 Tiffany King talked in a Q&A about asking her students, particularly her white male students, about their own desire at writing about Black women's bodies. Shifting the object of the gaze, she questioned their investment in degraded bodies and women's exposure and its recirculation. She suggested that this should be the first question that should be asked in the formation of their research. It is a question I have carried with me.

16 The oblique force field of violence that the photograph captures and collects is situated in a visual culture that celebrated lynchings through the public culture of photographs that were circulated as postcards, and that was part of the nature of whiteness: its structuring of the outdoors through gender and patriarchy; of the making of "white women" (Carby 1985; Feimster 2009); and the affective politics of property.

17 The anxiety that accompanied the imagination of a fortified genealogical account of whiteness was threatened by the "spill" of Blackness (even under conditions of segregation), in which a gaze could undo whiteness and womanhood. Whiteness was a spatiotemporal project that refuted the politics of emancipation (turning back time to conquer space, conquering space to turn back time). As Carby (1985, 269) comments, "Black disenfranchisement and Jim Crow segregation had been achieved; now, the annihilation of a black political presence was shielded behind a 'screen of defending the honor of [white] women.'" For discussions on gender and reconstruction, see Carby 1998; Feimster 2009.

18 In 2016–17, Sarah Jane Cervenak and J. Kameron Carter ran a speaker and working group series titled The Black Outdoors: Humanities Futures after Property and Possession.

19 Daniel Kato (2015) argued that lynching was legitimized by the state as a form of "constitutional anarchy," whereby the government operates in a version of the dual state in which it limits its powers to leave "anomalous zones" that are not subject to the law.

20 The Black Belt is recognized as a political formation that maps onto a geologic rift as an arc of Democratic voters in a heavily Republican state. Named in the nineteenth century by Dutch planters, the Black Belt refers to the rich, dark soils formed in the Cretaceous period that were seen as ideal for planting cotton, and thus it also refers to those who were unhomed by the Atlantic trade in enslaved persons to pick the fruits of that dark earth.

21 The song "Strange Fruit" was written by the son of Russian Jewish immigrants, Abel Meerpol, who was allegedly inspired by a photograph. Meerpol taught English at DeWitt Clinton High School in the Bronx, where James Baldwin was one of his students. The song started as an antilynching poem and was later set to music and performed by Billie Holiday and Nina Simone. Meerpol and his wife adopted Julius and Ethel Rosenberg's children after meeting them at a Christmas party at Du Bois's house.

22 Gary Brechin (1996) argues that eugenics in the United States was closely linked with the fledgling conservation movement, particularly through the writings and activities of Henry Fairfield Osborn at the American Museum of Natural History.

23 Wynter argues that "because of our role as the grammarians," we must integrate rather than stably replicate our "structural modes" to "make the transition from one Foucauldian *episteme*, from one founding and

behaviour-regulating narrative, to another" "so that we can now pose the questions whose answers can resolve the plight of the jobless archipelagoes, the NHI categories, and the environment" (Wynter 1994, 13).

24 "The dualism to which she subscribes dismisses the political significance of the 'private realm' and the bodies contained and policed there, allowing the personified public body to appear as both representative and universal (with little mention of the homogeneity of its appearance—propertied males racialized as white). The political person—naturalized and universalized as affluent, masculine European (or some approximation thereof)—shapes Arendt's color- and gender-blind analyses. Her model is premised on an inequality marked by assumptions of biologically determined superiority/inferiority; this model impedes critiques of white supremacy, heteropatriarchy, racial domination, and state violence" (J. James 2013, 308–9).

25 A recent article on America's national parks stated that less than 10 percent of the proportional population of park visitors were African American. Many groups now come together around the Black outdoors to refuse the legacy of the Enlightenment subject and the whiteness of nature activities. It is not new to suggest that geography is racialized; it is a practice that many people navigate every day in plotting their routes, hangouts, and outdoor spaces, but it is one that continues to code the production of space (see Finney 2014).

Chapter 13. Ghost Geologies

Portions of chapter 13 appeared in an earlier form as "Mine as Paradigm" in *e-flux Journal* (June 2021), https://www.e-flux.com/architecture/survivance/381867/mine-as-paradigm/.

1 Geologic drift: What kind of conceptual apparatus must be in place to think about straightening a river of which no part is straight (the Mississippi, for example)? Navigating rivers was part of "opening up" the territory; straightening rivers was a feature of the plantation and the control of irrigation and export.

2 Predation in biology is understood as the transfer of energy from the organism that is being consumed to the predator, to provide the energy to prolong its life and promote its reproduction.

3 The dirty streets of East London factories, with their relation to coal and sugar and the grinding weight of poverty, similarly made new mines of industrial relation.

4 Glissant and Césaire are not dialectical thinkers; they are expansive and recognized that they had to reinvent the terms of sovereignty and the composition of self, as well as the formation of social thought through the stratigraphic terms of dialectics.

5 Thomas describes this methodology as follows: "Exploring the constitution of the political subject not primarily through nationalism or through state-(and extra-state-) driven processes of subjectification, but through the cultivation of embodied affects that are shaped by the particular temporal conjunctures in which they emerge, enables us to interrogate the ways political affects can transcend the context of their emergence, allowing them to appear and resurface unpredictably. It can thus unbind sovereignty not only from territory, and therefore from the political centrality of the independent nation-state, but also from the teleologies of linear, progressive time" (2019, 5).

6 The inhuman is radically open and cannot be read solely through its social forms as signaling in any direction because we want the past to be a resource for more liberated futures, but this openness also has the potential to hold the unimaginable and the impossible closely.

7 An iteration of this colonial-contemporary earth praxis of white geology is the bodily and environmental pollution caused by the export of the French pesticide chlordecone (exported in the knowledge of its cancer-causing geochemical effects) to Martinique for use in banana plantations.

References

Agamben, Giorgio. 2005. *State of Exception*. Translated by Kevin Attell. Chicago: University of Chicago Press.

Agard-Jones, Vanessa. 2012. "What the Sands Remember." *GLQ* 18 (2–3): 325–46.

Agassiz, Louis. 1850a. "The Diversity of Origin of the Human Races." *Christian Examiner and Religious Miscellany* 49 (July): 110–45, 181–204.

Agassiz, Louis. 1850b. "Geographical Distribution of Animals." *Christian Examiner and Religious Miscellany* 48 (March): 184–85.

Agassiz, Louis. 1850c. "Remarks of Prof. Agassiz, after the Reading of this Paper; Zoological Evidence for the Diversity of the Races." In *Proceedings of the American Association for the Advancement of Science, Third Meeting Held at Charleston SC, March 1850*, 106–7. Charleston, SC: Walker and James.

Agassiz, Louis. 1854. "Sketch of the Natural Provinces of the World and Their Relations to the Different Types of Man." In Nott and Gliddon, *Types of Mankind*, 90–92.

Agassiz, Louis. 1857. *Contributions to Natural History of the United States of America*. Vol. 2. Boston: Little, Brown.

Agassiz, Louis. 1863a. "America the Old World." *The Atlantic Monthly* (March): 373–82.

Agassiz, Louis. 1863b. Transcript of letter to S. G. Howe, Nahant, August 1. Louis Agassiz correspondence and other papers (1807–1873), MS Am 1419 (150–53), Box 2, Houghton Library, Harvard University, Cambridge, MA.

Agassiz, Louis. 1874. *Methods of Study in Natural History*. Boston: James Osgood.

Agassiz, Louis, and Elizabeth Cabot Cary Agassiz. 1868. *A Journey in Brazil*. Boston: Ticknor and Fields.

Alabama. Board of Inspectors of Convicts. 1886. *Biennial Report of the Inspectors of Convicts to the Governor* (October 1884–October 1886). Montgomery, AL: The Board.

Alabama. Board of Inspectors of Convicts. 1886/88. *Biennial Report of the Board of Inspectors of Convicts to the Governor*. Montgomery, AL: The Board.

Alabama. Board of Inspectors of Convicts. 1888/90–1890/92. *Biennial Report of the Inspectors of Convicts to the Governor*. Montgomery, AL: The Board.

Alexander, M. Jacqui. 2005. *Pedagogies of Crossing: Meditations on Feminism, Sexual Politics, Memory, and the Sacred*. Durham, NC: Duke University Press.

Althusser, Louis. 2014 [1969]. *On the Reproduction of Capitalism: Ideology and Ideological State Apparatuses*. London: Verso.

Amin, Samir. 1976. *Unequal Development: An Essay on the Social Formations of Peripheral Capitalism*. New York: Monthly Review Press.

Ansfield, Bench. 2015. "Still Submerged: The Uninhabitability of Urban Redevelopment." In McKittrick, *Sylvia Wynter*, 124–41.

Ansted, David Thomas. 1854. "Slavery as an Economic Question." In *Scenery, Science, Art: Being Extracts from the Note-Book of a Geologist and Mining Engineer*, 294–311. London: John Van Voorst.

Appel, Hannah, Arthur Mason, and Michael Watts. 2015. *Subterranean Estates: Life Worlds of Oil and Gas*. Ithaca, NY: Cornell University Press.

Arendt, Hannah. 1951. *The Origins of Totalitarianism*. New York: Random House.

Arendt, Hannah. 1959a. *The Human Condition*. New York: Doubleday.

Arendt, Hannah. 1959b. "A Reply to Critics." *Dissent* (Spring 1959): 45–46. https://www.dissentmagazine.org/article/a-reply-to-critics/.

Arendt, Hannah. 1970. *On Violence*. New York: Houghton Mifflin Harcourt.

Arendt, Hannah. 2006 [1962]. "The Meaning of Love in Politics: A Letter by Hannah Arendt to James Baldwin." HannahArendt.net / Journal for Political Thinking. Uploaded September 15, 2006. http://www.hannaharendt.net/index.php/han/article/view/95/156. First sent to Baldwin November 21, 1962.

Armes, Ethel. 2011 [1910]. *The Story of Coal and Iron in Alabama*. Tuscaloosa: University of Alabama Press.

Armstrong, Meg. 1996. "'The Effects of Blackness': Gender, Race, and the Sublime in Aesthetic Theories of Burke and Kant." *Journal of Aesthetics and Art Criticism* 54 (3): 213–36.

Baldwin, James. 1955. *Notes of a Native Son*. Boston: Beacon Press.

Baldwin, James. 1962. "Letter from a Region in My Mind." *New Yorker*, November 9, 1962.

Baldwin, James. 1985. *The Price of the Ticket*. Boston: Beacon Press.

Barra, Monica P. 2020. "Good Sediment: Race and Restoration in Coastal Louisiana." *Annals of the American Association of Geographers* 111 (1): 266–82.

Bataille, Georges. 1991 [1949]. *The Accursed Share: An Essay on General Economy*. New York: Zone Books.

Bebbington, Anthony, and Jeffrey Bury, eds. 2013. *Subterranean Struggles: New Dynamics of Mining, Oil and Gas in Latin America*. Austin: University of Texas Press.

Berlant, Lauren. 2011. *Cruel Optimism*. Durham, NC: Duke University Press.

Berlant, Lauren. 2016. "The Commons: Infrastructures for Troubling Times." *Environment and Planning D: Society and Space* 34 (4): 393–419.

Bernard, Rachel E., and Emily H. G. Cooperdock. 2018. "No Progress on Diversity in 40 Years." *Nature Geoscience* 11:292–95.

Birmingham Age-Herald. 1888. "Alabama's Resources, as Seen by a Correspondent of the Boston Bulletin. Some Concluding Observations on Convict Labor—Ensley City—Prospects and Probabilities." June 1888.

Bishop, Elizabeth. 1979. "At the Fishhouses." In *The Complete Poems, 1927–1979*, 64–67. New York: Farrar, Straus and Giroux.
Blackmon, Douglas. 2008. *Slavery by Another Name: The Re-enslavement of Black Americans from the Civil War to World War II*. New York: Doubleday.
Blanchot, Maurice. 2010. *Political Writings, 1953–1993*. Translated by Zakir Paul. New York: Fordham University Press.
Blanchot, Maurice. 2014. *Into Disaster*. Translated by Michael Holland. New York: Fordham University Press.
Bonds, Anne. 2020. "Race and Ethnicity II: White Women and the Possessive Geographies of White Supremacy." *Progress in Human Geography* 44 (4): 778–88.
Bonham, Vence L., Sherrill L. Sellers, and Harold W. Neighbors. 2004. "John Henryism and Self-Reported Physical Health among High-Socioeconomic Status African American Men." *American Journal of Public Health* 94 (5): 737–38.
Bowler, Peter J. 1992. *The Fontana History of the Environmental Sciences*. London: Fontana Press / Harper Collins.
Bradford, Roark. 2008. *John Henry: Roark Bradford's Novel and Play*. Oxford: Oxford University Press.
Braidotti, Rosi. 2011. *Nomadic Theory: The Portable Rosi Braidotti*. New York: Columbia University Press.
Brand, Dionne. 1990. *No Language Is Neutral*. Toronto: McClelland and Stewart.
Brand, Dionne. 1999. *At the Full and Change of the Moon*. New York: Grove Press.
Brand, Dionne. 2001. *A Map to the Door of No Return*. Toronto: Vintage Canada.
Brand, Dionne. 2006. *Inventory*. Toronto: McClelland and Stewart.
Brand, Dionne. 2017. "An *Ars Poetica* from the Blue Clerk." *Black Scholar* 47 (1): 58–77.
Brand, Dionne. 2018. *The Blue Clerk: Ars Poetica in 59 Versos*. Durham, NC: Duke University Press.
Brassier, Ray. 2007. *Nihil Unbound: Enlightenment and Extinction*. New York: Palgrave-Macmillan.
Brathwaite, Edward Kamau. 1983. "Caribbean Culture: Two Paradigms." In *Missile and Capsule*, edited by Jürgen Martini, 9–54. Bremen: Universität Bremen.
Brechin, Gary. 1996. "Conserving the Race: Natural Aristocracies, Eugenics, and the U.S. Conservation Movement." *Antipode* 28 (3): 229–45.
Broeck, Sabine. 2018. *Gender and the Abjection of Blackness*. New York: State University of New York Press.
Brown, Edwin L., and Colin J. Davis, eds. 1999. *It Is Union and Liberty: Alabama Coal Miners and the UMW*. Tuscaloosa: University of Alabama Press.
Brown, Imani Jacqueline. 2020. "Black Ecologies: An Opening, an Offering." *MARCH: a journal of art and strategy* 1:86–96.
Bruce, La Marr Jurelle. 2019. "Shore, Unsure: Loitering as a Way of Life." *GLQ* 25 (2): 352–61.
Butler, Judith. 2002. "Is Kinship Always Already Heterosexual?" *differences* 13 (1): 14–44.

Byrd, Jodi A. 2011. *Transit of Empire: Indigenous Critiques of Colonialism*. Minneapolis: University of Minnesota Press.

Caillois, Roger. 1985. *The Writing of Stones*. Translated by Margaret Bray. Charlottesville: University Press of Virginia.

Campt, Tina M. 2017. *Listening to Images*. Durham, NC: Duke University Press.

Carby, Hazel V. 1985. "'On the Threshold of Woman's Era': Lynching, Empire, and Sexuality in Black Feminist Theory." *Critical Inquiry* 12 (1): 252–77.

Carby, Hazel V. 1998. *Race Men*. Cambridge, MA: Harvard University Press.

Carby, Hazel V. 2019. *Imperial Intimacies: A Tale of Two Islands*. London: Verso.

Cartwright, Samuel A. 1857. "Ethnology of the Negro or Prognathous Race: A Lecture." Lecture presented to the N. O. Academy of Sciences, African Continental Ancestry Group—Ethnology, New Orleans Academy of Sciences, November 30, 1857.

Cervenak, Sarah Jane, and J. Kameron Carter. 2017. "Untitled and Outdoors: Thinking with Saidiya Hartman." *Women and Performance: A Journal of Feminist Theory* 27 (1): 45–55.

Césaire, Aimé. 1969 [1956]. *Return to My Native Land*. Translated by John Berger and Anna Bostock. New York: Archipelago Books.

Césaire, Aimé. 2000 [1972]. *Discourse on Colonialism*. New York: Monthly Review Press.

Césaire, Aimé. 2010. "Earthquake." *New Yorker*, January 14, 2010.

Césaire, Aimé. 2013. *The Original 1939 Notebook of a Return to the Native Land*. Bilingual ed. Translated and edited by A. James Arnold and Clayton Eshleman. Middletown, CT: Wesleyan University Press.

Chamoiseau, Patrick. 1992. *Texaco*. Translated by Rose-Myriam Réjouis and Val Vinokurov. New York: Pantheon Books.

Chamoiseau, Patrick. 2018 [1997]. *Slave Old Man*. Translated by Linda Coverdale. New York: New Press.

Chandler, Nahum Dmitri. 2013. *X—The Problem of the Negro as a Problem for Thought*. New York: Fordham University Press.

Chandler, Nahum Dmitri. 2015. "Dry and Heavy: Or, Another Poetics and Another Writing—of History and the Future." *CR: The New Centennial Review* 15 (2): 1–22.

Cheddie, Janice. 2016. "Sasha Huber's *Rentyhorn*." *Third Text* 30 (5–6): 368–87.

Chen, Mel Y. 2012. *Animacies: Biopolitics, Racial Mattering, and Queer Affect*. Durham, NC: Duke University Press.

Clark, Nigel. 2011. *Inhuman Nature: Sociable Life on a Dynamic Planet*. London: Sage.

Clark, Nigel. 2017. "Politics of Strata." *Theory, Culture and Society* 34 (2–3): 211–31.

Clark, Nigel, and Kathryn Yusoff. 2017. "Geosocial Formations and the Anthropocene." *Theory, Culture and Society* 34:3–23.

Clark, Nigel, and Kathryn Yusoff. 2018. "Queer Fire: Ecology, Combustion and Pyrosexual Desire." *Feminist Review* 118 (1): 7–24.

Clark, Tiana. 2017. "ODE TO THE ONLY LIVING OBJECT THAT SURVIVED." *Obsidian: Literature and Arts in the African Diaspora* 43 (2): 79–80.

Coombes, P., and K. Barber. 2005. "Environmental Determinism in Holocene Research: Causality or Coincidence?" *Area* 37 (3): 303–11.
Coulthard, Glen Sean. 2014. *Red Skin, White Masks: Rejecting the Colonial Politics of Recognition*. Minneapolis: University of Minnesota Press.
Crawley, Ashon. 2020. *The Lonely Letters*. Durham, NC: Duke University Press.
Curtin, Mary Ellen. 2000. *Black Prisoners and Their World: Alabama, 1865–1900*. Charlottesville: University Press of Virginia.
Cuvier, Georges. 1827. *The Animal Kingdom Arranged in Conformity with Its Organization. With Additional Descriptions of All the Species Hitherto Named, and of Many Not Before Noticed, by Edward Griffith and Others*. Vols. 1–15. London: G. B. Whittaker.
Cuvier, Georges. 1849. *The Animal Kingdom, Arranged after Its Organization: Forming a Natural History of Animals, and an Introduction to Comparative Anatomy*. London: W. S. Orr and Co.
Cuvier, Georges. 2009 [1813]. *Essay on the Theory of the Earth*. Translated by Robert Kerr. Cambridge: Cambridge University Press.
Darwin, Charles. 1859a. *On the Origin of Species by Means of Natural Selection, or the Preservation of Favoured Races in the Struggle for Life*. London: John Murray.
Darwin, Charles. 1859b. "Letter 2503: C. R. Darwin to C. Lyell," October 11, 1859. Darwin Correspondence Project, University of Cambridge. Accessed September 29, 2023. https://www.darwinproject.ac.uk/letter/?docId=letters/DCP-LETT-2503.xml.
Darwin, Charles. 1862. "Letter 3439: Darwin to Charles Kingsley," February 6, 1862. Darwin Correspondence Project, University of Cambridge. Accessed September 29, 2023. https://www.darwinproject.ac.uk/letter/?docId=letters/DCP-LETT-3439.xml&query=3439.
Darwin, Charles. 1864. "Letter 4510: Darwin to A. R. Wallace," May 28, 1864. Darwin Correspondence Project, University of Cambridge. Accessed September 29, 2023. https://www.darwinproject.ac.uk/letter/?docId=letters/DCP-LETT-4510.xml.
Darwin, Charles. 1869. "Letter 6728: Charles Lyell to Darwin," May 5, 1869. Darwin Correspondence Project, University of Cambridge. Accessed September 29, 2023. https://www.darwinproject.ac.uk/letter/DCP-LETT-6728.xml.
Darwin, Charles. 1871. *The Descent of Man*. London: John Murray.
Darwin, Charles. 1881. "Letter 13230: Darwin to William Graham," July 3, 1881. Darwin Correspondence Project, University of Cambridge. https://www.darwinproject.ac.uk/letter/?docId=letters/DCP-LETT-13230.xml.
Davis, Angela Y. 1998. "From the Prison of Slavery to the Slavery of the Prison: Frederick Douglass and the Convict Lease System." In *The Angela Y. Davis Reader*, edited by Joy James, 74–95. Malden, MA: Blackwell.
Deleuze, Gilles. 1997. *Essays Critical and Clinical*. Minneapolis: University of Minnesota Press.
Deleuze, Gilles. 2006 [1986]. *Foucault*. London: Bloomsbury.

Deleuze, Gilles, and Félix Guattari. 1993 [1980]. *A Thousand Plateaus: Capitalism and Schizophrenia*. Minneapolis: University of Minnesota Press.

Deleuze, Gilles, and Félix Guattari. 1994. *What Is Philosophy*? London: Verso.

DeLoughrey, Elizabeth M. 2007. *Routes and Roots: Navigating Caribbean and Pacific Island Literatures*. O'ahu: University of Hawai'i Press.

DeLoughrey, Elizabeth. 2011. "Yams, Roots, and Rot: Allegories of the Provision Grounds." *Small Axe* 34:58–75.

Derrida, Jacques. 1999. *Monolingualism of the Other: or, The Prosthesis of Origin*. Translated by Patrick Mensah. Stanford, CA: Stanford University Press.

Diawara, Manthia, dir. 2010. *One World in Relation*. K'a Yéléma Productions. Color, 48 mins. Vimeo video, accessed September 21, 2023. https://vimeo.com/299644865.

Diaz, Natalie. 2020. *Postcolonial Love Poem*. Minneapolis, MN: Graywolf Press.

Dorninger, Christian, Alf Hornborg, David J. Abson, Henrik von Wehrden, Anke Schaffartzik, Stefan Giljum, John-Oliver Engler, Robert L. Feller, Klaus Hubacek, and Hanspeter Wieland. 2021. "Global Patterns of Ecologically Unequal Exchange: Implications for Sustainability in the 21st Century." *Ecological Economics* 179 (January), 106824.

Dott, R. 1996. "Lyell in America: His Lectures, Fieldwork and Mutual Influences, 1841–1853." *Earth Sciences History* 15 (2): 101–40.

Douglass, Frederick. 1854. *The Claims of the Negro, Ethnologically Considered: An Address before the Literary Societies of Western Reserve College, at Commencement, July 12, 1854*. Rochester, NY: Press of Lee, Mann.

Douglass, Patrice, and Frank Wilderson. 2013. "The Violence of Presence: Metaphysics in a Blackened World." *Black Scholar* 43 (4): 117–23.

Du Bois, W. E. B. 1891. *Enforcement of the Slave Trade Laws (American Historical Association, Annual Report)*. Washington, DC: Government Printing Office.

Du Bois, W. E. B. 1900. "The American Negro at Paris." *American Monthly Review of Reviews* 22 (5): 575–77.

Du Bois, W. E. B. 1901. "The Spawn of Slavery: The Convict Lease System in the South." *Missionary Review of the World* 14:737–45.

Du Bois, W. E. B. 1903. *The Souls of Black Folk: Essays and Sketches*. Chicago: A. C. McClurg.

Du Bois, W. E. B. 1910. "Reconstruction and Its Benefits." *American Historical Review* 15 (4): 781–99.

Du Bois, W. E. B. 1920. *Darkwater: Voices from Within the Veil*. New York: Harcourt, Brace and Howe.

Du Bois, W. E. B. 2013 [1935]. *Black Reconstruction in America, 1860–1880*. New Brunswick, NJ: Transaction Publishers.

Du Bois, W. E. B. 2019. *BLACK LIVES 1900: W. E. B. Du Bois at the Paris Exposition*. Edited by Jacqueline Francis and Stephen G. Hall. London: Redstone Press.

DuBose, John Witherspoon. 1886. *The Mineral Wealth of Alabama and Birmingham Illustrated*. Birmingham, AL: N. T. Green.

Dutt, Kuheli. 2020. "Race and Racism in the Geosciences." *Nature Geoscience* 13:2–3.
Dyson, J. F. 1886. *A New and Simple Explanation of the Unity of the Human Race and the Origin of Color*. Nashville, TN: Southern Methodist Pub. House.
Edelman, Lee. 2004. *No Future: Queer Theory and the Death Drive*. Durham, NC: Duke University Press.
Elhacham, Emily, Liad Ben-Uri, Jonathan Grozovski, Yinon M. Bar-On, and Ron Milo. 2020. "Global Human-made Mass Exceeds All Living Biomass." *Nature* 588 (December 9): 442–44.
Emerson, Ralph Waldo. 1836. *Nature*. Boston: James Munroe.
Emerson, Ralph Waldo. 1862. "American Civilization." *Atlantic* (April): 502–11.
Emerson, Ralph Waldo. 1904a. *Poems*. Boston: Houghton, Mifflin.
Emerson, Ralph Waldo. 1904b. "II. Civilization." In *The Complete Works of Ralph Waldo Emerson. Vol. 7: Society and Solitude*. Boston: Houghton, Mifflin.
Emerson, Ralph Waldo. 1909. *Nature*. New York: Duffield and Co.
Emerson, Ralph Waldo. 1963. *The Journals and Miscellaneous Notebooks of Ralph Waldo Emerson. Volume III: 1826–1832*, edited by William Gilman and Alfred Ferguson, 393. Entry 204, September 10, 1840. Cambridge, MA: Belknap Press of Harvard University Press.
Emerson, Ralph Waldo. 1969. *The Journals and Miscellaneous Notebooks of Ralph Waldo Emerson, 1838–1842*. Cambridge, MA: Harvard University Press.
Emerson, Ralph Waldo. 1982. *Emerson in His Journals, 1803–1882*. Edited by Joel Porte. Cambridge, MA: Belknap Press of Harvard University Press.
Eshun, Kodwo. 1998. *More Brilliant than the Sun: Adventures in Sonic Fiction*. London: Quartet Books.
Eshun, Kodwo. 2003. "Further Considerations on Afrofuturism." *CR: The New Centennial Review* 3 (2): 287–302.
Estes, Nick. 2019. *Our History Is the Future: Standing Rock versus the Dakota Access Pipeline, and the Long Tradition of Indigenous Resistance*. New York: Verso.
Evans, Chris, and Göran Rydén. 2018. "'Voyage Iron': An Atlantic Slave Trade Currency, Its European Origins, and West African Impact." *Past and Present* 239 (1): 41–70.
Fabian, Ann. 2010. *The Skull Collectors: Race, Science, and America's Unburied Dead*. Chicago: University of Chicago Press.
Fanon, Frantz. 1963. *The Wretched of the Earth*. Translated by Constance Farrington. London: Penguin Books.
Fanon, Frantz. 1986 [1952]. *Black Skin, White Masks*. 3rd ed. Translated by C. L. Markmann. London: Pluto Press.
Featherstonhaugh, George William. 1844. *Excursion through the Slave States: From Washington on the Potomac, to the Frontier of Mexico; with Sketches of Popular Manners and Geological Notices*. New York: Harper and Brothers.
Feimster, Crystal N. 2009. *Southern Horrors: Women and the Politics of Rape and Lynching*. Cambridge, MA: Harvard University Press.

Feldman, Glenn. 2011. "You Know What It Means to Have 9,000 Negroes Idle: Rethinking the Great 1908 Alabama Coal Strike." *Alabama Review* 64 (3): 175–223.

Feldman, Keith P. 2017. "Framed in Black." *PMLA* 132 (1): 156–63.

Ferreira da Silva, Denise. 2001. "Towards a Critique of the Socio-logos of Justice: The Analytics of Raciality and the Production of Universality." *Social Identities* 7 (3): 421–54.

Ferreira da Silva, Denise. 2007. *Toward a Global Idea of Race*. Minneapolis: University of Minnesota Press.

Ferreira da Silva, Denise. 2011. "Notes for a Critique of the 'Metaphysics of Race.'" *Theory, Culture and Society* 28 (1): 138–48. https://doi.org/10.1177/0263276410387625.

Ferreira da Silva, Denise. 2014. "Toward a Black Feminist Poethics: The Quest(ion) of Blackness toward the End of the World." *Black Scholar* 44 (2): 81–97.

Ferreira da Silva, Denise. 2017. "1 (life) ÷ 0 (blackness) = ∞ − ∞ or ∞ / ∞: On Matter beyond the Equation of Value." *eflux* 79 (February 2017). http://www.e-flux.com/journal/79/94686/1-life-0-blackness-or-on-matter-beyond-the-equation-of-value/.

Finney, C. 2014. *Black Faces, White Spaces: Reimagining the Relationship of African Americans to the Great Outdoors*. Chapel Hill: University of North Carolina Press.

Foner, Philip S., and Ronald L. Lewis, eds. 1980. *The Black Worker: A Documentary History from Colonial Times to the Present. Vol. 5: The Black Worker from 1900 to 1919*. Philadelphia: Temple University Press.

Forrest, Kimberly Y. Z., and Wendy O. Stuhldreher. 2011. "Prevalence and Correlates of Vitamin D Deficiency in US Adults." *Nutrition Research* 31 (1): 48–54.

Foucault, Michel. 1977. *History of Sexuality. Vol. 1: An Introduction*. London: Allen Lane.

Foucault, Michel. 1989. *Foucault Live: Collected Interviews, 1961–1984*. New York: Semiotext(e).

Foucault, Michel. 1994 [1966]. *The Order of Things: An Archaeology of the Human Sciences*. New York: Vintage Books.

Foucault, Michel. 2003 [1976]. *Society Must Be Defended: Lectures at the Collège de France, 1975–76*. Edited by Mauro Bertani and Alessandro Fontana. Translated by David Macey. New York: Picador.

Foucault, Michel. 2010. *The Birth of Biopolitics: Lectures at the Collège de France, 1978–1979*. Edited by Michel Senellart. Translated by Graham Burchell. London: Palgrave MacMillan.

Freud, Sigmund. 1922. *Beyond the Pleasure Principle*. London: Hogarth Press.

Frodeman, Robert. 2003. *Geo-Logic: Breaking Ground between Philosophy and the Earth Sciences*. Albany: State University of New York Press.

Fuentes, Marisa J. 2016. *Dispossessed Lives: Enslaved Women, Violence, and the Archive*. Philadelphia: University of Pennsylvania Press.

Furrow, Matthew. 2010. "Samuel Gridley Howe, the Black Population of Canada West, and the Racial Ideology of the 'Blueprint for Radical Reconstruction.'" *Journal of American History* 97 (2): 344–70.

Gabrys, Jennifer. "Becoming Planetary." *e-flux Architecture*, October 2018. https://www.e-flux.com/architecture/accumulation/217051/becoming-planetary.

Gardner, Tia-Simone. 2020. "Chronotopophobias." *Georgia*, July 31, 2020. https://www.georgiageorgia.org/chronotopophobias.

Garst, John. 2002. "Chasing John Henry in Alabama and Mississippi: A Personal Memoir of Work." *Progress Tributaries: Journal of the Alabama Folklife Association* 5:92–129.

Geoffroy Saint-Hilaire, Étienne, and Frédéric Cuvier. 1824. *Histoire naturelle des mammifères avec des figures originales, coloriées, dessinées d'après des animaux vivants.* Paris: A. Belin.

Gilmore, Glenda. 1996. *Gender and Jim Crow: Women and the Politics of White Supremacy in North Carolina, 1896–1920*. Chapel Hill: University of North Carolina Press.

Gilmore, Ruth W. 2002. "Fatal Couplings of Power and Difference: Notes on Racism and Geography." *Professional Geographer* 54:15–24.

Gilmore, Ruth W. 2006. "Race, Prisons and War: Scenes from the History of US Violence." *Socialist Register* 45:73–87.

Gines, Kathryn T. 2014. *Hannah Arendt and the Negro Question.* Bloomington: Indiana University Press.

Glissant, Édouard. 1989 [1981]. *Caribbean Discourse: Selected Essays*. Translated by J. Michael Dash. Charlottesville: University Press of Virginia.

Glissant, Édouard. 1997. *Poetics of Relation*. Translated by Betsy Wing. Ann Arbor: University of Michigan Press.

Glissant, Édouard. 2010 [1969]. *Poetic Intention*. Translated by Nathanaël. New York: Nightboat Books.

Glissant, Édouard. 2011. *The Overseer's Cabin*. Translated by Betsy Wing. Lincoln: University of Nebraska Press.

Glissant, Édouard. 2020. *Introduction to a Poetics of Diversity*. Translated by Celia Britton. Liverpool: Liverpool University Press.

Glissant, Édouard. 2021. *Treatise on the Whole-World*. Translated by Celia Britton. Liverpool: Liverpool University Press.

Gómez-Barris, Macarena. 2017. *The Extractive Zone: Social Ecologies and Decolonial Perspectives*. Durham, NC: Duke University Press.

Gougeon, Len. 1982. "Emerson and Abolition: The Silent Years, 1837–1844." *American Literature* 54 (4): 560–75.

Gould, Stephen J. 1981. *The Mismeasure of Man*. New York: W. W. Norton.

Gregory, J. W. 1928. *Human Migration and the Future: A Study of Causes, Effects and Control of Emigration*. London: Seeley, Service.

Gregory, J. W. 1931. *Race as a Political Factor*. London: Watts.

Green, Lesley. 2020. *Rock, Water, Life.* Durham, NC: Duke University Press.

Guano Islands Act of 1856. 1856. 48 USC Chapter 8: Guano Islands §1411. /uscode.house.gov/view.xhtml?path=/prelim@title48/chapter8&edition=prelim.

Grosz, Elizabeth. 2004. *The Nick of Time: Politics, Evolution, and the Untimely*. Durham, NC: Duke University Press.

Grosz, Elizabeth. 2008. *Chaos, Territory and Art: Deleuze and the Framing of the Earth*. Durham, NC: Duke University Press.

Grosz, Elizabeth. 2011. *Becoming Undone: Darwinian Reflections on Life, Politics, and Art*. Durham, NC: Duke University Press.

Grosz, Elizabeth. 2017. *The Incorporeal: Ontology, Ethics and the Limits of Materialism*. Durham, NC: Duke University Press.

Grosz, Elizabeth, Kathryn Yusoff, and Nigel Clark. 2017. "An Interview with Elizabeth Grosz: Geopower, Inhumanism, and the Biopolitical." *Theory, Culture and Society* 34 (2–3): 129–46.

Guattari, Félix. 2016. *Lines of Flight: For Another World of Possibilities*. London: Bloomsbury.

Gumbs, Alexis Pauline. 2018. *M Archive: After the End of the World*. Durham, NC: Duke University Press.

Guzmán, Patricio, dir. 2010. *Nostalgia for the Light*. New York: Icarus Films.

Hall, Stuart. 2017. *Familiar Stranger: A Life Between Two Islands*. Durham, NC: Duke University Press.

Haraway, Donna. 1984. "Teddy Bear Patriarchy: Taxidermy in the Garden of Eden, New York City, 1908–1936." *Social Text* 11:20–64.

Harris, Will. 2020. *Rendang*. Middletown, CT: Wesleyan University Press.

Harris, Wilson. 2017. *The Ghost of Memory*. London: Faber and Faber.

Hartman, Saidiya V. 1996. "Seduction and the Ruses of Power." *Callaloo* 19 (2): 537–60.

Hartman, Saidiya V. 1997. *Scenes of Subjection: Terror, Slavery, and Self-Making in Nineteenth-Century America*. Oxford: Oxford University Press.

Hartman, Saidiya V. 2002. "The Time of Slavery." *South Atlantic Quarterly* 101 (4): 757–77.

Hartman, Saidiya V. 2007. *Lose Your Mother: A Journey along the Atlantic Slave Route*. New York: Farrar, Straus and Giroux.

Hartman, Saidiya V. 2008. "Venus in Two Acts." *Small Axe* 26:1–14.

Hartman, Saidiya. 2016. "The Belly of the World: A Note on Black Women's Labors." *Souls* 18 (1): 166–73.

Hartman, Saidiya. 2018. "On Working with Archives: An Interview with Writer Saidiya Hartman." Creative Independent. Accessed August 18, 2023. https://thecreativeindependent.com/people/saidiya-hartman-on-working-with-archives/.

Hartman, Saidiya V. 2019. *Wayward Lives, Beautiful Experiments: Intimate Histories of Riotous Black Girls, Troublesome Women, and Queer Radicals*. New York: W. W. Norton.

Hartman, Saidiya V., and Frank B. Wilderson III. 2003. "The Position of the Unthought." *Qui Parle* 13 (2): 183–201.

Hartt, Charles Frederick. 1870. *Thayer Expedition: Scientific Results of a Journey in Brazil. by Louis Agassiz and His Travelling Companions. Geology and Physical Geography of Brazil*. Boston: Fields, Osgood.

Haymes, Stephen Nathan. 2018. "An Africana Studies Critique of Environmental Ethics." In *Racial Ecologies*, edited by Leilani Nishime and Kim D. Hester Williams, 34–49. Seattle: University of Washington Press.

Heringman, N. 2004. *Romantic Rocks, Aesthetic Geology*. Ithaca, NY: Cornell University Press.

Higgs, Robert. 1982. "Accumulation of Property by Southern Blacks before World War I." *American Economic Review* 72:725–37.

Hird, Myra, and Kathryn Yusoff. 2019. "Lines of Shite: Microbial-Mineral Chatter in the Anthropocene." In *Posthuman Ecologies: Complexity and Process after Deleuze*, edited by Rosi Braidotti and Simone Bignall, 265–82. New York: Rowman and Littlefield.

Hitchcock, Edward. 1840. *Elementary Geology*. Amherst, MA: J. S. and C. Adams.

Holt, D. 2015. "Heat in US Prisons and Jails: Corrections and the Challenge of Climate Change." Sabin Center for Climate Change Law, Columbia Law School, New York, NY, August 2015. https://www.icrrl.org/wp-content/blogs.dir/102/files/2021/06/Holt-2015-08-Heat-in-US-Prisons-and-Jails.pdf.

Howe, Samuel Gridley. 1863. Letter to [Louis] Agassiz. Portsmouth, 3 Aug 1863. Louis Agassiz correspondence and other papers, MS Am 1419 (415 and 416), Box 3. Houghton Library, Harvard University.

Hoxie, Frederick. 2007. "What Was Taney Thinking? American Indian Citizenship in the Era of *Dred Scott*." *Chicago-Kent Law Review* 82:329–59.

Hudson, Darrell L., Harold W. Neighbors, Arline T. Geronimus, and James S. Jackson. 2016. "Racial Discrimination, John Henryism, and Depression among African Americans." *Journal of Black Psychology* 42 (3): 221–43.

Hunter, T. 1997. *To 'Joy My Freedom: Southern Black Women's Lives and Labors after the Civil War*. Cambridge, MA: Harvard University Press.

Hutchinson, Henry N., John W. Gregory, and Richard Lydekker. 1898. *The Living Races of Mankind: A Popular Illustrated Account of the Customs, Habits, Pursuits, Feasts and Ceremonies of the Races of Mankind throughout the World*. Vol. 1. London: Hutchinson.

Hutchinson, Henry N., John W. Gregory, and Richard Lydekker. 1901. *The Living Races of Mankind: A Popular Illustrated Account of the Customs, Habits, Pursuits, Feasts and Ceremonies of the Races of Mankind throughout the World*. Vol. 2. London: Hutchinson.

Hutton, James. 1788. "Theory of the Earth." *Transactions of the Royal Society of Edinburgh* 1:209–304.

Ife, Fahima. 2021. *Maroon Choreography*. Durham, NC: Duke University Press.

Isaac, Gwyniera. 1997. "Louis Agassiz's Photographs in Brazil: Separate Creations." *History of Photography* 21 (1): 3–11.

Jackson, Shona N. 2012. *Creole Indigeneity: Between Myth and Nation in the Caribbean.* Minneapolis: University of Minnesota Press.

Jackson, Zakiyyah Iman. 2018. "'Theorizing in a Void': Sublimity, Matter, and Physics in Black Feminist Poetics." *South Atlantic Quarterly* 117 (3): 617–48.

Jamaica Almanac. 1811. Jamaican Family Search. Accessed January 20, 2024. http://www.jamaicanfamilysearch.com/Members/AL11Eliz.htm.

James, C. L. R. 1938a. *Black Jacobins*. London: Secker and Warburg.

James, C. L. R. 1938b. "British Barbarism in Jamaica: Support the Negro Worker's Struggle." *Fight* 1 (3): 1, 4.

James, Joy. 2013. *Seeking the Beloved Community: A Feminist Race Reader*. Albany: State University of New York Press.

James, S. A., S. A. Hartnett, and W. D. Kalsbeek. 1983. "John Henryism and Blood Pressure Differences among Black Men." *Journal of Behavioral Medicine* 6 (3): 259–78.

James, William. 1865. "Brazilian Expedition Diary and Sketchbook," November 10, 1865. William James papers, 1803–1941 and 1862–1910, MS Am 1092.9–1092.12. Houghton Library, Harvard University, Cambridge, MA.

James, William. 2006. *Brazil through the Eyes of William James: Letters, Diaries, and Drawings, 1865–1866* [*O Brasil no olhar de William James: Cartas, dírios e desenhos, 1865–1866*]. Edited by Maria Helena P. T. Machado. Translated by John M. Monteiro. Cambridge, MA: Harvard University Press.

Jemisin, N. K. 2019. *Broken Earth Trilogy*. London: Orbit.

Jenkins, Barry, dir. 2021. *The Underground Railroad*. Amazon Prime Video limited series. First aired May 14, 2021.

Johnson, Walter. 2017. "To Remake the World: Slavery, Racial Capitalism, and Justice." *Boston Review*, February 1, 2017. https://bostonreview.net/forum/walter-johnson-to-remake-the-world.

Kant, Immanuel. 1994 [1756]. "History and Physiography of the Most Remarkable Cases of the Earthquake which Towards the End of the Year 1755 Shook a Great Part of the Earth." In *Four Neglected Essays*, edited by Stephen R. Palmquist, n.p. Hong Kong: Philopsychy Press. http://staffweb.hkbu.edu.hk/ppp/fne/essay1.html.

Kant, Immanuel. 2009 [1755]. *Universal Natural History and Theory of the Heavens*. Translated by Ian Johnston. Arlington, VA: Richer Resources Publications.

Kato, Daniel. 2015. *Liberalized Lynching: Building a New Racialized State*. Oxford: Oxford University Press.

Keeler, Clarissa Olds. 1907. *The Crime of Crimes; or, The Convict System Unmasked*. African American Pamphlet Collection. Washington, DC: Pentecostal Era Company. https://www.loc.gov/item/07026922/.

Keeling, Kara. 2009a. *Queer Times, Black Futures*. New York: New York University Press.

Keeling, Kara. 2009b. "LOOKING FOR M—: Queer Temporality, Black Political Possibility, and Poetry from the Future." *GLQ* 15 (4): 565–82.

Kelley, Robin D. G. 1990. *Hammer and Hoe: Alabama Communists during the Great Depression*. Chapel Hill: University of North Carolina Press.

Kelley, Robin D. G. 2002. *Freedom Dreams: The Black Radical Imagination*. Boston: Beacon Press.

Kelly, Brian. 2001. *Race, Class and Power in the Alabama Coalfields, 1908–1921*. Urbana: University of Illinois Press.

King, Tiffany Lethabo. 2016a. "The Labor of (Re)reading Plantation Landscapes Fungible(ly)." *Antipode* 48:1022–39.

King, Tiffany Lethabo. 2016b. "New World Grammars: The 'Unthought' Black Discourses of Conquest." *Theory and Event* 19 (4): n.p. https://muse.jhu.edu/pub/1/article/633275.

King, Tiffany Lethabo. 2019. *Black Shoals: Offshore Formations of Blackness and Native Studies*. Durham, NC: Duke University Press.

Koch, Alexander, Chris Brierley, Mark M. Maslin, and Simon L. Lewis. 2019. "Earth System Impacts of the European Arrival and Great Dying in the Americas after 1492." *Quaternary Science Reviews* 207:13–36.

Kulik, Gary. 1981. "Black Workers and Technological Change in the Birmingham Iron Industry, 1881–1931." In *Southern Workers and Their Unions: Selected Papers, the Second Southern Labor History Conference 1977*, edited by G. Fink and M. E. Reed, 28–31. Westport, CT: Greenwood Press.

Lamarck, Jean Baptiste Pierre Antoine de Monet de. 1809. *Philosophie zoologique; ou, Exposition des considérations relatives à l'histoire naturelle des animaux*. Paris: n.p.

Lamarck, Jean Baptiste Pierre Antoine de Monet de. 1984 [1744–1829]. *Zoological Philosophy: An Exposition with Regard to the Natural History of Animals*. Chicago: University of Chicago Press.

Larsen, Svend Erik. 2006. "The Lisbon Earthquake and the Scientific Turn in Kant's Philosophy." *European Review* 14 (3): 359–67.

Last, Angela. 2015. "Fruit of the Cyclone: Undoing Geopolitics through Geopoetics." *Geoforum* 64 (August): 56–64.

Last, Angela. 2017. "We Are the World? Anthropocene Cultural Production between Geopoetics and Geopolitics." *Theory, Culture and Society* 34 (2–3): 147–68.

Last, Angela. 2018. "Open Space to Risk the Earth: The Nonhuman and Nonhistory." *Feminist Review* 118 (1): 87–92.

Lauden, Rachel. 1987. *From Mineralogy to Geology: The Foundations of a Science, 1650–1830*. Chicago: University of Chicago Press.

LeFlouria, Talitha. 2011. "'The Hand that Rocks the Cradle Cuts Cordwood': Exploring Black Women's Lives and Labor in Georgia's Convict Camps, 1865–1917." *Labor: Studies in Working-Class History of the Americas* 8 (3): 47–63.

LeFlouria, Talitha. 2015. *Chained in Silence: Black Women and Convict Labor in the New South*. Chapel Hill: University of North Carolina Press.

Letwin, Daniel L. 1998. *The Challenge of Interracial Unionism: Alabama Coal Miners, 1878–1921*. Chapel Hill: University of North Carolina Press.

Lewis, Cherry, and Simon Knell, eds. 2009. *The Making of the Geological Society of London*. London: Geological Society of London.

Lewis, Jovan Scott. 2020. *Scammer's Yard: The Crime of Black Repair in Jamaica*. Minneapolis: University of Minnesota Press.

Lewis, Ronald L. 1987. *Black Coal Miners in America: Race, Class and Community Conflict, 1780–1980*. Lexington: University Press of Kentucky.

Lichtenstein, Alex. 1993. "Black Labor and Technological Change at a National Historic Landmark: Sloss Furnaces, Birmingham, Alabama." *Radical History Review* 56:119–26.

Lichtenstein, Alex. 1996. *Twice the Work of Free Labor: The Political Economy of Convict Labor in the New South*. London: Verso.

Lorde, Audre. 1997. *Coal*. New York: W. W. Norton.

Lorde, Audre. 2007. "Poetry Is Not a Luxury." In *Sister Outsider: Essays and Speeches*, 36–44. Berkeley, CA: Crossing Press.

Louisiana. 1825. Civil Code of the State of Louisiana. Paris: Impr. de E. Duverger. https://babel.hathitrust.org/cgi/pt?id=hvd.32044004462917&seq=185&q1=537.

Lowe, Lisa. 2015. *The Intimacies of Four Continents*. Durham, NC: Duke University Press.

Lubrin, Canisia. 2017. *Voodoo Hypothesis: Poems*. Ontario: Wolsak and Wynn.

Lyell, Charles. 1830. *Principles of Geology: Being an Attempt to Explain the Former Changes of the Earth's Surface, by Reference to Causes Now in Operation*. Vol. 1. London: John Murray.

Lyell, Charles. 1845. *Travels in North America, in the Years 1841–2; with Geological Observations on the United States, Canada, and Nova Scotia*. New York: Wiley and Putnam.

Lyell, Charles. 1863. *The Geological Evidences of the Antiquity of Man with Remarks on the Origin of Species by Variation*. 3rd rev. ed. London: John Murray.

Lyell, Charles. 1997 [1830]. *Principles of Geology*. Edited by James A. Second. London: Penguin Books.

Malick, Terrence, dir. 2011. *The Tree of Life*. United States. 139 mins. Los Angeles and Minneapolis, MN: River Road Entertainment / Plan B Entertainment.

Malick, Terrence, dir. 2016. *Voyage of Time: Life's Journey*. Los Angeles: Broad Green Pictures / IMAX Corporation.

Mancini, Matthew J. 1978. "Race, Economics, and the Abandonment of Convict Leasing." *Journal of Negro History* 63 (4): 339–52.

Mancini, Matthew J. 1996. *One Dies, Get Another: Convict Leasing in the American South, 1866–1928*. Columbia: University of South Carolina Press.

Marcou, Jules. 1895. *Life, Letters, and Works of Louis Agassiz*. New York: Macmillan. https://hdl.handle.net/2027/pst.000029044597.

Mardorossian, Carine M. 2013. "'Poetics of Landscape': Édouard Glissant's Creolized Ecologies." *Callaloo* 36 (4): 983–94.

Marriott, David. 2018. *Whither Fanon? Studies in the Blackness of Being*. Stanford, CA: Stanford University Press.
Marx, Karl. 1976 [1867]. *Capital: A Critique of Political Economy*. Vol. 1. Translated by David Fernbach. London: Penguin Books in association with *New Left Review*.
Marx, Karl, and Friedrich Engels. 2012 [1888]. *The Communist Manifesto: A Modern Edition*. London: Verso.
Mbembe, Achille. 2017. *Critique of Black Reason*. Translated by Laurent Dubois. Durham, NC: Duke University Press.
McCalley, Henry. 1886. "The Mountain, Manufacturing and Mineral Region of Alabama." In DuBose, *The Mineral Wealth of Alabama and Birmingham Illustrated*, 17–42.
McKittrick, Katherine. 2006. *Demonic Grounds: Black Women and the Cartographies of Struggle*. Minneapolis: University of Minnesota Press.
McKittrick, Katherine. 2010. "Science Quarrels Sculpture: The Politics of Reading Sarah Baartman." *Mosaic: An Interdisciplinary Critical Journal* 43 (2): 113–30.
McKittrick, Katherine. 2013. "Plantation Futures." *Small Axe* 17:1–15.
McKittrick, Katherine, ed. 2015. *Sylvia Wynter: On Being Human as Praxis*. Durham, NC: Duke University Press.
McKittrick, Katherine. 2019. "Rift." In *Keywords in Radical Geography:* Antipode *at 50*, edited by *Antipode* Editorial Collective, 243–47. Hoboken, NJ: Wiley Blackwell. https://onlinelibrary.wiley.com/doi/epdf/10.1002/9781119558071.
McKittrick, Katherine. 2021. *Dear Science and Other Stories*. Durham, NC: Duke University Press.
McKittrick, Katherine, and Clyde Woods, eds. 2007. *Geographies and the Politics of Place*. Cambridge, MA: South End Press.
Meillassoux, Quentin. 2008. *After Finitude: An Essay on the Necessity of Contingency*. London: Continuum.
Meillassoux, Quentin. 2013. *Time without Becoming*. Edited by Anna Longo. Milan: Mimesis International.
Menand, Louis. 2001. *The Metaphysical Club*. London: Flamingo.
Merriam-Webster Dictionary. n.d. "Ore"noun (1)." Accessed January 18, 2024. https://www.merriam-webster.com/dictionary/ore.
Mikkelsen, Jon M., ed. 2013. *Kant and the Concept of Race: Late Eighteenth-Century Writings*. New York: State University of New York Press.
Mills, Catherine. 2010. *Regulating Health and Safety in the British Mining Industries, 1800–1914*. Burlington, VT: Ashgate.
Milner, John T. 1890. *White Men of Alabama—Stand Together, 1860 and 1890*. Box 96, Item 16, Alabama Department of Archives and History, Montgomery, AL.
Mintz, Sidney. 1985. *Sweetness and Power: The Place of Sugar in Modern History*. New York: Viking-Penguin.
Morgan, Jennifer L. 2004. *Laboring Women: Reproduction and Gender in New World Slavery*. Philadelphia: University of Pennsylvania Press.

Morrison, Toni. 1973. *Sula*. New York: Knopf.
Morrison, Toni. 1997. *Song of Solomon*. New York: Knopf.
Morrison, Toni. 1999 [1970]. *The Bluest Eye*. London: Vintage.
Morrison, Toni. 2007 [1987]. *Beloved*. London: Vintage.
Morrison, Toni. 2019. *The Source of Self-Regard*. New York: Knopf.
Morton, Samuel. 1847. "Hybridity in Animals, Considered in Reference to the Question of the Unity of the Human Species." *American Journal of Science and Arts*, 2nd ser., 3 (3): 39–50.
Morton, Samuel. 1849. *Catalogue of Skulls of Man and the Inferior Animals*. Philadelphia: Merrihew and Thompson.
Moten, Fred. 2003. *In the Break: The Aesthetics of the Black Radical Tradition*. Minneapolis: University of Minnesota Press.
Moten, Fred. 2007. "Uplift and Criminality." In *Next to the Color Line: Gender, Sexuality, and W. E. B. Du Bois*, edited by Alys Weinbaum and Susan Gilman, 317–49. Minneapolis: University of Minnesota Press.
Moten, Fred. 2008. "The Case of Blackness." *Criticism* 50 (2): 177–218.
Moten, Fred. 2016. *A Poetics of the Undercommons*. New York: Sputnik and Fizzle.
Moten, Fred. 2017. *Black and Blur*. Durham, NC: Duke University Press.
Moten, Fred. 2018. *Stolen Life*. Durham, NC: Duke University Press.
Mukherjee, Jenia, and Pritwinath Ghosh. 2020. "Fluid Epistemologies: The Social Saga of Sediments in Bengal." *Ecology, Economy and Society: The INSEE Journal* 3 (2): 135–48.
Muller, Christopher. 2018. "Freedom and Convict Leasing in the Postbellum South." *American Journal of Sociology* 142 (2): 357–405.
Nelson, Scott R. 2006. *Steel Drivin' Man: John Henry, the Untold Story of an American Legend*. Oxford: Oxford University Press.
Newfield, Christopher. 1996. *The Emerson Effect: Individualism and Submission in America*. Chicago: University of Chicago Press.
Norrell, R. 1986. "Caste in Steel: Jim Crow Careers in Birmingham, Alabama." *Journal of American History* 73 (3): 669–94.
Nott, Josiah C., and George R. Gliddon, with additional contributions from L. Agassiz, W. Usher, S. Morton, and H. S. Patterson. 1854. *Types of Mankind: Or, Ethnological Researches Based upon the Ancient Monuments, Paintings, Sculptures, and Crania of Races, and upon Their Natural, Geographical, Philological and Biblical History*. 6th ed. Philadelphia: J. B. Lippincott, Grambo.
Noxolo, Pat. 2016. "Locating Caribbean Studies in Unending Conversation." *Environment and Planning D: Society and Space* 34 (5): 830–35.
Nyong'o, Tavia. 2009. *The Amalgamation Waltz: Race, Performance and the Ruses of Memory*. Minneapolis: University of Minnesota Press.
Oshinsky, David M. 1996. *"Worse Than Slavery": Parchman Farm and the Ordeal of Jim Crow Justice*. New York: Free Press.
Owens, Patricia. 2017. "Racism in the Theory Canon: Hannah Arendt and 'the One Great Crime in Which America Was Never Involved.'" *Millennium* 45 (3): 403–24.

Parke, T. D. 1895. Report on Coalburg Prison (Birmingham). Thomas Duke Parke Papers, Acc. 76-15, Birmingham Public Library.

Patell, Cyrus R. K. 2001. *Negative Liberties: Morrison, Pynchon, and the Problem of Liberal Ideology*. Durham, NC: Duke University Press.

Patterson, Orlando. 1982. *Slavery and Social Death: A Comparative Study*. Cambridge, MA: Harvard University Press.

Pellow, David. 2021. "Struggles for Environmental Justice in US Prisons and Jails." *Antipode* 53:56–73.

Penal Code of Alabama; prepared by John Wesley Shepherd, and George Washington Stone and Adopted by the General Assembly at the Session of 1865–6, Together with the Other Criminal Laws Now in Force. 1866. Montgomery, AL: Reid and Screws, State Printers.

Philip, M. NourbeSe. 1994. "Dis Place—the Space Between." In *Feminist Measures: Soundings in Poetry Theory*, edited by Lynn Miller and Christianne Keller, 287–316. Ann Arbor: University of Michigan Press.

Philip, M. NourbeSe. 2017. *Black: Essays and Interviews*. Toronto: BookThug.

Povinelli, Elizabeth. 1995. "Do Rocks Listen? The Cultural Politics of Apprehending Australian Aboriginal Labor." *American Anthropologist* 97 (3): 505–18.

Povinelli, Elizabeth. 2006. *The Empire of Love: Toward a Theory of Intimacy, Genealogy, and Carnality*. Durham, NC: Duke University Press.

Povinelli, Elizabeth. 2011. *Economies of Abandonment: Social Belonging and Endurance in Late Liberalism*. Durham, NC: Duke University Press.

Povinelli, Elizabeth. 2016. *Geontologies: A Requiem to Late Liberalism*. Durham, NC: Duke University Press.

Povinelli, Elizabeth. 2017a. "An Interview with Elizabeth Povinelli: Geontopower, Biopolitics and the Anthropocene." Interview by Mathew Coleman and Kathryn Yusoff. *Theory, Culture and Society* 34 (2–3): 169–85.

Povinelli, Elizabeth. 2017b. "Bleak House." *Social Text 130* 35 (1): 131–36.

Quammen, David. 2018. *The Tangled Tree: A Radical New History of Life*. New York: Simon and Schuster.

Qureshi, Sadiah. 2004. "Displaying Sara Baartman, the 'Hottentot Venus.'" *History of Science* 42 (2): 233–57.

Raine, Anne. 2013. "Du Bois's Ambient Poetics: Rethinking Environmental Imagination in *The Souls of Black Folk*." *Callaloo* 36 (2): 322–41.

Rancière, Jacques. 2004. *The Politics of Aesthetics: The Distribution of the Sensible.* Translated by Gabriel Rockhill. London: Continuum.

Rankine, Claudia. 2014. *Citizen: An American Lyric.* Minneapolis, MN: Greywolf Press.

Reed, Roland, dir. 1938. *Steel: Man's Servant*. United States. Multiple cuts. United States Steel Corporation and Roland Reed Productions.

Reinhardt, O., and D. R. Oldroyd. 1982. "Kant's Thoughts on the Ageing of the Earth." *Annals of Science* 39 (4): 349–69.

Reinhardt, O., and D. R. Oldroyd. 1983. "Kant's Theory of Earthquakes and Volcanic Action." *Annals of Science* 40 (3): 247–72.

Reyes-Carranza, Mariana. 2021. "Racial Geographies of the Anthropocene: Memory and Erasure in Rio de Janeiro." *Politics* 43 (2): 250–66.

Rifkin, Mark. 2017. *Beyond Settler Time: Temporal Sovereignty and Indigenous Self-Determination*. Durham, NC: Duke University Press.

Roane, J. T. 2018. "Plotting the Black Commons." *Souls* 20 (3): 239–66.

Roane, J. T. 2020. "Tornado Groan: On Black (Blues) Ecologies." *Black Perspectives*, March 16, 2020. https://www.aaihs.org/tornado-groan-on-black-blues-ecologies/.

Roane, J. T. 2022. "Black Ecologies, Subaquatic Life, and the Jim Crow Enclosure of the Tidewater." *Journal of Rural Studies* 94:227–38.

Robinson, Cedric. 2000 [1983]. *Black Marxism: The Making of the Black Radical Tradition*. Chapel Hill: University of North Carolina Press.

Robinson, Cedric. 2007. *Forgeries of Memory and Meaning: Blacks and the Regimes of Race in American Theater and Film before World War II*. Chapel Hill: University of North Carolina Press.

Ross, RaMell, dir. 2018. *Hale County This Morning, This Evening*. United States. 76 mins.

Rudwick, Martin J. 1998. *Georges Cuvier, Fossil Bones, and Geological Catastrophes*. Chicago: University of Chicago Press.

Rudwick, Martin J. 2005. *Bursting the Limits of Time: The Reconstruction of Geohistory in the Age of Revolution*. Chicago: University of Chicago Press.

Rudwick, Martin J. 2008. *Worlds before Adam: The Reconstruction of Geohistory in the Age of Reform*. Chicago: University of Chicago Press.

Rusert, Britt. 2017. *Fugitive Science: Empiricism and Freedom in Early African American Culture*. New York: New York University Press.

Said, Edward W. 1994. *Culture and Imperialism*. New York: Vintage.

Saldanha, Arun. 2006. "Reontologising Race: The Machinic Geography of Phenotype." *Environment and Planning D: Society and Space* 24 (1): 9–24.

Saldanha, Arun. 2017. *Space after Deleuze*. London: Bloomsbury Academic.

Sandford, Stella. 2018. "Kant, Race, and Natural History." *Philosophy and Social Criticism* 44 (9): 950–77.

Schneider, Susan. 2012. "Louis Agassiz and the American School of Ethnoeroticism: Polygenesis, Pornography, and Other 'Perfidious Influences.'" In Wallace and Smith, *Pictures and Progress*, 211–43.

Scott, James C. 1992. *Domination and the Arts of Resistance: Hidden Transcripts*. New Haven, CT: Yale University Press.

Scott, James C. 2009. *The Art of Not Being Governed: An Anarchist History of Upland Southeast Asia*. New Haven, CT: Yale University Press.

Serres, Michel. 1995. *The Natural Contract*. Translated by Elizabeth MacArthur and William Paulson. Ann Arbor: University of Michigan Press.

Sexton, Jared. 2010. "People-of-Color-Blindness: Notes on the Afterlife of Slavery." *Social Text* 28 (2): 31–56.

Sexton, Jared. 2011. "The Social Life of Social Death: On Afro-Pessimism and Black Optimism." *Intensions* 5:1–47.

Sharma, Nandita. 2015. "Strategic Anti-essentialism: Decolonizing Decolonization." In *Sylvia Wynter: On Being Human as Praxis*, edited by Katherine McKittrick, 164–82. Durham, NC: Duke University Press.

Sharpe, Christina. 2010. *Monstrous Intimacies: Making Post-Slavery Subjects*. Durham, NC: Duke University Press.

Sharpe, Christina. 2012. "Blackness, Sexuality, and Entertainment." *American Literary History* 24 (4): 827–41. https://doi.org/10.1093/alh/ajs046.

Sharpe, Christina. 2014. "The Lie at the Center of Everything." *Black Studies Papers* 1 (1): 189–214.

Sharpe, Christina. 2016a. *In the Wake: On Blackness and Being*. Durham, NC: Duke University Press.

Sharpe, Christina. 2016b. "Lose Your Kin." *New Inquiry*, November 16, 2016. https://thenewinquiry.com/lose-your-kin/.

Sharpe, Christina. 2017. "'What Does It Mean to Be Black and Look at This?': A Scholar Reflects on the Dana Schutz Controversy." Interview by Siddhartha Mitter. *Hyperallergic*, March 17, 2017. https://hyperallergic.com/368012/what-does-it-mean-to-be-black-and-look-at-this-a-scholar-reflects-on-the-dana-schutz-controversy/.

Sharpe, Christina. 2019. "Beauty Is a Method." *e-flux* 105 (December). https://www.e-flux.com/journal/105/303916/beauty-is-a-method/.

Simpson, Audra. 2014. *Mohawk Interruptus: Political Life across the Borders of Settler States*. Durham, NC: Duke University Press.

Simpson, Leanne Betasamosake. 2014. "Land as Pedagogy: Nishnaabeg Intelligence and Rebellious Transformation." *Decolonization: Indigeneity, Education and Society* 3 (3): 1–25.

Simpson, Leanne Betasamosake. 2017. *As We Have Always Done: Indigenous Freedom through Radical Resistance.* Minneapolis: University of Minnesota Press.

Simpson, Leanne Betasamosake. 2021. *A Short History of the Blockade: Giant Beavers, Diplomacy, and Regeneration in Nishnaabewin*. Edmonton: University of Alberta Press.

Singh, Julietta. 2017. *Unthinking Mastery: Dehumanism and Decolonial Entanglements*. Durham, NC: Duke University Press.

Smallwood, Stephanie E. 2007. *Saltwater Slavery: A Middle Passage from Africa to American Diaspora.* Boston: Harvard University Press.

Smith, Danez. 2015. *Black Movie*. Minneapolis, MN: Button Poetry / Exploding Pinecone Press.

Smith, John Pye. 1847. *Introduction to Elementary Geology*. 8th ed. New York: M. H. Newman.

Smith, Shawn M. 2004. *Photography on the Color Line: W. E. B. Du Bois, Race, and Visual Culture*. Durham, NC: Duke University Press.

Smith, Shawn Michelle. 2012. "'Looking at One's Self through the Eyes of Others': W. E. B. Du Bois's Photographs for the Paris Exposition of 1900." In Wallace and Smith, *Pictures and Progress*, 274–98.

Smith, Tracy K. 2021. "I Sit Outside in Low Late-Afternoon Light to Feel Earth Call to Me." In *Such Color: New and Selected Poems*, 206. Minneapolis, MN: Greywolf Press.

Solomon, Marissa. 2019. "'The Ghetto Is a Gold Mine': The Racialized Temporality of Betterment." *International Labor and Working-Class History* 95:76–94.

Solomon, Rivers. 2017. *An Unkindness of Ghosts*. Brooklyn, NY: Akashic Books.

Spillers, Hortense J. 1987. "Mama's Baby, Papa's Maybe: An American Grammar Book." *Diacritics* 17 (2): 65–81.

Spillers, Hortense. 2003. *Black, White and in Color: Essays on American Literature and Culture*. Chicago: University of Chicago Press.

State of Alabama. 1891. *Journal of the Senate, Session of 1890–91*. Montgomery, AL: Smith, Allread and Co. State Printers.

Stoler, Ann Laura. 2013. *Imperial Debris: On Ruins and Ruination*. Durham, NC: Duke University Press.

Summers, Brandi T. 2019. *Black in Place: The Spatial Aesthetics of Race in a Post-Chocolate City*. Chapel Hill: University of North Carolina Press.

TallBear, Kim. 2013. *Native American: Tribal Belonging and the False Promise of Genetic Science*. Minneapolis: University of Minnesota Press.

Tamara Lanier v. President and Fellows of Harvard College, aka Harvard Corporation, Harvard Board of Overseers, Harvard University, the Peabody Museum of Archaeology and Ethnology. 2019. https://www.courthousenews.com/wp-content/uploads/2019/03/harvard-photos.pdf.

Tamara Lanier v. President and Fellows of Harvard College and Others. 2022. https://law.justia.com/cases/massachusetts/supreme-court/2022/sjc-13138.html.

Taney, Roger Brooke, and Supreme Court of the United States. 1857. "Judgment in the U.S. Supreme Court Case Dred Scott v. John F. A. Sandford; 3/6/1857; Dred Scott, Plaintiff in Error, v. John F. A. Sandford." Appellate Jurisdiction Case Files, 1792–2010, Records of the Supreme Court of the United States, Record Group 267; National Archives, Washington, DC.

Tettenborn, Éva. 2013. "'A Mountain Full of Ghosts': Mourning African American Masculinities in Colson Whitehead's *John Henry Days*." *African American Review* 46 (2–3): 271–84.

Thomas, Deborah. 2019. *Political Life in the Wake of the Plantation: Sovereignty, Witnessing, Repair*. Durham, NC: Duke University Press.

Thomas, Mary, and Kathryn Yusoff. 2017. "Geology." In *Gender: Matter*, edited by Stacy Alaimo, 123–37. Farmington Hills, MI: Macmillan.

Thomas, Mary, and Kathryn Yusoff. 2018. "Inhumanities." Paper presented at Deterritorialising the Future: A Symposium on Heritage in, of and after

the Anthropocene, University College London, August 10, 2020. https://mediacentral.ucl.ac.uk/Player/40828609.
Trouillot, Michel-Rolph. 1995. *Silencing the Past: Power and the Production of History*. Boston: Beacon Books.
United States Supreme Court, R. B. Taney, J. H. Van Evrie, and S. A. Cartwright. 1860. *The Dred Scott Decision: Opinion of Chief Justice Taney*. New York: Van Evrie, Horton. https://www.loc.gov/item/17001543/.
Van Evrie, John H. 1863. *Negroes and Negro "Slavery": The First an Inferior Race—The Latter Its Normal Condition*. New York: Van Evrie, Horton.
Van Evrie, J. H. 1866. Introduction to *The Negro's Place in Nature: A Paper Read before the London Anthropological Society*, by James Hunt, 3–4. New York: Van Evrie, Horton. https://www.loc.gov/item/12002987/.
Vermeulen, Heather V. 2018. "Thomas Thistlewood's Libidinal Linnaean Project." *Small Axe* 22:18–38.
Vuong, Ocean. 2019. *On Earth We're Briefly Gorgeous*. London: Jonathan Cape.
Walcott, Rinaldo. 2021. *The Long Emancipation: Moving toward Black Freedom*. Durham, NC: Duke University Press.
Wallace, Maurice O., and Shawn Michelle Smith, eds. 2012. *Pictures and Progress: Early Photography and the Making of African American Identity*. Durham, NC: Duke University Press.
Ward, Robert David, and William Warren Rogers. 1987. *Convicts, Coal and the Banner Mine Tragedy*. Tuscaloosa: University of Alabama Press.
Wardi, Anissa Janine. 2016. "August Wilson's Bioregional Perspective." *Callaloo* 39 (3): 680–94.
Warren, Calvin. 2018. *Ontological Terror: Blackness, Nihilism, and Emancipation*. Durham, NC: Duke University Press.
Waterhouse, Benjamin. 1805. "Lecture: The contents of the earth" (June 16). Benjamin Waterhouse papers, H MS c16, Box 02, Folder 41. Center for the History of Medicine (Francis A. Countway Library of Medicine), Harvard University, Boston, MA. https://id.lib.harvard.edu/ead/c/med00213c00091/catalog (accessed January 18, 2024).
Waterhouse, Benjamin. n.d. "Public Lecture on the Earth." Benjamin Waterhouse papers, H MS Center for the History of Medicine (Francis A. Countway Library of Medicine), Harvard University, Boston, MA. https://id.lib.harvard.edu/ead/c/med00213c00159/catalog (accessed January 18, 2024).
Weheliye, Alexander. 2002. "'Feenin': Posthuman Voices in Contemporary Black Popular Music." *Social Text* 71 (20, no. 2): 21–47.
Weheliye, Alexander. 2014. *Habeas Viscus: Racializing Assemblages, Biopolitics, and Black Feminist Theories of the Human*. Durham, NC: Duke University Press.
Weheliye, Alexander G. 2019. "Black Life/Schwarz-Sein: Inhabitations of the Flesh." In *Beyond the Doctrine of Man: Decolonial Visions of the Human*, edited by Joseph Drexler-Dreis and Kristien Justaert, 227–62. New York: Fordham University Press.

Weinbaum, Alys Eve. 2013. "Gendering the General Strike: W. E. B. Du Bois's *Black Reconstruction* and Black Feminism's 'Propaganda of History.'" *South Atlantic Quarterly* 112 (3): 437–63.

Wells, Ida B. 1892. *Southern Horrors: Lynch Law in All Its Phases*. New York: New York Age Print.

Whitehead, Colson. 2001. *John Henry Days*. New York: Anchor.

Whitehead, Colson. 2016. *The Underground Railroad*. London: Fleet.

Whyte, Kyle P. 2018. "Indigenous Science (Fiction) for the Anthropocene: Ancestral Dystopias and Fantasies of Climate Change Crises." *Environment and Planning E: Nature and Space* 1 (1–2): 224–42.

Wilderson, Frank B., III. 2003. "The Prison Slave as Hegemony's (Silent) Scandal." *Social Justice* 30 (2): 18–27.

Wilderson, Frank B., III. 2010. *Red, White and Black: Cinema and the Structure of U.S. Antagonisms*. Durham, NC: Duke University Press.

Wilderson, Frank B., III. 2020. *Afropessimism*. New York: Liveright.

Williams, Eric. 1944. *Capitalism and Slavery*. Chapel Hill: University of North Carolina Press.

Wilson, Bobby M. 2000. *Race and Place in Birmingham: The Civil Rights and Neighborhood Movements*. Lanham, MD: Rowman and Littlefield.

Wolfe, Patrick. 2006. "Settler Colonialism and the Elimination of the Native." *Journal of Genocide Research* 8 (4): 387–409.

Woods, Clyde A. 1998. *Development Arrested: The Blues and Plantation Power in the Mississippi Delta*. New York: Verso.

Woodward, C. Vann. 1981. *Origins of the New South, 1877–1913: A History of the South*. Baton Rouge: Louisiana State University Press.

Woodward, Horace. 1907. *The History of the Geological Society of London*. London: Geological Society.

Worger, William. H. 2004. "Convict Labour, Industrialists and the State in the US South and South Africa, 1870–1930." *Journal of Southern African Studies* 30 (1): 63–86.

Worthman, Paul B. 1969. "Black Workers and Labor Unions in Birmingham, Alabama, 1897–1904." *Labor History* 10 (3): 375–407.

Wright, Michelle. 2015. *Physics of Blackness: Beyond the Middle Passage Epistemology*. Minneapolis: University of Minnesota Press.

Wright, Richard. 2021. *The Man Who Lived Underground*. New York: Library of America.

Wright, Willie J. 2018. "As Above, So Below: Anti-Black Violence as Environmental Racism." *Antipode* 53:791–809.

Wynter, Sylvia. 1971. "Novel and History, Plot and Plantation." *Savacou: A Journal of the Caribbean Artists Movement* 5:95–102.

Wynter, Sylvia. 1984. "The Ceremony Must Be Found: After Humanism." *Boundary 2* 12 (3)–13 (1): 19–70.

Wynter, Sylvia. 1989. "Beyond the Word of Man: Glissant and the New Discourse of the Antilles." *World Literature Today* 63 (4): 637–48.

Wynter, Sylvia. 1994. "No Humans Involved: An Open Letter to My Colleagues." *Forum N.H.I.: Knowledge for the 21st Century* 1 (1): 1–17.
Wynter, Sylvia. 1996. "Is Development a Purely Empirical Concept or Also Teleological? A Perspective from We the Underdeveloped." In *Prospects for Recovery and Sustainable Development in Africa*, edited by Aguibou Yansané, 299–316. Westport, CT: Greenwood Press.
Wynter, Sylvia. 2000. "The Re-enchantment of Humanism: An Interview with Sylvia Wynter." Interview by David Scott. *Small Axe* 8:119–207.
Wynter, Sylvia. 2003. "Unsettling the Coloniality of Being/Power/Truth/Freedom: Towards the Human, after Man, Its Overrepresentation—An Argument." *CR: The New Centennial Review* 3 (3): 257–337.
Wynter, Sylvia. n.d. "Black Metamorphosis: New Natives in a New World." MG 502, Box 1. Institute of the Black World Records, Schomburg Center for Research in Black Culture, New York, NY.
Yusoff, Kathryn. 2005. "Arresting Visions: A Geographical Theory of Antarctic Light." PhD diss., University of London.
Yusoff, Kathryn. 2013. "Geologic Life: Prehistory, Climate, Futures in the Anthropocene." *Environment and Planning D: Society and Space* 31 (5): 779–95.
Yusoff, Kathryn. 2015a. "Geologic Subjects: Nonhuman Origins, Geomorphic Aesthetics and the Art of Becoming Inhuman." *cultural geographies* 22 (3): 383–407.
Yusoff, Kathryn. 2015b. "Queer Coal: Genealogies in/of the Blood." *PhiloSOPHIA* 5 (2): 203–29.
Yusoff, Kathryn. 2016. "Anthropogenesis: Origins and Endings in the Anthropocene." *Theory, Culture and Society* 33 (2): 3–28.
Yusoff, Kathryn. 2017. "Geosocial Strata." *Theory, Culture and Society* 34 (2–3): 105–27.
Yusoff, Kathryn. 2018a. *A Billion Black Anthropocenes or None*. Minneapolis: University of Minnesota Press.
Yusoff, Kathryn. 2018b. "Politics of the Anthropocene: Formation of the Commons as a Geologic Process." *Antipode* 50 (1): 255–76.
Yusoff, Kathryn. 2019. "Geologic Realism: On the Beach of Geologic Time." *Social Text* 37 (1): 1–26.
Yusoff, Kathryn. 2021. "The Inhumanities." *Annals of the American Association of Geographers* 111 (3): 663–76.
Yusoff, Kathryn, Elizabeth Grosz, Nigel Clark, Arun Saldanha, and Catherine Nash. 2012. "*Geo*power: A Panel on Elizabeth Grosz's *Chaos, Territory, Art: Deleuze and the Framing of the Earth*." *Environment and Planning D: Society and Space* 30 (6): 971–988.
Yusoff, Kathryn, and Mary Thomas. 2018. "The Anthropocene." In *Edinburgh Companion to Animal Studies*, edited by Lynn Turner, Undine Sellbach, and Ron Broglio, 52–64. Edinburgh: Edinburgh University Press.

Index